Very Large Volume Neutrino Telescope (VLVnT-2015)

EPJ Web of Conferences Volume 116 (2016)

Roma, Italy
14 –16 September 2015

Editors:

Antonio Capone
Giulia De Bonis
Irene Di Palma
Chiara Perrina

ISBN: 978-1-5108-2268-9

Printed from e-media with permission by:

Curran Associates, Inc.
57 Morehouse Lane
Red Hook, NY 12571

Some format issues inherent in the e-media version may also appear in this print version.

Printed by Curran Associates, Inc. (2016)

For additional information, please contact EDP Sciences – Web of Conferences at the address below.

EDP Sciences – Web of Conferences
17, Avenue du Hoggar
Parc d'Activité de Courtabœuf
BP 112
F-91944 Les Ulis Cedex A
France

Phone: +33 (0) 1 69 18 75 75
Fax: +33 (0) 1 69 28 84 91

contact-edps@webofconferences.org

Additional copies of this publication are available from:

Curran Associates, Inc.
57 Morehouse Lane
Red Hook, NY 12571 USA
Phone: 845-758-0400
Fax: 845-758-2633
Email: curran@proceedings.com
Web: www.proceedings.com

TABLE OF CONTENTS

EPJ Web of Conferences **116**, 00001 (2016)
DOI: 10.1051/epjconf/201611600001

Preface

VLVnT 2015 has been the 7th edition of the series of biannual Very Large Volume Neutrino Telescope Workshop. The character of the Workshop has largely changed since its first edition in 2003. At that time the IceCube and the ANTARES detectors were under construction while Baikal and AMANDA were taking data. The Workshop was launched with the aim to bring together astroparticle physicists, marine technologists and marine scientists in order to exchange ideas and expertise useful for the design, construction and operation, in the Mediterranean Sea, of a deep sea Neutrino Telescopes with an instrumented volume of multi-cubic kilometers. With the several editions of the Workshop it has been possible to follow the successful construction of IceCube at the South Pole and of ANTARES in the Mediterranean Sea, and gradually a growing fraction of the time was spent on data analysis, reconstruction techniques, calibrations, discussion of results. After a long period of preparation and of technology development the KM3NeT Collaboration has started the construction of the future Neutrino Telescope in the Mediterranean Sea. During the 6th edition of the Workshop in Stockholm (2013) the IceCube Collaboration presented the first direct evidence of a flux of high-energy cosmic neutrinos, during the 2015 edition of the Workshop this signal not only was confirmed with higher statistics but was the starting point for several other analyses also in close collaboration with ANTARES and in a multimessenger context.

Neutrino Telescopes have demonstrated to be also precision tools for fundamental physics: the measurement of neutrino oscillation parameters, the indirect search for dark matter, the possibility to study with future detectors like ORCA and PINGU the neutrino mass hierarchy are few example.

The Workshop was organized in plenary and parallel sessions.
Photon detection technique: Alexander Kappes, Dorothea Samtleben, Pasquale Migliozzi
Challenges in simulation and reconstruction: Jakob van Santen
Radio and acoustic detection techniques: Allan Hallgren, Robert Lahmann
Physics results: Chad Finley, Rosa Coniglione
Readout/DAQ/data transmission: Tommaso Chiarusi, Vladimir Aynutdinov
Calibration: Giorgio Riccobene, Dawn Williams
Computing models, data repositories, virtual observatory, ...: Gonzalo Merino, Christos Markou, Cristiano Bozza
Atmospheric neutrino phenomenology: maurizio spurio, Tyce DeYoung, Thomas Eberl, Spencer Klein
Installation/Infrastructure/Production techniques: Patrick Lamare, Jim Haugen
Multi-messengers search: Jon Dumm

On behalf of the guest editors, I would like to thank the Department of Physics of the "La Sapienza" University and the INFN that hosted the Workshop, the International Advisory Committee, the conveners of the parallel sessions and the Local Organizing Committee for their invaluable work.

Antonio Capone, for the guests editors

International Advisory Committee

Local Organizing Committee

List of participants

Ahlers Markus, UW-Madison
Ameli Fabrizio, INFN
Antochi Cristian Vasile, Università La Sapienza
Ardid Miguel, Universitat Politècnica de València
Arguelles Delgado Carlos, UW-Madison
Aversa Massimo, Hamamatsu Photonics Italia
Aynutdinov Vladimir, Institute for Nuclear Research RAS

Baret Bruny, CNRS
Barrios Martí Javier, IFIC – CSIC
Belolaptikov Igor, JINR
Biagioni Andrea, INFN
Bozza Cristiano, University of Salerno
Brunner Juergen, CPPM
Buis Ernst-Jan, TNO

Capone Antonio, La Sapienza University of Roma and INFN
Celli Silvia, La Sapienza University of Rome and INFN
Chiarusi Tommaso, INFN – Sezione di Bologna
Coenders Stefan, Technische Universität München
Coniglione Rosa, INFN
Cormack Andy, ET Enterprises Ltd
Cortopassi Alessandro, Caen SpA
Cowen Doug, Pennsylvania State University
Coyle Paschal Anthony, Centre de Physique des Particules de Marseille (CNRS-IN2P3)

De Angelis Alessandro, INFN Padova
De Bonis Giulia, Department of Physics, University of Roma "Sapienza"
De Sio Chiara, University of Salerno
de Wolf Els, Nikhef
DeYoung Tyce, Michigan State University
Di Capua Francesco, Università "Federico II" Napoli and INFN
Di Palma Irene, University of Rome La Sapienza
Dille Christian, Hamamatsu Photonics Deutschland GmbH
Distefano Carla, LNS–INFN
Dompè Valentina, Università degli Studi di Roma "La Sapienza"
Dornic Damien, CPPM
Drakopoulou Evangelia, N.C.S.R. DEMOKRITOS
Dumm Jon, Stockholm University
Dzhilkibaev Zhan, Institute for nuclear research

Euler Sebastian, Uppsala University

Favaro Matteo, INFN – National Institute for Nuclear Physics
Fedynitch Anatoli, KIT – Karlsruhe Institute of Technology (DE) / CERN
Fermani Paolo, Università di Roma "La Sapienza"
Filippidis Christos, Nat. Cent. for Sci. Res. Demokritos (GR)
Finley Chad, OKC – Stockholm University
Fusco Luigi Antonio, University of Bologna

Gal Tamas, ECAP, University of Erlangen
Garzelli Maria Vittoria, University of Hamburg
Gazizov Askhat, GSSI
Giommi Paolo, Agenzia Spaziale Italiana
Giordano Valentina, INFN
Giovenale Francesco, HAMAMATSU PHOTONICS ITALIA
Gladstone Laura, MIT
Gonzalez Javier, University of Delaware

Hallgren Allan, Uppsala University
Hanson Kael, University of Wisconsin – Madison
Hebecker Dustin, HU-berlin / DESY Zeuthen
Hoshina Kotoyo, Earthquake Research Institute, the University of Tokyo
Hotta Yuji, Hamamatsu Photonics K.K.
Hubbard Antonia, Yale University

Illuminati Giulia, La Sapienza University and INFN

Jero Kyle, University of Wisconsin – Madison
Jurkovic Martin, Technische Universität München

Kappes Alexander, Friedrich-Alexander-Univ. Erlangen (DE)
Karle Albrecht, Univ. of Wisconsin-Madison
Katz Uli, ECAP, University of Erlangen
Kieft Gerard, Nikhef
Kintscher Thomas, DESY
Kouchner Antoine, Universite de Paris VII (FR) – APC
Kuleshov Denis, Institute for Nuclear Research
Kulikovskiy Vladimir, LNS

Lahmann Robert, Friedrich-Alexander-Univ. Erlangen (DE)
Lamare Patrick, CPPM
Leonora Emanuele, INFN, section of Catania
Lisi Eligio, INFN

Manolopoulos Konstantinos, NCSR Demokritos
Maragos Nikos, NATIONAL CENTER FOR SCIENTIFIC RESEARCH "DEMOKRITOS"
Marinelli Antonio, Institute of Physics & INFN, Pisa
Markou Christos, Nat. Cent. for Sci. Res. Demokritos (GR)
Martineau Olivier, LPNHE Paris
Martini Agnese, Istituto Nazionale Fisica Nucleare Frascati (IT)
Mathieu Aurore, CPPM
Melis Karel, Nikhef
Migliozzi Pasquale, INFN – NA
Mollo Carlos Maximiliano, INFN

Nicolau Carlo Alessandro, INFN
Nieves Juan M, IFIC (CSIC-UV)

Papaleo Riccardo, INFN
Pausch Steffen, Nautilus Marine Service GmbH
Pellegrino Carmelo, Sezione di Bologna dell'INFN
Perrina Chiara, "La Sapienza" University of Roma – INFN

Pikounis Konstantinos, NCSR Demokritos
Pulvirenti Sara Rita, INFN – National Institute for Nuclear Physics

Rameez Mohamed, Universite de Geneve (CH)
Real Diego, IFIC – Instituto de Física Corpuscular
Riccobene Giorgio, INFN-LNS
Rongen Martin, RWTH Aachen
Ruhe Tim, TU Dortmund University

Samtleben Dorothea, Leiden University
Sandroos Joakim, Universität Mainz
Sanguineti Matteo, INFN Genova – Università di Genova
Schultz David, University of Wisconsin-Madison
Shaybonov Bair, Joint Institute for Nuclear Research
Sheifler Aleksei, Institute for Nuclear Research of the Russian Academy of Sciences
Simeone Francesco, INFN
Sperrin Graham, ET Enterprises Ltd
Spiering Christian, DESY
Spurio Maurizio, University of Bologna and INFN
Stellacci Simona Maria, University of Salerno
Stéphan Beurthey, CPPM

Tardy Christophe, POWERSEA

van der Meer Rob, ASTRON
van Eldik Christopher, ECAP, University of Erlangen-Nuremberg
Viola Salvatore, INFN – LNS
Vivolo Daniele, University of Naples "Federico II"

Weber Alfons, STFC/RAL and University of Oxford
Williams Dawn, University of Alabama

Yanez Juan Pablo, DESY

SAPIENZA
UNIVERSITÀ DI ROMA

INFN
Istituto Nazionale
di Fisica Nucleare

HELMHOLTZ
ASSOCIATION
Alliance for Astroparticle Physics

DESY

NIKHEF

HAMAMATSU
PHOTON IS OUR BUSINESS

Nautilus
MARINE SERVICE GmbH

ARAP
Associazione Romana per le Astro-Particelle

Very Large Volume Neutrino Telescope
VLVνT
workshop
Very Large Volume Neutrino Telescope
Physics Department, "La Sapienza" University - Roma
CAEN
HAMAMATSU
PHOTON IS OUR BUSINESS
Foto: Maurizio PERCIBALLI

Statement of Peer review

In submitting conference proceedings to EPJ Web of Conferences, I certify to the Publisher that I adhere to the **Policy on Publishing Integrity** of the journal in order to safeguard good scientific practice in publishing.

1. All articles have been subjected to peer review administered by the proceedings editors.
2. Reviews have been conducted by expert referees, who have been requested to provide unbiased and constructive comments aimed, whenever possible, at improving the work.
3. Proceedings editors have taken all reasonable steps to ensure the quality of the materials they publish and their decision to accept or reject a paper for publication has been based only on the merits of the work and the relevance to the journal.

Title, date and place of the conference

Very Large Volume Neutrino Telescope (VLVνT - 2015)
Physics Department of University "La Sapienza", Roma, Italy

Proceedings editor(s):

Antonio Capone, antonio.capone@roma1.infn.it
Giulia De Bonis, giulia.debonis@roma1.infn.it
Irene Di Palma, irene.dipalma@roma1.infn.it
Chiara Perrina, chiara.perrina@roma1.infn.it

Date and editor's signature

Roma, February 5, 2016

EPJ Web of Conferences **116**, 01001 (2016)
DOI: 10.1051/epjconf/201611601001

Multi-PMT Optical Module Designs for IceCube-Gen2

Alexander Kappes[a]

Erlangen Centre for Astroparticle Physics, Friedrich-Alexander-Universität Erlangen-Nürnberg, 91058 Erlangen, Germany

Abstract. IceCube-Gen2 is the planned next generation neutrino telescope at the South Pole incorporating a high-energy array for neutrino astronomy and a dense array (PINGU) aimed at the determination of the neutrino mass hierarchy. Here, we present alternative designs to IceCube-Gen2's single-PMT baseline optical module which are currently being developed. These designs feature up to 24 smaller photomultipliers and use glass and gel with enhanced UV transparency to increase the number of detected photons and provide additional information. Thereby, they have the potential to significantly enhance the performance of IceCube-Gen2.

1. Introduction

The design of the Digital Optical Module (DOM) used in the IceCube detector [1] has proven to be highly reliable and, thereby, has played a key role in providing the high-quality data that enabled the recent discovery of high-energy cosmic neutrinos [2]. For the next-generation neutrino telescope in the deep ice at the South Pole, IceCube-Gen2, about 10,000 additional optical sensors will be deployed. These are required to operate reliably without maintenance over the full livetime of the detector of 10+ years at depths down to 2,700 m with pressures up to 550 bar[1] and temperatures down to $-45\,\mathrm{C}^{\circ}$. In view of these harsh environmental conditions, the baseline design for IceCube-Gen2 is an advancement of the successful IceCube-DOM design. While all parts exposed to the high pressure are retained, developments in recent years allow to significantly simplify the readout electronics, thereby reducing complexity and power consumption [3].

Like the original DOM, this new design features a single large 10" photomultiplier (PMT) pointing downwards, resulting in a very asymmetric angular acceptance. In particular, in view of the recent success in observing cosmic neutrinos from the Southern hemisphere [2], i.e. detecting events entering the detector from above, this strongly downward peaked sensitivity is an undesirable feature[2]. In addition, the single-PMT design doesn't provide any information on the arrival directions of the detected photons, and the absorption of photons by the current DOM glass increases rapidly as the wavelength of

---

[a]e-mail: kappes@icecube.wisc.edu

[1] This pressure is reached during freeze-in.

[2] In fact, the IceCube DOM also has some sensitivity to photons produced above the optical module due to backscattering in the ice. However, scattered photons lose directional information on their origin.

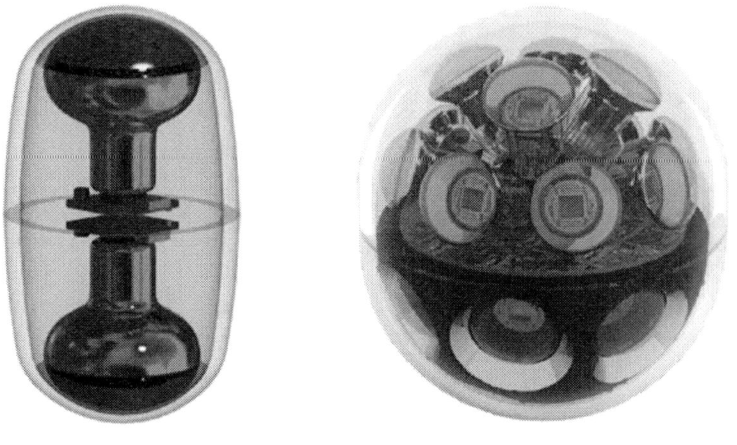

Figure 1. Renderings of the D-Egg (left, taken from [4]) and mDOM (right) concepts for IceCube-Gen2. In case of the latter, the top part of the black PMT holding structure visible in the lower half has been removed.

photons decreases below 400 nm, thereby losing the UV photons from the Cherenkov spectrum which rises with $1/\lambda^2$. Therefore, in parallel to the development of this baseline design, new detector concepts are being explored that aim at improving on these and other features. Two of them are the so-called D-Egg [4] and mDOM [5] concepts which use multiple PMTs and glass with improved UV transparency. Another concept, discussed in [6], uses large surfaces of wavelength shifters to significantly enhance the number of detected photons at low costs.

2. The D-Egg and mDOM concepts

The idea of optical sensors with multiple PMTs was first proposed in 1979 in the framework of the DUMAND project [7]. The D-Egg concept, depicted in Fig. 1 left, incorporates two oppositely oriented 8" PMTs (R5912-100 HQE) housed in an ellipsoidal pressure vessel. At its widest part, the pressure vessel measures 300 mm (~12") and, hence, is about 1" narrower than the standard IceCube DOM. This allows for drilling narrower holes into the ice, thereby saving fuel costs and drilling time. First pressure tests up to 700 bar including penetrator, a dummy PMT and optical gel were successful. The observed bubble formation in the gel is under investigation. The glass of the pressure vessel as well as the optical gel for coupling the PMTs to the glass have been developed to provide increased UV transparency. For readout of each PMT, the analog waveforms will be continuously sampled with a fast ADC.

The mDOM concept (Fig. 1 right), which is based on the KM3NeT optical module [8], pushes the multi-PMT concept even further and incorporates 24 3" PMTS (HamamatsuR12199-02) into a pressure vessel of 358 mm (14") diameter. The large number of PMTs allows to gain information on the direction of a light source already on the single module level and coincidences between PMTs can, e.g, be used for suppression of background. Each PMT has a base attached to it that, in addition to containing electronics that generates the high voltage via a Cockcroft Walton circuit[3], performs advanced signal processing. For readout of the 24 PMTs, a multi-level comparator design with up to 63 thresholds is envisioned which samples the signal of each PMT already on the PMT base. Each comparator will provide the leading-edge time and the time over threshold.

[3] Developed by Nikhef for the KM3NeT optical module.

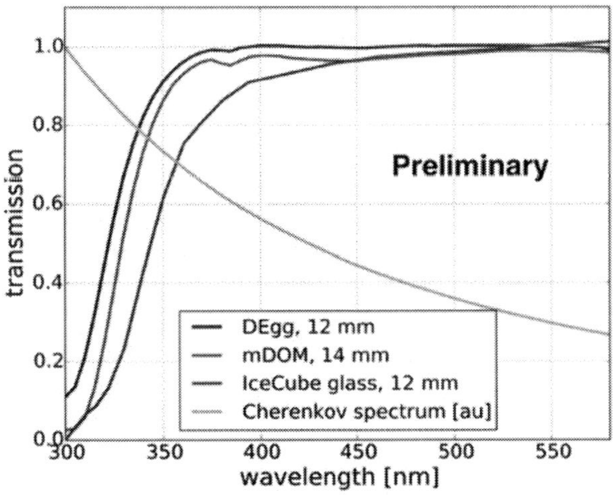

Figure 2. Transmissivity of different pressure vessel glasses as function of wavelength. The measurement for the D-Egg and mDOM glasses were performed in air and losses due to reflection deducted. For illustrative purposes, the $1/\lambda^2$ Cherenkov spectrum is also shown with an arbitrary scale.

2.1 UV sensitivity

Due to the Cherenkov spectrum increasing with $1/\lambda^2$, already minor extensions of the sensitivity to shorter wavelengths increases the number of detected photons significantly. The achievable gain is weakened by the fact that both scattering and absorption length rise below $\sim$ 400 nm. However, a sizable gain remains even for larger distances between photon source and optical module.

For the IceCube DOM, the limiting factor here is the transmissivity of the pressure vessel which starts to significantly absorb photons already at 400 nm (see Fig. 2) and reaches 50% transmissivity around 345 nm. The glass of the D-Egg, specifically developed to provide better UV transparency by reducing its Fe_2O_3 concentration, features a significantly lower cutoff wavelength and reaches 50% around 320 nm. In particular, it also maintains its maximum transmissivity to much lower wavelengths. The mDOM glass, which is the standard glass used by the manufacturer (NAUTILUS Marine Service GmbH) for their deep-see pressure vessels, reaches 50% transmissivity around 330 nm, thereby being only slightly worse than the special UV-transparent glass. Both glasses therefore represent attractive alternatives to the one used in the current optical modules. Their contamination with radioactive elements still needs to be determined carefully though, as this has a large impact on the dark count rate of the PMTs in the otherwise very radioactive pure environment in the deep ice.

2.2 Effective areas

In addition to an increased UV transparency, both new concepts also feature increased photocathode area (D-Egg: $\sim 2 \times 320\,\text{cm}^2 = 640\,\text{cm}^2$; mDOM: $\sim 24 \times 50\,\text{cm}^2 = 1200\,\text{cm}^2$) compared to the current IceCube DOM with $\sim 500\,\text{cm}^2$. Also, both the Gen2 baseline design and the D-Egg concept foresee the usage of super-bialkali photocathode material which provides up to $\sim$ 20% higher photon detection efficiency. Whether this also poses a cost-effective option for the 3" PMTs used in the mDOM has to be evaluated. Figure 3 shows the effective areas for the D-Egg and mDOM concepts including quantum efficiency. Compared to the single-PMT baseline design (not shown), the D-Egg shows a significantly reduced angular dependence with a maximum reduction of about 20%. Owing to its large number of PMTs, the mDOM features an even more uniform angular acceptance with a maximum variation of

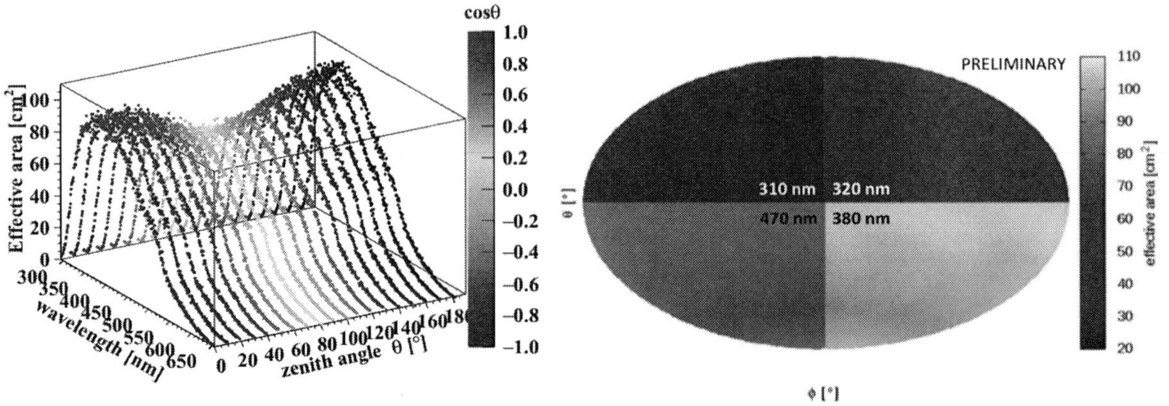

Figure 3. Left: Effective area of the D-Egg including PMT quantum efficiency as a function of wavelength for different photon zenith angles (module orientation as in Fig. 1 left). Taken from [4]. — Right: Hammer projection of the mDOM photon acceptance including PMT quantum efficiency represented as effective area as a function of direction of the incident photons in zenith and azimuth angle (module orientation as in Fig. 1 right). Values are shown for different photon wavelengths. Due to the symmetry in the orientation of the PMTs, each shown quadrants is representative of the total solid angle for the respective wavelength.

12%. Both new concepts therefore improve significantly on the amount of effective area per optical module and its uniformity in solid angle. Detailed measurements and simulations currently underway will allow to quantify the impact of these improvements on the detector performance.

3. Conclusions

Parallel to the advancement of the proven IceCube DOM design, several alternative designs for the next generation neutrino telescope at the South Pole, IceCube-Gen2, are being investigated. These designs improve on several aspects of the Gen-2 baseline design, in particular the uniformity of the angular acceptance and the overall photon acceptance. The aim is to develop prototypes within the next two years which can be tested under real conditions in the deep ice at the South Pole in advance of the installation of the actual IceCube-Gen2 detector. A first mechanical design for both sensor types has been completed and the development of specific read-out electronics has commenced. Conceptual prototypes are planned for 2016. Simulations currently undertaken address the impact of the utilization of the new module concepts on the performance of IceCube-Gen2 which is expected to benefit from the increased effective area and superior information on the recorded photons. In the end, a combination of different module types may turn out to be most effective.

References

[1] F. Halzen, S.R. Klein, Rev. Sci. Instrum. **81**, 081101 (2010), `arXiv:1007.1247`

[2] M. Aartsen et al. (IceCube), Phys. Rev. Lett. **113**, 101101 (2014), `arXiv:1405.5303`

[3] M.G. Aartsen et al. (IceCube-Gen2), *Generation 2 IceCube Digital Optical Module and DAQ*, in *Proc. of the 34th International Cosmic Ray Conference* (The Hague, The Netherlands, 2015), `arXiv:1510.05228`

[4] M.G. Aartsen et al. (IceCube-Gen2), *A dual-PMT optical module (D-Egg) for IceCube-Gen2*, in *Proc. of the 34th International Cosmic Ray Conference* (The Hague, The Netherlands, 2015), `arXiv:1510.05228`

[5] M.G. Aartsen et al. (IceCube-Gen2), *Multi-PMT optical modules for IceCube-Gen2*, in *Proc. of the 34th International Cosmic Ray Conference* (The Hague, The Netherlands, 2015), `arXiv:1510.05228`

[6] D. Hebecker, *A wavelength-shifting optical module (WOM) for in-ice neutrino detectors*, these proceedings

[7] V.V. Borog et al., *Measuring module for registration of Cherenkov radiation in the water*, in *Proc. of the 16th International Cosmic Ray Conference* (1979), Vol. 8, p. 380

[8] H. Löhner et al. (KM3NeT), Nucl. Inst. Meth. **A718**, 513 (2013)

EPJ Web of Conferences **116**, 01002 (2016)
DOI: 10.1051/epjconf/201611601002

The KM3NeT Digital Optical Module

Daniele Vivolo[1,2] for the KM3NeT Collaboration

[1] University of Naples "Federico II", Department of Physics, via Cintia, 80126 Naples, Italy
[2] INFN – Istituto Nazionale di Fisica Nucleare, Section of Naples, via Cintia, 80126 Naples, Italy

Abstract. KM3NeT is a European deep-sea multidisciplinary research infrastructure in the Mediterranean Sea. It will host a km3-scale neutrino telescope and dedicated instruments for long-term and continuous measurements for Earth and Sea sciences. The KM3NeT neutrino telescope is a 3-dimensional array of Digital Optical Modules, suspended in the sea by means of vertical string structures, called Detection Units, supported by two pre-stretched Dyneema ropes, anchored to the seabed and kept taut with a system of buoys. The Digital Optical Module represents the active part of the neutrino telescope. It is composed by a 17-inch, 14 mm thick borosilicate glass (Vitrovex) spheric vessel housing 31 photomultiplier tubes with 3-inch photocathode diameter and the associated front-end and readout electronics. The technical solution adopted for the KM3NeT optical modules is characterized by an innovative design, considering that existing neutrino telescopes, Baikal, IceCube and ANTARES, all use large photomultipliers, typically with a diameter of 8" or 10". It offers several advantages: higher sensitive surface (1260 cm^2), weaker sensitivity to Earth's magnetic field, better distinction between single-photon and multi-photon events (photon counting) and directional information with an almost isotropic field of view. In this contribution the design and the performance of the KM3NeT Digital Optical Modules are discussed, with a particular focus on enabling technologies and integration procedure.

1. Introduction

The KM3NeT project aims at the realization of a deep-sea multidisciplinary research infrastructure in the Mediterranean Sea. In particular, a multi-km^3-scale neutrino telescope, with dedicated instruments for long-term and continuous measurements in earth and sea sciences, is currently under construction. The KM3NeT research facility will be realized as an installation distributed over several sites, with common detector technology, management, data handling and operation control. In particular, a large volume neutrino telescope, called ARCA (Astroparticle Research with Cosmics in the Abyss), will be installed in the Italian site of Portopalo di Capopassero in Sicily, while a denser and more compact version of the detector, ORCA (Oscillation Research with Cosmics in the Abyss) in the KM3NeT-Fr site of Toulon (France), will be dedicated to the study of low energy atmospheric neutrino oscillations [1].

KM3NeT is a three-dimensional array of optical detectors contained in glass spheres that are designed to resist the hostile conditions (high hydrostatic pressure and water salinity) of the deep-sea environment. Such modules, called Digital Optical Modules (DOMs), are suspended in the sea by means of vertical string structures, called Detection Units (DUs), supported by two pre-stretched

Dyneema ropes, anchored to the seabed and kept taut by a system of buoys. Each DU is equipped with 18 DOMs, for a total length of more than 700 m, and hosts the optical detectors readout electronics and all the devices required for calibration, positioning and acoustics measurements as well as sea and Earth science instrumentation like video cameras, conductivity-temperature-depth probes, Doppler current profilers, chemical analysers and seismographs [2].

Currently, the KM3NeT collaboration has started the construction of the first phase of the project, consisting of 24 DUs of the ARCA detector and 7 DU of ORCA.

2. The Digital Optical Module

The Digital Optical Module represents the active part and hence the real heart of KM3NeT.

An exploded view is shown in Fig. 1. The DOM is composed by a 17-inch, 14 mm thick borosilicate glass (Vitrovex) spheric vessel housing 31 PMTs with 3-inch photocathode diameter, the associated frontend and readout electronics, calibration components and accessory instrumentation for monitoring purposes. The vessel provides an adequate mechanical resistance to the extreme compressive stresses of deep-sea environment (the ambient hydrostatic pressure is in the range of 250–400 bar) while keeping a transparency of more than 95% at a wavelength of $\lambda = 350$ nm.

The multi-PMT solution, which had already been considered in the framework of the DUMAND project [3], represents one of the main differences of the KM3NeT DOM with respect to the standard concept of optical modules in large-volume neutrino telescopes, which are traditionally equipped with single large PMTs.

First of all, an overall sensitive surface of $1260\,\mathrm{cm}^2$ is achieved, thus optimizing the detection efficiency. This is a crucial factor, considering the extremely low interaction rate of astrophysical neutrinos. Small surface PMTs exhibit a weak sensitivity to Earth's magnetic fields, with the appealing consequence that no shielding is required and hence integration complexity and costs are reduced. A segmented detection surface makes possible an improved distinction between single-photon from multi-photon events. Indeed, in a DOM a two-photon event can be unambiguously recognized if the two photons hit separate tubes, a condition that occurs in 85% of cases for photons arriving from the same direction. Moreover, the almost isotropic angular coverage of the PMTs, which are oriented from vertically downwards to 58 deg upwards, permits the measure of unscattered light from sources with arbitrary orientation from the DOMs [4].

In addition, a strong reduction of the failure risk is achievable. Indeed, for a small diameter device the integrated anode charge is significantly smaller, thus making it less subject to ageing and decreasing its failure rate to something around 10^{-4} per year.

2.1 Mechanics

A nylon support structure, manufactured by 3D printing with the selective laser sintering method, has the function to hold the 31 PMTs inside the DOM, defining their positions, their spatial orientations and their distances from the glass vessel. All the PMT locations are tapered and a silicone O-ring tightly seals around the PMT just below its head [4].

The DOM vessel is made of two half spheres. The lower hemisphere houses 19 PMTs, power board and readout electronics, a LED flasher pointing upwards for timing calibration between DOMs (called nanobeacon, see Fig. 1) and a pressure gauge to monitor the air pressure inside the DOM before deployment. The upper one contains 12 PMTs and an acoustic sensor attached to the glass sphere for positioning purposes. The photomultipliers are glued to the DOM glass vessel by means of a two-components transparent silicon optical gel (Wacker 612), filling the cavity between the support structure and the glass, and assuring optical contact. In order to reduce the unwanted reflections, a refractive index of ≈ 1.40 is required, close to both the refractive indexes of the glass vessel (1.47) and of the PMT

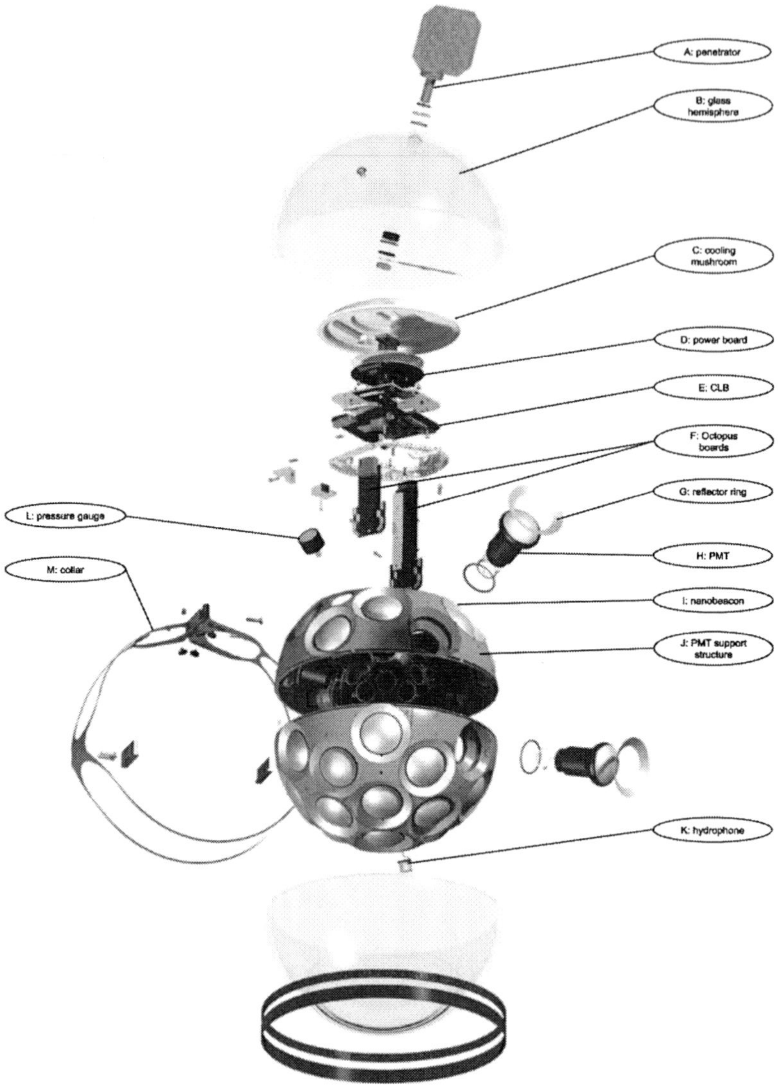

Figure 1. Exploded view of a KM3NeT DOM. The main components are indicated and referenced in the text.

window (1.51–1.54), with an attenuation length greater than 40 cm for wavelengths above 350 nm. In addition, sufficiently elastic support structure and gel provide a precious help in the absorption of the shocks and the vibrations induced by transportation and deployment, contributing to accommodate the shrinkage of the glass vessel under the high hydrostatic pressure.

The DOM contains also a passive cooling system, based on the heat conduction mechanism, aimed at keeping the temperature of the electronic components as low as possible, thus maximizing their lifetime. In order to optimize the transfer of the heat generated by the electronics to the seawater, a metallic structure is required, with a large contact surface with the inner surface of the glass vessel. The cooling system is made by a mushroom-shaped Aluminium structure coupled to an Aluminium bar and is dimensioned to maintain temperatures below 30 °C for an overall power dissipation inside the glass vessel of up to 20 W.

2.2 Photomultiplier tubes

The choice of the optical sensors is crucial. The overall performance of the telescope, in terms of efficiency and track reconstruction resolution, is intrinsically related to the basic performance of the adopted photodetectors. The adopted solution is represented by a 3-inch PMT. In particular, the PMTs used in the current KM3NeT phase are the Hamamatsu R12199-02.

In order to meet the KM3NeT requirements, the PMTs need to fulfil several requisites [2], including some geometry constraints. The reduced space available inside the DOM imposes an overall limitation on the length of the device (less than 122 mm), while the outer convex surface of the PMT must match as precisely as possible the curvature of the inner surface of the glass sphere.

The performance specifications for the PMTs are summarized in the following. High quantum efficiency is required ($> 23\%$ at 404 nm and $> 20\%$ at 470 nm), with an inhomogeneity of cathode response of less than 10%. Taking into account the noise level of the electronic circuit for PMT readout, the value of the gain of each photomultiplier is fixed at 3×10^6, with a peak-to-valley ratio of at least 3 for an acceptable resolution. Considering that for a 3-inch PMT the optical background in the sea water is of the order of few tens of kHz, the dark count rate of the adopted PMTs (given by the combination of thermal electron emission from photocathode and dynode chain and radioactive decay in the glass and in the optical gel, mainly in the form of ^{40}K decay) must be less than 3 kHz.

Since the time resolution of the optical sensor has a crucial impact on the reconstructed muon track angular resolution, a small Transit Time Spread is preferable, compared with the timescale of the absolute time calibration of the telescope. For the KM3NeT case, a TTS of 2 ns (expressed in terms of the standard deviation σ) has been set as an upper limit.

3. DOM integration

DOM integration is a delicate process, managed by a dedicated procedure. All the design features are optimized for functionality and all the components must pass characterization tests and fulfil acceptance criteria before use.

The two hemispheres of the DOM are constructed separately. At the end of the integration process, the DOM is closed connecting the two halves and making a slight underpressure in the sphere (0.2 bar below atmospheric pressure). Also the support structure is segmented in two halves, each one made of two parts. After the 3-D printing process the structures are white, therefore, in order to reduce the optical background they must be painted black before being glued together.

After mounting the PMTs in the support structure, they are connected to frontend and readout electronics and then glued to the glass vessel. In order to transmit the optical data through the fibre leading to the outside of the DOM, a splicing of optical fibres is required.

During the integration process, several tests are performed: the mounted penetrator is subject to a helium leak test, the fibre splice and resulting optical power are measured and a functional test of all systems is performed before applying the optical gel [2]. After the DOM closure, an extensive final acceptance test is performed in a light-tight dark box.

4. Conclusions

The multi-PMT solution represents one of the main differences of the KM3NeT DOM with respect to the standard concept of optical modules in large-volume neutrino telescopes, which are traditionally equipped with single large PMTs, providing strong advantages in terms of photon counting, background rejection capabilities and direction sensitivity.

After a dedicated prototyping program, aimed at validating its design and performance, the KM3NeT collaboration has started a mass production of DOMs for both ARCA and ORCA detectors in four DOM integration sites.

References

[1] R. Coniglione, these proceedings (2015)
[2] R. Bruijn and D. van Eijk (for the KM3NeT collaboration), *PoS(ICRC2015)*, **1157**
[3] V.V. Borog et al., Proceedings of the 16th ICRC, **10**, 380–384
[4] KM3NeT Technical Design Report, `http://www.km3net.org/TDR/TDRKM3NeT.pdf`

EPJ Web of Conferences **116**, 01003 (2016)
DOI: 10.1051/epjconf/201611601003

The optical module of Baikal-GVD

A.D. Avrorin[1], A.V. Avrorin[1], V.M. Aynutdinov[1], R. Bannash[7], I.A. Belolaptikov[2], D.Yu. Bogorodsky[3], V.B. Brudanin[2], N.M. Budnev[3], I.A. Danilchenko[1], G.V. Domogatsky[1], A.A. Doroshenko[1], A.N. Dyachok[3], Zh.-A.M. Dzhilkibaev[1], S.V. Fialkovsky[5], A.R. Gafarov[3], O.N. Gaponenko[1], K.V. Golubkov[1], T.I. Gress[3], Z. Honz[2], K.G. Kebkal[7], O.G. Kebkal[7], K.V. Konischev[2], A.V. Korobchenko[2], A.P. Koshechkin[1], F.K. Koshel[1], A.V. Kozhin[4], V.F. Kulepov[5], D.A. Kuleshov[1], V.I. Ljashuk[1], M.B. Milenin[5], R.A. Mirgazov[3], E.R. Osipova[4], A.I. Panfilov[1], L.V. Pan'kov[3], E.N. Pliskovsky[2], M.I. Rozanov[6], E.V. Rjabov[3], B.A. Shaybonov[2], A.A. Sheifler[1], M.D. Shelepov[1], A.V. Skurihin[4], A.A. Smagina[2], O.V. Suvorova[1], V.A. Tabolenko[3], B.A. Tarashansky[3], S.A. Yakovlev[7], A.V. Zagorodnikov[3], V.A. Zhukov[1], and V.L. Zurbanov[3]

[1] Institute for Nuclear Research, 60th October Anniversary pr. 7A, Moscow 117312, Russia
[2] Joint Institute for Nuclear Research, Dubna 141980, Russia
[3] Irkutsk State University, Irkutsk 664003, Russia
[4] Skobeltsyn Institute of Nuclear Physics MSU, Moscow 119991, Russia
[5] Nizhni Novgorod State Technical University, Nizhni Novgorod 603950, Russia
[6] St. Petersburg State Marine University, St. Petersburg 190008, Russia
[7] EvoLogics GmbH, Berlin, Germany

Abstract. The Baikal-GVD neutrino telescope in Lake Baikal is intended for studying astrophysical neutrino fluxes by recording the Cherenkov radiation of the secondary muons and showers generated in neutrino interactions. The first stage of Baikal-GVD will be equipped with about 2300 optical modules. We describe the design of the optical module, the front-end electronics and the laboratory characterization and calibration before deployment.

1. Introduction

The objective of the Baikal Project is the creation of a kilometer-scale high-energy neutrino observatory: the Gigaton Volume Detector (GVD) in Lake Baikal [1–4]. The first phase of GVD will consist of 12 independent (i.e. connected to the shore by 27 individual electro-optical cables) clusters of strings. In April 2015 the first cluster of Baikal-GVD was deployed in Lake Baikal and put into operation. It comprises a total of 192 optical modules (OM) arranged at eight 345 m long strings, as well as an acoustic positioning system. Seven side strings are located at 40 m distances from a central one. Each string comprises 24 OMs spaced by 15 m at depths of 900 m to 1250 m below the surface. An optical module contains a large area photomultiplier R7081-100 that detects Cherenkov radiation produced by relativistic charged particles passing through the water. The information from the ensemble of

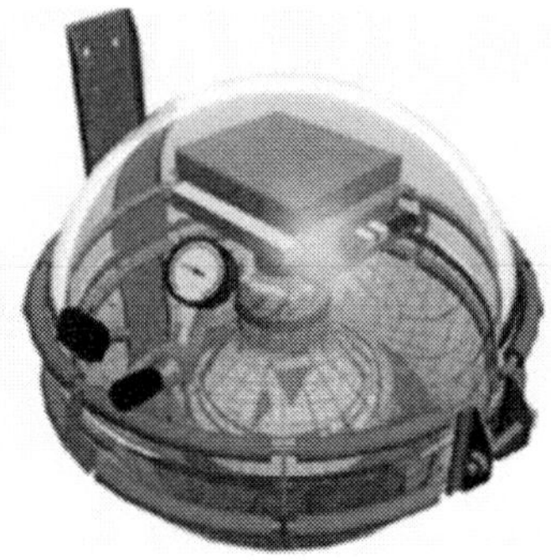

Figure 1. The Baikal-GVD optical module.

OMs allows reconstruction of direction and energy of the parent neutrino. Optical modules are the key elements of a neutrino telescope. The present paper presents the design and performance of the components of the optical module, and describes selected results of the tests of 87 OMs.

2. The Optical Module Design

A sketch of a GVD optical module is presented in Fig. 1. The OM contains a photomultiplier tube (PMT) enclosed in a transparent, nearly spherical pressure housing (VITROVEX) with 42 cm diameter. The optical contact between the photocathode region of the tube and the pressure sphere is provided by optically transparent silicone gel. A high permittivity alloy cage surrounds the PMT, shielding it against the Earth's magnetic field. A vacuum valve allows evacuating the sphere down to 0.7 atm. The OM is equipped with one deep-underwater connector (SubConn Low Profile 5-contacts). It is used for analog pulse transmission, slow control (2-wire RS-485) and OM power supply (12 VDC). The OM electronics unit is mounted directly on the PMT base.

After testing different options for the photomultiplier, Hamamatsu R7081-100 was selected as a light sensor for the OM. This PMT has a hemispherical SBA (super bialkali) photocathode with 10" diameter and a quantum efficiency at peak up to 35%.

The block diagram of the optical module electronics is presented in Fig. 2. The OM electronics includes a controller, a high voltage (HV) power supply unit, a fast two-channel amplifier, and a LED flasher. The OM controller is intended for communication to a Central electronics Module, for HV regulation and monitoring, for PMT noise measurements, and for time and amplitude calibration with LEDs. It is designed by SNIIP-AUNIS Ltd (Russia) on the basis of the SiLabs C8051F121 microcontroller.

Slow control data to and from the OMs are transferred via an underwater RS-485 bus. The unit has an instruction set for the control of OM electronics: HV control (to switch HV on/off, to set the PMT voltage and readout the HV value), LED flasher control (to set LED intensities, the delay between LED pulses, and the period of flashes), the counter control (setting counter threshold, time window, off-duty factor and the size of circular memory data buffer). Also a set of procedures is foreseen for the laboratory calibration of the counter threshold and PMT voltage.

The PMT power supply is provided by a passive HV divider circuit with a resistance of 18 MΩ and a HV unit (SHV 12-2.0 K 1000 P produced by TRACO Electronic AG) with positive polarity. The tube gains have been adjusted to about 10^7.

The PMT amplifier is comprised of two channels. The first channel with an amplification factor of 14 forms a signal with negative polarity that is transmitted to the ADC board. The second channel with an amplification factor of 21 produces positive pulses that are intended for PMT noise monitoring. The

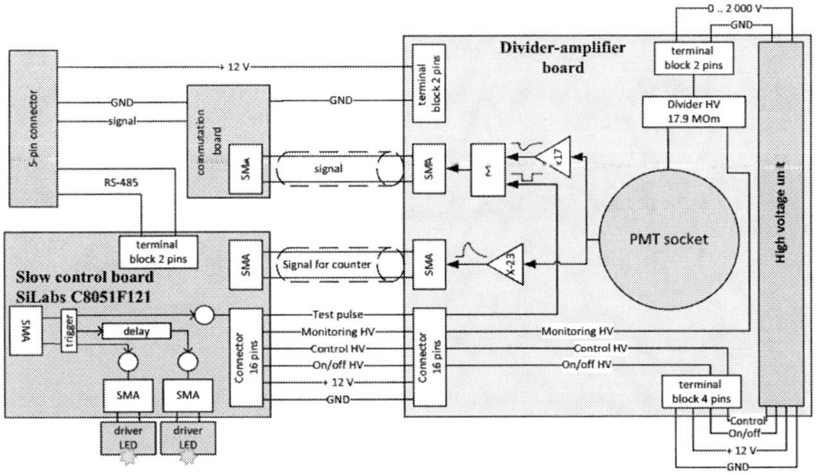

Figure 2. Block diagram of the OM electronics.

outer cascades are implemented as emitter-follower amplifiers and provide operation with 50 Ω load. The maximum amplitude for both channels is limited to 4 V.

The LED flasher is intended for time and amplitude calibration of OM channels during long-term exposition. It includes two LEDs Kingbright L7113 with a dominant wavelength about 470 nm. The LED pulse has a width of ~ 6 ns (FWHM). The control system of the flasher has to provide independent tuning of two LED luminosities in a wide dynamic range (up to about 10^8 photons per LED flash) and has to have a minimal level of cross-talk between two LED channels (less than 1%).

The OM counter with programmable threshold is intended for PMT count rate monitoring. Count rate data is accumulated in the circular buffer and transmitted to the shore for each OM.

3. OMs calibration and characterization

Before OM deployment in Lake Baikal a series of test procedures is foreseen. There are tests of all OM electronic components, stress tests, and check-up of the OM in various modes of operation (final OM tests before transportation to Baikal). Final OM tests are performed by means of a digital storage oscilloscope (LeCroy HDO 4034, 350 MHz bandwidth, 0.4 ns samples, 4 channels). Four OMs are housed in a screened dark box and are connected to the oscilloscope inputs with 90 m coaxial cables, identical to the underwater OM cables. The test procedures are performed after at least 2 hours exposition in the darkness with PMT high voltage switched on. Internal LEDs of the OMs are used as calibration light sources. The OM test procedures are fully automatized and comprise a set of measurements of the time and amplitude parameters of the OMs.

The first stage of the OM test procedure is an adjustment of the PMT power supply voltages to provide OM channel gains about 10^8. An OM analog channel comprises PMT, preamplifier and 90 m coaxial cable connecting OM and Central electronics Module. Taking into account signal amplification with the preamplifier ($k_{amp} = 14$) and pulse attenuation in the cable ($k_{att} = 0.7$), a 10^8 channel gain corresponds to a PMT gain about 10^7.

The OM channel gains were derived on the basis of the single photoelectron distributions (SPE spectra) of the PMTs obtained with LED sources. Intensities of the LEDs were adjusted to provide a detection probability of SPE signals of 10%. The SPE pulse detection thresholds were about 0.2 SPE pulse amplitude. The oscilloscope input was triggered by the synchronization signal of the LED pulse generator. The check for pedestal and noise contributions was done with the LED light output disabled.

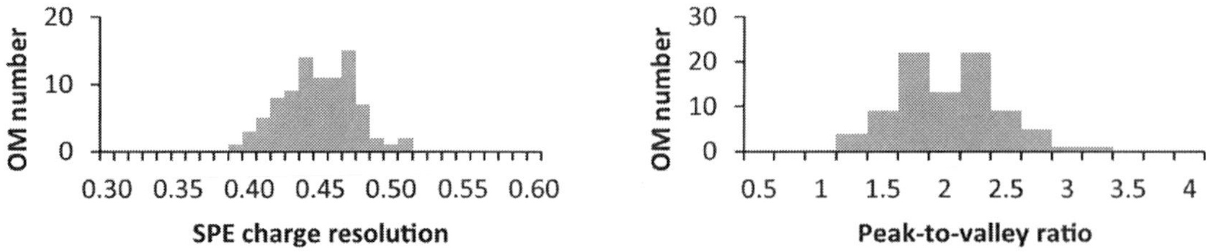

Figure 3. The distributions of the OMs on SPE charge resolution (left) and on the peak-to-valley ratio (right).

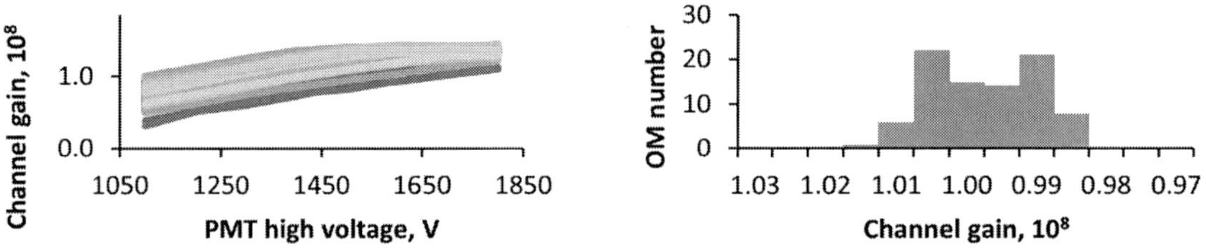

Figure 4. The dependences of the channel gains on high voltage for the set of 87 OMs (left), and OMs distribution on the fitted channel gain (right).

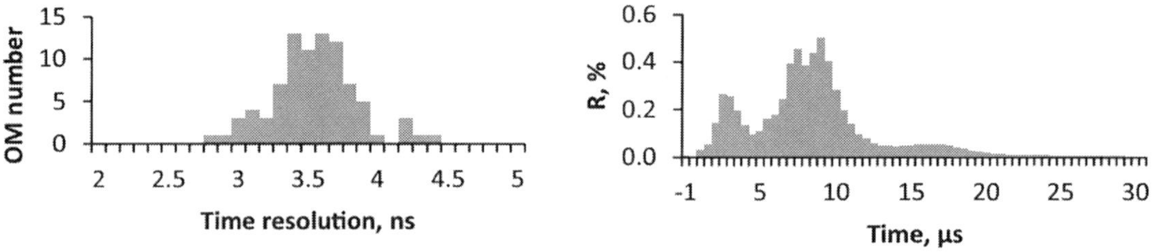

Figure 5. The distribution of the OMs on the time resolution (left) and the typical time dependence of the afterpulse charges for primary pulses with one photoelectron (right).

The distributions of the OMs on SPE charge resolution (left) and on the peak-to-valley ratio are presented in Fig. 3. The SPE charge resolution is defined as one standard deviation of the SPE spectrum. The contribution of multi-electron pulses in the SPE spectrum overestimates the SPE charge resolution by about 20%.

The results of the high voltage adjustment are presented in the Fig. 4. The channel gain of 10^8 is provided by divider voltages between 1150 V and 1750 V for the investigated set of PMTs. Adjusted channel gains are about 1×10^8.

The OM time resolution (the standard deviations of hit times of the SPE pulses) is presented in Fig. 5 (left). Afterpulse measurements were made with LED pulses of about 5 ns width. The total charge of the afterpulses (N_{pe} afterpulses) was measured in the range from 300 ns up to 30 μs after the main pulse, in steps of 500 ns and for various LED intensities. N_{pe} afterpulses grows almost linearly with the flash brightness. We characterize the OM afterpulses by the parameter $R = N_{pe\ \text{afterpulses}} / N_{pe\ \text{main pulse}} \times 100\%$. The typical dependence of R on the time after the main pulse is presented in Fig. 5 (right). For most of the OMs, the integral charges of the afterpulses correspond to about 0.10 ... 0.15 SPE per primary photoelectron.

4. Conclusion

For the OM mass production a fully automatized test facility was designed. The test of the OM performance with this facility provides information about the basic OM parameters: SPE spectrum, time resolution, systematic shifts of the hit time in dependence on the pulse amplitude, saturation curve, and afterpulse probability. These data are used as input for the detailed simulation of the Baikal-GVD physics events.

This work was supported by the Russian Found for Basic Research (grants 13-02-12221, 14-02-00175, 14-02-00972).

References

[1] A. Avrorin et al., *The prototyping/early construction phase of the BAIKAL-GVD project*, NIM **A742**, 82

[2] V. Aynutdinov et al., *The prototype string for the km3-scale Baikal neutrino telescope*, NIM **A602**, 227

[3] V. Aynutdinov et al., *The gigaton volume detector in Lake Baikal*, NIM **A639**, 30

[4] A. Avrorin et al., *Status and recent results of the Baikal-GVD project, Phys. of Part. and Nucl.* **46**, 211

EPJ Web of Conferences **116**, 01004 (2016)
DOI: 10.1051/epjconf/201611601004

VSiPMT a new photon detector

F. Di Capua[1,2,a], G. Barbarino[1,2], F.C.T. Barbato[1,2], L. Campajola[1,2], R. de Asmundis[2], G. De Rosa[1,2], P. Migliozzi[2], C.M. Mollo[2], and D. Vivolo[1,2]

[1] Physics Department, University of Naples, Napoli, Italy
[2] INFN, Sezione di Napoli, Napoli, Italy

Abstract. Photon detection is a key factor to study many physical processes in several areas of fundamental physics research. Focusing the attention on photodetectors for particle astrophysics, the future experiments aimed at the study of very high-energy or extremely rare phenomena (e.g. dark matter, proton decay, neutrinos from astrophysical sources) will require additional improvements in linearity, gain, quantum efficiency and single photon counting capability. To meet the requirements of these class of experiments, we propose a new design for a modern hybrid photodetector: the VSiPMT (Vacuum Silicon PhotoMultiplier Tube). The idea is to replace the classical dynode chain of a PMT with a SiPM, which therefore acts as an electron detector and amplifier. The aim is to match the large sensitive area of a photocathode with the performances of the SiPM technology.

1. Introduction

Photodetectors are widely used in many areas of fundamental physics research. Focusing on astroparticle physics experiments, up to the present the photon detection capability of photomultiplier tubes (PMTs) seems to be unrivalled. Anyway, they have several disadvantages: fluctuations in the first dynode gain make single photon counting difficult; the linearity is strongly related to the gain; the transit time spreads over large fluctuations; the need of voltage dividers increases failure risks and the power consumption.

The future generation of experiments will require further improvement in linearity, gain, and sensitivity (quantum efficiency and single photon counting capability). So, even though PMTs represent a certified technology since 1936 [1], an alternative solution should be found.

The solid state photodetectors, APDs and SiPMs, represent a technological progress that goes beyond PMTs' limits. In particular, SiPM technology has achieved a very high level of performances but the size limited by the thermal noise partially compromises their application in astroparticle physiscs experiments.

The realization of a new photon detector with photoelectron multiplication based on SiPMs, having a photocathode of at least 1 inch, will have a tremendous impact on astroparticle physics experiment.

---

[a]e-mail: `dicapua@na.infn.it`

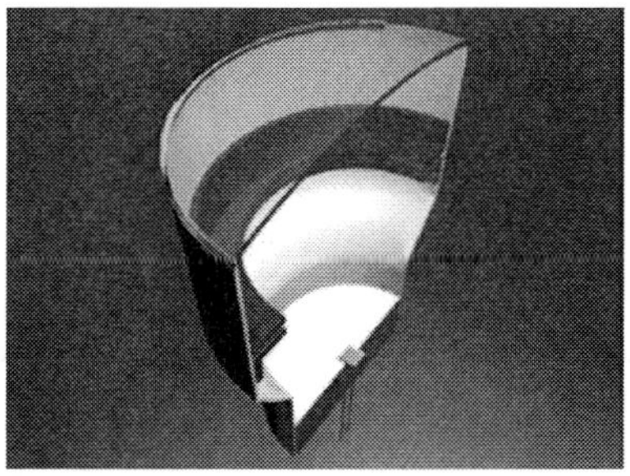

Figure 1. A cutaway of the VSiPMT showing the interior composition of the device.

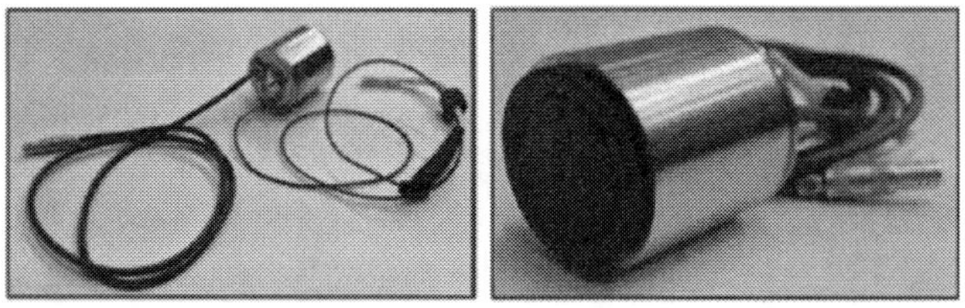

Figure 2. The VSiPMT prototype by Hamamatsu.

2. A new high gain photodetector: The VSiPMT

The central point of the challenge is finding a way to detect photons from large surfaces volumes as typically needed in many astroparticle experiments. An interesting solution is represented by VSiPMT (Vacuum Silicon PhotoMultiplier Tube), a new design for a modern hybrid photodetector based on the combination of a SiPM with an hemispherical PMT standard envelope.

The main idea, born in Naples in 2007 is to replace the classical dynode chain of a PMT with a SiPM, the latter acting as an electron multiplier at low voltage [2], see Fig. 1.

For this purpose a special no-windowed SiPM is necessary to allow the electrons to enter into the Silicon.

The VSiPMT feasibility study started with testing of a special non-windowed SiPM by Hamamatsu to an electron beam. The results[4] obtained encouraged Hamamatsu Photonics to realize two VSiPMT prototypes (Fig. 2).

The replacement of the divider makes the device compact and with three connections only: two cables to supply power to the photocathode and to the SiPM and one output for signal readout.

The prototypes have an envelope with a $7 \times 7\,\mathrm{mm}^2$ borosilicate glass entrance window and a GaAsP photocathode with a 3 mm diameter and a spectral response in the range between 300 and 750 nm.

3. The VSiPMT prototype characterization

The prototypes underwent many tests in our laboratories [6], in order to achieve a full characterization of the devices.

A picosecond pulsed laser emitting in blue region (407 nm) is sent trough a system of optical fibres to the VSiPMT, its intensity of light is varied by using differently calibrated filters and is continuously monitored by a power meter.

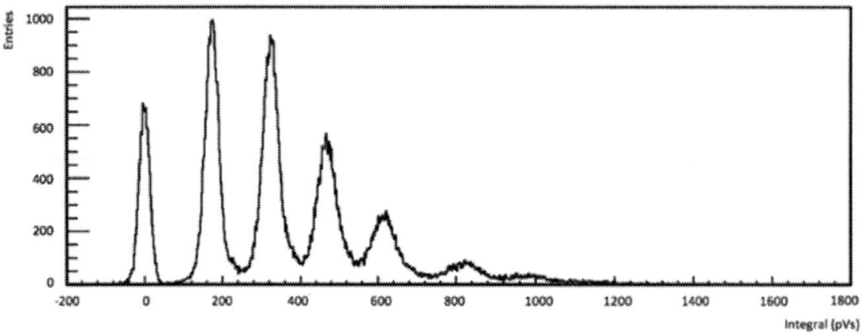

Figure 3. A typical VSiPMT charge spectrum.

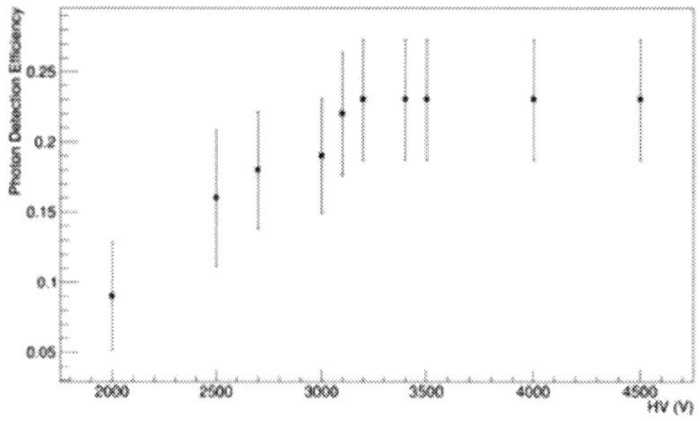

Figure 4. Results of the photon detection efficiency test in function of the HV. The PDE errors are obtained by the propagation of the errors on N_{PE} and N_{PH}.

3.1 Signal quality

A typical charge spectrum of VSiPMT is shown in Fig. 3. The remarkable separation between the peaks demonstrate the excellent photon counting capability of such device. The peak-to-valley ratio has been measured to be about 60 in the single photon condition. The device exhibits an excellent gain stability (better than 2%) in time.

3.2 Detection efficiency and stability

The device detection efficiency has been measured illuminating the device with few photons per pulse ($\sim$70) by using an optical fibre. Defining the Photon Detection Efficiency as

$$PDE = \frac{N_{PE}}{N_{PH}}$$

where N_{PE} is the number of fired cells and N_{PH} is the number of photons per laser pulse hitting the photocathode. As shown in Fig. 4 the PDE become stable around $-3\,\mathrm{kV}$ with a plateau where a PDE value of about 23% has been found.

The PDE value measured is in agreement with expectation formula

$$PDE_{VSiPMT} = QE \times \epsilon_{SiPM} \times \epsilon_{coll}$$

where QE is the photocathode quantum efficiency, ϵ_{SiPM} is the SiPM electron detection efficiency and ϵ_{coll} is the collection efficiency of the photoelectrons on the SiPM surface ($QE \sim 0.38$, $\epsilon_{SiPM} \sim 0.61$, $\epsilon_{coll} \sim 1$).

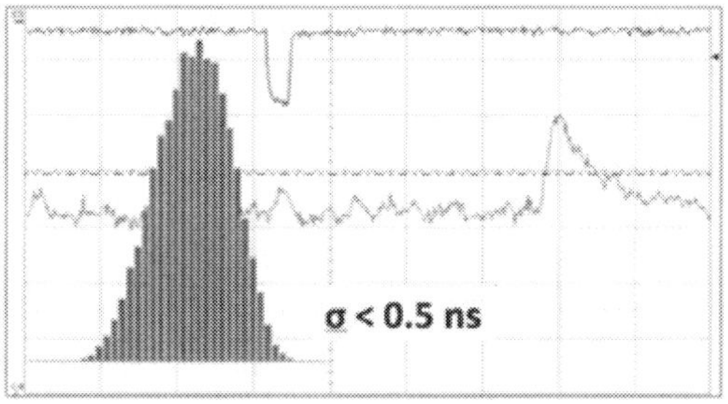

Figure 5. Printscreen of the oscilloscope. The blue squarewave is the trigger signal from the oscilloscope. The green waveform is the VSiPMT spe signal. The x-axes unit is 20 ns/div while the y-axes unit for the green line is 5 mV/div. The histogram is the distribution of the arrival time of the VSiPMT signal with respect to the laser trigger.

In VSiPMT, the plateau region is linked to the energy threshold of the photoelectrons to enter into the the silicon bulk and, consequently, to produce a signal. The high voltage in such device is required only to drive photoelectrons to the SiPM surface and to give them the right energy to trigger the geiger avalanche.

3.3 Gain

As for a SiPM the gain in this device is the ratio between the single pixel charge output signal and the electron charge. As expected we found that the gain is totally independent of the high voltage and we measured for SiPM bias voltage set to 72.5 V a value of 5.6×10^5.

3.4 Transit Time Spread

We expect the TTS to be smaller for the VSiPMT than for a standard PMT. For a standard PMT, indeed, there are two contributions to the TTS: one is due to the different trajectories of the photoelectrons going from the photocathode to the first dynode, the other is due to the different trajectories of the seconday electrons between the dynode chain. In the VSiPMT the dynode chain is replaced by a SiPM, thus for such a device the TTS is simply given by first contribution. Taking into account an eventual laser jitter, the TTS upper limit has been estimated to be <0.5 ns (Fig. 5).

4. Summary of VSiPMT performances and conclusions

The VSiPMT project born in the University of Naples in 2007 to provide good alternative to PMTs. The simple substitution of the dynode chain with a SiPM allows to reach a very high gain totally provided by the pixels working in geiger mode.

The devices show outstanding properties and performances beyond expectations, such as: excellent photon counting capability, fast response, low power consumption and great stability. VSiPMT linearity follows the same behaviour as for SiPM with the possibility to play with the number of cells in order to vary the dynamic range.

All these attractive features lead us to go ahead with a second phase aimed at the realization of a new optimized and usable version of the prototypes.

We are now attempting to realize a prototype with a larger photocatode and study an optimized configuration for a better electron focusing in order to fire full SiPM and, consequently, improve the dynamic range of the device.

Moreover in the future a SiPM optimized for electron multiplier could be designed. In fact in contrast with the photon detection, the precise and very short range of the photoelectrons incoming in SiPM, in which few electron-hole pairs are enough to induce the micro-pixel Geiger discharge, permits the reduction of the thickness of the junction p+nn+ layers with great advantage for lowering the dark current, increasing at same time the efficiency and the time resolution. Definitive and commercial VSiPMT will be constructed in synergetic collaboration with specialized factories and semiconductor research laboratories.

References

[1] H. Iams, B. Salzberg, The secondary emission phototube, Proceedings of the Institute of Radio Engineers, 1935, 0731–5996

[2] G. Barbarino et al., A new high-gain vacuum photomultiplier based upon the amplification of a Geiger-mode p-n junction, Nucl. Instrum. Meth. A, 2008, **594**, 326–331

[3] G. Barbarino et al., VSiPMT for underwater neutrino telescopes, Nucl. Instrum. Meth. A, 2013, **725**, 162–165

[4] G. Barbarino et al., Proof of feasibility of the Vacuum Silicon PhotoMultiplier Tube (VSiPMT), JINST, 2013, **8**, P04021

[5] C. Joram et al., Proof of principle of G-APD based hybrid photodetectors, Nucl. Instrum. Meth. A, 2010, **621**(1–3) 171–176

[6] G. Barbarino et al., A new generation photodetector for astroparticle physics: the VSiPMT, Astroparticle Physics Journal, 2015, **67**, 18–25

EPJ Web of Conferences **116**, 01005 (2016)
DOI: 10.1051/epjconf/201611601005

The effects of Earth's magnetic field on 3-inch diameter photomultipliers used in KM3NeT neutrino telescope

V. Giordano[a], S. Aiello, E. Leonora, and N. Randazzo on behalf of the KM3NeT Collaboration

INFN sezione di Catania, via Santa Sofia 64, 95125 Catania, Italy

Abstract. The KM3NeT neutrino telescope will be the largest underwater neutrino telescope and will be located in the abyss of the Mediterranean Sea. In neutrino telescopes the key element of the detector is the optical module and for KM3NeT it consists of 31 PMTs stored inside a transparent pressure-resistant glass sphere of 17-inch that serves as mechanical protection while ensuring good light transmission. Since the PMTs installed into an underwater neutrino telescope can change their orientation because of movements of the detector structure due to sea currents, the influence of Earth's magnetic field has been investigated. Magnetic shielding by means of a mu-metal cage is used to reduce magnetic effects and to make the response of the PMT sufficiently orientation independent. In order to quantify the effect on magnetic field, we compared measurements on variation of gain, transit time spread and detection efficiency for a 3-inch PMT in shielded and unshielded condition at 3 PMT inclinations. Data shows that variations are sufficiently low especially for timing properties.

1. Introduction

The KM3NeT Observatory [1] is a large scale neutrino telescope to be built in the deep waters of the Mediterranean Sea. In neutrino telescope the key element of detection is the optical module that consists of one or more photomultipliers (PMTs) stored inside a transparent pressure-resistant glass sphere that serves as mechanical protection while ensuring good light transmission. The KM3NeT collaboration has developed an innovative design of an optical module composed by 31 PMTs of 3-inch diameter and the readout electronics housed together in a 17-inch glass sphere (Digital Optical Module-DOM) [2]. Since PMTs are installed into an underwater neutrino telescope, they can change their orientation because of the movements of the detector structure due to sea currents so the influence of Earth's magnetic field should be investigated. Indeed the performance of a PMT could be subject to significant variation due to magnetic fields. Detection efficiency, transit time spread (TTS) and gain were measured simultaneously while varying the PMT's orientations with respect to the Earth's magnetic field, both with and without a mu-metal cage as magnetic shielding.

---

[a] e-mail: valentina.giordano@ct.infn.it

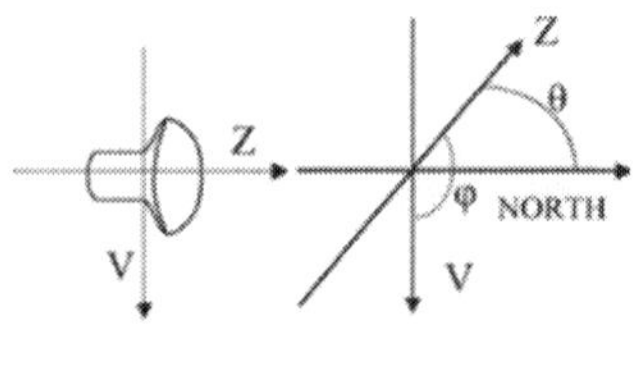

Figure 1. On the left: PMT axis (Z) and vertical axis (V). On the right: angle of inclination ϕ (zenith angle) and angle of rotation θ (azimuth angle).

2. Experimental setup

We tested one PMT with a 3-inch photocathode diameter model R12199-02 produced by Hamamatsu [3]. In the laboratory used for these measurements, the value of the magnetic field was locally measured and its value was about 40 μTesla with good uniformity over 1 m of area. In order to characterize the PMT response to an injected light source while varying its orientation relative to the Earth's magnetic field, a light-tight dark box that can rotate with respect to vertical axes (step of 1°) and to change its inclination (step of 10°) was constructed. Orientation of the PMT is shown in Fig. 1. It is possible to shield a 3-inch PMT against the magnetic field using a passive magnetic shield. The shield used in this work was a wire cage, made of 1 mm diameter wire of mu-metal, a nickel-iron alloy with very high relative magnetic permeability of 10^5. The cage was composed of two parts: an hemispherical part (which surrounds the entire PMT) and a flat part. The pitch of the grid was $(68 \cdot 68)$ mm^2, giving a shadow effect on the photocathode of less than 1% of the cathode area. The reduction factor of the magnetic field inside the cage was about 4. A laser source of 410 nm wave length and 60 ps pulse width was used to illuminate the photocathode of the PMT in the dark box. The variables measured changing the orientation (θ, Φ) were: transit time (TT), transit time spread (TTS), gain and detection efficiency, defined as the ratio between the number of detected pulses and those emitted by the laser. For the detection efficiency a second PMT was used to count photons of the laser source. In order to compare results obtained with the different photomultipliers, the PMT started its rotation from the same position with respect to the box and to the Earth's magnetic field. The PMT was measured in three inclinations of the box: horizontally oriented ($\Phi = 90°$), 50° downwards and 50° upwards. For each inclination, the PMT under test was rotated over 360° in steps of 45°. All measurements were made first with the PMT unshielded and then repeated with the mu-metal magnetic shield. The PMT was powered by a passive base.

3. Results

For all sets of measured parameters the minimum, maximum, average values and maximum variation $((max.value - min.value)/min.value)$ are calculated for the unshielded and shielded condition of the PMT. Two examples of the behaviour of the detection efficiency and gain variations with θ at constant $\Phi = -50°$ are shown in Fig. 2 where the blue lines represent the unshielded condition and the red lines the shielded ones. As shown in Fig. 2, higher fluctuations in gain and detection efficiency are found with rotations in the unshielded (naked) condition, while in the shielded condition the variation with θ is smaller. In Tables 1, 2 and 3 these values are reported for each inclination Φ of dark box and for the two conditions of PMT, unshielded and shielded. The measured values for the relative transit time are not reported in tables since the maximum variation found for PMT was less than 1% in unshielded condition, i.e the relative TT is only weakly dependent on the magnetic field. Values of the TTS in full width half maximum (FWHM) are shown on Table 1. We find that also TTS seems to be weakly affected by the magnetic field since generally its maximum variation on unshielded condition has a maximum of 8.10% reaching 5.01% with the mu-metal cage. Time properties can be affected by the geometry of the

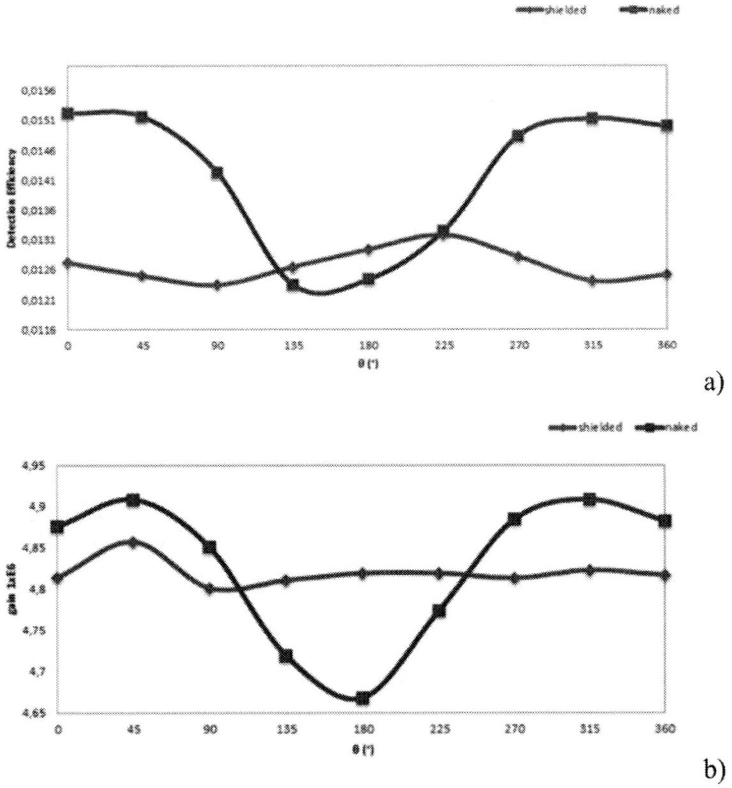

Figure 2. Detection Efficiency (a) and Gain (b) for a 3-inch PMT, 360° rotated, with an inclination $\Phi = -50°$ in shielded (red line) and unshielded condition (blue line).

Table 1. Transit Time Spread Measurements.

TTS FWHM (ns)	$\Phi = 90°$		$\Phi = -50°$		$\Phi = 50°$	
	unshielded	shielded	unshielded	shielded	unshielded	shielded
Minimum Value (ns)	3.64	3.55	3.63	3.52	3.58	3.49
Maximum Value (ns)	3.79	3.61	3.75	3.58	3.87	3.67
Average Value (ns)	3.72	3.58	3.71	3.55	3.71	3.59
Maximum Variation [%]	4.04	1.69	3.30	1.49	8.10	5.01

PMT, as has been reported for 10-inch PMTs [4]. We can say that in a 3-inch PMT, time properties are not highly influenced by the effect of the natural magnetic field. Concerning the gain of the PMT, Table 2 shows that the maximum variation was not higher than 5.1%. It should be noticed that the base voltage partition of PMTs will change in the final DOM configuration, so the voltage between the photocathode and the first dynode will change value. Moreover the final nominal gain will be of $3 \cdot 10^6$ instead than $5 \cdot 10^6$ but for these sets of measurements of gain we can say that the effects of the magnetic field are low and not as high as for the large area PMTs [4]. Concerning the measurements of the detection efficiency in Table 3, it seems to be the parameter with higher variation between shielded and unshielded condition. The largest deviation is measured for $\Phi = -50°$ in which unshielded variations are at a maximum of 23.30%. However also in this case the mu-metal cage turns out to reduce variations reducing maximum variation to the value of 6.75%.

Table 2. Gain Measurements.

Gain	$\Phi = 90°$		$\Phi = -50°$		$\Phi = 50°$	
	unshielded	shielded	unshielded	shielded	unshielded	shielded
Minimum Value [10^6]	4.59	4.75	4.67	4.77	4.63	4.77
Maximum Value [10^6]	4.72	4.83	4.91	4.86	4.80	4.90
Average Value [10^6]	4.67	4.78	4.83	4.81	4.72	4.82
Maximum Variation [%]	2.98	1.62	5.10	1.79	4.10	2.83

Table 3. Relative Detection Efficiency Measurements.

Rel. Detection Eff.	$\Phi = 90°$		$\Phi = -50°$		$\Phi = 50°$	
	unshielded	shielded	unshielded	shielded	unshielded	shielded
Minimum Value [10^{-3}]	11.86	12.34	12.34	12.34	12.34	11.90
Maximum Value [10^{-3}]	13.28	13.17	15.22	13.17	13.11	13.20
Average Value [10^{-3}]	12.58	12.36	14.17	12.67	12.35	12.48
Maximum Variation [%]	11.93	6.76	23.30	6.75	11.90	11.10

4. Conclusions

The KM3NeT collaboration aims at constructing a cubed kilometer sized telescope in the abyss of the Mediterranean Sea using optical modules comprising: 31 PMTs of 3-inch stored in a glass sphere of 17-inch. Since the orientation of photomultipliers is crucial, we have investigated the possible effects of the Earth's magnetic field on the performance of the PMT. We choose as dark box for testing the 3-inch PMT, a dark box which can rotate in θ and Φ. We chose 3 inclinations of Φ and we variate θ at steps of 45°. We shield PMT from magnetic field using a mu-metal cage and we measure several PMT's quantities at several angles determining their variations. We investigated variations on time properties of the PMT (TT and TTS) and found no significant variation. The maximum variation on the gain was also very low for the 3-inch PMT, while higher variations were found for the detection efficiency. Using a mu-metal shielding, the variations were drastically reduced for all the quantities measured. For the 3-inch PMTs the impact of the magnetic field was found to be smaller than those of large area PMTs. The larger effect was found on detection efficiency but not so high to justify the use of a magnetic shielding in the DOM made with 3-inch PMTs.

References

[1] KM3NeT (2010), ISBN 978-90-6488-033-9, `www.km3net.org/TDR/TDRKM3NeT.pdf`

[2] S.A. Martinez et al., The European Physical Journal C **74** (2014)

[3] *Hamamatsu photonics web site* (2015), `https://www.hamamatsu.com`

[4] S. Aiello et al., IEEE Transaction on Nuclear Science **59**, issue 4 (2012)

EPJ Web of Conferences **116**, 01006 (2016)
DOI: 10.1051/epjconf/201611601006

A Wavelength-shifting Optical Module (WOM) for in-ice neutrino detectors

Dustin Hebecker[1,2,a], Markus Gerhard Archinger[3], Sebastian Böser[3], Jannes Brostean-Kaiser[1], Esther Del Pino Rosendo[3], Vincenzo Di Lorenzo[3], Michael DuVernois[4], Peter Johannes Falke[5], Carl-Christian Fösig[3], Timo Karg[2], Lutz Köpke[3], Marek Kowalski[1,2], Andreas Looft[1], Krystina Sand[3], and Delia Tosi[4]

[1] Institut für Physik, Humboldt-Universität zu Berlin, 12489 Berlin, Germany
[2] DESY, 15738 Zeuthen, Germany
[3] Institute of Physics, University of Mainz, Staudinger Weg 7, 55099 Mainz, Germany
[4] Dept. of Physics and Wisconsin IceCube Particle Astrophysics Center, University of Wisconsin, Madison, WI 53706, USA
[5] Physikalisches Institut, Universität Bonn, Nussallee 12, 53115 Bonn, Germany

Abstract. We report on the development status of a single-photon sensor that employs wavelength-shifting and light-guiding techniques to maximize the collection area while minimizing the dark noise rate. The sensor is tailored towards application in ice-Cherenkov neutrino detectors embedded in inert and cold, low-radioactivity and UV transparent ice as a detection medium, such as IceCube-Gen2 or MICA. The goal is to decrease the energy threshold as well as to increase the energy resolution and the vetoing capability of the neutrino telescope, when compared to a setup with optical sensors similar to those used in IceCube. The proposed sensor captures photons with wavelengths between 250 nm and 400 nm. These photons are re-emitted with wavelengths above 400 nm by a wavelength shifting coating applied to a 90 mm diameter polymer tube. The tube guides the light towards a small-diameter PMT via total internal reflection. By scaling the results from smaller laboratory prototypes, the total efficiency of the proposed detector for a Cherenkov spectrum is estimated to exceed that of a standard IceCube optical module. The status of the prototype development and the performance of its main components will be discussed.

1. Motivation and concept

IceCube has seen its first astrophysical neutrinos [1]. Building on this success the planned IceCube-Gen2 facility [2] will allow for collecting higher statistics of high energy astrophysical neutrinos and a determination of the neutrino mass hierarchy. Deploying new optical modules in the ice gives a unique chance to improve the optical modules in noise rate and effective area. We have developed a concept that achieves this by a wavelength-shifter (WLS) coated light guiding tube connected to small PMTs. The WLS absorbs UV light (250–400 nm) and re-emits photons isotropically in the blue (> 400 nm).

[a]e-mail: dustin.hebecker@desy.de

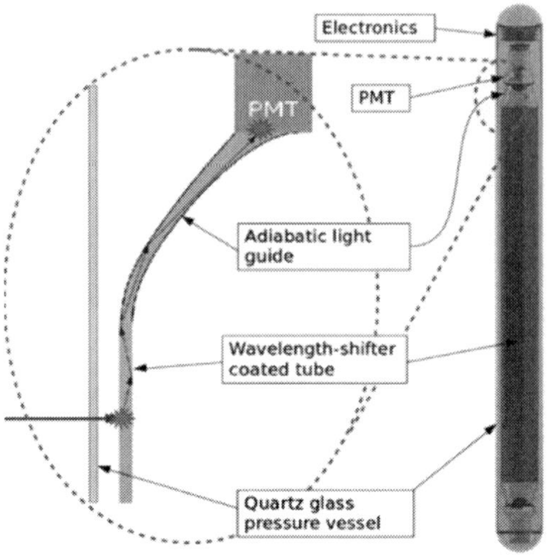

Figure 1. WOM. The working principle is sketched in a cross section. Light is absorbed by a thin layer of WLS and re-emitted isotropically at a lower wavelength. With a probability of $\approx 75\%$ the re-emitted light is captured by total internal reflection and guided towards the PMTs where it can be detected. Also shown an artists view of the full module.

Combined with the high transparency of the ice [3] the WLS allows us to access the Cherenkov photons with wavelengths $< 350\,\text{nm}$, with significant gains due to the steeply falling Cherenkov spectrum $\propto \frac{1}{\lambda^2}$. The shifted photons are detected by small and hence less noisy PMTs at the ends of the tube. This concept, the Wavelength-shifting Optical Module (WOM), provides a low cost single-photon detector, with an increased photosensitive area and reduced overall noise. In large volume detectors like IceCube-Gen2 [2] and MICA [4], this allows the energy threshold to be lowered while keeping the detector volume and amount of deployed modules constant. In addition, the large sensitive area and signal to noise ratio allows for good vetoing capabilities making this concept interesting also for experiments like SHiP [5]. The WOM as shown in Fig. 1 consists of a pressure vessel, a wavelength-shifting and light-guiding tube, an adiabatic light guide and PMTs with their readout electronics. The pressure vessel protects the sensitive components inside from the outer pressure. The pressure vessel is 1.3 m long, 114 mm in diameter and made from quartz glass in order to be transparent to UV-light. The wavelength-shifting and light-guiding tube within the pressure vessel has a diameter of 90 mm and is made of acrylic glass. The thin air layer around the tube and high refractive index difference facilitates the light guiding of the shifted light by total internal reflection (theoretical maximum 74.6%). The light captured in the tube propagates to the upper or lower end. Adiabatic light guides, glued to the ends, direct the light virtually lossless towards the PMTs. Several low noise PMTs are currently under consideration. The readout electronics will be an adapted version of the Gen2 Digital Optical Module (Gen2-DOM [6]) electronics.

2. Efficiency

We have developed a WLS paint and investigated its properties when coated on a transparent light guiding tube. The key value for the quality is the capture and transport efficiency ϵ defined as the ratio of shifted photons leaving on either side of a tube coated with a WLS paint to the amount of UV photons the tube is exposed to. This value has been measured in different ways. They all have in common that the test tube is illuminated with a small light beam perpendicular to the tube axis and the light coming out at one end is detected. The other end is blackened to reduce reflections. Since we plan for reading

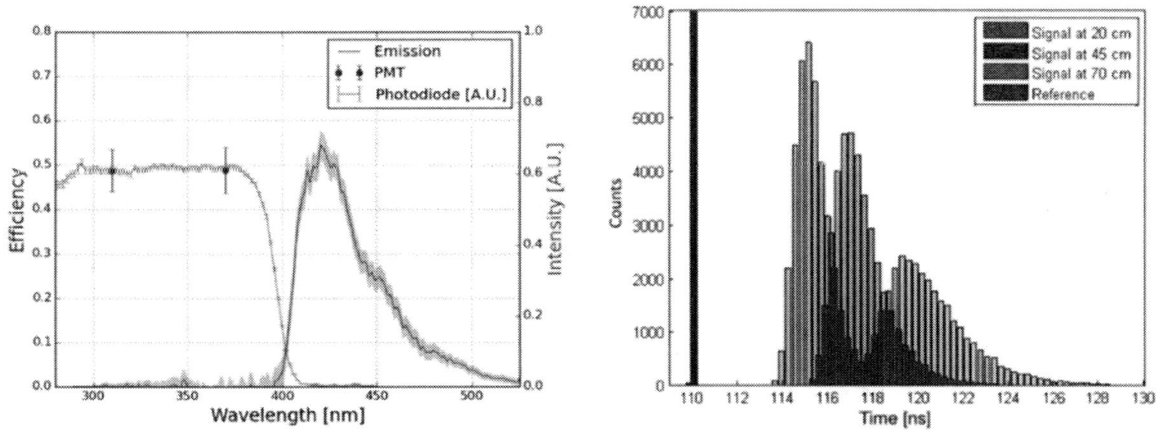

Figure 2. Efficiency ϵ of a custom made WLS paint coated on a PMMA tube for light capturing and transport measured with a photodiode and a PMT. The photodiode spectrum is scaled to match the PMT results and allows the reader to see the flatness of the spectrum. Also the emission spectrum of the same paint is shown (left). For different distances between illumination point on the tube and the PMT the broadening of the arrival time distributions is shown (right).

out both ends, we correct ϵ for this by multiplying with a factor of two. This efficiency ϵ is used as a measure to optimize the paint, for selection of the right substrate and for estimating the final detector efficiency.

2.1 WLS paint development

The paint performing best in efficiency ϵ, adhesiveness and surface quality[1], is made with a mixture of 77.31%[2] Toluene, 22.29% Paraloid B72[3], 0.13% Bis-MSB and 0.27% P-Terphenyl [7]. The efficiency and emission spectra of this paint are shown in Fig. 2 (left). To apply the paint to the substrate a dip-coating process is used, where the substrate is immersed in WLS paint and withdrawn with a well defined velocity. The higher the withdraw velocity the thicker the paint film on the substrate. With the above mixture a velocity of 9.3 cm/min has given the best absorption and surface quality.

2.2 Timing

Photon transit times through the device are affected by light paths of different lengths in the light guide and by the WLS decay time. This has been investigated with a sub-nanosecond input pulse as shown in Fig. 2 (right). For the tube of 90 cm length and homogeneous illumination an overall FWHM of $\approx$ 10 ns is expected.

2.3 Theoretical expectations

The efficiency ϵ had been measured with a photodiode setup in the first development stages as described in [8]. The measured values have been a factor $>$ 2 lower than expected when considering a geometrical capture efficiency of 74.6%, 96% WLS-quantum efficiency [9] and Fresnel losses. Measurements with

[1] Evaluated by visual inspection.

[2] Percentage by weight.

[3] Ethyl-methacrylate copolymer.

a PMT matched the expectation, however, if reasonable transport losses were assumed. Subsequent measurements of the photodiode showed limited angular acceptance which was not taken into account in the original measurement. A simulation of the setup including this effect produces consistent results between the PMT and photodiode measurements.

3. Glass housing

The quartz glass pressure housing has undergone static pressure testing up to 5250 psi (362 bar). After the test some optical defects were seen in the glass, but they may have been present prior to the testing. Repeated tests did not result in additional, or modified, glass features. No cracking was observed. Feedthrough connectors of the Seacon type previously employed in IceCube DOMs can be mounted on the quartz glass housing as needed.

We plan to structure the inside of the pressure vessel to further increase the efficiency. As the light passes from ice to quartz glass to air, it passes different refractive indices. As quartz glass has a greater refractive index than ice, nearly all the light passes through the first surface. Air has a lower refractive index therefore the light with angles $\theta \geq 48.75° = \arcsin(\frac{n_{air}=1}{n_{ice}=1.33})$ does not enter the detector at the inner surface of the pressure vessel when the module is deployed in ice. This not only reduces the efficiency of the detector but also decreases the angular acceptance. In order to counteract both of these effects we investigate the use of lenticular lens arrays to cover the inside of the pressure vessel. A first calculation indicates that this can give an efficiency gain of up to 37%.

4. Conclusion and outlook

The WOM is a low cost, low noise and high efficiency optical module. It is mainly developed to be used in IceCube-Gen2 and similar large volume neutrino Cherenkov detectors. However the concept is flexible and can be adapted for other types of detectors like in the surrounding background tagger of the SHiP experiment [5]. For the realization of this detector a high efficiency WLS paint has been developed. It shifts light from the range of 250 – 400 nm to > 400 nm. Efficiency measurements with PMTs showed capture and transport efficiencies of $\approx 50\%$ while photodiodes measurements show lower results. First simulations show consistency between both measurements when accounting for the photodiode's angular efficiency. An adiabatic light guide has been produced that acts as an adapter between the WLS tube and the small PMTs. The quartz glass pressure vessel in the necessary elongated shape has passed pressure testing. The assembly of a first complete prototype is progressing well, and gives hope that this novel concept can be tested in-situ soon.

References

[1] IceCube Collaboration: M.G. Aartsen et al., Science **342**, 1242856 (2013)
[2] IceCube-Gen2 Collaboration, `arXiv:1412.5106` (2014)
[3] M. Ackermann et al., J. Geophys. Res., **111**, D13203 (2006)
[4] S. Böser, M. Kowalski, L. Schulte et al., Astropart. Phys. **62**, 54-65 (2015)
[5] SHiP Collaboration, `arXiv:1504.04956` (2015)
[6] M. A. DuVernois for the IceCube-Gen2 Collaboration, PoS (ICRC2015) 1148
[7] Dustin Hebecker, `https://www-zeuthen.desy.de/~hebecked/Publications_etc./Master_Thesis/Dustin_hebecker_master_thesis.pdf` Master Thesis, University of Bonn (2014)
[8] L. Schulte, Markus Voge, Akos Hoffmann et al., ICRC, `arXiv:1307.6713` (2013)
[9] X. Hua-Lin et al., Chinese Physics C, **34**, No. 11, Nov. (2010)

EPJ Web of Conferences **116**, 02001 (2016)
DOI: 10.1051/epjconf/201611602001

Energy reconstruction of high energy muon and neutrino events in KM3NeT

Evangelia Drakopoulou[1,2,a], Christos Markou[1,b], Ekaterini Tzamariudaki[1,c], Konstantinos Pikounis[1,2,d] for the KM3NeT Collaboration

[1] N.C.S.R. Demokritos, Patriarchou Grigoriou and Neapoleos 27, Agia Paraskevi, Greece
[2] National Technical University of Athens, Heroon Polytechniou 9, Zografou Campus, Greece

Abstract. KM3NeT will be a European deep-sea infrastructure of neutrino telescopes covering a volume of several cubic kilometers in the Mediterranean Sea aiming to search for high energy neutrinos from galactic and extragalactic sources. This analysis focuses on muons coming from neutrino charged-current interactions. In large water Cherenkov detectors the reconstructed muon is used to approximate the neutrino direction and energy, thus providing information on the astrophysical neutrino source. Muon energy estimation is also critical for the differentiation of neutrinos originating from astrophysical sources from neutrinos generated in the atmosphere which constitute the detector background. We describe a method to determine the muon and neutrino energy employing a Neural Network. An energy resolution of approximately 0.27 has been achieved for muons at the TeV range.

1. Introduction

The KM3NeT Collaboration aims to deploy a neutrino telescope with volume of several cubic kilometers at the Mediterranean Sea that will search for neutrinos from galactic and extragalactic sources. The first phase of the neutrino telescope is called the KM3NeT-ARCA detector and consists of two building blocks covering a total volume of approximately 1 km^3. The basic detection units are the optical modules (OMs) consisting of 17 inch diameter pressure-resistant glass spheres, each equipped with 31 photomultipliers (PMTs) of 3-inch photocathode diameter and their readout electronics. The PMTs are arranged in vertical strings covering a roughly cylindrical instrumented volume.

The reconstruction of the muon and consequently the neutrino energy is critical for neutrino telescopes. The signal corresponds to astrophysical ν_μ events with an energy spectrum extending to higher energies compared to the energy spectrum of the atmospheric ν_μ events which are the background. The energy estimation thus allows the discrimination of astrophysical ν_μ from atmospheric ν_μ events that refer to neutrinos produced by charged Kaons or pions during the interaction of

[a] e-mail: liliadrak@inp.demokritos.gr
[b] e-mail: cmarkou@inp.demokritos.gr
[c] e-mail: katerina@inp.demokritos.gr
[d] e-mail: pikounis@inp.demokritos.gr

cosmic rays in the atmosphere. Moreover, the determination of the energy is a major parameter in the identification of the neutrino sources as it enhances many analyses performed in Cherenkov detectors, such as searches for point sources of neutrinos or diffuse extraterrestrial neutrinos.

Monte-Carlo simulations have been performed for one block (with a volume of $\simeq 0.5\,\mathrm{km}^3$) of the KM3NeT-ARCA detector including the neutrino interactions in the medium, generation and propagation of secondary particles, the light generation and propagation in water and finally the PMT response to the incident photons. The light from the background of ${}^{40}K$ decays is also considered. 10^{10} neutrinos have been generated according to a spectrum $E^{-1.4}$, instead of E^{-2} that is theoretically expected, in order to get reasonable statistics in the whole energy range from 10^2 GeV to 10^8 GeV. A cylindrical volume, the can, encloses the detector volume and is used for the simulation of the Cherenkov photons. The radius and the height of the can is approximately three times the absorption length of light in sea water larger than the detector volume. A larger volume, the generation volume, is used for neutrino generation and corresponds to the can volume expanded at the maximum lepton range in the medium (rock or sea water) for the maximum value of the energy range that is generated. Prior to the estimation of energy, the direction of the particle is reconstructed using the standard algorithm for track reconstruction in KM3NeT. In the final event sample, events are selected (well reconstructed tracks) if the maximum log-likelihood value per degree of freedom found by the fit (Λ) satisfies the relation $\Lambda > -5.8$ [1].

2. Energy reconstruction using a neural network

Muons lose energy via ionization and by stochastic processes, such as bremsstrahlung, pair production, and photonuclear interactions. The total average energy loss of the muon is:

$$\frac{-dE}{dx} = a(E) + b(E) \cdot E_\mu$$

where $a \simeq 0.274\,\mathrm{GeV\,m^{-1}}$ accounts for the energy loss due to ionization and $b \simeq 0.000349\,\mathrm{m^{-1}}$ is due to the stochastic energy loss. Muons with $E_\mu > 1$ TeV lose energy dominantly stochastically, while for lower energies ionization dominates. In this study a method to derive muon (and neutrino energy) from the light collected as the muon passes through the instrumented volume has been developed.

Since the reconstruction of the muon energy is based on the collection of light in PMTs, muons should travel an adequate distance inside the instrumented volume before an attempt to estimate the energy is made. Therefore, a containment selection has been established aiming to distinguish muon tracks which are well contained in the detector, thus depositing the light in the instrumented volume, from muons that pass outside or cross the borders of the detector. Muons which do not travel an adequate distance inside the detector deposit only a fraction of light in the instrumented volume thus not allowing a reliable estimation of the muon energy. In this case a lower limit of the muon energy is calculated.

This method employs a Multi-Layer Percepton Neural Network with appropriate input variables that show a strong dependence on the muon energy and can thus be used for the energy reconstruction. These variables are:

i. The number of optical modules with hits used for the track direction reconstruction per track length.
ii. The number of photomultipliers with recorded hits (N_{PMTs}) that are used for the track reconstruction, weighted according to the vertical distance from the reconstructed track (D_i). This weight refers to the ratio D_i/D_{max} where D_{max} is a maximum distance of 500 m.
iii. The ratio of N_{PMTs} to the number of PMTs that could have been hit by Cherenkov photons according to the direction of the reconstructed track but have not recorded pulses.
iv. The total Time over Threshold (ToT) (as a measure of charge) of all PMTs that are used for the track reconstruction.

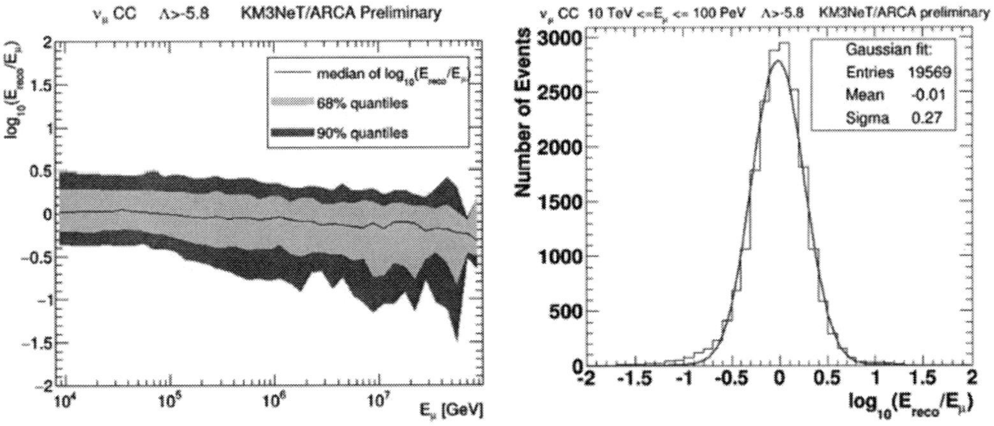

Figure 1. Left plot: The median of $\log_{10}(E_{\text{reco}}/E_\mu)$ (with 68% and 90% quantiles) as a function of E_μ for events that satisfy the containment selection. Right plot: The distribution of $\log_{10}(E_{\text{reco}}/E_\mu)$ for events with $E_\mu \geq 10\,\text{TeV}$ that satisfy the containment selection. A Gaussian fit is applied.

The input variables are inserted in the Neural Network which is trained and tested using half the sample of ν_μ events while the energy is determined for the other half of events to avoid bias.

2.1 Results on muon and neutrino energy reconstruction

The results on the muon energy reconstruction for reconstructed events with $\Lambda > -5.8$ that satisfy the containment selection are shown in Fig. 1. Figure 1 (left plot) depicts the median of $\log_{10}(E_{\text{reco}}/E_\mu)$ with 68% and 90% quantiles as a function of E_μ where E_μ is the MC muon energy at the can level and E_{reco} is the reconstructed muon energy. A very good energy reconstruction has been achieved for high energy muons with $10\,\text{TeV} \leq E_\mu \leq 100\,\text{PeV}$ resulting in a median of 0.02 for $E_\mu \geq 10\,\text{TeV}$ while for muons with energies $E_\mu > 10\,\text{PeV}$ the calculation suffers from limited statistics. The energy resolution that has been attained for these events, that correspond to approximately 90% of all the reconstructed events with $\Lambda > -5.8$, is $\simeq 0.27$ in $\log_{10} E_\mu$ as it is shown in Fig. 1 (right plot). For the rest 10% of events that do not satisfy the containment selection a lower limit of the energy is calculated. In this case the energy is underestimated leading to larger tails in the left side of the distribution of $\log_{10}(E_{\text{reco}}/E_\mu)$ for all reconstructed events. However, due to the low percentage of these events (10%) in the whole sample, the energy resolution is $\simeq 0.28$.

The neutrino energy cannot be reliably reconstructed for events that cross the detector and the neutrino interaction vertex lies outside the instrumented volume. In such cases most photons from the hadronic activity are not detected, thus resulting in an underestimation of the neutrino energy. Only a lower limit of the neutrino energy can then be reported that corresponds to the reconstructed muon energy. In order to make sure that the muon and neutrino energy is reliably reconstructed for events with the neutrino vertex close or inside the instrumented volume we consider events with the reconstructed vertex inside a fiducial volume which is slightly smaller than the volume of the detector. This criterion alone is not sufficient to reject events that have interacted outside the detector. Therefore we apply selection criteria originally developed for atmospheric muon rejection. This selection keeps most events that have interacted inside this fiducial volume and rejects events with high activity on border strings. Moreover, events with activity upstream of the reconstructed vertex that is compatible with the track hypothesis are examined and rejected if they have given pulses in PMTs at the edge of the detector. The results on muon and neutrino energy reconstruction (obtained using the aforementioned Neural Network) for this subsample of events, are shown in Fig. 2 (left and right plot respectively).

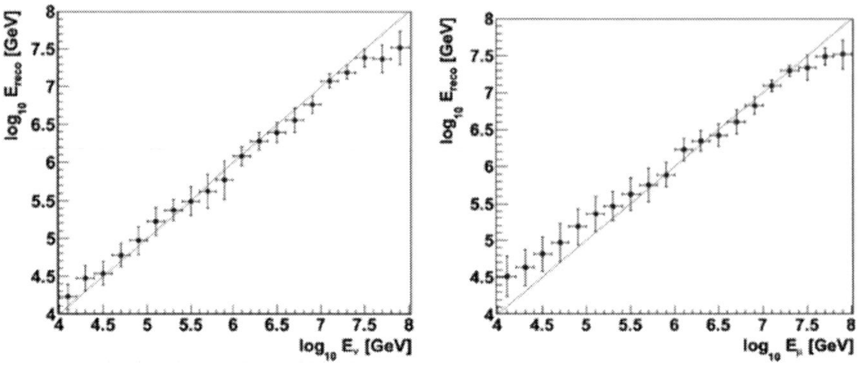

Figure 2. Left (right) plot: The reconstructed energy ($\log_{10} E_{\text{reco}}$) with respect to the muon (neutrino) energy for events (with $E_\mu \geq 10\,\text{TeV}$) that satisfy the containment selection and the interaction vertex is estimated to be inside the fiducial volume. The mean with RMS error is depicted.

The overestimation of the muon energy (left plot) is due to the fact that most photons from the hadronic shower are detected. A very good linear relation between the reconstructed energy with the neutrino energy can be observed (right plot) leading to a reliable reconstruction of the neutrino energy for tracks that interact well inside the instrumented volume.

The systematics uncertainties for this energy reconstruction have been studied assuming a variation of $\pm 10\%$ on the estimation of the absorption and scattering length of light in sea water and on the effective area of the PMTs. The relative deviation in the mean reconstructed energy $\Delta E/E$ corresponds to $\pm 3\%$, $\pm 0.1\%$ and $\pm 1\%$ respectively.

This method for the muon energy reconstruction has been applied to an alternative detector configuration with a larger average horizontal distance between the detection units (strings) leading to an instrumented volume of $\simeq 0.8\,\text{km}^3$ (compared to $\simeq 0.5\,\text{km}^3$ for the previous detector configuration). The containment selection and the input variables have been appropriately adjusted in this case and the Neural Network has been trained again. For high energy events with $10\,\text{TeV} \leq E_\mu \leq 100\,\text{PeV}$ satisfying the containment selection an energy resolution of $\simeq 0.25$ in $\log_{10} E_\mu$ has been achieved increasing to $\simeq 0.26$ when all reconstructed events with $\Lambda > -5.8$ are taken into account.

Conclusions

The presented work has been partially financed by the European Social Fund (NSRF 2007–2013) under the Thalis initiative GRBNeT project-MIS 360381. A method to estimate the muon and neutrino energy in the KM3NeT-ARCA detector, has been presented. A very good energy resolution of $\simeq 0.27$ in $\log_{10} E_\mu$ for high energy muons with $10\,\text{TeV} \leq E_\mu \leq 100$ PeV has been attained which improves to $\simeq 0.25$ in $\log_{10} E_\mu$ when a larger instrumented volume is considered.

Reference

[1] A. Trovato, E. Drakopoulou, P. Sapienza, proceedings of ICRC2015 (2015), 8, to be published

EPJ Web of Conferences **116**, 02002 (2016)
DOI: 10.1051/epjconf/201611602002

The Run-by-Run Monte Carlo simulation for the ANTARES experiment

L.A. Fusco[1,2,a], A. Margiotta[1,2] on behalf of the ANTARES Collaboration

[1] Dipartimento di Fisica e Astronomia dell'Universitá, Viale Berti-Pichat 6/2, 40127 Bologna, Italy
[2] INFN - Sezione di Bologna, Viale Berti-Pichat 6/2, 40127 Bologna, Italy

Abstract. The ANTARES neutrino telescope is the largest and longest-operated underwater neutrino telescope. Data acquisition conditions in a marine environment are not stable in time: biological and physical phenomena follow a seasonal evolution producing a periodical change of the rates registered at the neutrino telescope. Variations in the sea current velocity also affect the measured baseline value and the burst fraction on short time scales.

Monte Carlo simulations of the detector response to charged particles in the proximity of the telescope should reproduce the conditions of the medium and of the acquisition setup as much as possible. An efficient way to account for their variability is to extract related information directly from the data runs. A Run-by-Run simulation procedure has been developed to follow the time evolution of data acquisition in ANTARES.

1. Introduction

The ANTARES detector is at present the largest neutrino telescope in the Northern Hemisphere [1]. It has been operated continuously since 2007, and in its full configuration since 2008. It is located at a depth of 2475 m in the Mediterranean Sea at (42° 48′ N, 6° 10′ E), 40 km off-shore from Toulon in the Gulf of Lion, Southern France.

The neutrino detector consists of 12 slender detection lines, kept taut by a top buoy, with 25 storeys each. A storey hosts three optical modules (OMs), glass spheres housing 10-inch photomultiplier tubes (PMTs) and a local control module (LCM) containing the off-shore electronics. The distance between storeys on each line is 14.5 m and the first storey is located ~100 m from the sea-bed. PMTs in the storey are oriented 45° downwards in order to optimise their acceptance to Cherenkov light coming from upgoing particles. The length of a line is 450 m and the horizontal distance between neighbouring lines is 60–75 m.

A Junction Box (JB), on the sea bed, connects the lines to the shore station with an electro-optical cable providing power, transferring data to shore and distributing a clock signal responsible for the synchronisation of the different detector elements. At the shore station, filtering and triggering of data

[a] e-mail: `lfusco@bo.infn.it`

take place. Data stream is subdivided into "runs" lasting from 2 to 12 hours depending on the data taking setup.

Monte Carlo simulations are necessary to understand the response of the detector to physics events, to check the correct functioning of the apparatus, to optimise selection cuts for the rejection of background events and to evaluate the level of purity of the selected sample of events. The simulation chain can be subdivided into two main steps:

1. Physics simulation: neutrino interactions and muon bundles are generated in the proximity of the detector. Particles are propagated through the active volume and Cherenkov light is produced along their paths.
2. Data acquisition simulation: the PMT behaviour and the behaviour of the data acquisition electronics are simulated. Filtering and triggering algorithms are applied. The optical background due to environmental light (mainly bioluminescent bacteria and ^{40}K decay) is also added.

2. Physics simulation

2.1 Event generation

The instrumented volume of the apparatus is a cylinder which contains all the PMTs. A larger cylinder, called the *can*, surrounding the first one, extending it by three light attenuation lengths. The *can* defines the active volume within which Cherenkov photon emission can lead to a signal in a PMT. Outside this volume, only particle energy losses during propagation are considered.

The dedicated *GENHEN* [2] package is suitable for the full range of neutrino studies in ANTARES from neutrino oscillations to high energy astrophysics, being the majority of detectable neutrinos in a range of energies from tens GeV, limited by the energy threshold of muon detection, to hundreds PeV, where the absorption of neutrinos in the Earth, which strongly attenuates the upward neutrino flux, must be taken into account.

Atmospheric muon bundles are generated in ANTARES with the *MUPAGE* software [3]. *MUPAGE* is based on the parametrisation of the angular and energy distributions of muons under-water or in ice as a function of the muon bundle multiplicity [4]. These parametrisations are taken from complete simulations and data collected with the MACRO experiment at Gran Sasso, extrapolated either under the sea or under an ice layer. The usage of parametric formulas allows the production of an extremely large number of events with a reasonable CPU time consumption, the main limitation being the primary cosmic ray flux that cannot be modified.

2.2 Particles and light propagation

Long-lived particles produced by the physics generators are tracked through the water in the *can* volume using a GEANT-based [5] package, denoted in ANTARES as *KM3* [2]. The composition and density of the sea water and its optical properties as measured at the experimental site are considered. All relevant physics processes are activated (energy loss, multiple scattering, radiative processes and photo-nuclear interactions).

Thanks to the homogeneity of the sea-water, a photon-by-photon simulation for the Cherenkov light is not required, differently from what is done to describe the photon propagation through the Antarctic ice where optical properties show a dependence on the ice stratification. This means that a set of "scattering tables" containing the probability of each photon to give a hit on a PMT as a function of 5 parameters: the distance from the muon; 3 angles defining the direction of the photons with respect to the muon and to the PMT; the photon arrival time; and considering the diffusion and absorption phenomena, can be created in advance. Muons are propagated in the can volume by MUSIC [6]. Steps of 1 m for muon tracks are considered, the ionisation energy loss and the Cherenkov light emission are calculated. The probability for each Cherenkov photon to reach PMTs is extracted from the scattering tables. Also the probability of a catastrophic energy loss is evaluated. Similar scattering tables are created for electromagnetic showers.

Hadronic showers are treated differently. In this case a large number of charged particles is produced at the interaction vertex. The computation of scattering tables for each particle would require an event-by-event simulation, and a huge amount of CPU time, because of the high variability in the hadronic shower composition. To overcome this problem, a Multi Particle Approximation approach is followed. Each hadron is considered as equivalent to an electron. The electron scattering tables are used in association with opportune weights, evaluated for each hadron after many complete photon tracking simulations. The light yield from hadronic showers obtained with this technique is compatible with what is achieved with a detailed simulation.

3. Data acquisition simulation

The final step of the simulation is performed by the *TriggerEfficiency* program. This software adds the optical background to the hits generated by physics events, reproducing a sort of data stream, simulates the behaviour of the front-end electronics and applies the trigger algorithms for candidate event selection. The optical background can be generated and added to MC events according to a Poisson distribution using a fixed background rate specified by the user or, in alternative, extracting information on rates directly from real data.

Two front-end ARS (Analogue Ring Sampler) chips are connected to each PMT, integrating the analogue signal from the PMT over a typical time window of 25 ns. This behaviour is simulated by summing up the number of detected photons in that window. The two ARS operate in a token ring scheme. After the integration time of the first ARS chip, the second takes over. Only after the integration time of the second ARS chip, the dead time of the first plays a role. The time resolution for single photo-electron signals is 1.3 ns and decreases for higher amplitudes. To simulate this effect, the hit times are smeared using a Gaussian function with a width $\sigma = 1.3\ \text{ns}/\sqrt{N_\gamma}$, where N_γ is the number of simultaneously detected photons.

Different level of triggers are considered in the ANTARES DAQ:

- "Level zero" (L0) trigger selects hits that have a charge greater than a low threshold – typically 0.3 photo-electrons (p.e.). The amplitude measurement is simulated by smearing the integrated number of photons with an empirical function that produces a (roughly Gaussian) smearing of about 30%. The dynamic range of the charge integration has a saturation level corresponding to about 20 photo-electrons. MC events are triggered with the same trigger algorithm used for real data.
- The "first level trigger" (L1) is built up of coincidence hits in the same storey within a 20 ns time window or single hits with a large charge amplitude, greater than a "high threshold" tuneable from 2.5 p.e. to 10 p.e.
- A trigger logic algorithm, "level 2" trigger (L2), is then applied to data and operates on L1 hits.

All trigger definitions are reproduced inside *TriggerEfficiency*.

3.1 Run-by-run approach

Environmental conditions in a marine environment are not stable and constant in time. Variations affect data acquisition in an under-sea neutrino telescope like ANTARES. Biological [7] and physics phenomena [8] show evolving trends on seasonal timescales producing a periodical change of the background optical rates registered at the detector. Short term variations are also present as the optical rates are modified by the sea current velocity. In addition to environmental effects, the detector elements might not collect data continuously, because of temporary or permanent malfunctioning of the optical modules or lack of connection to some part of the apparatus. Finally, it is possible to modify the trigger algorithms applied to the data taking along the life of the detector.

A reliable Monte Carlo simulation of the detector should reproduce all these effects. An efficient way to account for the variations of the optical background is to extract related information directly from the data. The *TriggerEfficiency* program takes this information from raw data files. It considers the counting rate in short segments of the data stream, about 100 ms long, and simulates the corresponding optical background according to the measured charge distribution of hits. DAQ conditions (the status of each detector element, the active triggers, the data filtering parameters) for each run are stored in the ANTARES database, which is accessed by *TriggerEfficiency*, and used during the simulation.

The resulting output has the same format of raw data and is processed using the available reconstruction algorithm to extract physics information from data. This approach has significantly improved the data/MC agreement, and allowed for a better monitoring of the time evolution of the data acquisition.

4. Conclusions

The Run-by-Run scheme is a reliable tool to produce Monte Carlo simulations for an under-sea neutrino telescope like ANTARES, following the real evolution of environmental conditions. Extracting information from the raw data and the database, on OM-per-OM basis, it allows to reproduce the behaviour and the environment of the detector even on short time scales. All data analyses currently produced by the experiment [9] are based on this simulation chain and good agreement is found between simulations and real data. ANTARES data acquisition will go on at least until the end of 2016 and constant efforts are ongoing to keep the application of this simulation strategy on track with data acquisition.

References

[1] M. Ageron et al., Nucl. Instrum. Meth. A, **656**, 11–38 (2011)
[2] A. Margiotta, Nucl. Instr. Meth. A **725**(2013) 98
[3] G. Carminati et al., Computer Physics Communications, **179**(12), 915–923 (2008)
[4] Y. Becherini et al., Astrop. Phys., **25**, 1 (2006)
[5] GEANT program manual, CERN program library long writeup, W5013 (1993)
[6] P. Antonioli et al., Astrop. Phys., **7**, 357 (1997)
[7] M. Circella, Nucl. Instr. and Meth. A **602**, 1–6 (2009)
[8] H. van Haren et al., Deep-Sea Research I **58**, 875–884 (2011)
[9] C.W. James, Proceedings of the 34th ICRC, The Hague, The Netherlands (2015)

EPJ Web of Conferences **116**, 02003 (2016)
DOI: 10.1051/epjconf/201611602003

CORSIKA modifications for faster background generation

Kyle Jero[a]

University of Wisconsin Physics Department, 1150 University Ave, Madison, WI 53706, USA

Abstract. CORSIKA is a simulation program for extensive air showers initiated by high energy cosmic particles. These air showers create the majority of the muons and neutrinos which neutrino that telescopes detect and are considered a background signature in searches for astrophysical neutrinos. This contribution will discuss changes to CORSIKA which allow for faster high energy background simulation. The theory, implementation, application, and performance of these modifications will be presented.

1. Introduction

When estimating the background from air showers present in neutrino telescopes it is necessary to use simulations. Often the air shower simulation CORSIKA [1] is used. To accurately model the final products of an air shower, CORSIKA must track every interaction, decay, and particle present in these fabricated air showers. Since the number of secondary particles increases with energy this leads to a computation time which rises with the energy of the initial incident particle. The computation time required increases even more if rare showers are required, such as in [2].

In the specific case of [2] high energy neutrinos with no accompanying muons were the focus of study. Because neutrinos and muons are largely the decay products of secondary particles in air showers, they tend to be produced one power steeper than the cosmic ray spectrum. This makes obtaining statistics for high energy neutrinos difficult because showers which could produce a relevant neutrino often do not. Herein however, also lies an advantage. Since these neutrinos are the byproduct of secondary hadronic particles, the maximum neutrino energy obtainable in a given shower is limited by the highest energy hadronic particle. Thus, if, at any point in shower development no hadron or neutrino above an energy of interest exists, then the shower being simulated is not relevant. Such showers do not need to be simulated to completion.

2. Modifications

When CORSIKA was developed in 1985 it was designed to handle primaries of all energies on contemporary systems. This meant keeping the memory footprint low was important. As such, a last in first out (LIFO) buffer was implemented. In this configuration, after the initial secondaries were produced, the most recently generated secondary was simulated next. Since new particles are always

[a] e-mail: `kjero@icecube.wisc.edu`

placed at the front of the buffer, the most recently generated child was simulated next. This process continues repeatedly until no new particles are added to the buffer after the simulation of an existing particle. At this point, a sibling of the previously simulated particle is simulated. Again, the generation of children repeats until no new particles are created. This process repeats until the shower completes. Effectively what is happening is the complete simulation of one branch of the shower before another begins. This completes the shower in a depth-first manner. However, this type of shower development is not optimal for searching for the highest energy hadron since most energetic hadron could not be simulated until nearly the end of the simulation of the shower.

Another option that exists for memory management is the first in first out (FIFO) buffer. In such a configuration, after the direct secondaries are generated, the first direct secondary would be simulated and its children placed at the end of the buffer. This continues until all the direct secondaries have been simulated. At this point the children of the direct secondaries, which are now at the front of the buffer, would be simulated and their children would be placed at the end of the buffer. This process would continue until all particles in the buffer have been simulated. A shower simulated in this way is propagated breadth-first and keeps particles in the buffer which are roughly the same number of decays/interactions after the primary particle. While this type of shower development requires more memory than a LIFO to complete the same task, it is near optimal for searching for the highest energy hadron. This is because every particle is kept at the same stage of shower development. Further, high energy neutrinos appear soon after the initial interaction since it is difficult for hadrons to keep large amounts of energy through decays/interactions.

The only way to make the process of finding these neutrinos faster is to first simulate the hadrons which can create a neutrino above threshold. This can be done simply by operating the buffer as a dual ended insert and single ended removal buffer, making it a hybrid between the FIFO and LIFO. By putting particles which are above threshold at the front of the buffer and particles which are below threshold at the end of the buffer the relevant hadrons are inspected first while the rest of the shower is held. If a neutrino above threshold is produced, then the full shower can be completed. In this way only the necessary computations are completed to find out if the shower is relevant or not.

A version of CORSIKA with the implementation of the hybrid buffer and a threshold energy was the final outcome of the modifications. The threshold works by stopping showers and starting the next if no particles are above the energy threshold are in the buffer and no neutrino above threshold has been created. Showers which are stopped save the primary particles information to the output file, while fully propagated showers appear as normal output. This modification has been submitted to and accepted by the CORSIKA maintainers. It will appear in the next major release of CORSIKA.

3. Performance

By stopping showers once they have crossed a threshold energy for neutrino production and saving their full output only if they posses a relevant neutrino we improve the run time and file size requirements of our simulation. Since the reduction truncates events and does not change the underlying algorithms, the savings are proportional to the amount of truncation achieved. This can be seen in Fig. 1 on top of a minimum floor value.

4. Applications

This modification was first used to obtain statistics at high neutrino energy for [2]. Here, a dataset of showers with neutrinos above 100 TeV allowed a model fit of suppressed atmospheric neutrino sources as well as a confirmation of the conventional atmospheric neutrino veto.

One detail about neutrino vetos that was not in the scope of [2] is the question of what primaries create neutrinos which are not accompanied by enough muon energy at depth to consider them vetoed.

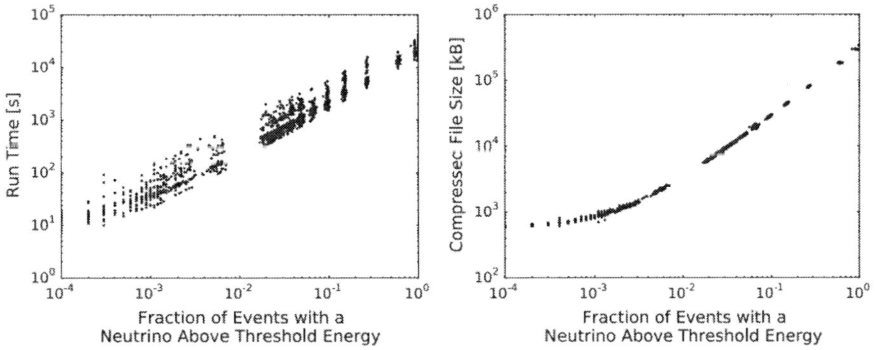

Figure 1. The file size (left) and run time (right) as a function of the fraction of events above a threshold energy. Both show a linear dependence with an increasing fraction of events on top of a minimum floor value to run and save basic information when no events meet the criterion.

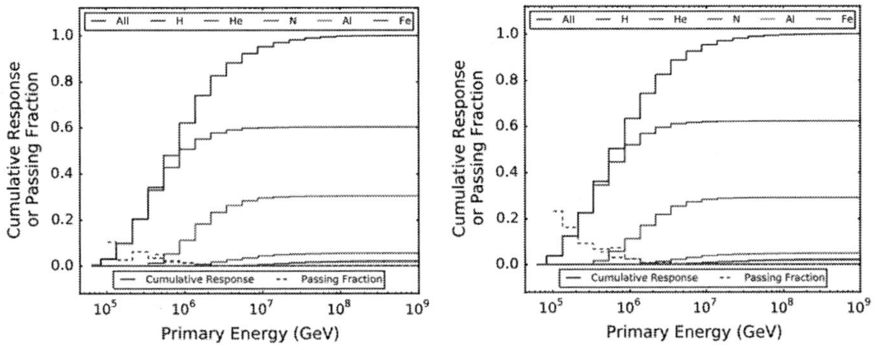

Figure 2. The cumulative response(solid) and passing fraction(dashed) of 50 to 55 TeV neutrinos from various primaries are presented for two different zenith bands, $1<\text{Cos}(\theta)<.8$ (left) and $.8<\text{Cos}(\theta)<.6$ (right). The cumulative response shows how much a given primary type and energy contributes to the flux of neutrinos between 50 and 55 TeV in the zenith band. The passing fraction shows how often a shower of a given type and energy contributes has a neutrino between 50 and 55 TeV in the zenith band but less than 1 TeV of muons at the depth of IceCube. The largest contribution to the passing fraction always comes from primaries near the neutrino energy.

This question is particularly relevant for surface vetos of neutrinos since it gives the relevant energy of primaries one must be able to reliably detect at the surface to reject the background of atmospheric neutrinos [3, 4]. To do this requires a connection between a neutrino and its parent primary, thus the question must be answered using Monte Carlo. Using the modification, a simulation sample was produced with a neutrino energy threshold of 50 TeV and the SIBYLL 2.1 interaction model [4]. From this data, the cumulative response to a given neutrino energy from possible parent primary types and energies was calculated assuming the composition model of [5] and is shown in the plots of Fig. 2.

These plots show that most of the contribution to neutrinos from 50–55 TeV comes from hydrogen primaries between 1–2 PeV. Additionally plotted is the fraction of neutrinos which have less than 300 GeV of accompanying muons (the passing fraction). For nearly vertical events, the largest contribution comes from neutrinos which take nearly all their parent's energy. As the zenith angle increases the muon energy decreases due to increased overburden leading to more events from higher energy primaries contributing to the passing fraction.

Another interesting potential use for these simulations is to determine the significance of events with a contained vertex from IceCube's high energy starting event search [6]. In the most recently released starting event data [7] a very vertical muon neutrino was found. This event starts after passing through more than 10 layers of DOMs and deposits 400 TeV of energy in the detector. Because of energy and zenith only charmed decays could potentially produce this event. Producing a target simulation with the modification and the DPMJET II.5 [8] hadronic model a background of one event per 391 years was found to be accompanied by a muon bundle with less than 300 GeV of total energy. This gives the event a conservative estimate of a 3σ deviation from background since DPMJET II.5 is known to exceed the limits on charmed atmospheric muons and neutrinos [9, 10].

5. Conclusion

Modifications to CORSIKA's memory management were made which make "breadth-first" shower propagation possible. With this it is possible to stop showers when particles in the shower have dropped below a threshold energy. In the case of this modification neutrinos were the focus, and a neutrino energy threshold was also implemented. The modified code will be available for use in the next public release of CORSIKA. Already this modification has assisted in determining the significance of a starting event from IceCube's high energy starting event sample and can play a role in determining the backgrounds for a possible surface veto above IceCube.

References

[1] D. Heck et al., Forschungszentrum Karlsruhe Technik und Umwelt Wissenschaftliche Berichte **FZKA 6010** (1998), `https://web.ikp.kit.edu/corsika/physics_description/corsika_phys.pdf`

[2] Thomas K. Gaisser, Kyle Jero, Albrecht Karle, and Jakob van Santen, Phys. Rev. D. **90**, 023009 (2014), `1405.0525`

[3] K. Jero, D. Tosi, *IceTop as Veto for IceCube*, in *Proceedings, 34th International Cosmic Ray Conference (ICRC 2015)* (2015), `http://pos.sissa.it/archive/conferences/236/1086/ICRC2015_1086.pdf`

[4] E.J. Ahn, R. Engel, T.K. Gaisser, P. Lipari, T. Stanev, Phys. Rev. D **80**, 094003 (2009), `http://link.aps.org/doi/10.1103/PhysRevD.80.094003`

[5] T.K. Gaisser, Astroparticle Physics **35**(12), 801 (2012), ISSN 0927-6505, `http://www.sciencedirect.com/science/article/pii/S0927650512000497`

[6] M.G. Aartsen et al. ((IceCube Collaboration)), Phys. Rev. Lett. **113**, 101101 (2014), `http://link.aps.org/doi/10.1103/PhysRevLett.113.101101`

[7] C. Kopper, W. Giang, and N. Kurahashi for the IceCube Collaboration, *Observation of Astrophysical Neutrinos in Four Years of IceCube Data*, in *Proceedings, 34th International Cosmic Ray Conference (ICRC 2015)* (2015), `1510.05223`

[8] J. Ranft, ArXiv High Energy Physics - Phenomenology e-prints (1999), `hep-ph/9911213`

[9] M.G. Aartsen, R. Abbasi, M. Ackermann, J. Adams, J.A. Aguilar, M. Ahlers, D. Altmann, C. Arguelles, T.C. Arlen, J. Auffenberg et al., Phys. Rev. D. **89**, 062007 (2014)

[10] A. Fedynitch, R. Engel, T.K. Gaisser, F. Riehn, T. Stanev, *Calculation of conventional and prompt lepton fluxes at very high energy*, in *European Physical Journal Web of Conferences* (2015), Vol. 99 of *European Physical Journal Web of Conferences*, p. 08001, `1503.00544`

EPJ Web of Conferences **116**, 03001 (2016)
DOI: 10.1051/epjconf/201611603001

A compact array calibrator to study the feasibility of acoustic neutrino detection

M. Ardid[a], F. Camarena, I. Felis, A. Herrero, C.D. Llorens, J. Martínez-Mora, and M. Saldaña

Institut d'Investigació per a la Gestió Integrada de les Zones Costaneres (IGIC) – Universitat Politècnica de València, C/ Paranimf 1, 46730 Gandia, Spain

Abstract. Underwater acoustic detection of ultra-high-energy neutrinos was proposed already in 1950s: when a neutrino interacts with a nucleus in water, the resulting particle cascade produces a pressure pulse that has a bipolar temporal structure and propagates within a flat disk-like volume. A telescope that consists of thousands of acoustic sensors deployed in the deep sea can monitor hundreds of cubic kilometres of water looking for these signals and discriminating them from acoustic noise. To study the feasibility of the technique it is critical to have a calibrator able to mimic the neutrino "signature" that can be operated from a vessel. Due to the axial-symmetry of the signal, their very directive short bipolar shape and the constraints of operating at sea, the development of such a calibrator is very challenging. Once the possibility of using the acoustic parametric technique for this aim was validated with the first compact array calibrator prototype, in this paper we describe the new design for such a calibrator composed of an array of piezo ceramic tube transducers emitting in axial direction.

1. Introduction

The study of Ultra-High-Energy (UHE) neutrinos by means of the detection of the acoustic pulse generated by the energy deposition after interacting with a nucleus in water was proposed already in 1950s [1]. This phenomenon induces a local heating in a very short period of time leading to a sub-millisecond pressure pulse signal with bipolar shape and very directive flat disk-like pattern, with the bipolar pulse being emitted perpendicularly to the shower axis direction.

The feasibility of the acoustic detection neutrino technique is currently being studied in the Mediterranean Sea [2, 3] and could be implemented in the new deep-sea Cherenkov neutrino telescope under construction, the KM3NeT telescope [4], which will have a volume of several cubic kilometres. Thus, the acoustic detection is a promising technique to cover the detection of UHE neutrinos, being possible to combine these two neutrino detection techniques for a hybrid underwater neutrino telescopes, especially considering that the optical neutrino detection technique needs acoustic sensors for monitoring the position of the optical sensors [5, 6]. A key tool to study the viability of the acoustic detection is an acoustic calibrator able to imitate the bipolar pulse signal. This will be needed to calibrate the sensors, as well as, to improve and train the classification and discernment of the neutrinos bipolar

[a] e-mail: mardid@fis.upv.es

pulse signals from noise or other background transient signals. The proposed design, described in this paper, is a compact calibrator composed of an array of piezo-ceramic tube transducers emitting in axial direction. The emission of the low-frequency acoustic bipolar pulse is generated by parametric emission at high-frequency. For this, specific electronics adapted to the ceramics in order to feed them efficiently are required, as well as, to allow and control the different functionalities the calibrator has to provide.

2. Compact array calibrator approach

The use of parametric acoustic sources to generate neutrino-like signals with cylindrical transducers has been achieved and validated in previous studies [7]. The acoustic parametric effect occurs when two intense monochromatic beams with two close frequencies travel together through the medium. Under these conditions, in the region of nonlinear interaction, secondary harmonics of these frequencies are produced at lower frequency with a beam pattern directivity similar to the primary beam. Since this technique offers the advantage of generating a parametric signal much more directive, it allows designing a compact array with fewer units with respect to classical solutions, and thus reducing costs and facilitating the deployment and operation issues. The array calibrator under development will be constituted of a few tube piezo ceramics aligned with optimal distance separation for obtaining an opening angle of about 1° and phased emission of the multi-element array. The array of transducers will emit at high frequency (hundreds of kHz) in the perpendicular direction to the axis and produce a bipolar pulse (with lower frequency spectrum, tens of kHz) due to the interaction of parametric signals generated for each element.

The first phase of the calibrator design presented in this paper consists of the study and selection of transducers, their characterization, and the emission of neutrino-like signals with a single transducer using the parametric technique. In a second phase, a study on matching layer materials and transmission loses has to be studied and finalizing the studies of the parametric emission for single and group of transducers, moulding them, designing the electronics and testing them. The last phase will be completing the integration of the system and testing it for long distances. Afterwards, the calibrator will be ready for in situ tests and use either installed in the deep-sea neutrino telescope or from the surface of the sea in a sea campaign with a vessel.

The array calibrator will be able to work in different operation modes; at linear low frequency range by emitting long non-directive signals (easy to detect), at high frequency range by emitting long parametric directive signals, and at high frequency range by emitting the transient and directive parametric bipolar signal. This will allow to use the array transmitter for different functionalities, such as training and tuning the acoustic detector, cross-checking the detector hydrophones, and other marine applications. On the other hand, the different operation modes will allow us to plan a sea campaign in order to facilitate the signal detection in the three steps with increasing difficulty.

3. Transducer characterization

Two types of piezo-ceramic tube transducers from UCE ultrasonic Co. LTD have been selected as candidates for the emitters of the array calibrator. The choice was based on resonance frequencies, power emission, dimensions, and costs. Both are able to emit high frequency signals within high power emission and they have a reasonable power level at lower frequencies as well. The first candidate is the piezo-ceramic UCE-534541 with dimensions: outer diameter of 5.3 cm, inner diameter of 4.5 cm and height of 4.1 cm. The primary resonance frequency is at 490 kHz, but it has also a secondary resonance at low frequency, around 35 kHz. The second candidate is the piezo-ceramic UCE-343020 with dimensions: outer diameter of 3.4 cm, inner diameter of 3 cm and height of 2 cm. The primary resonance frequency is at 890 kHz, with a secondary resonance frequency around 75 kHz.

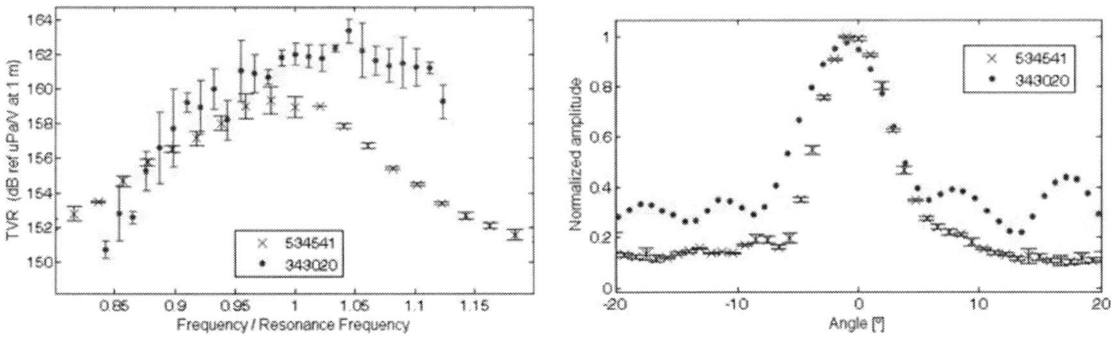

Figure 1. Left: TVR of UCE-534541 at $F_R = 490\,kHz$ (blue x) and for UCE-343020 at $F_R = 890\,kHz$ (red dot). Right: Directivity at 490 kHz for the UCE-534541 (blue x) and at 890 kHz for UCE-343020 (red dot).

The piezo-ceramics have been characterized both at high and low resonance frequencies in a water tank of $87.5 \times 113 \times 56.5\,cm^3$ with fresh-water at the laboratory. The calibration of the ceramics has been performed by measuring the Transmitting Voltage Response (TVR) and directivity in emission. The signals used for calibration have been tone bursts with the different frequency ranges of interest of each transducer. The hydrophones used for the calibration are omnidirectional, model RESON-TC40348 for high frequency and model RESON-TC4034 for low frequency. Figure 1 shows the characterization of both ceramics, the left plot shows the sensitivity obtained at high frequency with resonance frequency (F_R) of 490 kHz for the UCE-534541 and 890 kHz for the UCE-343020. The TVR of the UCE-534541 is 159 dB (re µPa/V at 1m) at F_R (490 kHz) with a directivity of $\pm 5°$. On the other hand, the TVR of the UCE-343020 is 162 dB (re μ Pa/V at 1m) at F_R (890 kHz) with a directivity of $\pm 7°$. The TVR at lower frequencies, 10 kHz to 100 kHz, is in the range 132–143 dB (re µPa/V at 1m) for both ceramics [8], so it will be possible to use these ceramics as well for generating low frequency signals directly, which is very useful for calibration purposes.

4. Parametric bipolar pulse emission

The experiment of generating the bipolar pulse was carried out in the same water tank. The piezo-ceramic was connected to a linear 55 dB RF amplifier ENI 1040L to feed the emitter and generate a more powerful signal in order to achieve the non-linear parametric effect. The chosen receiver was the RESON-TC4034, this transducer is an omnidirectional, broad-band hydrophone with enough sensitivity to detect the primary beam (high frequency) and better sensitivity to lower frequencies for the bipolar pulse detection. Additionally it was connected to a charge amplifier CCA 1000 (Teledyne RESON©) which amplifies the received signal, especially the low frequency range. Here, we present the study performed with the UCE-343020 ceramic, hence, the signal for emission has been derived for parametric generation at 890 kHz as described in ref. [7]. The distance between the emitter and the receiver was 30 cm. The results with the UCE-534541 are reported in ref. [8], where the shapes of the primary beam and the secondary bipolar pulse can be observed in Fig. 3 of this reference.

Since the amplitude of the secondary bipolar signal is about 100 times smaller than the primary beam, it is necessary to use a band pass ([5–80] kHz) filter to be able to observe this signal. In order to check that the secondary beam has a non-linear behaviour, the amplitude of the parametric signal, as a function of the voltage applied to the transducer, was measured. Figure 2-left shows the amplitude behaviour of the received signal, without filtering (original) and with the band pass filtering (the parametric bipolar signal, secondary beam). Linear behaviour can be observed for the amplitude of the primary beam, whereas, for the secondary received signal (parametric bipolar signal) the amplitude behaviour is proportional to the square of the amplitude of the input signal, showing the nonlinearity of

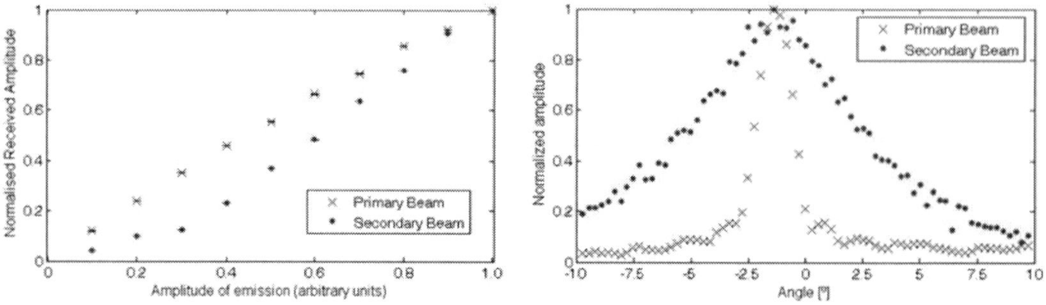

Figure 2. Left: Amplitude of the received primary beam and of the secondary bipolar pulse. Right: Directivity obtained for both signals.

the effect. On the other hand, in Fig. 2-right it can be observed that, despite the frequency content of both signals is so different, the directivity achieved is quite similar.

5. Conclusions and future steps

The results obtained on the piezo-ceramics characterization show that both could be good candidates for being part of the future calibrator since they have an optimal sensitivity at both high and low frequency and narrow beam directivity at high frequencies which will lead, in turn, to a very directive low-frequency (parametric) pulse. Moreover, the first studies on the acoustic neutrino-like signals generation by using the parametric technique show that the signal reproduction is achieved.

The next steps will consist in completing the studies of the parametric emission for single and group of transducers, and moulding them. In addition, the electronics design will be finalized and tested. After this, the complete integration of the system and the tests at long distances will be performed before doing sea campaign tests.

We acknowledge the financial support of the Spanish Ministerio de Economía y Competitividad, Grants FPA2012-37528-C02-02, and Consolider MultiDark CSD2009-00064, of the Generalitat Valenciana, Grants ACOMP/2015/175 PrometeoII/2014/079 and of the European FEDER funds.

References

[1] G.A. Askaryan, J. A. Energy **3**, 921 (1957)
[2] J.A. Aguilar, et al., Nucl. Instrum. and Meth. A **626**, 128 (2009)
[3] G. Riccobene, Nucl. Instrum. and Meth. A **604**, S149 (2009)
[4] The KM3NeT Collaboration, KM3NeT Technical Design Report (2010) ISBN 978-90-6488-033-9, available on `www.km3net.org`
[5] M. Ardid, Nucl. Instrum. and Meth. A **604**, S203 (2009)
[6] M. Ardid, Nucl. Instrum. and Meth. A **602**, 174 (2009)
[7] M. Ardid et al., Sensors, **12**, 4113 (2012), pp. 4113–4132
[8] M. Saldaña et al., New design of an acoustic array calibrator for underwater neutrino telescopes. In Proceedings of the 2nd Int. Electron. Conf. Sens. Appl., 15–30 November 2015; Sciforum Electronic Conference Series, Vol. 2, 2015, E011; doi: 10.3390/ecsa-2-E011

EPJ Web of Conferences **116**, 03002 (2016)
DOI: 10.1051/epjconf/201611603002

Fiber optic hydrophones for acoustic neutrino detection

E.J. Buis[1,a], E.J.J. Doppenberg[1], R. Lahmann[2], P.M. Toet[1], and J. de Vreugd[1]

[1] TNO Technical Sciences, PO Box 155, 2600 AD Delft, The Netherlands
[2] Erlangen Centre for Astroparticle Physics, University of Erlangen, Erwin-Rommel-Str. 1, 91058 Erlangen, Germany

Abstract. Cosmic neutrinos with ultra high energies can be detected acoustically using hydrophones. The detection of these neutrinos may provide crucial information about then GZK mechanism. The flux of these neutrinos, however, is expected to be low, so that a detection volume is required more than a order of magnitude larger than what has presently been realized. With a large detection volume and a large number of hydrophones, there is a need for technology that is cheap and easy to deploy. Fiber optics provide a natural way for distributed sensing. In addition, a sensor has been designed and manufactured that can be produced cost-effectively on an industrial scale. Sensitivity measurements show that the sensor is able to reach the required sea-state zero level. For a proper interpretation of the expected bipolar signals, filtering techniques should be applied to remove the effects of the unwanted resonance peaks.

1. Introduction

The detection of ultra high energy neutrinos, with an energy above 10^{18} eV, may provide crucial information about the GZK mechanism [1, 2], but could also point to violent and distant astrophysical objects. The expected flux of these cosmic neutrinos, however, is expected to be low, so that large scale neutrino telescopes are needed for this emerging field of astroparticle physics.

Acoustic signals in water induced by energetic neutrino interactions [3, 4] may provide a handle to detect these cosmic particles using hydrophones. Such an acoustics detection method profits from the fact that sound can travel for some kilometers with only small attenuation in the relevant frequency range. A large scale neutrino telescope can therefore be established using hydrophone arrays in the deep sea.

In this paper we discuss the developments on a hydrophone sensor in view of its application in a very large network. We explore the feasibility of manufacturing hydrophone sensors, according to the requirements on the sensitivity and dynamic range, on an industrial scale.

[a]e-mail: ernst-jan.buis@tno.nl

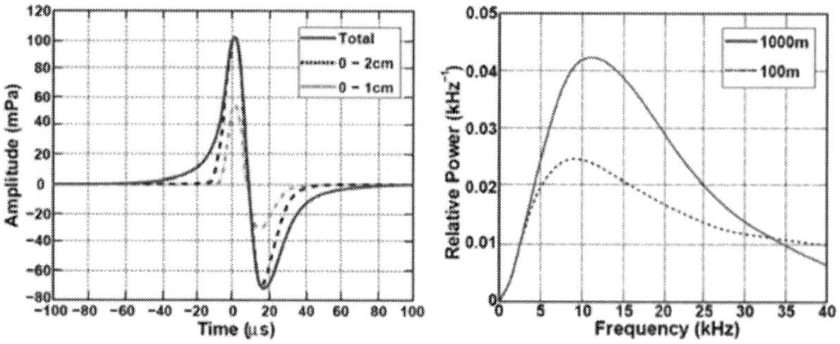

Figure 1. Left: Expected signal from a 1 EeV neutrino shower at a distance of 1 km (with relative contributions from the shower axis). Right panel: (relative) power spectral density function of an acoustics signal sensed at a distance of 100 m and 1 km. Adapted from [6].

2. Acoustic neutrino detection

The acoustic signals of particle interactions in water have been studied through extensive and detailed simulations, which have been validated by experimental measurements using particle and laser beams [5, 6]. The acoustic signal expected expected for a 1 EeV neutrino interaction at a 1 km distance from the sensor is shown in Fig. 1. The shape of the signal is the well-known bipolar pulse, which is characteristic for a neutrino event. In the right panel of the same figure the power spectral density is plotted, which shows that the spectrum peaks around 10 kHz, but extends up to 40 kHz and higher.

As the acoustic source, i.e. the particle shower, can be approximated by a thin line source, the acoustic pattern that is formed in the water has a shape of a pancake: the pattern is, according the length of the particle shower, about 10–20 m thick, but may extend to several km's in radial direction. The radial exention of the accoustic pressure field is due large absorption length of sound, which is about 1 km at 25 kHz, depending on salinity and temperature [7]. The distinct acoustic pattern will prove crucial in event reconstruction to discriminate a neutrino event from background as well as to obtain the direction of the neutrino.

3. Fiber based neutrino telescope

It is clear from the various discussions in literature that a(n acoustic) neutrino telescope suitable for the study of ultra-high energy neutrinos is required to have a detection volume of at least 100 km^3 or more [8]. In addition, such a volume should be equipped with at least 1000 hydrophones for a proper event reconstruction. Such a detection volume demands a truly Very Large Volume neutrino Telescope (VLVνT).

3.1 Detection concept and technology

It has been advocated that fiber optic hydrophone technology is a promising means to establish a sensitive, cost-effective and large scale sensor network [9–11], that could provide the basis of a future neutrino telescope. This technology is based on fiber lasers and hydrophone sensors that are integrated on a single optical fiber as shown in Fig. 2. The sensors transform the acoustic pressure into strain in the fiber. At the position of each sensor a fiber laser is inserted in the optical fiber, which emits light of a well-known wavelength. The strain causes a wavelength shift of the laser light and is sensed using an interrogator. The interrogator houses an interferometer to determine the wavelength shift, but also a pump laser to pump all the fiber lasers in the string. The readout scheme of the interrogator is

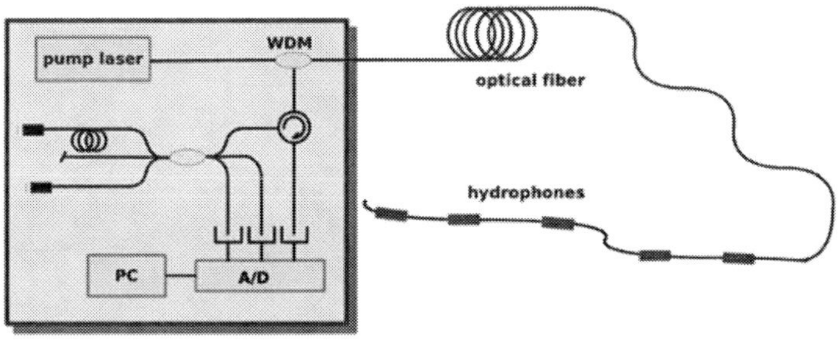

Figure 2. Concept of the optical fiber sensor network: An optical fiber with several fiber lasers and sensors are read out using an interrogator. Once deployed in the deep-sea, several of these strings make up a neutrino telescopes.

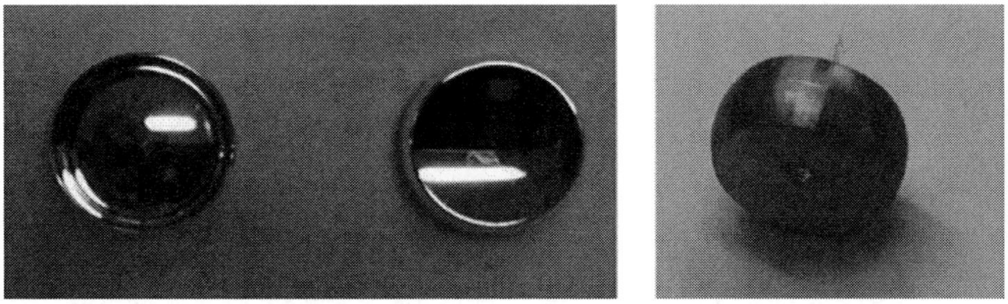

Figure 3. Left photograph: two halves of the sensor. In the photograph on the right the sensors is assembled and glued to an optical fiber.

plotted in Fig. 2. Note that the distance between the sensors and the interrogator can be large, so that the interrogator can be located on shore or on any other convenient location.

3.2 Sensor development for a VLVνT

An attempt was made to develop a sensor that is cost effective and can be mass produced. The design consisted of two halves made out of aluminum as shown in Fig. 3. The aluminum parts were constructed using a turning machine, but could in future be produced through the method of die cutting. Such a construction method allows for a mass scale production of the sensor.

During assembly, the two halves were glued and an optical fiber was guided through a hole in the center of the sensor as shown in the right panel of Fig. 3. At the time of the measurements, no fiber lasers of the size that match the dimensions of the sensor were available, so that simple and less sensitive fiber bragg gratings (FBG) were used instead. As a concequence, no pulses could be recorded at the mPa level (Fig. 1), but a characterization of the sensor could be done nevertheless on the overall response as a function of the acoustic frequency to study the mechanical resonances.

4. Results

The sensor as described in the previous section has been tested in an anechoic basin at TNO. A B&K8103 type hydrophone from Bruël and Kjaer was taken as a reference hydrophone that has a flat response at the relevant acoustic frequencies. The response of the hydrophone was determined as a function of the acoustic frequency and is shown in Fig. 4 (left panel). The graph shows two large resonance peaks that are due to the eigen frequencies of the aluminum housing as well as the fiber that is attached to it. A filter has been applied to remove these peaks of which the result is shown in the right

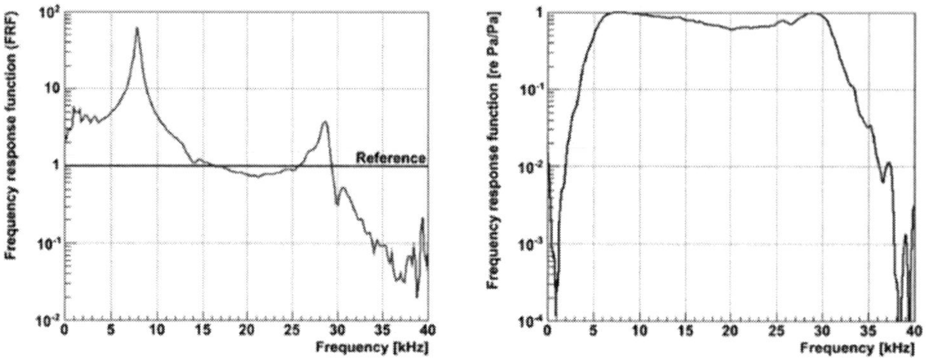

Figure 4. Measured response of a sensor (shown in Fig. 3 prior (left) and after filtering (right).

panel of the figure. Here, the Box-Jenkins method was applied (see e.g. [12]). It can be seen that the filtering is quite effective and that the peaks disappear and the response is relatively flat in the region of interest.

5. Conclusions

Within the field of astroparticle physics, the study of ultra-high energy neutrinos may come within reach when neutrino telescopes can be constructed using large scale fiber optic networks. Such a network, when equiped with a large number of hydrophones, would allow to detect the cosmic neutrinos through the acoustic signals that the may induce in the deep sea.

We have designed, realized and characterized a new hydrophone sensor based on fiber lasers that can be produced cost-effectively on a mass-scale. It was found that the sensor exhibits two resonance peaks that are related to the aluminum housing of the sensor as well as the fiber that is attached to it. Filtering has been applied to obtain a flat response of the sensor, which is required for a proper interpretation of the data.

References

[1] K. Greisen, Physical Review Letters **16**, 748 (1966)
[2] G.T. Zatsepin, V.A. Kuz'min, ZhETF Pisma Redaktsiiu **4**, 114 (1966)
[3] G.A. Askaryan, Zemlia i Vselennaia **1**, 13 (1979)
[4] J.G. Learned, Phys. Rev. D **19**, 3293 (1979)
[5] L. Sulak, et al., Nuclear Instruments and Methods **161**, 203 (1979)
[6] S. Bevan, et al. (Acorne Collaboration), Astroparticle Physics **28**, 366 (2007)
[7] M.A. Ainslie, J.G. McColm, Acoustical Society of America Journal **103**, 1671 (1998)
[8] E. Waxman, Science **315**, 63 (2007)
[9] P.E. Bagnoli, et al., Nuclear Instruments and Methods in Physics Research A **567**, 515 (2006)
[10] E. Marconi, et al., Nuclear Instruments and Methods in Physics Research A **572**, 490 (2007)
[11] E.J. Buis, et al., Journal of Instrumentation **9**(03), C03051 (2014)
[12] L. Ljung, ed., *System Identification (2Nd Ed.): Theory for the User* (Prentice Hall PTR, Upper Saddle River, NJ, USA, 1999), ISBN 0-13-656695-2

EPJ Web of Conferences **116**, 03003 (2016)
DOI: 10.1051/epjconf/201611603003

The ARIANNA Hexagonal Radio Array – performance and prospects

Allan Hallgren[a] for the ARIANNA collaboration

Department of Physics and Astronomy, Uppsala University, Box 516, 751 20 Uppsala, Sweden

Abstract. The origin of the highest energy cosmic rays at $\sim 10^{20}$ eV is still unknown. Ultra-high energy neutrinos from the GZK process should provide information on the sources and their properties. A promising and cost effective method for observing GZK-neutrinos is based on detection of Askaryan radio pulses with antennas installed in ice. The ARIANNA project aims at instrumenting a 36*36 km^2 large area on the Ross Ice Shelf with an array of radio detection stations. The deployment of a test system for ARIANNA, the Hexagonal Radio Array (HRA), was completed in December 2014. The three first stations were installed in 2012. Solar panels are used to drive the < 10 W stations. The system hibernated at sunset in April and all stations returned to operation in September. The site is essentially free of anthropogenic noise. Simple cuts eliminate background and provides for efficient selection of neutrino events. Prospects for the sensitivity of the full ARIANNA array to the flux of GZK neutrinos are shown.

1. Introduction

The origin of the very highest energy cosmic rays remains enigmatic still more than 100 years after the discovery of cosmic rays by V. Hess in 2012. A cut-off in the spectrum is observed and may correspond to interactions with the cosmic microwave background as suggested by Greisen, Zatsepin and Kuzmin (GZK) [1, 2]. Production of neutrinos with energies around 10^{18} eV would then occur. Detection of GZK-neutrinos requires surveillance of a large amount of target material. Observation of radio pulses from the Askaryan effect [3, 4] using antennas embedded in Antarctic ice or firn promises a cost effective technique to achieve the goal of GZK-neutrino detection. ARIANNA (Antarctic Ross Ice-Shelf ANtenna Neutrino Array) is a proposed detector system for this purpose. Determination of the flux of GZK-neutrinos, their spectral properties, directional and temporal distributions would provide crucial information on the high energy universe. Further, physics beyond the standard model may be revealed if the flux permits to determine e.g. the cross section at these extreme energies. Even if no GZK-neutrinos are found limiting the flux below most current models will provide important information on the origin of the highest energy particles observed.

The detection mechanism relies on coherent emission of a sub-nanosecond Cherenkov pulse at radio wavelengths. The particle shower caused by the neutrino-interaction develops a negative charge excess as it travels through the medium. Constructive interference of Cherenkov radiation generates a

[a] e-mail: `allan.hallgren@physics.uu.se`

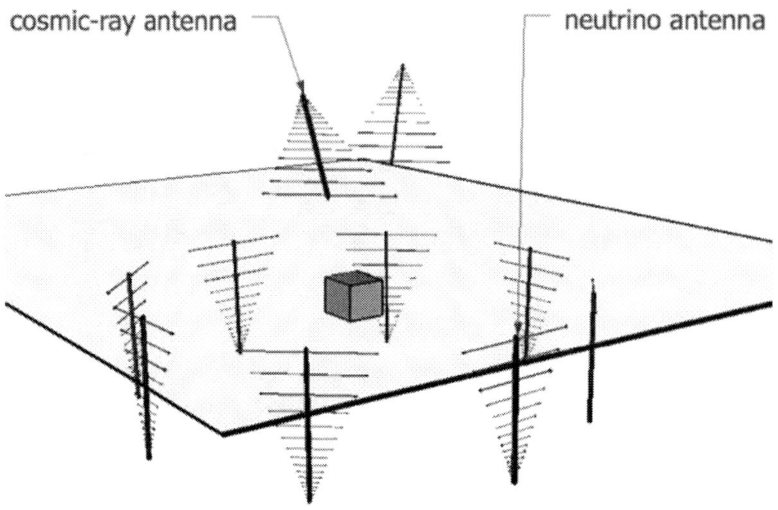

Figure 1. Conceptual design of an ARIANNA station with eight downward pointing antennas just below snow surface and two upward tilted antennas for identification of radio signals from cosmic rays. All antennas are connected to the station central readout system. Power and communication systems are not shown. The prototype stations in the Hexagonal Radio Array is a reduced version with 4 downward directed antennas.

sub-nanosecond pulse in radio-wavelengths as predicted by G. Askaryan. This radio pulse generation and its properties have been experimentally verified with good agreement using electron bunches stopped in e.g. blocks of ice [5]. The signal generation and pulse properties behave as expected, allowing for simulations of all the variations of pulses that a detector will have to efficiently identify.

2. The ARIANNA development work

The full ARIANNA system would comprise 1236 independent stations distributed over an area of 36 km*36 km on the immense Ross Ice-Shelf [6]. Each station has in the conceptual design eight downward directed antennas deployed just below the snow surface and two upward antennas for suppression of background from radio pulses induced by cosmic rays, Fig 1. The flat ice to sea-water interface reflects the downward directed part of the Askaryan radio pulse. Most of the neutrino events recorded by ARIANNA will be seen through the reflected signal. The proposed location is ~120 km from the McMurdo station, a US driven base for research in Antarctica. This remote site is essentially free of anthropogenic radio noise due to the distance and to screening by mountain ridges.

2.1 The Hexagonal Radio Array

The Hexagonal Radio Array (HRA) serves as a prototype system for ARIANNA. The seven HRA stations contain a reduced number of four antennas per station. Three stations were installed in 2012, the remaining four were deployed in December 2014. The equipment for the last four stations was prepared and planned for deployment in 2013 but due to US Government shutdown the installation was postponed. Only some maintenance work was possible in 2013. Data taken by HRA-3 thereafter, Jan-April 2014, have been analysed for content of neutrino events. The closing date for this analysis was set by hibernation of the solar powered stations. No ν-event was found, but more importantly demonstrated that three simple cuts were sufficient to remove all background events [6]. Autocorrelation, where a pulse is correlated with a time shifted copy of itself, is a simple test that removes much of the thermal noise, showing strong correlation only at a time shift of zero. Pulses from ν-events, as simulated through the electronics chain, show negative correlation when shifted a few nanoseconds and can thus be separated. The second cut analyse the frequency power distribution, neutrino events have a broader distribution than noise events that typically have power in only few of the frequency bins. The third and final selection criterion compares the recorded event to a library of events simulated with different observational conditions and the best correlation is used for selecting neutrino candidate events. The efficiency of the analysis was above 80% for a typical GZK spectrum and the flux limit obtained from three stations in three months of data was $E^2\phi(E) \sim 2 \cdot 10^5$ GeVcm^{2}s^{1}sr^1 [6]. This result is promising. To asses rarer types of background more stations and live-time is needed and is forthcoming from the now completed HRA.

The full HRA system also serves to test and verify the functionality of all technical systems [7]. An important challenge is the power system. Solar power panels of are used together with lithium batteries for buffering. The first generation of batteries used have proven unreliable, and some of the stations were run with non-functional batteries or without batteries. If no battery is available the fraction of the day the station is alive gradually decreases from mid-January to mid-April. If a working battery is present full functionality can be maintained until mid-April when the Sun is only a few degrees above horizon and the power generated becomes too low. All stations came back to operation during September when the Sun came back over horizon. The resulting live-time fraction achieved is thus $\sim$60% for stations with working buffering capacity [8]. This live-time fraction is included in the sensitivity estimates for ARIANNA below.

A more reliable Li-battery with lower but sufficient capacity (12 V, 20 Ah) has been found and extensively tested in the lab. Additionally a better charge control unit has been evaluated, in particular the noise generated when switching is substantially reduced. The batteries and charge controller will therefore be exchanged on all installed stations in the coming 15/16 field season. The smaller dimensions allow these components to be included in an RF-tight box. This is likely to remove a large part of noise seen when solar power generation is marginal.

The trigger rate has been verified to be low in all stations. After proper adjustment the rate is $\sim$few*10^{-2} Hz. The electronics gain is known to rise with decreasing temperature, thus trigger level adjustments are required to maintain the rates. The electronics box is buried in the firn and the temperature changes only slowly. Periods of strong winds are seen as increased trigger rate, but these events are all eliminated in the template correlation analysis [9].

Tests are done with RF pulses emitted downwards from known positions at surface and detected in the array after bounce on the ice-water interface. The RF pulse has thus propagated $\sim$1.1 km through the ice. The reflected pulses maintain their polarization. The time differences between the signals' received in the different antennas permit determination of the direction of the incoming radiation. A likelihood fit to the direction gives a resolution of $\sim 1^\circ$ [9]. The direction of the incoming neutrino can be determined from the direction to the event paired with the measured polarisation. The likelihood can be used in the event selection as shown in Fig. 2. This was not used in the analysis of HRA-3 data [6] but will allow further noise suppression.

One station was installed with upward tilted antennas to verify the signal amplitudes recorded for cosmic ray events. This station detected signals from a coronal mass ejection from the Sun on Dec 20, and the subsequent auroras activity a few days later. Noise from the Galactic centre is also visible in forced readouts with a daily modulation and with the expected power ratio between upward and downward antennas [8]. These observations demonstrate that the system sensitivity to RF signals is according to expectations. The characteristics of the radiation, e.g. the frequency content, show that this type of background can be readily discarded from neutrino candidate events.

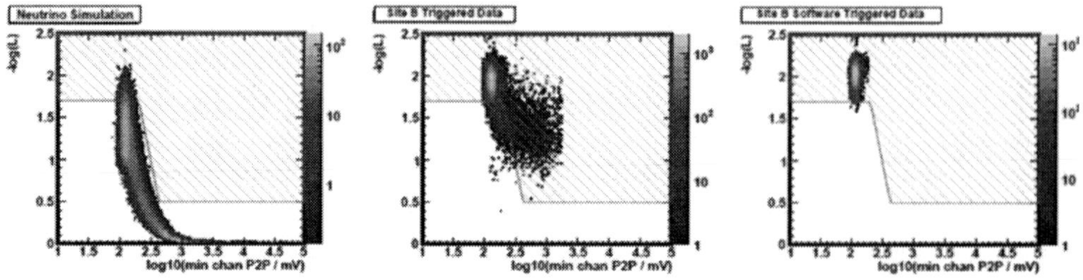

Figure 2. The directional fit likelihood plotted versus the peak to peak amplitude for simulated neutrino events (left), triggered data in HRA (middle) and pure thermal noise events (right). All plots are on events surviving the cut on distribution of power [9]. The passing fraction is 96% for neutrino events if the indicated cut is used.

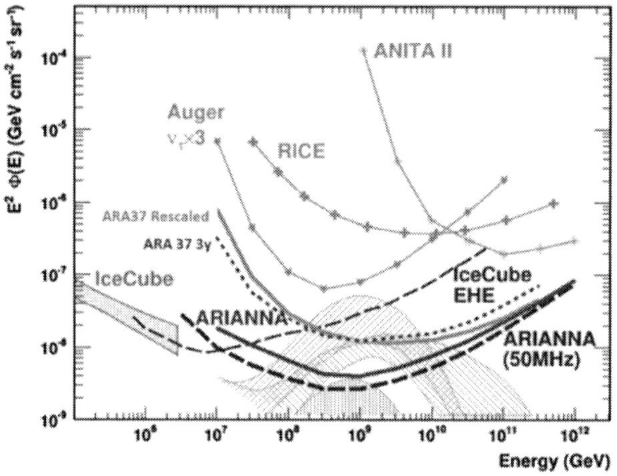

Figure 3. Projected 5 calendar year sensitivities for ARIANNA (36*36 stations) and 50 MHz ARIANNA option. For ARA37 the projected 3 year sensitivity from [10] is given as well as a rescaled 5 year estimate, see text for explanation. The 5 year sensitivity of the IceCube EHE analysis is shown as well as a selection of current limits and range of predictions for the flux of GZK-neutrinos.

3. Projected performance of ARIANNA

The 90% CL sensitivity of the full 36*36 station ARIANNA array has been estimated in the standard configuration with 100 MHz antennas and with an alternative 50 MHz antenna, as shown in Fig. 3. The performance of the 50 MHz option, including noise, will be tested in the near future as it offers an interesting possibility to enhance the sensitivity at lower energies. This sensitivity corresponds to $N_{bck} = 0$ and $N_{90} = 2.3$ events. To compare to the sensitivity of the proposed ARA37 array two curves are shown. The curve labelled "ARA37 3yr" is taken from [10]. The curve labelled "ARA37 Rescaled" is for a 5 calendar year period but includes reduction in analysis efficiency and live time derived from [10]. This curve is further recalculated to be comparable to the ARIANNA curves by using e.g. the same cross sections and statistical methods for determination of the differential sensitivity [11].

The Askaryan RF pulse is polarized with the shower direction included in the E-field plane. The ARIANNA antennas measure polarization and this combined with the direction to the interaction point (described above) allows the ν-direction to be resolved. Simulations show that the expected resolution is ~2.5–3 degrees [6]. The sky coverage starts at $-90°$ declination and has an energy dependent cut-off somewhat below horizon. Due to the ARIANNA position at about 78° S a larger part of the sky is visible than would be possible for a detector located at 90° S. This location results in that a point source of GZK-neutrinos will be seen with a modulated viewing angle, allowing for additional information on cross section and spectral properties.

The ARIANNA HRA has now demonstrated a cost effective possibility to build an instrument capable of measuring the flux of GZK-neutrinos, often dubbed "guaranteed". This would provide conclusive information on the yet enigmatic origin of the cosmic rays with the highest energies observed. It is time to start building!

References

[1] K. Greisen, Phys. Rev. Lett. **16**, 748 (1966)
[2] G.T. Zatsepin and V.A. Kuzmin, JETP Lett. **4**, 78 (1966)
[3] G.A. Askaryan, JETP **14**, 441 (1962)
[4] G.A. Askaryan, JETP **21** 658 (1965)
[5] P. Gorham et al. (ANITA Collaboration), Phys. Rev. Lett. **99**, 171101 (2007)
[6] S.W. Barwick et al. [`arxiv:1410.7352`], Astroparticle Phys. **70** 12 (2015)
[7] S.W. Barwick et al. [`arxiv:1410.7369`], *IEEE TNS* **62**, 2202 (2015)
[8] A. Nelles and C. Persichelli for the ARIANNA Collaboration, *PoS(ICRC2015)* 1087 (2015)
[9] C. Reed for the ARIANNA Collaboration, *PoS(ICRC2015)* 1149 (2015)
[10] P. Allison et al. (ARA collaboration), [`arxiv:1507.08991v2`]
[11] C. Reed, S.W. Barwick, Private communication

EPJ Web of Conferences **116**, 03004 (2016)
DOI: 10.1051/epjconf/201611603004

Acoustic neutrino detection investigations within ANTARES and prospects for KM3NeT

Robert Lahmann[a]

Friedrich-Alexander-Universität Erlangen-Nürnberg, Erlangen Centre for Astroparticle Physics (ECAP), Erwin-Rommel-Straße 1, 91058 Erlangen, Germany

Abstract. The acoustic neutrino detection technique is a promising approach for future large-scale detectors with the aim of measuring the small expected flux of cosmogenic neutrinos at energies exceeding 1 EeV. It suggests itself to investigate this technique in the context of underwater Cherenkov neutrino telescopes, in particular KM3NeT, because acoustic sensors are present by design to allow for the calibration of the positions of the optical sensors. For the future, the KM3NeT detector in the Mediterranean Sea will provide an ideal infrastructure for a dedicated array of acoustic sensors. In this presentation results from the acoustic array AMADEUS of the ANTARES detector will be discussed with respect to the potential and implications for acoustic neutrino detection with KM3NeT and beyond.

1. Introduction: Acoustic neutrino detection

Measuring acoustic pressure pulses in huge underwater acoustic arrays is a promising approach for the detection of ultra-high-energy (UHE, $E_\nu \gtrsim 10^9$ GeV) neutrinos. These are expected to be produced in interactions of cosmic rays with the cosmic microwave background [1]. The pressure signals are produced by the particle showers that evolve when neutrinos interact with nuclei in water. The resulting energy deposition of the hadronic cascade within a cylindrical volume of a few centimetres in radius and several metres in length leads to a local heating of the medium. According to the thermo-acoustic model [2, 3], the accelerated expansion of the heated volume forms a pressure pulse of bipolar shape which propagates in the surrounding medium. Coherent superposition of the elementary sound waves, produced over the volume of the energy deposition, leads to a propagation within a flat disk-like volume (often referred to as *pancake*) in the direction perpendicular to the axis of the particle shower. After propagating several hundreds of metres in sea water, the pulse has a characteristic frequency spectrum that is expected to peak around 10 kHz [4–6]. At these frequencies, the attenuation length of sound in sea water is about one to two orders of magnitude larger than that of Cherenkov light.

In this article, acoustic neutrino detection activities with a present (AMADEUS) and future (KM3NeT) experiment will be described and it will be pointed out how the experience with the former can be used for measurements with the latter.

[a]e-mail: robert.lahmann@fau.de

2. Lessons learned from the AMADEUS project

The AMADEUS project [7] was conceived to perform a feasibility study for a potential future large-scale acoustic neutrino detector in the Mediterranean Sea. For this purpose, a dedicated array of acoustic sensors was integrated into the ANTARES neutrino telescope [8]. The detector is located in the Mediterranean Sea at the geographic position of 42°48′ N, 6°10′ E at a water depth of about 2500 m, roughly 40 km south of the town of Toulon at the French coast. ANTARES comprises 12 vertical structures, the *detection lines*. Each detection line holds up to 25 *storeys* that are arranged at equal distances of 14.5 m along the line. A standard storey holds three *optical modules*, each one consisting of a photomultiplier tube inside a water-tight pressure-resistant glass sphere. A 13th line, the *Instrumentation Line (IL)*, is equipped with instruments for monitoring the environment.

Within the AMADEUS system, acoustic sensing is integrated in the form of *acoustic storeys* with six acoustic sensors per storey, arranged at distances of roughly 1 m from each other. The AMADEUS system comprises a total of six acoustic storeys: three on the IL and three on the 12th detection line (Line 12). In April 2013, the IL was re-deployed at a new position at 150 m distance from Line 12. Until then, the distance between the two lines was 220 m. The vertical distances between the two topmost acoustic storeys on the IL is increased from the standard of 14.5 m to about 110 m.

To assess the feasibility of acoustic neutrino detection in a natural body of water, ambient noise and transient background at the site of the installation have to be investigated. The ambient noise is broadband and is mainly caused by agitation of the sea surface [9], i.e. by rain, wind, breaking waves, spray, and cavitations. It is predominantly the ambient background that determines the energy threshold for neutrino detection. Transient noise signals have short duration and an amplitude exceeding the ambient noise level. When of bipolar shape, these signals can mimic pulses from neutrino interactions. Sources of such signals can be anthropogenic, e.g. shipping traffic, or marine fauna. In particular dolphins emit short signals with a spectrum similar to that of acoustic emissions from neutrino interactions.

Two main lessons have been learned from the operation of the AMADEUS system. First, the ambient background is low and stable which is favourable for the operation of an acoustic neutrino detector. Second, the transient background has a high level. It is essential to reduce this background completely in order to measure the low expected rate of cosmogenic neutrinos.

A strong background reduction is already achieved with methods developed with AMADEUS [10]. Machine learning algorithms are used to identify bipolar pulses. Furthermore, events that are spatially and temporally clustered are discarded, as neutrinos are unlikely to produce several signals from one position within a short period of time. For further reduction it is essential to use the pancake shape of the signal as a selection criterion. This is not possible with AMADEUS due to its small size; such investigations are a main focus for the KM3NeT detector.

3. Future activities in KM3NeT

KM3NeT is the next generation Cherenkov neutrino detector currently under construction in the Mediterranean Sea [11]. After its first phase of the construction, namely KM3NeT Phase 1, the detector will consists of 24 *detection units* (DUs) deployed off-shore Capo Passero in Italy (KM3NeT-IT), and 7 more DUs, deployed off-shore Toulon in France (KM3NeT-FR). Each DU holds a total of 18 *optical modules* [12], distributed equidistantly along flexible strings which for the KM3NeT-IT site have a length of about 700 m. Each optical module contains 31 three-inch-photomultiplier tubes (PMTs) together with their readout electronics within a pressure-resistant glass spheres. When fully implemented, KM3NeT will eventually consist of several hundreds of DUs at multiple sites.

In order to determine the relative positions of the optical modules with a precision of not worse than 20 cm, the detector will be equipped with an acoustic positioning system. The system employs acoustic

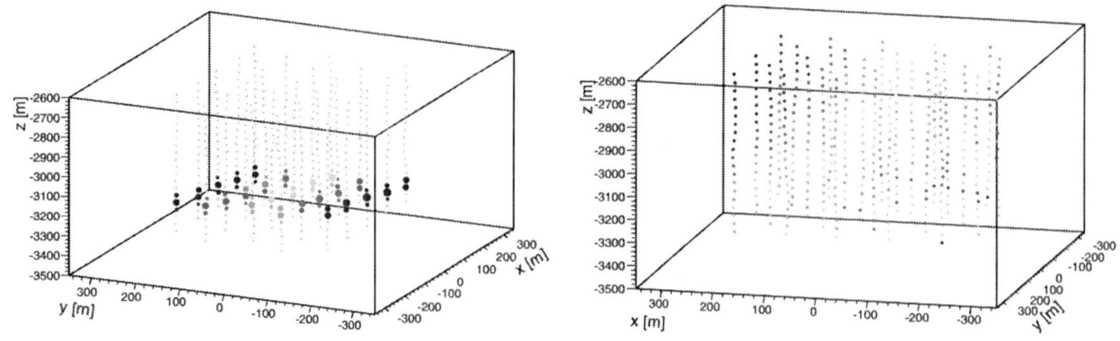

Figure 1. The signature of a neutrino event (left) and a spherical event originating from the surface of the sea, simulated in a KM3NeT-like acoustic detector with 25 lines spaced 90 m apart. Each dot represents an optical module with an integrated acoustic sensor, while the color indicates the arrival time of the acoustic signal (red: early, blue: late, grey: not triggered). The size of the dot represents the signal amplitude. The neutrino interacted at a distance of 1.8 km from the detector center with an energy of 10^{21} eV.

transceivers on the sea floor and acoustic receivers (hydrophones) in each storey and at the bases of the DUs. By performing multiple time-delay measurements and using these to triangulate the positions of the individual hydrophones, the hydrophone positions can be reconstructed relative to the positions of the emitters.

The design of the KM3NeT positioning system is based on experience with the systems developed for ANTARES and the former NEMO experiment, see [13] and references therein. Two types of acoustic sensors, both based on the piezo-electric effect, are used:

1. custom designed internal sensors. These are compact units of a piezo ceramic and a preamplifier, glued to the inside of the glass sphere of the optical module near its "South Pole";
2. external hydrophones, installed on the bases of the DUs.

Sampling is done at about 200 k samples per second and all data is transmitted to shore. This way, algorithms for the position calibration running on an on-shore computer farm can be adapted to in-situ conditions that may affect the data quality. Furthermore, the data can be used for additional analyses, in particular acoustic detection of neutrinos, or marine science investigations.

The crucial questions for acoustic neutrino detection are: can neutrinos be classified background-free? How low an energy threshold can be reached? And what volume needs to be instrumented to detect a predicted flux of cosmogenic neutrinos? These questions can be addressed by implementing the measurements for the ambient and transient noise from AMADEUS into simulations for acoustic neutrino detection with KM3NeT and developing suitable filter and reconstruction algorithms.

To demonstrate the neutrino classification based on the shape of the acoustic emission pattern, a simulated neutrino signal and a background event for a detector configuration similar to that of KM3NeT-IT in Phase 1 are shown in Fig. 1 [14]. As can be seen, the size of the detector allows for a discrimination of the characteristic pancake shape against the spherical sound waves emitted from background events. First studies indicate that for an ambient noise and transient background as measured at the ANTARES site, the remaining background of neutrino-like events after suppression based on signal shape and clustering alone, as discussed in Sect. 2, can be reduced significantly when taking the emission pattern into account [14].

As a potential future extention to KM3NeT, the use of optical fibre-based hydrophones is under investigation [15]. To survey large volumes of water for acoustic signals from neutrino interactions, optical fibre-based hydrophones could potentially have several advantages over conventional hydrophones based on piezo ceramics. Investigations of adapting this technique to neutrino detection in combination with an under-water Cherenkov telescope are planned in the context of the KM3NeT experiment.

4. Summary and conclusions

First generation acoustic arrays such as AMADEUS have been used to investigate neutrino detection methods and provide input for simulations. Ambient noise could be shown to be low and stable while transient background is higher than expected and will require sophisticated methods of suppression. KM3NeT provides an excellent framework for a second generation acoustic detection test setup – the detector is equipped with acoustic sensors for the purpose of position calibration which can also be used for neutrino detection investigations. Software tools developed for AMADEUS can be easily applied to KM3NeT. New concepts, in particular fibre-based hydrophones, can be used in the long run to further increase the instrumented volume of KM3NeT beyond the volume instrumented for optical detection of Cherenkov light.

The AMADEUS project is supported by the German federal ministry for education and research (BMBF) through grants 05A08WE1 and 05A11WE1.

References

[1] V.S. Berezinsky and G.T. Zatsepin, PL **B 28**, 423 (1969)
[2] G.A. Askariyan et al., Nucl. Inst. and Meth. **164**, 267 (1979)
[3] J.G. Learned, Phys. Rev. **D 19**, 3293 (1979)
[4] S. Bevan et al. (ACoRNE Coll.), Astropart. Phys. **28**, 366 (2007)
[5] S. Bevan et al. (ACoRNE Coll.), Nucl. Inst. and Meth. **A 607**, 398 (2009)
[6] V. Niess and V. Bertin, Astropart. Phys. **26**, 243 (2006)
[7] J.A. Aguilar et al. (ANTARES Coll.), Nucl. Inst. and Meth. **A 626-627**, 128 (2011)
[8] M. Ageron et al. (ANTARES Coll.), Nucl. Inst. and Meth. **A 656**, 11 (2011)
[9] R.J. Urick, *Ambient Noise in the Sea* (Peninsula publishing, 1986), ISBN 0-932146-13-9
[10] M. Neff, Ph.D. thesis, Univ. Erlangen-Nürnberg (2013)
[11] The KM3NeT homepage, `http://www.km3net.org/`
[12] D. Vivolo, *The KM3NeT Digital Optical Module*, in *Proceedings of VLVnT 2015, Rome, Italy, Sept. 14–16, 2015* (2016), to be published in EPJ Web of Conf.
[13] S. Viola et al., *Acoustic positioning system for KM3NeT*, in *Proceedings of ICRC2015, The Hague, Netherlands, July 30–Aug. 6* (2016), to be published in: Proc. of Science
[14] D. Kießling et al., *Simulation chain and signal classification for acoustic neutrino detection in seawater*, in *Proceedings of ARENA 2014, Annapolis, MD (USA), June 9–12* (2014), to be published in: AIP Conf. Proc.
[15] E.J. Buis et al., *Fiber optic hydrophones for acoustic neutrino detection*, in *Proceedings of VLVnT 2015, Rome, Italy, Sept. 14–16, 2015* (2016), to be published in EPJ Web of Conf.

EPJ Web of Conferences **116**, 03005 (2016)
DOI: 10.1051/epjconf/201611603005

The Giant Radio Array for Neutrino Detection

Olivier Martineau-Huynh[1], Kumiko Kotera[2], Mauricio Bustamente[3,4], Didier Charrier[5], Sijbrand De Jong[6], Krijn D. de Vries[7], Ke Fang[8], Zhaoyang Feng[9], Chad Finley[10], Quanbu Gou[9], Junhua Gu[10], Jordan C. Hanson[3,4], Hongbo Hu[9], Kohta Murase[12], Valentin Niess[13], Foteini Oikonomou[12], Nicolas Renault-Tinacci[4], Julia Schmid[14], Charles Timmermans[5], Zhen Wang[9], Xiangping Wu[11], Jianli Zhang[11], and Yi Zhang[9]

[1] LPNHE, CNRS-IN2P3 and Universités Paris VI & VII, 4 place Jussieu, 75252 Paris, France
[2] Institut d'Astrophysique de Paris, Sorbonne Universités, UPMC Univ. Paris 6 and CNRS, UMR 7095, 98 bis bd Arago, 75014 Paris, France
[3] Center for Cosmology and AstroParticle Physics, The Ohio State University, Columbus, OH 43210, USA
[4] Department of Physics, The Ohio State University, Columbus, OH 43210, USA
[5] SUBATECH, IN2P3-CNRS, Université de Nantes, Ecole des Mines de Nantes, Nantes, France
[6] Nikhef/Radboud University, Nijmegen, The Netherlands
[7] Vrije Universiteit Brussel, Dienst ELEM, 1050 Brussels, Belgium
[8] Department of Astronomy and Astrophysics, University of Chicago, Chicago, IL 60637, USA
[9] Key Laboratory of Particle Astrophysics, Institute of High Energy Physics, Chinese Academy of Sciences, Beijing 100049, China
[10] Oskar Klein Centre and Dept. of Physics, Stockholm University, 10691 Stockholm, Sweden
[11] National Astronomical Observatory, Chinese Academy of Sciences, Beijing 100012, China
[12] Department of Physics, Department of Astronomy & Astrophysics, Pennsylvania State University, University Park, PA, USA
[13] Clermont Université, Université Blaise Pascal, CNRS/IN2P3, Laboratoire de Physique Corpusculaire, BP. 10448, 63000 Clermond-Ferrand, France
[14] Laboratoire AIM, Université Paris Diderot/CEA-IRFU/CNRS, Service d'Astrophysique, CEA Saclay, 91191 Gif-sur-Yvette, France

Abstract. High-energy neutrino astronomy will probe the working of the most violent phenomena in the Universe. The Giant Radio Array for Neutrino Detection (GRAND) project consists of an array of $\sim 10^5$ radio antennas deployed over $\sim 200\,000\,\text{km}^2$ in a mountainous site. It aims at detecting high-energy neutrinos via the measurement of air showers induced by the decay in the atmosphere of τ leptons produced by the interaction of cosmic neutrinos under the Earth surface. Our objective with GRAND is to reach a neutrino sensitivity of $5 \times 10^{-11} E^{-2}\,\text{GeV}^{-1}\,\text{cm}^{-2}\,\text{s}^{-1}\,\text{sr}^{-1}$ above 3×10^{16} eV. This sensitivity ensures the detection of cosmogenic neutrinos in the most pessimistic source models, and up to 100 events per year are expected for the standard models. GRAND would also probe the neutrino signals produced at the potential sources of UHECRs.

1. Detection method

Cosmic ν_τs can produce τ particles underground through charged current interaction. τs travel to the surface of the Earth and decay in the atmosphere, generating Earth-skimming extensive air showers (EAS)[1] [1, 2]. Coherent electromagnetic radiation is associated to the shower development at frequencies of a few to hundreds of MHz at a detectable level for showers with $E \gtrsim 10^{17}$ eV.

The strong beaming of the electromagnetic emission combined with the transparency of the atmosphere to radio waves will allow the radio-detection of EAS initiated by τ decays at distances up to several tens of kilometers. Radio antennas are thus ideal instruments for this purpose. Furthermore, they offer practical advantages (limited unit cost, easiness of deployment, ...) that allow the deployment of an array over very large areas, as required by the expected low neutrino rate.

Remote sites, with low electromagnetic background, should obviously be considered for the array location. In addition, mountain ranges are preferred, first because they offer an additional target for the neutrinos, and also because mountain slopes are better suited to the detection of horizontal showers compared to flat areas, parallel to the showers trajectories.

GRAND antennas are foreseen to operate in the 30–100 MHz frequency band. An extension of the antenna bandwidth up to 200 or 300 MHz would enable to observe the Cherenkov ring associated with the air shower, that represents a sizable fraction of the total electromagnetic signal and may provide an unambiguous signature for background rejection.

2. GRAND layout and neutrino sensitivity

We present here a preliminary evaluation of the potential of GRAND for the detection of cosmic neutrinos. We initially simulated the response of a setup of 40 000 antennas deployed on a square layout of 60 000 km^2 in a remote mountainous area (Tianshan mountains, XinJiang, China).

Simulation method. We perform a 1D tracking of a primary ν_τ, simulated down to the converted τ decay. Standard rock with a density of 2.65 g/cm^3 is assumed down to sea level and the Earth core is modeled with the Preliminary Reference Earth Model [3]. The simulation of the Deep Inelastic Scattering of the neutrinos is performed with Pythia6.4, using the CTEQ5d probability distribution functions (PDF) combined with [4] for cross section calculations. The propagation of the produced τ is simulated using randomized values from parameterizations of GEANT4.9 PDFs for τ path length and proper time. Photonuclear interactions in GEANT4.9 have been extended above PeV energies following [5]. The τ decay is simulated using the TAUOLA package.

For the time being only an analytical treatment of EAS radio detection has been performed. ZHAireS simulations [6] of horizontal showers have been used to parametrize the distance at which radio antennas may detect EAS radio emissions. Assuming an aggressive (conservative) detection threshold $\epsilon_a = 30\ \mu$V/m ($\epsilon_c = 100\ \mu$V/m), it was found that a conical detection volume centered on the shower axis could be considered, with a half-angle between 1.4° (0.9°) for a shower energy $E_{sh} = 10^{17}$ eV and 4.7° (3.8°) for $E_{sh} = 3 \times 10^{20}$ eV and a distance to the τ decay point between 14 and 90 km (45 km) for $E_{sh} = 10^{17}$ eV, and between 14 and 300 km (200 km) for $E_{sh} = 3 \times 10^{20}$ eV. We considered that a shower is detected if a cluster of 8 neighbouring antennas or more is inside this detection volume and in direct view of the τ decay point.

Note that the Cherenkov effect is not taken into account in this study. This will be done in the up-coming full Monte-Carlo study of the GRAND neutrino sensitivity.

---

1 Other neutrino flavors can be neglected as the electron range in matter at these energies is too short and the muon decay length too large compared to flight distances underground.

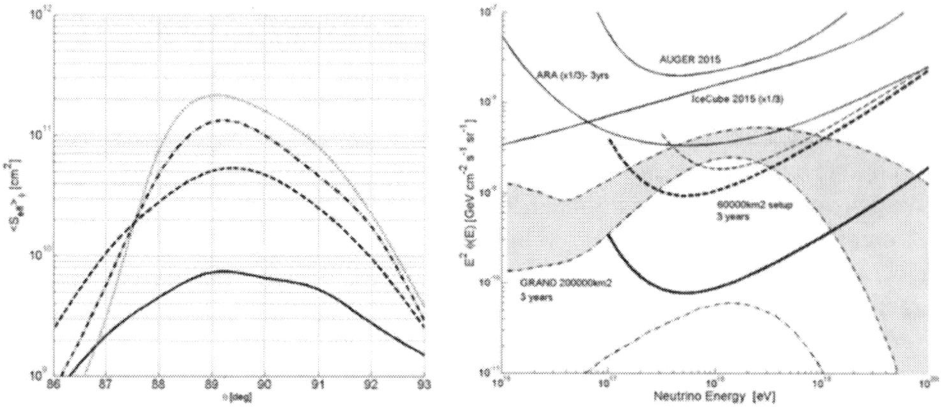

Figure 1. Left: effective area of the 60 000 km^2 simulated set-up averaged over azimuth angle ϕ as a function of zenith angle θ for various initial ν_τ energies (plain line: $10^{17.5}$ eV, dashed: $10^{18.5}$ eV, dash-dotted: $10^{19.5}$ eV, dotted:$10^{20.5}$ eV). Zenith angles > 90° correspond here to down-going trajectories. Right: differential sensitivity limit of the 60 000 km^2 simulated setup (dashed line, thin: conservative, thick: aggressive), of the projected 3 times larger GRAND array (thick line), and for other instruments (thin lines), and estimated theoretical cosmogenic neutrino fluxes for one neutrino flavor [8].

Results and implications. Assuming a 3-year observation on an area of 60 000 km^2 with no neutrino candidate, a 90% C.L. integral limit of 6.6×10^{-10} (1.3×10^{-9}) GeV^{-1} cm^{-2} s^{-1} can be derived for an E^{-2} neutrino flux in our aggressive (conservative) scenario.

This preliminary analysis also demonstrates that mountains constitute a sizable target for neutrinos, with ~50% of down-going events, corresponding to neutrinos interacting inside the mountains (Fig. 1). It also appears that specific parts of the array (large mountains slopes facing another mountain range at distances of 30–80 km) are associated with a detection rate well above the average. We thus realized that a factor of 10 improvement in sensitivity – corresponding to cosmogenic neutrinos event rates between 1 to 100 per year – may be reached with a factor 3 increase in the detector area only, provided the detector is composed of several sub-arrays of smaller size (few 10 000 km^2) deployed solely on favorable sites. This is the envisioned GRAND setup.

The expected angular resolution on the arrival direction of the events detected was computed analytically following [7]. A mean resolution of 0.05° should be achievable for a 3 ns precision on the antenna trigger timing. Energy estimation will be challenging, mainly because the τ decay point is unknown. The very specific ground pattern of the electric field amplitude and polarization associated with ν-induced EAS should provide a very efficient tool for the discrimination of background (very inclined EAS induced by UHECRs or events of anthropic origin for example). This shall be tested on an engineering array of size ~1000 km^2. This array will obviously be too small to perform a neutrino search, but cosmic rays should be detected above 10^{18} eV. Their reconstructed properties will enable us to validate this stage, if found to be compatible with expectations.

3. High-energy neutrino astronomy with GRAND and other science cases

UHECRs are likely produced in extragalactic sources, given the strength of Galactic magnetic fields and the lack of correlations with the Galactic plane. Some fraction of their energy is converted to high-energy neutrinos through the decay of charged pions produced by interactions with ambient matter and radiation. This can happen in the source environment or during the flight in the intergalactic medium (*cosmogenic* neutrinos). The range of expected cosmogenic neutrino fluxes can be calculated precisely, and depends mostly on parameters inherent to the cosmic rays (Fig. 1) [8]. The calculated level of high-energy neutrino fluxes produced at the sources depends on the modeling of the acceleration region, but robust flux estimates have recently been derived for some sources [9–12], arguing that many scenarios can be tested with increased sensitivities.

The sensitivity of GRAND should guarantee the detection of cosmogenic EeV neutrinos. For reasonable source scenarios (Fig. 1), GRAND aims at collecting up to 100 events per year above 10^{17} eV. The ideal way to identify high-energy neutrino sources would be to observe a point source – a *transient* source is a likely candidate due to the power and density constraints from UHECR observations (e.g., [13]). GRAND opens this possibility with its excellent spatial resolution.

Given its effective area, GRAND would also be the largest UHECR observatory on ground, and should provide valuable information to help resolve these puzzles if precision of reconstructed parameters is satisfactory. Determination of the UHECR composition in particular is a key issue. Detailed simulation studies will be performed to estimate performances reachable by GRAND.

The large array size and the frequency range of GRAND – if extended to 300 MHz – could be profitable for other physics goals (such as the study of the Epoch of Reionization [14] or the search for Fast Radio Bursts [15]). We are presently assessing the technical cost of pushing the detector towards these directions.

4. Conclusion

The GRAND project aims at building a next-generation neutrino telescope composed of radio antennas deployed over $\sim 200\,000\,\mathrm{km}^2$. Preliminary simulations indicate that a sensitivity guaranteeing the detection of cosmogenic neutrinos may be achievable. GRAND would also be a powerful instrument for UHECR detection at energy $\gtrsim 10^{19}$ eV. Work is ongoing to determine precisely the achievable scientific goals and the corresponding technical constraints.

We thank J. Alvarez-Muniz and W. Carvalho for helpful comments and discussions. The GRAND and GRANDproto projects are supported by the Institut Lagrange de Paris, the France China Particle Physics Laboratory, the Natural Science Fundation of China (Nos.11135010, 11375209) and the Chinese Ministery of Science and Technology.

References

[1] D. Fargion, International Cosmic Ray Conference **2**, 396 (1999), astro-ph/9906450
[2] X. Bertou, et al, Astroparticle Physics **17**, 183 (2002), astro-ph/0104452
[3] A. Dziewonski, D. Anderson, Physics of the Earth and Planetary Interiors **25**(4), 297 (1981)
[4] R. Gandhi, C. Quigg, M.H. Reno, I. Sarcevic, Phys. Rev. D **58**(9), 093009 (1998), hep-ph/9807264
[5] S. Dutta, et al., European Physical Journal C **18**, 405 (2000), hep-ph/9905475
[6] J. Alvarez-Muniz, W.R. Carvalho, Jr., E. Zas, Astropart. Phys. **35**, 325 (2012), 1107.1189
[7] D. Ardouin et al., Astroparticle Physics **34**, 717 (2011), 1007.4359
[8] K. Kotera, D. Allard, A.V. Olinto, J. Cos. and Astro. Phys. **10**, 13 (2010), 1009.1382
[9] P. Baerwald, M. Bustamante, W. Winter, Astroparticle Physics **62**, 66 (2015), 1401.1820
[10] K. Murase, P. Mészáros, B. Zhang, Phys. Rev. D **79**(10), 103001 (2009), 0904.2509
[11] K. Fang, K. Kotera, K. Murase, A.V. Olinto, Phys. Rev. D **90**(10), 103005 (2014)
[12] K. Murase, Y. Inoue, C.D. Dermer, Phys. Rev. D **90**(2), 023007 (2014), 1403.4089
[13] K. Murase, H. Takami, ApJ Letters **690**, L14 (2009), 0810.1813

[14] S. Zaroubi, *The Epoch of Reionization*, in *Astrophysics and Space Science Library*, edited by T. Wiklind, B. Mobasher, V. Bromm (2013), Vol. 396 of *Astrophysics and Space Science Library*, p. 45

[15] S.R. Kulkarni, E.O. Ofek, J.D. Neill, Z. Zheng, M. Juric, The Astrophysical Journal **797**, 70 (2014), 1402.4766

EPJ Web of Conferences **116**, 03006 (2016)
DOI: 10.1051/epjconf/201611603006

Characterisation and testing of the KM3NeT acoustic positioning system

S. Viola[1,a], F. Simeone[2,b], and M. Saldaña[3,c] for the KM3NeT Collaboration

[1] Istituto Nazionale di Fisica Nucleare, Laboratori Nazionali del Sud
[2] Istituto Nazionale di Fisica Nucleare, Sezione di Roma
[3] Instituto de Investigación para la Gestión Integrada de las Zonas Costeras, Universitat Politècnica de València, Gandia, Spain

Abstract. In underwater neutrino telescopes, the search of point-like sources through the Cherenkov detection technique requires a precise knowledge of the positions of thousands of optical sensors, spread in a volume of a few cubic kilometres. In KM3NeT the optical sensors are hosted in 700 m high semi-rigid structures, called detection units, which move under the effects of underwater currents. These movements are continuously monitored through an underwater positioning system based on acoustic emitters and receivers. In this work, the tests performed on the key elements of the positioning system are presented.

1. Introduction

The KM3NeT neutrino telescope has been designed to measure the visible Cherenkov photons originated by charged particles propagating at velocities greater than the speed of light through the seawater by means of an array of photomultipliers (PMTs). The charged particle tracks are reconstructed measuring the time of arrival of the Cherenkov photons on PMTs, placed at known positions. In each detection unit (DU) of the telescope, PMTs are located within 18 digital optical modules (DOMs), that host 31 PMTs each one [1]. The position of each DOM is calculated acoustically by means of a Long Base Line of acoustic beacons placed in fixed positions and an array of digital acoustic receivers (DARs) installed in each DOM and at the base of the DUs and of the calibration units (CUs) [2].

2. LBL acoustic beacon characterization

The acoustic beacon, used for the KM3NeT acoustic positioning system is composed by a piezo-ceramic transducer and an electronics board integrated in an cylindrical hard-anodized aluminium vessel by the MSM company ([3]). The transducer is a Free Flooded Ring (FFR SX30) manufactured by Sensor Technology Ltd [4]. The electronic board is specifically designed to fulfil the requirements of the

---

[a] e-mail: `sviola@lns.infn.it`
[b] e-mail: `francesco.simeone@roma1.infn.it`
[c] e-mail: `masalcos@upv.es`

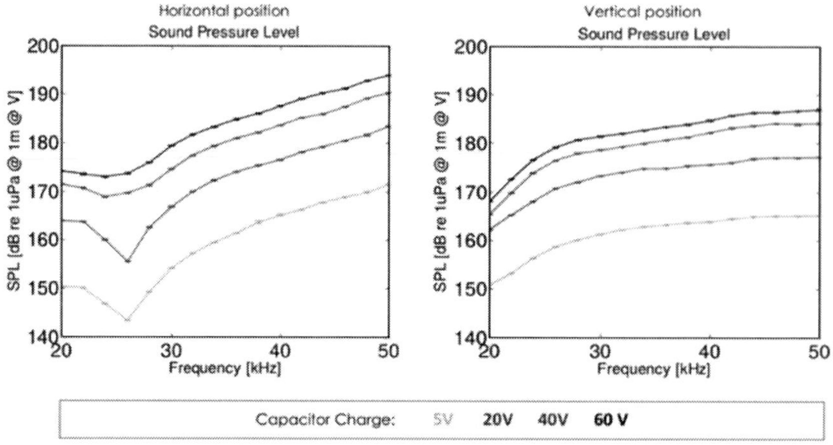

Figure 1. Sound Pressure Level (SPL) of an acoustic beacon at axial direction (left side) and radial direction (right side) for different capacitor charges.

telescope positioning system, enabling the transducer communication and the signal emission control and amplification [5]. The acoustic signal emission parameters can be configured by commands through a RS-232 serial interface allowing signal reconfiguration from shore. The acoustic beacon (MAB-100) for the LBL positioning system is a broadband range acoustic emitter (20 kHz–60 kHz) able to work at rating depths up to 400 bar in underwater environments. It provides the emission of short intense signals (Sound Pressure Levels of 180 dB re 1 μPa at 1 m at 34 kHz) with a variation of $\pm$ 6 dB in the frequency range of 20 kHz to 60 kHz. The radial beam pattern is omnidirectional with $\pm$ 2 dB for each work frequency and the axial beam pattern is toroidal with $\pm$ 10 dB of variation at 60° and $\pm$ 5 dB at 180°. The signal emission is activated by responding to an external trigger, in addition, it has two operation emission modes: automatic (continue) and triggered emission. The electronics delay of the object, that is the time from the reception of the trigger to the acoustic signal emission, is provided by the manufacturer. Typical values are 4.2 ± 0.8 μs. The Acoustic Beacons were characterized in a water tank of $87.5 \times 113 \times 56.5$ cm^3 with fresh-water. The calibration was performed in terms of measuring the Transmitting Voltage Response (TVR) and directivity in emission. During the calibration, the distance between emitter (Acoustic Beacon) and receiver (calibrated hydrophone) was fixed at 20 cm. The signals sent for the calibration were tone bursts of 5 wavelength cycles with frequencies ranging from 20 kHz to 60 kHz in steps of 2 kHz. Measurements were performed with different capacitor charge (5 V, 20 V, 40 V and 60 V) both in radial and axial direction (Fig. 1). The hydrophone used for the calibration was the omnidirectional transducer model ITC-1042 [7]. The signal generation and acquisition process was conducted by a National Instruments PCI-Technology controller NI8176, which also controlled the positioning [6].

3. Digital Acoustic Receivers characterisation

In the KM3NeT acoustic positioning system the acoustic signals produced by the acoustic emitters of the LBL are received by two kinds of acoustic receivers: an "internal" digital piezoelectric-sensor, installed inside the DOM and a digital "external" hydrophone, installed at the base of the DUs and on the CUs, where no DOM are forseen.

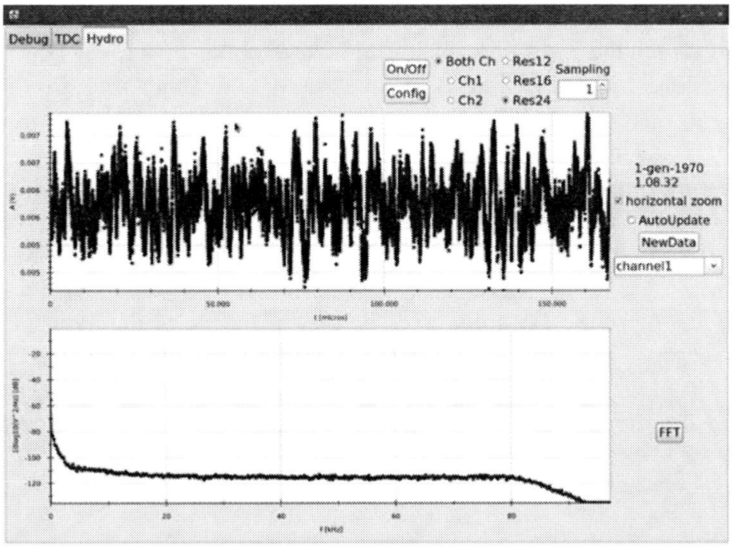

Figure 2. Oscillogram and power spectral density of the electronic noise of a typical "internal" digital acoustic receiver.

3.1 The internal Digital Acoustic Receiver

The "internal" Digital Acoustic Receiver is a digital piezoelectric-sensor based on a ceramic material produced by Ferroperm [8], readout by 24-bit Sigma-Delta ADC (model AD7764BRUZ -TSSOP28) [9]. The sensor is glued to the inside to the glass sphere of each DOM. The performance of the digital piezo sensors were investigated in the laboratory. Power spectral density of the electronic noise was measured thanks to a proper GUI (see Fig. 2) that provides in real time spectrum and oscillogram of the acquired signal. The noise spectrum shows a very flat behaviour in the range 10 kHz to 80 kHz. Beyond that frequency the anti-alias filter sets in. Below $\approx$ 5 kHz a strong rise of the noise can be observed before the effect of a high pass filter, which is intended to prevent saturation of the sensor due to strong low frequency noise. The RMS of the noise was measured to be about 1250 LSB ($1\text{LSB} = 4V/2^{24}$) for the full range and about 300 LSB for the frequency range above 5 kHz. A test with the final configuration for DU-1 of KM3NeT was performed at Nikhef laboratory with an open DOM inside a light-tight box. It was observed that turning on the PMT power results in spikes at distinct frequencies. However, in the range of frequencies between 20 and 40 kHz where the positioning system will be operating, no disturbances are observed so that no adverse effects on the performance of the position reconstruction are expected. The sensor's functionality check is also performed after installation in the DOM, during the final DOM calibration measurement. A function generator produces sinusoidal signals at arbitrary amplitude (always the same for all DOM tests) at 30 kHz frequency. The electric signal is amplified by an audio amplifier and fed into a piezo-electric, not-magnetic tweeter, placed close to the piezo-sensor. One minute of continuous data taking is analysed on-line looking at noise spectrum the presence of the peak produced by the emitter.

3.2 The external acoustic Digital receivers: Hydrophones

External hydrophones are hosted on the base of the DUs and on CUs, where no DOM is foreseen. These hydrophones will be used to calibrate the LBL, measuring the relative distances among LBL elements. The selected hydrophone is the DG0330 manufactured by CO.L.MAR. s.r.l. [10]. It consists of a spherical piezo-ceramic element, read-out by an analogue board splitting the signal in two lines with different gains (+46 dB and +26 dB respectively). The double gain feature was implemented to avoid signal saturation when the hydrophone is placed close (about 1 m) to the beacon. Two hydrophone productions are available, without and with an analogue signal high-pass filtering stage. The filter frequency is 700 Hz, to reject the low frequency ambient sea-noise and thus improve Signal to Noise ratio in the detection of beacon pulses (20–40 kHz). The unfiltered version can be also used for acoustic noise monitoring and marine bio-acoustics. The hydrophones are delivered from the company after a pressure-crush test at 400 bars (4000 m water equivalent) and shallow water calibration (at 1 bar). A sub batch of few hydrophones was also calibrated at high pressure (350 bars). Once received at INFN, the hydrophones were tested with a dedicated set-up, current and intrinsic noise were measured for the two channels using proprietary software. The time response of the hydrophone as a function of the frequency was measured and reported in acceptance test sheets. A functional test (communication and noise spectrum visualization) is carried out before the closure of the DU Base container and line installation. For all hydrophones the sensitivity curves and the time response were stored in the database.

4. Conclusions

Measurements in the water tank demonstrate that the acoustic pulses produced by the acoustic beacon can reach a SPL of 180 dB re 1 μPa in the range of frequencies from 20 kHz to 50 kHz. It means that beacon signals in this range can be easily recognised by the acoustic receivers of the telescope since that an acoustic pulse with a SPL of 180 dB re 1 μPa at 1 m can effectively propagate in water up to distance of 2 km with about 110 dB re 1 μPa at 0° (depending on the frequency). "Internal" acoustic receiver functionality in deep sea was demonstrated analysing acoustic data after the deployment of the first detection unit at the KM3NeT-IT site. Sweep sine signals from 38 kHz to 39 kHz, emitted by a LBL beacon (capacitor charge 50 V) have been used as test signals for the receivers located along the DU at distances between 220 m and 690 m from the beacon. The data analysis showed that beacon pulses were properly detected by all the acoustic receivers.

References

[1] KM3NeT website, www.km3net.org
[2] S.Viola , Proceedings of 34th ICRC, PoS(ICRC2015)1169 (2015)
[3] Mediterráneo Señales Marítimas (MSM) website, http://en.mesemar.com/
[4] Sensor Technology std website: http://sensortech.ca
[5] C D Llorens et al., Journal of Instrumentation, **7**, C01001 (2012)
[6] PXI-8176 Controller website, http://sine.ni.com/nips/cds/view/p/lang/it/nid/212342
[7] ITC hydrophone data sheet, http://www.channeltechgroup.com/publication/view/model-itc-1042-spherical-omnidirectional-transducer
[8] FERROPERM website, www.ferroperm-piezo.com
[9] Analog Devices website, www.analog.com
[10] CO.L.MAR. srl website, www.colmaritalia.it

EPJ Web of Conferences **116**, 04001 (2016)
DOI: 10.1051/epjconf/201611604001

Search for spatial and temporal collective effects in the ANTARES neutrino telescope data

Alexis Coleiro[a], Rodrigo Gracia Ruiz[b], Antoine Kouchner[c] on behalf of the ANTARES collaboration

APC, Université Paris Diderot, CNRS/IN2P3, CEA/Irfu, Observatoire de Paris, Sorbonne Paris Cité, 10 rue Alice Domon et Léonie Duquet, 75205 Paris Cedex 13, France

Abstract. We investigate potential collective effects in the spatial and temporal domains in ANTARES data sets. On the one hand, we apply a two-point correlation analysis to look for inhomogeneities in the arrival directions of the high energy muon neutrino candidates detected between 2007 and 2012. This enables us to provide constraints on models of a population of point sources too faint to be detected by a likelihood-based method. On the other hand, we perform a search for ANTARES neutrino events in temporal coincidence with IceCube High-Energy Starting Events located within 45° from the Galactic Center. This study, also based on a two-point correlation function, is sensitive to transient emission and does not a prior on either the burst timing structure or on the electromagnetic emission. Therefore, it provides an effective way to acquire information on the possible origin of the IceCube astrophysical signal from transient sources.

1. Introduction

High energy neutrinos are expected to be produced in sources of cosmic rays. Since they are neutral, weakly interacting and traveling straight from their source without suffering from absorption, neutrinos are unique messengers to further understand the particle acceleration processes in astrophysical sources.

One way of looking for an astrophysical neutrino signal, is to search for clustering in the arrival directions of the reconstructed events. This can be achieved with spatial autocorrelation analysis applied to discrete data ensembles. In the present analysis the method is performed with five years of data and used to search for a signal coming from sources that are too faint to be detected by likelihood analyses.

Furthermore, when dealing with transient emission, the atmospheric background can be significantly reduced using a time-dependent approach. It was pointed out recently that two of the IceCube High-Energy Starting Events (HESE, see [1]) occurred close to the Galactic Center within 1 day of each other with a p-value of 1.6% (see [4]). This was interpreted as the possible signature of a transient point source of very-high energy neutrinos in this part of the sky. Consequently, we carry out a model-independent

---

[a]e-mail: coleiro@apc.in2p3.fr
[b]e-mail: rgracia@apc.in2p3.fr
[c]e-mail: kouchner@apc.univ-paris7.fr

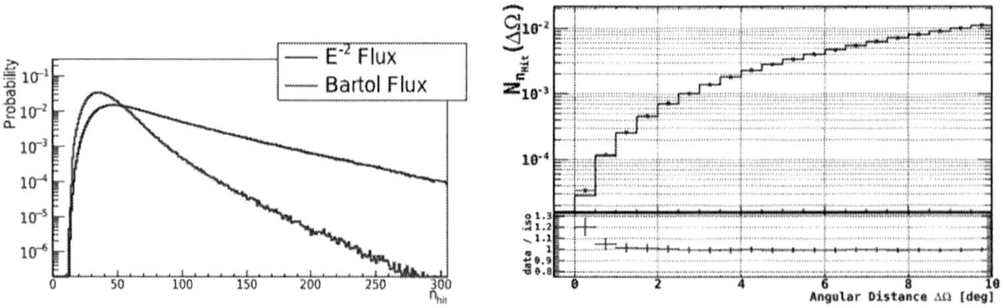

Figure 1. Left: Distributions of the n_{hit} energy estimator for simulated atmospheric neutrinos, following a Bartol flux ($\sim E^{-3.7}$), and simulated cosmic neutrinos following an E^{-2} spectrum. Right: Cumulative autocorrelation distribution for the 2007-2012 ANTARES data set (red points) and for the average randomized ensemble (black line).

analysis based on the timing properties of both the ANTARES and IceCube HESE events to constrain a potential transient neutrino signal in the Galactic Center region.

2. Search principles

2.1 Spatial autocorrelation

The autocorrelation analysis allows to find inhomogeneities within a discrete data set by studying the two point correlation distribution, which is defined as the distribution of the number of pairs of events as a function of their mutual angular distance $\Delta\Omega$. As it was shown in [2] and [3], weights based on some energy estimator $\bar{E}$ can be applied to the events in order to discriminate between atmospheric neutrinos, which dominate at lower energies, and cosmic neutrinos, of which the spectral distribution in energy is harder (see Fig. 1, left). Formally, the cumulative autocorrelation distribution can be defined as

$$\mathcal{N}_{n_{Hit}}(\Delta\Omega) = \sum_{i=1}^{N} \sum_{j=i+1}^{N} \omega_{ij} \cdot \left[1 - H(\Delta\Omega_{ij} - \Delta\Omega)\right], \tag{1}$$

where H is the Heaviside step function and $\omega_{ij} = \omega_i \cdot \omega_j$ are weights assigned to the couple of events i and j. Each of the individual weights ω_i is defined as $\omega(\bar{E}_i) = \int_0^{\bar{E}_i} f(\bar{E}) d\bar{E}$, and $f(\bar{E})$ is the normalized distribution of the energy estimator, which can be obtained from Monte Carlo simulations of the atmospheric background. As it is shown in [2], the selected choice for the energy estimator is the number of hits used in the event reconstruction, n_{hit}. Figure 1 left, shows a comparison between the n_{hit} distribution for atmospheric and astrophysical neutrinos.

A comparison of the autocorrelation function resulting from measured events with the one corresponding to a randomized sample will allow to detect possible clusters of events if a statistically significant excess with respect to isotropy is present in the data.

2.2 Time correlation

In the time correlation study, we extend the two-point correlation function to the time domain. Thus, we define the two-point cumulative distribution as

$$\mathcal{N}(\Delta t) = \sum_{i=1}^{N_{IC}} \sum_{j=1}^{N_{ANT}} \omega_{ij} \cdot \left[1 - H\left(\Delta t_{ij} - \Delta t\right)\right], \tag{2}$$

where H is the Heaviside step function depending on the absolute value of the temporal distance Δt. ω_{ij} are the weights assigned to each couple of IceCube (IC) and ANTARES (ANT) events, labelled respectively i and j. Each weight ω_{ij} is computed according to a normal distribution centered on the IceCube HESE events and defined as $\omega_{ij} = \exp\left(\frac{-\Delta\Omega_{ij}^2}{2\sigma_i^2}\right)$, where $\Delta\Omega_{ij}$ is the angular distance between each couple of events, and σ_i is the standard deviation of the ith IceCube HESE angular error distribution.

3. Spatial autocorrelation

3.1 The ANTARES data set and the reference autocorrelation distribution

For the present analysis, a data set recorded by the ANTARES neutrino telescope between 2007 and 2012 has been used. The sample contains 5243 up-going neutrino candidates that satisfy selection criteria optimized in order to obtain the best average upper limit on the flux of neutrinos coming from point like sources and extends the data set of the previous analysis (see [2]) by about 50%. The reference autocorrelation distribution has been built as the average of the autocorrelation distributions derived from about 10^4 scrambled data sets. Each of them was obtained by keeping the local coordinates of one neutrino candidate, but assigning it a time randomly selected from another event. This process allows to build an isotropic sky map with the same coverage as the original data set, and in which the non uniformity in the data taking conditions is taken into account.

3.2 Results and discussion

The spatial autocorrelation analysis showed that the largest difference between the distributions corresponds to an excess of the ANTARES neutrino candidates with respect to the isotropic ensemble, at scales $< 0.5°$ (see Fig. 1, right). A statistical comparison following the same method described in [2] was made, and led to the conclusion that this corresponds to a $\sim 2.3\sigma$ excess. This difference is therefore not significant. Upper limits are currently being derived on the cumulative flux of unresolved sources.

4. Time correlation with IceCube HESE events

4.1 Data samples

Among the three-year HESE data set, consisting of thirty-seven events (see [1]), nine of them are located within 45° from the GC, of which eight occur between May 2010 and November 2012. Moreover, the ANTARES data set, between May 2010 and November 2012, has been optimized through Monte-Carlo simulations to reach a constant neutrino candidate rate over time. That was performed by taking into account variations of the mean counting rate of the optical modules due to the evolution of the data taking conditions in the sea. The final ANTARES data set consists of 4337 events.

4.2 Reference distribution

To detect correlated structures in the data set, a reference cumulative correlation distribution has been built by averaging 10^4 pseudo-experiments. For each pseudo-experiment, the event times are generated following the time dependent background fluctuations while the event local coordinates are drawn from the local coordinates distribution of the data.

A timing correlation between ANTARES events and IceCube HESE events is searched for by comparing, bin-by-bin, the cumulative two-point distribution of the data with the background distribution. Time correlation features will thus appear as differences between these two distributions.

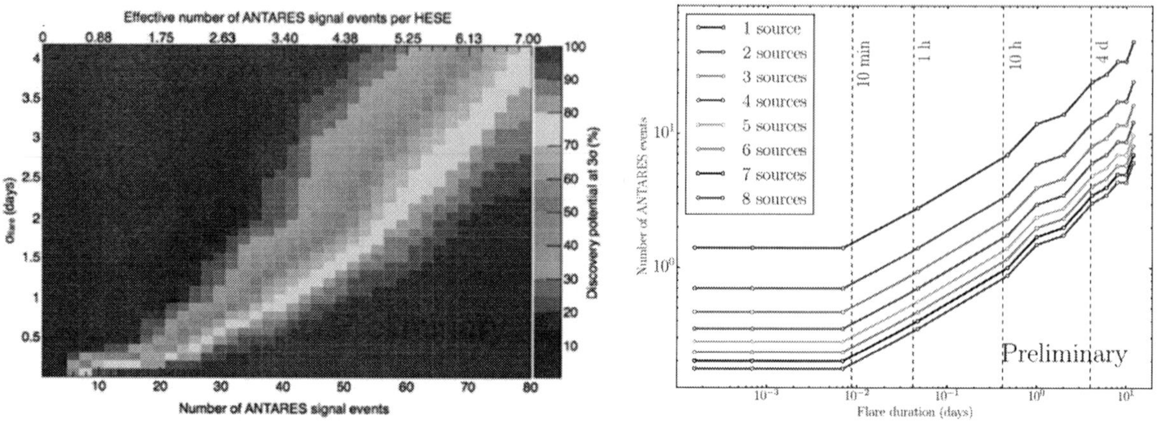

Figure 2. Left: Discovery potential at 3σ with respect to the flare duration (σ_{flare}) and the total number of signal events. The effective number of signal events per IC HESE is also given in the upper part. Right: 90% confidence level upper limit on the number of ANTARES events temporally correlated with the IceCube HESE events as a function of both the flare duration and the assumed number of sources.

In order to estimate the detection power of the analysis, we performed pseudo-experiments in which one or more event(s) is (are) replaced with one or more signal event(s) that would have come from a transient astrophysical source. The probability for a 3σ effect is provided with respect to the different combinations of flare duration and number of generated signal events (see Fig. 2, left plot).

4.3 Results and discussion

For the unblinded ANTARES data set, a $\sim 0.9\sigma$ effect is observed after having performed a statistical comparison similar to the spatial autocorrelation one. Thus, no significant correlation has been found. Consequently, we set upper limits on such a time correlation. Figure 2 (right plot) provides the 90% confidence level upper limit on the number of ANTARES events temporally correlated with the IceCube HESE events as a function of both the flare duration and the number of sources.

References

[1] M. G. Aartsen et al., Phys. Rev. Lett., **113**, 10, 2014

[2] S. Adrian-Martinez et al., JCAP05, 0001, 2014

[3] ANTARES Collaboration, ICRC 2013 Proc., `arXiv:1312.4308`

[4] Y. Bai et al., Phys. Rev. D **90**, 6, 2014

EPJ Web of Conferences **116**, 04002 (2016)
DOI: 10.1051/epjconf/201611604002

Indirect detection of Dark Matter with the ANTARES Neutrino Telescope

Miguel Ardid[a] on behalf of ANTARES Collaboration

Institut d'Investigació per a la Gestió Integrada de les Zones Costaneres (IGIC), Universitat Politècnica de València, C/ Paranimf 1, 46730 Gandia, Spain

Abstract. One of the main objectives of the ANTARES neutrino telescope is the search for neutrinos produced in self-annihilation of Dark Matter (DM) particles. The analysis for different sources of DM (Sun, Galactic Center, Earth, . . .) or DM models (SUSY, Secluded) will be described and the results presented. The specific advantages of neutrino telescopes in general and of ANTARES in particular will be explained. As an example, the indirect search for DM towards the Sun performed by neutrino telescopes currently leads to more stringent limits on the spin-dependent WIMP-nucleon cross section with respect to existing direct detection experiments.

1. Introduction

The concept of Dark Matter (DM) has been introduced in order to explain the observations in cosmology and astrophysics which seem to indicate that about 84% of the matter in the Universe is non-baryonic, non-relativistic and does not interact electromagnetically but only significantly through gravity. One of the most favourite models for DM is the so called Weakly Interacting Massive Particles (WIMPs) that form a halo in which the visible baryonic part of galaxies is embedded. Different candidates for WIMPs attract a great deal of interest, especially those provided by theories based on supersymmetry (SUSY). In some of these models, secondary high-energy neutrinos could be produced from the self-annihilation products, or decays of them. In this paper the indirect searches for DM by detecting high-energy neutrinos coming from the GC, Sun and Earth with the ANTARES neutrino telescope are presented.

2. Method

2.1 The ANTARES neutrino telescope

The ANTARES neutrino telescope [1] is located in the Mediterranean Sea, at a depth of 2.5 km, about 40 km off Toulon (France). It consists of 885 Optical Modules (OMs) arranged in a three-dimensional array. The operation principle is based on the detection by these OMs of the Cherenkov light induced by relativistic charged particles, mostly muons produced in interactions of high energy neutrinos in the

[a] e-mail: mardid@fis.upv.es

vicinity of the detector. The OMs are installed along 12 lines anchored to the sea floor and kept vertical by buoys at the top of them. The length of the lines is 450 m and the distance between the lines is 60–75 m. The position [2] and time [3] information of the photons detected by the OMs can be used to reconstruct the muon, and thus neutrino, direction. The installation of the detector was completed in 2008. In these DM searches data recorded between 2007 and 2012 are used. During this time, the detector consisted of 5 lines for most of 2007 and of successively 8, 9, 10 and 12 lines.

2.2 Signal and background estimation

The signal, i.e. the flux of muon neutrinos as a function of their energy arriving at the Earth's surface from the source is computed using Monte Carlo packages and previously obtained results (WimpSim [4] for the Sun and Earth, and [5] for GC). With this, neutrinos for the relevant channels and for DM masses ranging from 25 GeV to 10 TeV are derived. Propagation effects, i.e. oscillations among the three neutrino flavours, neutrino absorption and regeneration, are taken into account by these Monte Carlo packages. There are two sources of background in the ANTARES detector: the muons and neutrinos produced by cosmic rays interacting in the atmosphere. In order to reduce the first one, only upgoing events are selected, since muons cannot traverse the Earth. However, some of these muons (a tiny fraction of the total, but non-negligible number given the flux involved) are misreconstructed as upgoing. This is why further cuts are needed on the quality of the reconstructed track in order to reject them. On the other hand, the atmospheric neutrinos are irreducible background. For the Sun and Galactic Center cases, the background estimations are done by scrambling real data using random times, which reduces the impact of systematic uncertainties. For the Earth, since we look always to vertical tracks, this approach is not possible and background is estimated from simulations.

2.3 Optimization

A blind strategy procedure has been followed for the analysis in order to avoid selection biases. This means that the cuts have been selected before looking at the source region. The criterion to choose the cuts has been to optimize (minimize) the average upper limit which can be set for any given DM mass and channel. This has been done using Feldman-Cousins method [6] and using an average effective area that takes into account the livetime of each detector configuration [7].

The optimization is done by scanning the average upper limit for different values of quality reconstruction parameter and search cone angles around the source direction for the binned searches and using a likelihood procedure for unbinned analyses. For each WIMP mass and channel, an optimized set of cuts is found using the blind procedure. Then, real data is looked at and since there were not any significant excess in the searches, limits to DM were derived.

3. Searches and results

3.1 Indirect search for Dark Matter towards the Galactic Centre

A binned search for high-energy neutrinos coming from the direction of the Galactic Centre has been performed with the event selection criteria to maximise the sensitivity to possible signals produced by the self-annihilation of WIMPs accumulated around the centre of the Milky Way with respect to the atmospheric background for different possible channels: bb, WW, $\tau\tau$, $\mu\mu$, $\nu\nu$ [8]. Since the number of neutrinos observed is compatible with background expectations, the 90% C.L. upper limits in terms of the neutrino flux and the velocity averaged annihilation cross-section $<\sigma v>$ are derived for the different WIMP self-annihilation channels. These limits for $<\sigma v>$ are shown to be the most stringent for a neutrino telescope over the WIMP masses 25 GeV $< M_{WIMP} <$ 10TeV. As an example, the 90%

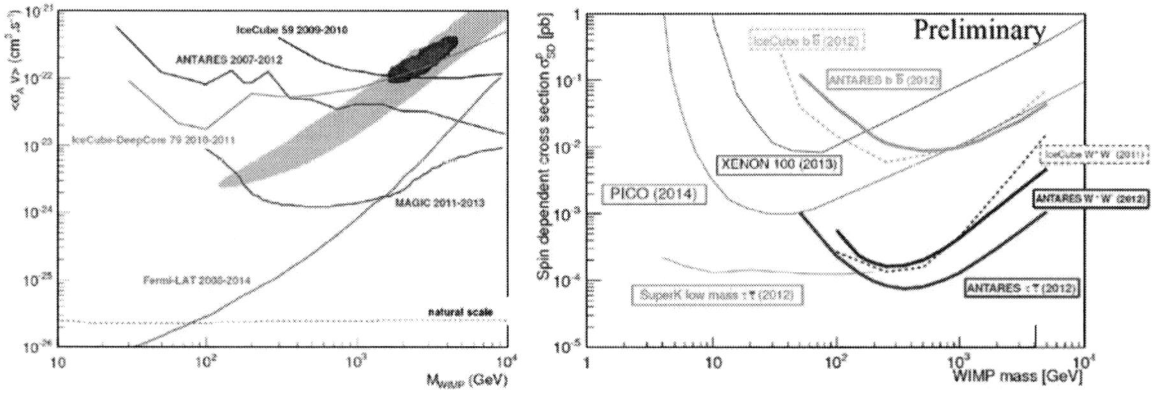

Figure 1. Left: The 90% C.L. upper limit on the WIMP velocity averaged self-annihilation cross-section, $<\sigma v>$ as a function of the WIMP mass in the range 25 GeV $< M_{WIMP} <$ 10TeV for the WIMP self-annihilation into $\tau^+\tau^-$, for ANTARES 2007-2012 GC search (red). This is compared to the limits from other experiments. The interpretation of observed electron/positron excesses as DM self-annihilations is also shown as filled areas orange (PAMELA) and green (PAMELA, Fermi-LAT and H.E.S.S). The dashed line indicates the natural scale for which a WIMP is a thermal relic of the early Universe. Figure from [8]. Right: The 90% C.L. upper limits on the spin-dependent WIMP-proton scattering cross section as a function of the WIMP mass obtained and compared to other results from neutrino telescopes and direct detection experiments. Figure from [9].

C.L. upper limit for $< \sigma v>$ WIMP self-annihilation into $\tau^+\tau^-$ is presented in Fig. 1-left and compared to the limits obtained from other experiments and regions of interest proposed.

3.2 Indirect search for Dark Matter towards the Sun

An indirect search for DM towards the Sun has been performed using the unbinned method [9]. Since no excess of data has been observed in the direction of the Sun, limits have been calculated in terms of muon neutrino fluxes and spin-dependent WIMP-nucleon scattering cross sections. Since the Sun is composed mainly of hydrogen indirect searches are more sensitive to the spin-dependent scattering cross section and can surpass even direct detection experiments, such as PICO [10], with the ANTARES limits becoming very competitive for high WIMP mass ($\geq$ 1 TeV), see Fig. 1-right. Direct detection experiments, such as LUX [11], are definitively more competitive for the spin-independent cross section (see Fig. 2).

3.2.1 Search for Secluded Dark Matter

Secluded DM is a special scenario where DM annihilates into a pair of non-Standard Model mediators, ϕ, which subsequently decay into SM particles. It was suggested to explain some experimental observations, such as the positron-electron ratio observed by satellite detectors. In case of meta-stable mediators, the decay products could either be detected directly in the Earth or the final spectrum of neutrino fluxes could be modified significantly. In this search [12, 13] two different cases are studied: a) detection of di-muons from mediator decays [14], either directly or indirectly through mediators which decay into di-muons that, in turn, decay into neutrinos, and b) the mediator that directly decays into neutrinos [15]. These results [12, 13] –the first experimental limits established directly in neutrino telescopes– are more restrictive than those derived in direct detection searches for the case of spin-dependent interaction for a wide range of lifetimes of the meta-stable mediator.

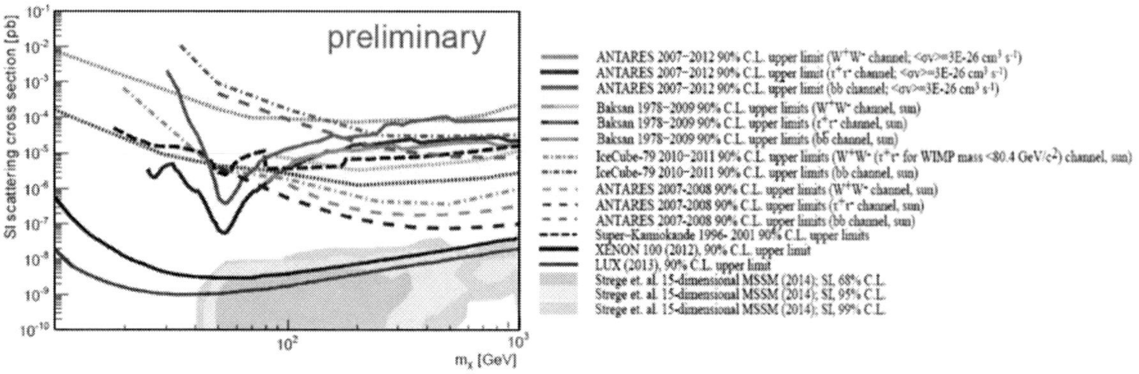

Figure 2. 90% CL upper limits on σSI as a function of the WIMP mass for $<\sigma v> = 3{\cdot}10^{-26}\,\mathrm{cm^3 s^{-1}}$ and WIMP pair annihilation to 100% into either $\tau^+\tau^-$, W^+W^- or bb, for ANTARES (Earth) 2007 – 2012, compared to other results from neutrino telescope, direct detection experiments and MSMM predictions. Figure from [16].

3.3 Indirect search for Dark Matter towards the Earth Centre

WIMPs could also scatter with Earth matter and be trapped. An indirect search for DM can be performed with ANTARES by looking for an excess of the neutrino flux from the Earth's centre [16]. The spectrum of the neutrino flux from the Earth would depend on the WIMP mass, the annihilation channel, the spin independent scattering cross section and the thermally averaged annihilation cross section of the WIMPs. No significant excess over the background expectation has been found and 90% CL upper limits on the WIMP self-annihilation rate were set as a function of the WIMP mass for different channels. These were translated to limits on the spin independent scattering cross section. Figure 2 shows the results of this search and compare them to other searches with neutrino telescopes and direct detection experiments. Although direct detection experiments put more stringent constrains, there might be some differences between experiments using different targets and technologies. In this sense, this kind of search can be complementary to direct detection.

We acknowledge the financial support of the Spanish Ministerio de Economía y Competitividad, Grants FPA2012-37528-C02-02, and Consolider MultiDark CSD2009-00064, of the Generalitat Valenciana, Grants ACOMP/2015/175 PrometeoII/2014/079 and of the European FEDER funds.

References

[1] M. Ageron, et al., ANTARES Collaboration, Nucl. Instrum. and Meth. A **656**, 11 (2011)

[2] M. Ardid, for the ANTARES Collaboration, Nucl. Instrum. and Meth. A **602**, 174 (2009). S. Adrián-Martínez et al., ANTARES Collaboration, J. Instrum. **7**, T08002 (2012)

[3] J.A. Aguilar et al., ANTARES Collaboration, Astropart. Phys. **34**, 539 (2011)

[4] Mattias Blennow, Joakim Edsjo, Tommy Ohlsson, JCAP **01**, 021 (2008). J. Edsjo, WimpSim Neutrino Monte Carlo, `http://www.fysik.su.se/ edsjo/wimpsim/`

[5] M. Cirelli, et al., JCAP **03**, 051 (2011)

[6] G.J. Feldman, R.D. Cousins, Phys. Rev. D **57**, 3873(1998)

[7] S. Adrián-Martínez et al., ANTARES Collaboration, JCAP **11**, 032 (2013)

[8] S. Adrián-Martínez et al., ANTARES Collaboration, JCAP **10**, 068 (2015)

[9] C. Tönnis, for the ANTARES collaboration, PoS (ICRC2015) 1207

[10] C. Amole et al., PICO Collaboration, Phys. Rev. Lett. **114**, 231302 (2015)

[11] D.S. Akerib et al., LUX Collaboration, Phys. Rev. Lett., **112**, 091303 (2014)

[12] M. Ardid and C. Tönnis, for the ANTARES collaboration, PoS (ICRC2015) 1212
[13] S. Adrián-Martínez. Ph. D. Thesis. Universitat Politècnica de València (2015)
[14] P. Meade, S. Nussinov, M. Papucci, T. Volansky. J. High Energy Phys. **06**, 29 (2010)
[15] N.F. Bell, K. Petraki. JCAP **04**, 003 (2011)
[16] A. Gleixner and C. Tönnis, for the ANTARES collaboration, PoS (ICRC2015) 1110

EPJ Web of Conferences **116**, 04003 (2016)
DOI: 10.1051/epjconf/201611604003

IceCube results from point-like source searches using 6 years of through-going muon data

Stefan Coenders[a] for the IceCube Collaboration[b]

Technische Universität München, Boltzmannstr. 2, 85748 Garching, Germany

Abstract. The IceCube Neutrino Observatory located at the geographic South Pole was designed to study and discover high energy neutrinos coming from both galactic and extra-galactic astrophysical sources. Track-like events induced by charged-current muon-neutrino interactions close to the IceCube detector give an angular resolution better than 1° above TeV energies. We present here the results of searches for point-like astrophysical neutrino sources on the full sky using 6 years of detector livetime, of which three years use the complete IceCube detector. Within 2000 days of detector livetime, IceCube is sensitive to a steady flux substantially below $E^2\partial\phi/\partial E = 10^{-12}\,\mathrm{TeV cm^{-2}\,s^{-1}}$ in the northern sky for neutrino energies above 10 TeV.

1. Introduction

IceCube is a cubic-kilometer neutrino detector installed in the ice at the geographic South Pole [1] between depths of 1450 m and 2450 m. Detector construction started in 2005 and finished in 2010. Neutrino reconstruction relies on the optical detection of Cherenkov radiation emitted by secondary particles produced in neutrino interactions in the surrounding ice or the nearby bedrock. During construction, partial detector configurations were operational taking data with 40 strings from April 2008, 59 strings from May 2009, and 79 strings from May 2010 on before IceCube was fully operational in May 2011 using 86 strings. Partial detector configurations are denoted by *IC*, followed by the number of active strings, e.g., *IC59* for the 59 string season.

The observation of high-energy astrophysical neutrinos by IceCube using events starting in the detector volume [2], and with up-going muons mainly created in charged-current muon-neutrino interactions outside the detector [3], adds impetus to the searches for astrophysical neutrino sources. The first detection of a cosmic neutrino point-like source will represent a major breakthrough and will provide a *smoking gun* for cosmic ray acceleration. In hadronic interactions, neutrinos are produced alongside photons and charged particles, but due to their neutral charge and low interaction cross-section, neutrinos are neither deflected by magnetic fields, nor absorbed by interstellar media. Here, we present the results of searches for time-independent neutrino emission using both a scan of the full-sky and a dedicated search using known high-energy γ-ray objects.

---

[a]e-mail: coenders@icecube.wisc.edu
[b]http://icecube.wisc.edu

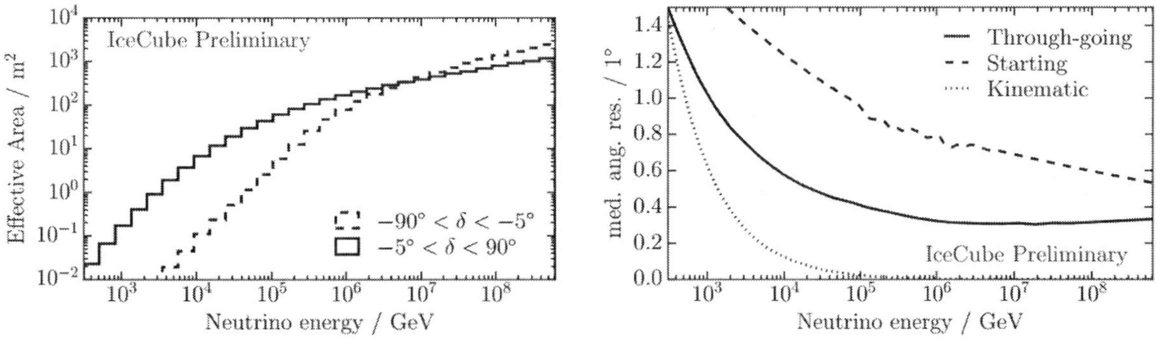

Figure 1. Effective area (left) and angular uncertainty (right) for $\nu_\mu + \bar{\nu}_\mu$ in this analysis. The effective area is shown for up-going (solid) and down-going (dashed) events separately. Angular resolutions are shown for tracks traversing (solid) and starting (dashed) in the detector. The intrinsic kinematic angle of muon and parent neutrino in ν_μ charged current interactions is shown as well (dotted).

2. Point source searches in IceCube

The analysis presented here is a search for neutrino point-like sources using a sample chiefly made up of muon events entering the detector, but also identifying starting muon tracks identified with veto techniques [2]. Muons are created by charged-current $\nu_\mu + \bar{\nu}_\mu$ interactions and travel several kilometers. Thus, IceCube collects large amounts of neutrino interactions far outside the detector, increasing the statistics largely compared to starting event searches.

Using a multivariate selection technique, approximately 70 000 neutrinos are collected in the northern sky each year, and a similar number of muons in the southern sky. The event sample in the northern sky consists of atmospheric neutrinos that form an irreducible background. In the southern sky, the largest background are bundles of atmospheric muons. The atmospheric spectrum is softer than the one expected for astrophysical neutrinos. Thus, the background rate is reduced by requiring only the highest-energy tracks to pass the event selection, yielding an event sample that is efficient for neutrino sources from above PeV up to EeV energies. More details about the event selection can be found at [4–6]. The graphs in Fig. 1 show the effective area (left) and median angular resolution (right) of $\nu_\mu + \bar{\nu}_\mu$ events detected by IceCube versus neutrino energy after the event selection. Above TeV energies, the muon direction is well correlated with the parent neutrino interaction. Overall, a neutrino reconstruction accuracy better than 0.8° is achieved above TeV energies.

Starting events using veto techniques as in [2] provide a complementary sample compared to through-going muons in the southern sky. Using an active veto, the atmospheric background of incoming muons is significantly removed, which allows energy regions as low as $\sim$ 100 TeV to be probed while retaining an angular resolution below 1°. These energy regions are not accessible using through-going muons in the southern sky, due to the large background.

The event sample is tested for clustering of events by maximizing the unbinned likelihood

$$\mathcal{L} = \prod_i \prod_{j \in i} \left(\frac{n_s^j}{N} S\left(\left|\mathbf{x}_j - \mathbf{x}_S\right|, E_j, \sigma_j; \gamma\right) + \left(1 - \frac{n_s^j}{N}\right) B\left(\delta_j, E_j\right) \right) \tag{1}$$

for all events j (total sum N) using partial and full detector configurations ($j \in \{\mathrm{IC40}, \mathrm{IC59}, \mathrm{IC79}, \mathrm{IC86}\}$), the probability of an event to be signal $\mathcal{S}$ or background $\mathcal{B}$ is used to optimize for the number of associated source events n_s^j in that sample and the source spectral index γ. Signal events at position $\mathbf{x}_j$ are assumed to cluster at the source position $\mathbf{x}_S$ given their reconstruction uncertainty σ_j. The background probability is estimated using an off-source region of events at the same declination δ_j. In addition to spatial clustering, the energy is used to distinguish signal

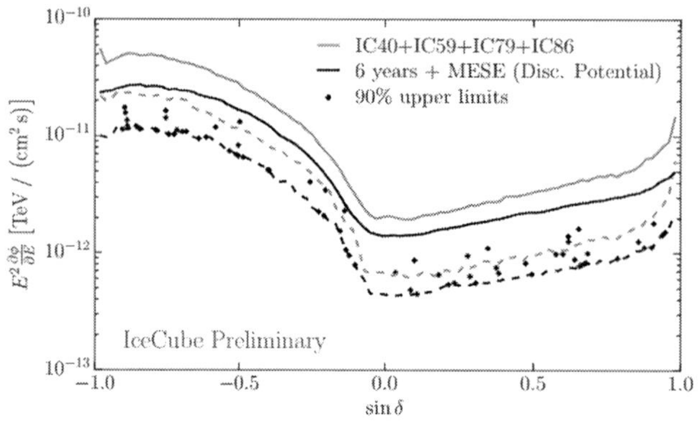

Figure 2. IceCube sensitivity (dashed) and discovery potential (solid) for point-like $\nu_\mu + \bar{\nu}_\mu$ sources using six years of detector exposure. In addition, medium energy starting events (MESE) are used in the southern sky. Black crosses indicate 90% upper limits for selected source candidates assuming an unbroken E^{-2} power-law. Grey lines show the same properties of IceCube's previous point source search using four years of exposure [5].

from background. Astrophysical sources are expected to produce high-energy neutrinos, whereas the atmospheric background has a soft energy spectrum. For each spot on the sky, the test statistic $\log \Lambda = -2\log \frac{\mathcal{L}(n_s,\gamma)}{\mathcal{L}(n_s=0)}$ with respect to the null hypothesis of no clustering ($n_s = 0$) is calculated.

The sensitivity and discovery potential of the six year event sample is shown in Fig. 2. Compared to the previous analysis [5], a gain is observed for all points in the sky. The effect of including starting events can be seen particularly in the extent that the bound is lowered in the southern hemisphere meaning that IceCube reaches the sensitivity of ANTARES [7]. However, the accessible energy range in the southern sky is different for both experiments, where IceCube covers the region above 100 TeV, while ANTARES is much more sensitive at lower energies, for example cut-off E^{-2}-spectra [8]. Dedicated searches using solely events starting in IceCube give additional sensitivity in that regime [9].

Using the test statistic of the unbinned likelihood (1), the entire sky is probed for significant clustering above background expectation. For this *all-sky scan*, in both northern and southern sky the point with most clustering is reported. To account for the *look elsewhere effect*, this procedure is redone for pure background data-sets, where the right ascension of all events is scrambled uniformly. Hence, the *pre-trial* p-value of these points is trial-corrected to obtain the true significance.

In order to reduce the large effect of the trial-factor from the unbiased *all-sky scan*, a source list of prominent γ-ray sources was selected *a-priori* to be analyzed individually. This source list consists of sources of different types, for example pulsar wind nebulae in the galactic plane, and extra-galactic active galactic nuclei like BL Lac.

3. Results

In the analysis of over 600 000 events distributed over the full sky within 6 years of IceCube livetime, no significant clustering was observed. Every cluster seen is compatible with expectations from pure background fluctuations.

In the northern sky, the most significant clustering (*hot spot*) is located at $\alpha = 249.6°$, $\delta = 63.6°$. The likelihood is fit to $n_s = 29$ and $\gamma = 2.1$ corresponding to a pre-trial p-value of $-\log_{10} \mathrm{p} = 5.75$. After accounting for the trial factor, the p-value is corrected to 35%. In the southern sky, the hot spot is located at $\alpha = 300.4°$, $\delta = -33.2°$ and yields a post-trial p-value of 87% at $n_s = 19$, $\gamma = 2.3$.

The analysis of the promising sources found that the source PKS1406-076 (a flat spectrum radio quasar) is associated with the highest neutrino clustering ($n_s = 12.9$). The observation is compatible with background (p-value 5.1%) and upper limits are set for all sources. The limits are indicated with crosses at the source declination in Fig. 2. The 90% upper limit for PKS1406-076 yields $2.32 \times 10^{-12}\,\mathrm{TeV}\,/\,\mathrm{cm}^2\,\mathrm{s}$.

From the observed γ-ray spectra of sources, models of the expected neutrino spectra can be built. Two models from [10] and [11] are tested for the Crab Nebula. The model from [10] uses the connection between γ-rays and neutrinos produced in pion decays while the model in [11] is derived from Monte Carlo simulation of inelastic p-p scattering in the source. For the model of [11], the IceCube upper limit is two thirds of the model prediction, whereas the limit for the model from [10] is still 30% above the prediction.

Note, that the sensitivity of IceCube with respect to that model matches the prediction in this analysis, but an observed fluctuation ($n_s = 6.8$) at the position of the Crab Nebula gives a higher upper limit. Within the next years, IceCube will be able to tighten the bounds on that model as well and thus probe the amount of hadronic acceleration which is occurring in addition to purely leptonic processes.

4. Conclusion

The analysis presented here, based on 6 years exposure and more than 600 000 events in a through-going track sample, did not find any clustering above the background expectation. The previous search [5] was extended by an additional two years of livetime in the full detector configuration, significantly improving the discovery potential of the analysis. For the first time, the sensitivity to steady neutrino sources significantly surpasses $10^{-12}\,\mathrm{TeV}\,/\,\mathrm{cm}^2\,\mathrm{s}$ over most of the northern sky.

IceCube puts stringent upper limits on the neutrino flux coming from various sources in the northern sky, constraining models of hadronic acceleration due to the non-detection of neutrinos from the sources. For some of the sources, IceCube has capabilities to test models and probe the amount of hadronic versus leptonic acceleration.

References

[1] A. Achterberg et al. (IceCube), Astropart.Phys. **26**, 155 (2006)
[2] M.G. Aartsen et al. (IceCube), Phys.Rev.Lett. **113**, 101101 (2014)
[3] M.G. Aartsen et al. (IceCube), Phys. Rev. Lett. **115**, 081102 (2015)
[4] M.G. Aartsen et al. (IceCube), Astrophys.J. **779**, 132 (2013)
[5] M.G. Aartsen et al. (IceCube), The Astrophysical Journal **796**, 109 (2014)
[6] S. Coenders (IceCube), PoS(ICRC2015)1047 (2015)
[7] S. Adrian-Martinez et al. (ANTARES), Astrophys. J. **786**, L5 (2014)
[8] S. Adrian-Martinez et al. (IceCube, ANTARES) (2015), 1511.02149
[9] S. Euler (IceCube), IceCube results from low-energy point-source searches in the southern hemisphere, these proceedings (2015)
[10] A. Kappes et al., The Astrophysical Journal **656**, 870 (2007)
[11] E. Amato, D. Guetta, P. Blasi, Astron.Astrophys. **402**, 827 (2003)

EPJ Web of Conferences **116**, 04004 (2016)
DOI: 10.1051/epjconf/201611604004

Low-energy point source searches with IceCube

Sebastian Euler[1,a], David Altmann[2,3,b], Rickard Ström[1,c] for the IceCube Collaboration[d]

[1] Department of Physics and Astronomy, Uppsala University, 751 20 Uppsala, Sweden
[2] Deutsches Elektronen-Synchrotron (DESY), 15735 Zeuthen, Germany
[3] Erlangen Centre for Astroparticle Physics (ECAP), Friedrich-Alexander-Universität Erlangen-Nürnberg, 91058 Erlangen, Germany

Abstract. Due to the overwhelming background of atmospheric muons, the traditional IceCube point source search in the Southern Hemisphere is mainly sensitive to neutrinos with energies above 100TeV. A new approach focuses on events starting inside the instrumented volume. By utilizing different veto techniques we are able to significantly reduce the energy threshold and can now for the first time explore the entire Southern Hemisphere at neutrino energies as low as 100GeV. We present the results of two analyses targeting slightly different energy ranges. Both use one year of data taken with the completed IceCube detector in 2011/12.

1. Introduction

The standard IceCube neutrino point source search [1] is most sensitive in the Northern Hemisphere, where the Earth shields the massive background from atmospheric muons. In the Southern Hemisphere, this analysis is limited to the highest energies above 100TeV, where the atmospheric background is sufficiently low. The Southern Hemisphere is, however, an interesting target, in particular at lower energies, as it contains the Galactic center and a large part of the inner Galaxy, with many potential neutrino sources [2, 3].

Events starting inside the IceCube detector [4] are an unambiguous signature of a neutrino interaction. By utilizing different veto techniques and focusing on starting events it is possible to reduce the background to a tolerable level and thus significantly lower the energy threshold for neutrino point source searches in the Southern Hemisphere. Here we present two such analyses, one aimed at an energy range of a few TeV to 100 TeV (STeVE – Starting TeV Events), and the other one aimed at even lower energies above 100GeV (LESE – Low Energy Starting Events).

[a]e-mail: `sebastian.euler@icecube.wisc.edu`
[b]e-mail: `david.altmann@icecube.wisc.edu`
[c]e-mail: `rickard.strom@icecube.wisc.edu`
[d]`http://www.icecube.wisc.edu`

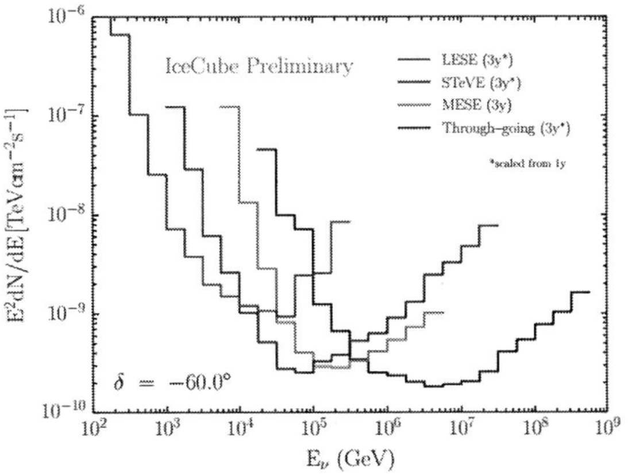

Figure 1. Differential sensitivities of four IceCube point source searches [1, 6] targeting different energy ranges, corresponding to 3 years of operation.

2. Data samples

Both analyses presented here use data from the first year of the completed 86-string IceCube detector. Due to slightly different run selections, there is a small difference in the final livetimes, 329days for LESE and 337days for STeVE. Both analyses rely on veto methods to reject the atmospheric muon background, using for example timing information of hits in the outer layers of IceCube and in IceTop. A cornerstone of both selections is the use of the *FiniteReco* algorithm [5], which reconstructs the neutrino interaction vertex based on the hit pattern seen in the detector. Events with a vertex outside the detector are rejected. For the final step of the event selection both analyses employ Boosted Decision Trees (BDT). The final LESE sample contains 6191 events and the STeVE sample 10178 events, with both samples still being dominated by atmospheric muons leaking in through the veto and mimicking starting neutrinos. The median angular resolution of both selections is about 2°, slightly worse than for the standard IceCube point source search, due to the lower energies of the events and the resulting shorter track lengths. Figure 1 shows the differential sensitivities of IceCube analyses aimed at different energy ranges: the LESE and STeVE analyses at the lowest energies, the MESE (Medium Energy Starting Events) analysis at somewhat higher energies [6] and the standard point source search (labeled "Through-going") at the highest energies [1]. Together, these analyses cover an energy range of 7 orders of magnitude from 100GeV to the EeV-range. Figure 2 shows the effective areas for the same set of analyses, compared to an analysis by ANTARES [7].

3. Analysis method

Both the LESE and the STeVE analysis use an unbinned maximum likelihood method [8] to search for a spatial clustering of events. The full likelihood $\mathcal{L}$ is given by

$$\mathcal{L}(n_s, \gamma) = \prod_i^N \left(\frac{n_s}{N} S_i(\Delta\psi, \sigma, E; \gamma) + \left(1 - \frac{n_s}{N}\right) B_i(\delta, E) \right).$$

For a given event and position in the sky we evaluate the compatibility with being signal or background, using the PDFs S_i and B_i. The signal PDF S_i depends on the angular difference $\Delta\psi$ between the reconstructed event direction and the considered source position, the angular uncertainty *gma* and reconstructed energy E of the event, and the spectral index of the source γ. The background PDF B_i

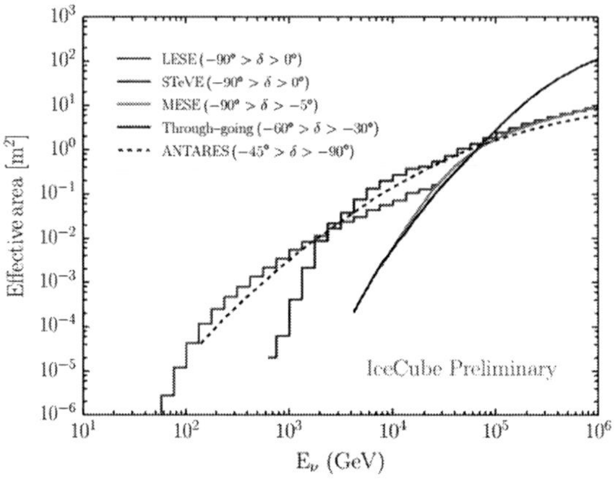

Figure 2. Effective areas of the analyses shown in Fig. 1 and of a corresponding ANTARES analysis [7].

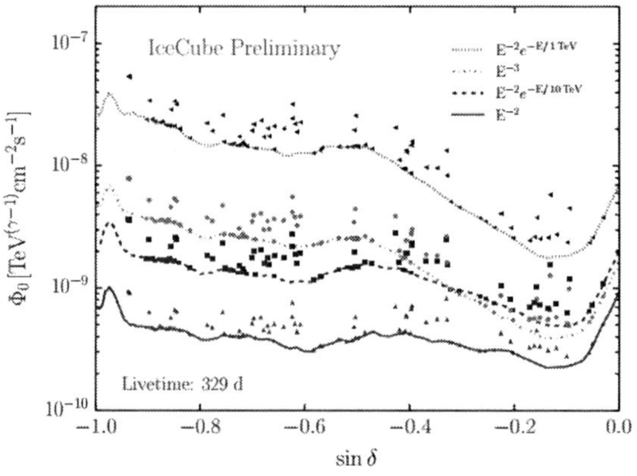

Figure 3. Median sensitivities as a function of declination (lines) and source upper limits (markers) at 90% C.L. obtained with the LESE analysis for various energy spectra.

depends only on reconstructed declination δ and reconstructed energy E. The likelihood is the product over all N events. It is maximized with the number of signal events n_s and the spectral index γ as free parameters. The test statistic is then given by the ratio of the likelihood for the null hypothesis ($n_s = 0$) and that for the best fit values.

When testing many points in the sky, trial factors need to be taken into account. To convert the pre-trial p-values obtained directly from the test statistic distribution into post-trial p-values, we generate an ensemble of 10000 pseudo-experiments with the R.A. directions scrambled, and count the fraction of these background-only experiments that yield an equal or greater pre-trial p-value.

4. Results

Both analyses evaluate the test statistic for all 84 Southern Hemisphere sources in the TeVCat [9] catalog (the "Default" and "Newly announced" catalogs, as of May 2015). We also test 12 sources not in TeVCat, but included in earlier IceCube point source searches. After trial factor correction, none of these 96 sources show a significant over-fluctuation. The obtained 90% upper limits on the neutrino flux, calculated using the Neyman method [10], are shown in Figs. 3 and 4 together with the sensitivities as a function of declination. Figure 3 shows the LESE results for a few different spectral hypotheses of the

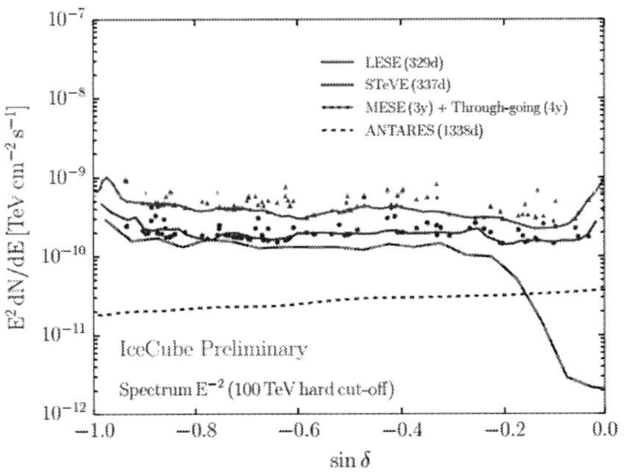

Figure 4. Median sensitivities and source upper limits for an E^{-2} spectrum with a hard cut-off at 100TeV for LESE and STeVE, compared to a combination of MESE [6] and the standard point source search [1], as well as ANTARES [11].

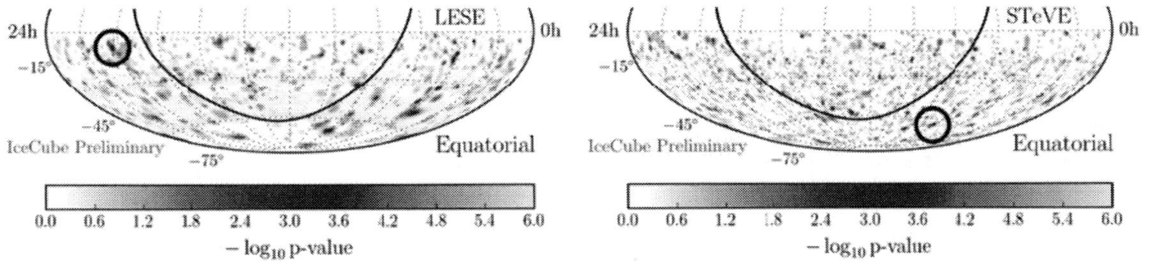

Figure 5. Sky maps of the pre-trial p-values obtained with LESE (left) and STeVE (right).

form $dN/dE = \Phi_0(E/\mathrm{TeV})^{-\gamma}\exp(-E/E_{\mathrm{cut}})$ with $\gamma = (2,3)$ and $E_{\mathrm{cut}} = (1\mathrm{TeV}, 10\mathrm{TeV}, \infty)$. Figure 4 shows limits and sensitivities of both LESE and STeVE for an E^{-2} spectrum with a hard cut-off at 100TeV, compared to the sensitivities of the combined MESE [6] and standard point source analysis [1], and a corresponding ANTARES analysis using 6 years of data [11]. The most significant source in the LESE analysis is QSO 2022-077 (also known as 2EG J2023-0836) at R.A. 306.5° and Dec. −7.6°, with a post-trial p-value of 14.8%. For STeVE, the most significant source is SNR G000.9+00.1 at R.A. 266.8° and Dec. −28.2°, with a post-trial p-value of 35.4%.

In addition to the pre-defined source list, both analyses perform a scan of the whole Southern Sky on a HEALPix [12] grid with a bin width of about 0.5°, well below the angular resolution for both samples. For each bin, the test statistic is evaluated in the same way as for the source list. The resulting pre-trial sky maps are shown in Fig. 5 for LESE (left) and STeVE (right). Again, no significant clustering of events is found. The hottest spot is marked with a circle in both maps. In the case of LESE, it is found at R.A. 305.2° and Dec. −8.5°, close to the most significant source, with a post-trial p-value of 88.1%. The STeVE analysis finds the hottest spot at R.A. 93.1° and Dec. −64.3°, with a post-trial p-value of 74.9% In both maps, the black line marks the Galactic plane.

5. Summary and outlook

Complementary to the standard IceCube point source search in the Southern Hemisphere, which is mainly sensitive to neutrinos with energies above 100TeV, we have performed two analyses searching for neutrino point sources at energies as low as 100GeV. The rejection of the large background of atmospheric muons is possible by making extensive use of veto information from the outer layers of IceCube. The data are analyzed using the same unbinned likelihood method as the standard point source analysis. No significant neutrino emission has been found. The obtained upper limits are about one order of magnitude weaker than those by ANTARES. Both analyses presented here use data from the first year of the completed IceCube detector with 86 strings. Three more years of data are available and ready to be incorporated. In addition, searches are planned for neutrino emission from extended sources and in particular from the Galactic plane, using the same data samples. In the long term, a unification of the different point-source searches will improve the overall sensitivity.

References

[1] M.G. Aartsen et al., Astrophys. J. **796**(2), 109 (2014), http://stacks.iop.org/0004-637X/796/i=2/a=109

[2] M.D. Kistler, J.F. Beacom, Phys. Rev. D **74**, 063007 (2006), http://link.aps.org/doi/10.1103/PhysRevD.74.063007

[3] A. Kappes, J. Hinton, C. Stegmann, F.A. Aharonian, Astrophys. J. **656**(2), 870 (2007)

[4] A. Achterberg et al., Astroparticle Physics **26**(3), 155 (2006), ISSN 0927–6505, http://www.sciencedirect.com/science/article/pii/S0927650506000855

[5] S. Euler, Ph.D. thesis, RWTH Aachen University (2014), http://publications.rwth-aachen.de/record/444944

[6] J. Feintzeig, Ph.D. thesis, University of Wisconsin, Madison (2014), http://search.proquest.com/docview/1641120849

[7] S. Adriàąn-Martìnez et al., The Astrophysical Journal **760**(1), 53 (2012), http://stacks.iop.org/0004-637X/760/i=1/a=53

[8] J. Braun et al., Astropart.Phys. **29**(4), 299 (2008), ISSN 0927-6505, http://www.sciencedirect.com/science/article/pii/S0927650508000303

[9] S.P. Wakely, D. Horan, *TeVCat: An online catalog for Very High Energy Gamma-Ray Astronomy*, in *Proceedings of the 30th International Cosmic Ray Conference* (2007), Vol. 3, p. 1341

[10] J. Neyman, Philosophical Transactions of the Royal Society of London. Series A, Mathematical and Physical Sciences **236**(767), 333 (1937), ISSN 00804614, http://www.jstor.org/stable/91337

[11] S. Adriàąn-Martìnez et al., The Astrophysical Journal Letters **786**(1), L5 (2014), http://stacks.iop.org/2041-8205/786/i=1/a=L5

[12] K.M. Gòrski et al., Astrophys. J. **622**(2), 759 (2005), http://stacks.iop.org/0004-637X/622/i=2/a=759

EPJ Web of Conferences **116**, 04005 (2016)
DOI: 10.1051/epjconf/201611604005

First combined search for neutrino point-sources in the southern sky with the ANTARES and IceCube neutrino telescopes

J. Barrios-Martí[1,a] on behalf of the ANTARES collaboration, and C. Finley[2,b] on behalf of the IceCube collaboration

[1] Institut de Física Corpuscular (UV-CSIC), C/Catedrático José Beltrán 2, 46980 Paterna, Spain
[2] Oskar Klein Centre and Dept. of Physics, Stockholm University, 10691 Stockholm, Sweden

Abstract. A search for cosmic neutrino point-like sources using the ANTARES and IceCube neutrino telescopes over the Southern Hemisphere is presented. The ANTARES data were collected between January 2007 and December 2012, whereas the IceCube data ranges from April 2008 to May 2011. An unbinned maximum likelihood method is used to search for a localized excess of muon events in the southern sky assuming an $E^{-}2$ neutrino source spectrum. A search over a pre-selected list of candidate sources has also been carried out for different source assumptions: spectral indices of 2.0 and 2.5, and energy cutoffs of 1 PeV, 300 TeV and 100 TeV. No significant excess over the background has been found, and upper limits for the candidate sources are presented compared to the individual experiments.

1. Introduction

Neutrinos offer unique insight into the Universe due to the fact that they interact only weakly. This also implies that their detection is challenging. The high-energy field is currently led by the IceCube [1] and ANTARES [2] experiments. While the instrumented volume of ANTARES is significantly smaller than that of IceCube, its geographical location provides a better view of the southern sky for neutrino energies below 100 TeV. The complementarity of the detectors for southern sky sources allows for a gain in sensitivity by combining the analysis of data from both experiments in a joint search for point sources. The improvement with this combination depends on the details of the fluxes, in particular the energy spectrum and a possible energy cut-off of the signal.

2. Neutrino data samples

The data sample corresponds to all events from the southern sky which were included in the latest published ANTARES point-source analysis [3] combined with the events in the three-year IceCube point-source analysis [4]. The ANTARES sample corresponds to data recorded from 2007 January 29

[a] e-mail: javier.barrios@ific.uv.es
[b] e-mail: cfinley@fysik.su.se

to 2012 December 31. The total number of events in this sample amounts to 5516, of which 4136 are from the Southern Hemisphere. The estimated contamination of mis-reconstructed atmospheric muons is of 10%. The IceCube data was recorded from 2008 April 5 to 2011 May 13, with a total number of 146 018 events in the southern sky. These data were collected within different configurations (40, 59 and 79 strings). These events are predominantly well-reconstructed atmospheric muons rather than atmospheric neutrinos, because the Earth cannot be used to filter out the muon background for down-going events.

3. Search method

An unbinned maximum likelihood has been performed to search for excesses of events. In order to estimate the significance of a cluster, the likelihood takes into account the energy and directional information of each event. The data sample to which an event belongs is also taken into account, due to the differences in detector response and background. The likelihood, as a function of the total number of signal events, n_s, can be expressed as

$$L(n_s) = \prod_{j=1}^{4} \prod_{i=1}^{N^j} \left[\frac{n_s^j}{N^j} S_i^j + \left(1 - \frac{n_s^j}{N^j}\right) B_i^j \right], \tag{1}$$

where j indicates one of the four data samples (ANTARES, IC40, IC59 or IC79), i indicates an event belonging to the j-th sample, S_i^j is the value of the signal probability distribution function (PDF) for the i-th event in the j-th sample, B_i^j indicates the value of the background PDF, N^j is the total number of events in the j-th sample, and n_s^j is the number of signal events fitted for in the j-th sample. Since a given evaluation of the likelihood refers to a single source hypothesis at a fixed sky location, the number of fitted signal events n_s^j in each sample is related to the total fitted number of signal events n_s by the relative contribution of each sample, $n_s^j = n_s \cdot C^j(\delta, \frac{d\Phi}{dE})$, which allows us to only fit in the likelihood the total number of signal events. This contribution corresponds to the fraction of signal events expected for the j sample compared to the total for a given source spectrum and declination.

The signal and background PDFs for the IceCube and ANTARES samples have slightly different definitions. The signal PDFs are defined as

$$S^{\mathrm{ANT}} = \frac{1}{2\pi\sigma^2} \exp\left(-\frac{\Delta\Psi(\vec{x}_s)^2}{2\sigma^2}\right) P_s^{\mathrm{ANT}}(N^{hits}, \sigma), \quad S^{\mathrm{IC}} = \frac{1}{2\pi\sigma^2} \exp\left(-\frac{\Delta\Psi(\vec{x}_s)^2}{2\sigma^2}\right) P_s^{\mathrm{IC}}(\mathcal{E}, \sigma|\delta) \tag{2}$$

where $\vec{x}_s = (\alpha_s, \delta_s)$ indicates the source direction in equatorial coordinates, $\Delta\Psi(\vec{x}_s)$ is the angular distance of a given event to the source, σ is the angular error estimate, and $\mathrm{P}_s^{\mathrm{ANT}}(N^{hits}, \sigma)$ is the probability for a signal event to be reconstructed with an angular error estimate of σ and a number of hits taken in the event reconstruction N^{hits}. The number of hits is a proxy for the energy of the event [5]. In the case of the IceCube signal PDF, the main difference lies in the use of the reconstructed energy, $\mathcal{E}$, and the declination dependence of the probability for a signal event to be reconstructed with a given σ and $\mathcal{E}$.

The background PDFs are obtained from the experimental data itself. The definitions of the PDFs are:

$$B^{\mathrm{ANT}} = \frac{B^{\mathrm{ANT}}(\delta)}{2\pi} P_b^{\mathrm{ANT}}(N^{hits}, \sigma), \quad B^{\mathrm{IC}} = \frac{B^{\mathrm{IC}}(\delta)}{2\pi} P_b^{\mathrm{IC}}(\mathcal{E}, \sigma|\delta), \tag{3}$$

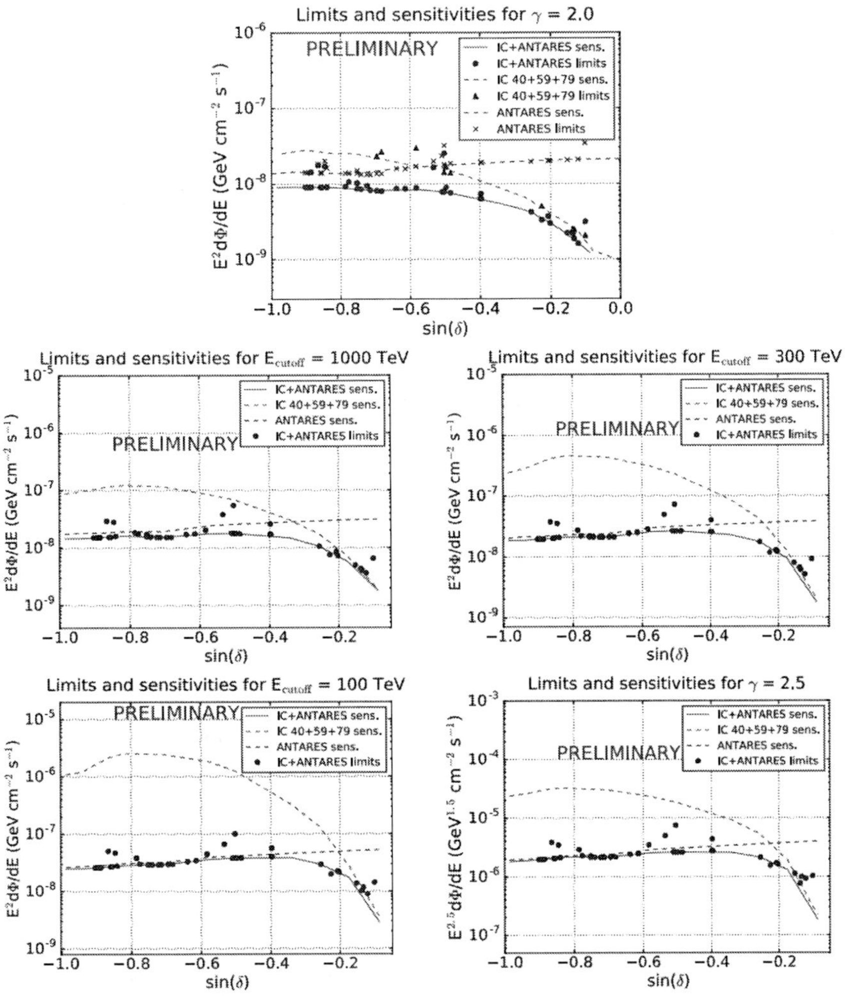

Figure 1. Point source sensitivities and limits for the following energy spectra: unbroken E^{-2} spectrum (top), E^{-2} with a square-root exponential cut-off at $E = 1\,\text{PeV}$ (middle left), $E = 300\,\text{TeV}$ (middle right), $E = 100\,\text{TeV}$ (bottom left) and $E^{-2.5}$ unbroken power-law (bottom right). Green points indicate the limits on the southern sky candidate sources given in Ref. [4] and [3]. The green line indicates the sensitivity for the combined search. Blue and red curves/points indicate the sensitivities for the individual IceCube and ANTARES analyses, respectively.

where B(δ) is the per-solid-angle rate of observed events as a function of the declination in the corresponding sample. $P_b^{\text{ANT}}(\mathcal{N}^{hits}, \sigma)$ and $P_b^{\text{IC}}(\mathcal{E}, \sigma|\delta)$ characterize the distributions for background event properties, in analogy with the definitions of P_s^{ANT} and P_s^{IC} for signal events given above.

The test statistic, *TS*, is determined from the likelihood (Eq. (1)) as $TS = \log L(\hat{n}_s) - \log L(n_s = 0)$, where $\hat{n}_s$ is the value that maximizes the likelihood. The p-value is calculated as the probability of the background to produce a TS of equal or higher value and becomes smaller for larger TS.

Two different searches for point-like neutrino sources have been performed. In the candidate list search, a possible excess of neutrino events is looked for at the location of 40 pre-selected neutrino source candidates.These candidates correspond to all sources in the southern sky considered in the previous candidate-source list searches performed in the ANTARES and IceCube point-source analyses [3, 4]. The second search is a "full sky" search, looking for a significant point-like excess anywhere in the southern sky.

4. Results

No significant event clusters are found over the expected background. The most significant cluster in the full southern sky search is located at equatorial coordinates $\alpha = 332.8°, \delta = -46.1°$, with best-fit $\hat{n}_s = 7.9$ and pre-trial p-value of 6.0×10^{-7}, with a post-trial significance of 24% (0.7σ in the one-sided sigma convention). The direction of this cluster is consistent with the second most significant cluster in the previous ANTARES point-source analysis (but also less significant).

No statistically significant excess is found in the candidate source list search. The most significant excess for any object in the list corresponds to HESS J1741-302 with a pre-trial p-value of 0.003. To account for trial factors, the search is performed on the same list of sources using pseudo data-sets. 11% of randomized data sets have a smaller p-value for some source than that found for the real data; the post-trial significance of the source list search is thus 11% (1.2σ in the one-sided sigma convention).

Figure 1-top shows the sensitivities and limits for this search (assuming an E^{-2} spectrum) in comparison with the previously published ANTARES and IceCube analyses of the same data. The point source sensitivity in a substantial region of the sky, centered approximately at the declination of the Galactic Center ($\delta = -30°$), can be seen to have improved by up to a factor of two. Similar gains in other regions of the sky can be seen for different energy spectra in Fig. 1.

5. Conclusion

We have presented the first combined point-source analysis of data from the ANTARES and IceCube detectors. We have calculated the sensitivity to point sources and, with respect to an analysis of either data set alone, found that up to a factor of two improvement is achieved in different regions of the southern sky, depending on the energy spectrum of the source. Two joint analyses of the data sets have been performed: a search over the whole southern sky for a point-like excess of neutrino events, and a targeted analysis of 40 pre-selected candidate source objects. The largest excess in the southern sky search has a post trial p-value of 0.24 (significance of 0.7σ). In the source list search the candidate with the highest significance corresponds to HESS J1741-302, with a post-trial p-value of 0.11 (significance of 1.2σ). Both of the results are compatible with the background-only hypothesis and no significant excess is found. Flux upper limits for each of the source candidates have been calculated for E^{-2} and $E^{-2.5}$ power-law energy spectra, as well as for E^{-2} spectra with cut-offs at energies of 1 PeV, 300 TeV, and 100 TeV. Because of their complementary nature, with IceCube providing more sensitivity at higher energies and ANTARES at lower energies, a joint analysis of future data sets will continue to provide the best point-source sensitivity in critical overlap regions in the southern sky, where neutrino emission from Galactic sources in particular may be found.

JBM acknowledges the support of the Spanish MINECO through project FPA2012-37528-C02-01, the MultiDark Consolider Project CSD2009-00064, the Generalitat Valenciana via Prometeo-II/2014/079, and of Universitat de València, Atracció de Talent. CF acknowledges the support of the Swedish Research Council (VR).

References

[1] Achterberg, A., Ackermann, M., Adams, J., et al., APh, **26**, 155 (2006)
[2] Ageron M., Aguilar J. A., Samarai I. Al et al., Nucl. Instrum. Meth., A **656**, 11 (2011)
[3] Adrián-Martínez S., Albert A., André M., et al., ApJ, **786**, L5 (2014)
[4] Aartsen M. G., Abbasi R., Abdou Y. et al., ApJ, **779**, 132 (2013)
[5] Adrián-Martínez S., Samarai I. Al., Albert A. et al., ApJ, **760**, 53 (2012)

EPJ Web of Conferences **116**, 04006 (2016)
DOI: 10.1051/epjconf/201611604006

KM3NeT/ARCA sensitivity and discovery potential for neutrino point-like sources

A. Trovato[a] on behalf of the KM3NeT Collaboration

INFN-LNS, via S. Sofia 62, 95123 Catania, Italy

Abstract. KM3NeT is a large research infrastructure with a network of deep-sea neutrino telescopes in the abyss of the Mediterranean Sea. Of these, the KM3NeT/ARCA detector, installed in the KM3NeT-It node of the network, is optimised for studying high-energy neutrinos of cosmic origin. Sensitivities to galactic sources such as the supernova remnant RXJ1713.7-3946 and the pulsar wind nebula Vela X are presented as well as sensitivities to a generic point source with an E^{-2} spectrum which represents an approximation for the spectrum of extragalactic candidate neutrino sources.

1. Introduction

KM3NeT is a new Research Infrastructure, consisting of a network of deep-sea neutrino telescopes in the Mediterranean Sea. In response to the IceCube discovery of cosmic neutrinos [1] and the reported non-zero value for the neutrino mixing angle θ_{13}, KM3NeT 2.0 will consist of two detectors with different granularity: KM3NeT/ARCA [2] at the KM3NeT-It site (Capo Passero, Italy) with two building blocks dedicated to high-energy neutrino astronomy and KM3NeT/ORCA [3], a single building block located at the KM3NeT-Fr site (Toulon, France), dedicated to the study of the neutrino mass hierarchy. Thanks to its geographical location in the Northern hemisphere, KM3NeT/ARCA can observe most of the Galactic Plane, including the Galactic Centre. The detection of galactic point sources is one of the main goals of the experiment. Under the hypothesis of 100% hadronic mechanisms acting inside transparent sources, models for galactic neutrino emission are well constrained by TeV gamma-ray observations and allow the calculation of realistic predictions of the detector discovery potential.

Neutrinos are detected by measuring the Cherenkov light induced by charged secondary particles emerging from neutrino interactions. KM3NeT will consist of arrays of light detectors deployed in the deep, clear water of the Mediterranean Sea which acts as target, Cherenkov radiator and shield for atmospheric muons. The light detectors are photo-multiplier tubes (PMTs) arranged in glass spheres that resist the water pressure (digital optical modules, DOM [4]). The optical modules are arranged along flexible strings with a total height of about 700 m for KM3NeT/ARCA. This detector will consist of two building blocks of 115 strings each, with 18 DOMs per string and a spacing of 36 m between

[a] e-mail: atrovato@lns.infn.it

DOMs. The footprint will be roughly circular with an average distance between strings of about 90 m. Each block will have a volume of about $\sim$0.5 km^3.

2. Simulation codes and analysis

The software used in this work [5] provides a complete simulation for neutrinos of all flavours with energy in the range $10^2 - 10^8$ GeV, including their interaction in the medium [6, 7], the propagation of the emerging secondary particles [8], the light generation and propagation in water and the detector response. The depth and the optical water properties measured at the KM3NeT-It Capo Passero site are used [9]. Background light due to the presence of ^{40}K in salt water has been also simulated.

The neutrino induced events can be divided into two classes, track-like and cascade-like events, each class requiring a specific event reconstruction. Only the results of the track analysis will be reported here. The reconstruction algorithm for track-like events used in this work is described in [10, 11]. The angular resolution, calculated as the median angle between the reconstructed muon direction and the generated neutrino track, is about 0.2° at 10 TeV.

In point source searches, the simulated events are analysed through statistical techniques to look for an excess around the source position. Indeed, the neutrino signal from a cosmic source needs to be identified among the large background of atmospheric muons and neutrinos, produced by the interaction of primary cosmic rays with the atmosphere. The atmospheric neutrino event rate is calculated assuming the Honda et al. [12] conventional atmospheric model, the Enberg et al. [13] prompt component and adding a correction due to the cosmic rays knee on the neutrino spectrum as described in [14]. Atmospheric muons are generated with the MUPAGE event generator [15]. Atmospheric muons can be rejected by selecting only upward-going events, which can be only produced by neutrinos, the only particles that can traverse the whole Earth without being absorbed. This rejection requires a reliable determination of the zenith angle. For this reason beside a zenith angle cut, a cut on the goodness of fit criterion of the reconstruction is applied. Atmospheric neutrinos are an unavoidable background but a key for a possible discrimination against cosmic neutrinos is their different spectrum, which is softer than the expected spectrum of cosmic neutrinos.

The discovery potential, that is the signal flux required to obtain an observation at a given significance level (e.g. 5σ or 3σ) with 50% probability [16], has been taken as a figure of merit of the telescope performance. The method used considers two hypotheses: the hypothesis that data consist only of background and the hypothesis that signal events are present in the data in addition to the background. The likelihood of obtaining the data given each hypothesis is calculated, and the ratio of likelihoods, or equivalently the log of the likelihood ratio, serves as test statistic [11, 17]. Having Probability Density Functions (PDF) that describe the distribution of signal and background events as a function of a given variable x, $PDF_{sig}(x)$ and $PDF_{bkg}(x)$, the likelihood ratio can be written as:

$$LR = \sum_{i=1}^{n} log \frac{\frac{n_{sig}}{n} \times PDF_{sig}(x_i) + \left(1 - \frac{n_{sig}}{n}\right) \times PDF_{bkg}(x_i)}{PDF_{bkg}(x_i)} \tag{1}$$

where n is the total number of recorded events in a given period of time and n_{sig} is the expected number of signal events in the sample of n events. For each sample LR is maximised as function of n_{sig} and the maximum value of LR is used as test statistic. The variable x in eq. 1 is chosen in different ways in the analyses described in the following and will be discussed in each case.

3. Results

The operation of gamma-ray telescopes have revealed in the last decades a large number of TeV γ-ray sources in our Galaxy associated with Supernova Remnants (SNR) and Pulsar Wind Nebulae (PWN)

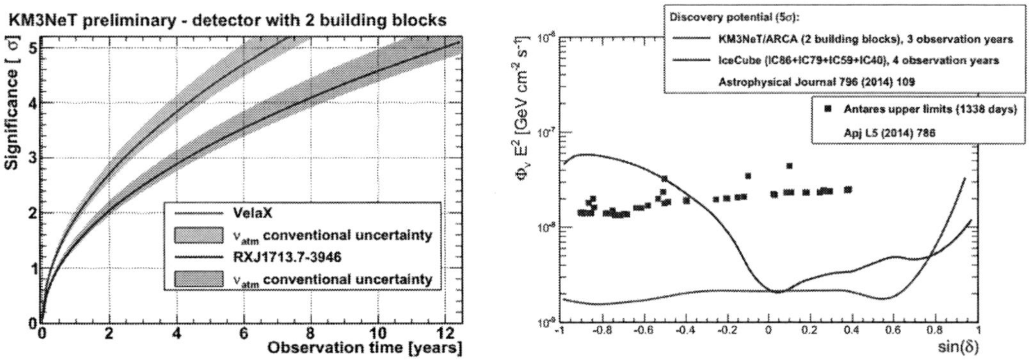

Figure 1. Left: Significance of the RXJ1713.7-3946 (blue line) and Vela X (red line) observation as a function of observation time for the KM3NeT/ARCA detector. Right: KM3NeT/ARCA 5σ discovery flux (red line) per flavor for point sources with an E^{-2} spectrum as a function of their declination calculated for 3 years of observation time. For comparison also the IceCube 4 years discovery flux [23] and the upper limit for the ANTARES detector [24] are shown.

that are also promising neutrino sources. In particular, the SNR RXJ1713.7-3946 and the Pulsar Wind Nebulae (PWN) Vela X are at present two of the most intense galactic objects in the high-energy gamma-ray band and are used here to evaluate the KM3NeT/ARCA performance. In both cases the expected neutrino spectra are calculated from the measured γ-ray spectrum under the hypotheses of a transparent source and 100% hadronic emission.

The young shell-type SNR RX J1713.7-3946 has been observed by H.E.S.S. [18] and its gamma energy spectrum is measured up to about 100 TeV. Having a declination of $-39^\circ 46'$ it is visible by KM3NeT for about 80% of the time. This SNR has been simulated as an extended homogeneous neutrino source with a 0.6° radius and an energy spectrum parametrised as [19]:

$$\Phi(E) = 16.8 \times 10^{-15} \left[\frac{E}{\mathrm{TeV}}\right]^{-1.72} e^{-\sqrt{E/2.1TeV}} \mathrm{GeV}^{-1}\mathrm{s}^{-1}\,\mathrm{cm}^{-2}. \tag{2}$$

Vela X is one of the nearest PWN centered at RA $= 08^h35^m$, DEC $= -45^\circ 36'$. The VHE γ-ray emission from a circular region of 0.8° around the source centre is reported by H.E.S.S. [20]. The corresponding neutrino spectrum derived using the Vissani prescription [21] is parametrised as:

$$\Phi(E) = 7.2 \times 10^{-15} \left[\frac{E}{\mathrm{TeV}}\right]^{-1.36} e^{-(E/7\,\mathrm{TeV})}\,\mathrm{GeV}^{-1}\mathrm{s}^{-1}\mathrm{cm}^{-2}. \tag{3}$$

The source has been simulated as a flat spatial distribution within a disk of 0.8° radius.

Under the assumed hypotheses, the significance of the observation as a function of the observation time for KM3NeT/ARCA has been calculated and is shown in the left panel of Fig. 1. In both cases a Boosted Decision Tree (BDT) from the ROOT TMVA package [22] has been used and the distributions of the BDT output for the signal and background events are used as PDFs in Eq. (1). An observation with 3σ significance is expected with the KM3NeT/ARCA detector after about 2.5 and 4.5 years for Vela X and RXJ1713.7-3946, respectively. The major uncertainty in this calculation is due to the contribution of conventional component of the neutrino flux. This uncertainty, shown as a shaded area in Fig. 1, is taken into account assuming a $\pm 25\%$ variation in the normalization of the Honda et al. flux model [12].

The discovery flux is also evaluated for a generic E^{-2} point source as an approximation for neutrino extragalactic sources. The right panel in Fig. 1 shows the 5σ discovery flux as a function of the point source declination. In this case the PDFs in Eq. (1) are functions of the angular distance of the reconstructed track from the source and the number of hits, N_{hit}, in a space-time correlation with the fitted track. The parameter N_{hit} is a rough energy estimator. The 3 years observation time has been chosen in such a way to have a comparable exposure with respect to the IceCube results in [23]. The different trend of the KM3NeT and IceCube curves is mainly due to the geographical location of the detectors. KM3NeT can reach an unprecedented sensitivity in the Southern hemisphere and good sensitivity everywhere else compared to IceCube.

4. Conclusions

The results presented in this paper represent the status of the point source analysis and show that at least the most intense galactic ones are within reach for KM3NeT. In particular, the expectation for RXJ1713.7-3946 and Vela X are reported. The discovery potential for point sources with E^{-2} spectrum shows that KM3NeT has a very large field of view thus not only complementing, but also overlapping to a large extent the IceCube one. KM3NeT can thus provide important contributions to the new born field of neutrino astronomy.

References

[1] M.G. Aartsen et al. (IceCube), Phys. Rev. Lett. **113**, 101101 (2014)
[2] R. Coniglione et al. (KM3NeT), These proceedings (2015)
[3] D. Samtleben et al. (KM3NeT), These proceedings (2015)
[4] R. Bruijn et al. (KM3NeT), PoS(ICRC15)1157 (2015)
[5] A. Margiotta et al. (ANTARES), NIM A **725**, 98 (2013)
[6] G. Ingelmann, A. Edin and J. Rathsman, Comput. Phys. Commun. **101**, 108 (1997)
[7] G. Barr, Ph.D. thesis, University of Oxford (1987)
[8] P. Antonioli et al., Astropart. Phys. **7**, 357 (1997)
[9] G. Riccobene et al., Astropart. Phys. **27**, 1 (2006)
[10] A. Trovato et al. (KM3NeT), PoS(ICRC15)1114 (2015)
[11] A. Trovato, Ph.D. thesis, Università degli Studi di Catania (2014)
[12] M. Honda et al., Phys. Rev. D **75**, 043006 (2007)
[13] R. Enberg, M.H. Reno and I. Sarcevic, Phys. Rev. D **78**, 043005 (2008)
[14] M.G. Aartsen et al. (IceCube), Phys. Rev. B **89**, 062007 (2014)
[15] Y. Becherini et al., Astropart. Phys. **25**, 1 (2006)
[16] G.C. Hill, J. Hodges, B. Hughey, A. Karle and M. Stamatikos, Proceedings of PHYSTAT05 p. 108 (2006)
[17] J. Braun et al., Astropart. Phys. **29**, 299 (2008)
[18] F. Aharonian et al. (H.E.S.S.), Astron. Astroph. **464**, 235 (2007), erratum in A&A **531**, 2 (2011)
[19] S.R. Kelner, F.A. Aharonian and V.V. Bugayov, Phys. Rev. D 74, 063007 (2006)
[20] F. Abramowsky et al. (H.E.S.S.), Astron. Astrophys. **548**, A38 (2012)
[21] F.L. Villante and F. Vissani, Phys. Rev. D **78**, 103007 (2008)
[22] A. Hoecker et al., PoS ACAT **040** (2007), `arXiv:physics/0703039`
[23] M.G. Aartsen et al. (IceCube), Astrophys. J. **796**, 109 (2014)
[24] S. Adrían-Martínez et al. (ANTARES), Astrophys. J. Lett. **L5**, 786 (2014)

EPJ Web of Conferences **116**, 04007 (2016)
DOI: 10.1051/epjconf/201611604007

Neutrino fluxes from the Galactic plane and the ANTARES limit

Luigi Antonio Fusco[1,2,a] on behalf of the ANTARES Collaboration

[1] Dipartimento di Fisica e Astronomia dell'Università, Viale Berti-Pichat 6/2, 40127 Bologna, Italy
[2] INFN – Sezione di Bologna, Viale Berti-Pichat 6/2, 40127 Bologna, Italy

Abstract. The existence of cosmic neutrinos has been reported by the IceCube Collaboration. Though this measurement is consistent with an isotropic neutrino flux, a sub-dominant galactic component coming from extended regions such as the Galactic Plane cannot be excluded. The ANTARES detector, located in the Mediterranean Sea, is currently the largest and longest operated under-water neutrino telescope; its effective area and good exposure to the Southern Sky allow to constrain an enhanced muon neutrino emission from extended sources such as the Galactic Plane. ANTARES data from 2007 to 2013 have been analysed and upper limits on the neutrino production from the central region of our galaxy have been set.

1. Introduction

The ANTARES detector [1] is currently the largest and longest operated under-water neutrino telescope. Its main scientific goal is to detect high energy neutrinos of cosmic origin. The IceCube Collaboration has reported the observation of a cosmic neutrino signal in the High Energy Starting Events (HESE) analysis. The purely atmospheric expectation is rejected at a level of 5.7σ with three years of data [2]. This signal is compatible with an isotropic cosmic flux which is equally distributed in the three neutrino flavours [3]. However, some possible enhancement of the neutrino flux from the Southern hemisphere with respect to the Northern one has been underlined, for example in [4]. The low energy extension of the IceCube HESE analysis [5] shows a steepening of the energy spectrum of the cosmic signal producing more events at low energy. Even if the IceCube volume is much larger, the effective area of ANTARES in the region between 20 and 50 TeV for the Southern Sky is similar [6]: some IceCube-like signal events could be observed by ANTARES too.

Because of the nature of the IceCube analysis, based on vetoing techniques to detect downward-going events, the signal is mainly distributed in the shower channel: the directional resolution for these events is poor. Since ANTARES is located in the Mediterranean Sea, the Southern sky is accessible to the detector via up-going tracks, for which sub-degree good angular resolution can be achieved [15]. A diffuse neutrino flux in the shower channel for IceCube might appear as an ensemble of individual point

[a] e-mail: lfusco@bo.infn.it

sources in ANTARES, or a region with an enhanced diffuse emission. The first possibility is addressed in [7], while the latter will be presented in this contribution.

2. Neutrinos from the Milky Way

Neutrinos can be produced close to galactic Cosmic Ray (CR) accelerators such as supernova remnants, when high energy protons or nuclei interact with the surrounding matter or photon field. Charged pions from those interactions decay into leptons (including neutrinos), while π^0, produced in equal amount at the interaction point, give γ-rays: the expected all-flavour neutrino flux from these interactions is equal to the γ one. The neutrino energy spectrum directly follows that of CRs at the acceleration site – $E^{-\Gamma}$ with $\Gamma = 2.0$ for Fermi acceleration scenarios [8] – since the decay usually takes place before the primary particle can interact.

A contribution to a cosmic neutrino flux from the Milky Way is expected from interactions of CRs propagating in the inner region of our Galaxy. The neutrino spectrum should have the same spectral behaviour as propagating CRs, which is softer than that at sources. Data from the Fermi/LAT detector provide the best observation of this diffuse γ flux in the Galactic Plane [9], though no observation of the neutrino counterpart is available. Assuming that a certain fraction of the observed diffuse γ flux in the central region of the Galaxy originates from hadronic mechanisms, the neutrino yield from CR propagation can be calculated. Different predictions are available such in [10–12]. Each of these provides a different description of the expected neutrino flux, with an overall normalisation that can vary by one order of magnitude. A rather hard spectrum is expected, described, at least in part of the energy range, by a power law with spectral index $\Gamma \sim 2.4 \div 2.5$.

3. The IceCube signal

The one-flavour IceCube cosmic neutrino signal was initially [2] fitted by a power law spectrum with spectral index Γ equal to 2:

$$E^2 \frac{d\Phi}{dE} = (0.95 \pm 0.3) \times 10^{-8}\ \mathrm{GeV\,s^{-1}\,sr^{-1}\,cm^{-2}}. \tag{1}$$

while also an $E^{-2.3}$ spectrum was reported to be compatible with the observed signal, mainly because of the absence of events above 2 PeV and the lack of an enhanced neutrino detection in the region of the Glashow resonance (~6 PeV). The further extension of the IceCube data set and refinements in the analysis for the observation of lower energy neutrinos resulted in a steepening of the energy spectrum; in [5] the best fit is:

$$\frac{d\Phi}{dE} = \left(2.06^{+0.4}_{-0.3}\right) \times 10^{-18} \left(\frac{E}{100\ TeV}\right)^{-2.46\pm0.12} \mathrm{GeV^{-1}\,s^{-1}\,sr^{-1}\,cm^{-2}} \tag{2}$$

for 25 TeV < E_ν < 1.4 PeV, rejecting the E^{-2} hypothesis with a significance of 3σ. This flux is observed as an all sky flux by IceCube. This can be related to the bias in efficiency of the analysis to shower events, for which IceCube has a rather bad angular resolution (10 to 20 degrees). An excess of events close to the galactic plane is reported for example in [13]. This accumulation could point towards a diffuse emission region of cosmic neutrinos. In addition, as the Southern sky contains most of the Galactic Plane, the soft spectrum derived from IceCube measurement and such a possible enhancement of the neutrino flux in the Southern hemisphere hint towards a possible neutrino emission from the Galactic Plane.

4. Data analysis

Assuming that part of the IceCube HESE signal comes from a region of the sky with angular size $\Omega \sim 0.1$–0.2 sr, the corresponding neutrino flux can be computed from the IceCube effective area [4]. Such a signal, coming from an enhanced neutrino emission scenario can be searched for in ANTARES comparing the event rate from this area of the sky to what is observed where no signal is expected. A rectangular region having galactic coordinates $|\,l\,| < 40°$ and $|\,b\,| < 3°$, enclosing the central part of the Fermi/LAT diffuse galactic plane flux, is chosen to represent the Galactic Plane area. Upward-going neutrino candidates collected by the ANTARES neutrino telescope from 2007 to 2013 are considered for this analysis.

4.1 Atmospheric background

The main background in neutrino telescopes is given by downward-going atmospheric muons reaching the detector. Since only neutrinos can traverse the Earth, neutrino telescopes look at upward-going events to reject this background. Wrongly reconstructed atmospheric muons, mimicking neutrino induced events, can be rejected by a selection on the track quality parameter Λ and the estimate for the angular error β [15].

Atmospheric neutrinos from the decay of short-lived particles in extensive air showers represent the surviving background in cosmic neutrino searches. A *conventional* component, coming from the decay of pions and kaons, here described by the *Honda* et al. flux [16], can be modelised as a power law with spectral index Γ asymptotically going to 3.7. The *prompt* component, expected from charmed hadrons, yields a harder neutrino energy spectrum ($\Gamma \sim 2.7$). The *Enberg* et al. [17] prescription is used in this work. Since signal and background spectra are described by power laws with different indices, a cut on the reconstructed energy [18] provides a selection for cosmic neutrinos.

4.2 Background estimation from data

A signal from a diffuse emission region in the Southern Sky is searched for by comparing the number of events from the chosen signal region (on-zone) to that coming from regions with no expected signal having the same exposure (off-zones). This choice avoids simulation related biases in the estimation of the signal intensity after event selection. Off-zones are defined as fixed regions in equatorial coordinates which have identical size and shape as the on-zone but have no overlap with it. In local coordinates, off-zones have the same sidereal-day periodicity as the on-zone and span the same fraction of the sky, but with some fixed delay in time. This approach has been already used in ANTARES to search for events from the Fermi bubbles [19]. 9 off-zones are selected for the Galactic plane. Data from the signal region are blinded until the event selection has not been completely defined on Monte Carlo simulations, while data from off-zones can be used to estimate the agreement between data and Monte Carlo, with the expected fluxes introduced in the previos section, as well as the relative agreement between data from different off-zones. Good agreement between data and Monte Carlo, as well as between data from different off-zones, is found.

4.3 Cut optimisation

The optimisation of the event selection to enhance the possible cosmic signal against the atmospheric background is made on the basis of track quality parameters to reject wrongly reconstructed atmospheric muons and on the energy estimation to select cosmic neutrinos over the atmospheric background. The optimal selection cut, which maximises the sensitivity, is computed on the basis of the Model Rejection Factor (MRF) procedure [20] for the $\Gamma = 2.4$. The final selection cut is $\Lambda > -5.0$, $\beta < 0.5°$,

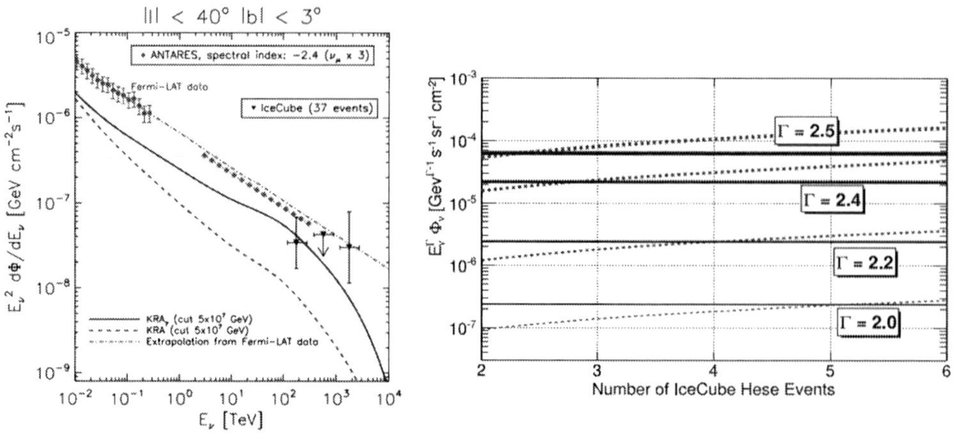

(a) (b)

Figure 1. a) ANTARES upper limit (magenta dots) compared to the expected neutrino fluxes from [12] (blue – solid and red -dashed line for different model assumptions) and the power law extrapolation of Fermi/LAT data from [10] to the IceCube data from that part of the sky. b) ANTARES upper limit for different energy spectra (black – solid lines) compared to the cosmic flux (dashed lines) that would produce a certain number of IceCube HESE events from a limited region of the sky.

$E_{ANN} > 10$ TeV, where Λ and β are the track quality parameters described in ref. [15], while E_{ANN} is the energy estimator from the Artificial Neural Network algorithm presented in [18].

5. Results and outlook

Considering a signal flux with an energy spectrum $\sim E^{-2.4}$ ($E^{-2.5}$) the 90% confidence level sensitivity is 2.0 (6.0) $10^{-5}\,\mathrm{GeV^{-1}cm^{-2}s^{-1}sr^{-1}}$. For comparison, assuming an $E^{-2.4}$ spectrum, the normalisation of an IceCube-like flux producing 3 or more events from a region in the sky of 0.1 sr is larger than 2.0 $10^{-5}\,\mathrm{GeV^{-1}cm^{-2}s^{-1}sr^{-1}}$ and any of these scenarios can be rejected. Figure 1 – left, compares the obtained sensitivity in the Galactic Plane region to the expected neutrino flux from CR propagation of [12] and the extrapolated Fermi γ spectrum of [10]. This sensitivity holds in the energy range 3 TeV – 300 TeV, which contains the central 90% of the expected signal.

For the Galactic Plane 3.7 events are observed on average for the background regions and two are detected from the on-zone after the final selection. An underfluctuation of the background is thus present in the signal region. Since the Galactic Plane area does not show an excess of events with respect to the background only evaluation, an upper limit at 90% confidence level on the signal flux can be set and it corresponds to the ANTARES sensitivity for the analysed spectral indices. This upper limit is reported in Fig. 1 right panel, compared to the expected neutrino flux that would induce a certain number of IceCube HESE events. These fluxes have been computed on the basis of the effective areas reported in [2]. Any model producing more than 3 events in the IceCube HESE sample from the Galactic Plane region is excluded at 90% confidence level for a spectral index larger than 2.4. The upper limits for harder spectral assumptions with the same selection criteria are also shown.

The lack of signal events in ANTARES data significantly constrains the possible Galactic origin of the IceCube Southern sky excess. Adding cascade events in the analysis is foreseen. For these events the angular resolution of ANTARES is about 3 degrees, better than the one of IceCube [21] and the combined search in track and showers can improve the possiblity to observe such a diffuse flux from our Galaxy and to explain the Southern sky excess in IceCube data.

References

[1] M. Ageron et al., Nucl. Instrum. Meth. A, **656**, 11–38 (2011)
[2] M.G. Aartsen et al., Phys. Rev. Lett., **113**, 101101 (2014)
[3] M.G. Aartsen et al., Phys. Rev. Lett., **114**, 171102 (2015)
[4] M. Spurio, Phys. Rev. D, **90**, 103004 (2014)
[5] M.G. Aartsen et al., Phys. Rev. D, **91**, 022001 (2015)
[6] S. Adrián Martínez et al., Astrophys. J. **760**: 53 (2012)
[7] J. Barrios Martí, Proceedings of the 34th ICRC, The Hague, Netherlands, ID 1077 (2015)
[8] E. Fermi, Phys. Rev., **75**, 1169 (1949); E. Fermi, Astroph. J., **119**, 1 (1954)
[9] M. Ackermann et al., ApJ, **750**, 3 (2012)
[10] A. Neronov, D. Semikoz, C. Tchernin, Phys. Rev. D, **89**, 103002 (2014)
[11] Y.Q. Guo, H.B. Hu & Z. Tian, `arXiv:1412.8590` (2014)
[12] D. Gaggero et al., ApJ Letters **815**: L25 (2015)
[13] A. Neronov & D. Semikoz, Astropart. Phys. **75**: 60-63 (2016)
[14] A. Marinelli et al., Proceedings of the 34th ICRC, The Hague, Netherlands, ID 1126 (2015)
[15] S. Adrián Martínez et al., ApJ Letters, **786**, L1 (2014)
[16] M. Honda et al., Phys. Rev. D, **75**, 043006 (2007)
[17] R. Enberg et al., Phys. Rev. D, **78**, 043005 (2008)
[18] J. Schnabel, Nucl. Instr. and Meth. A, **725**, 106–109 (2013)
[19] S. Adrián Martínez et al., Eur. Phys. J. C, **74**, 2701 (2014)
[20] G.C. Hill & K. Rawlins, Astrop. Phys., **19**, 393 (2003)
[21] T. Michael, Proceedings of the 34th ICRC, The Hague, Netherlands, ID 1078 (2015)

EPJ Web of Conferences **116**, 04008 (2016)
DOI: 10.1051/epjconf/201611604008

Predicted sensitivity of the KM3NeT/ARCA detector to a diffuse flux of cosmic neutrinos

R. Coniglione[1], L.A. Fusco[2,3,a], and D. Stransky[4] for the KM3NeT Collaboration

[1] INFN – Laboratori Nazionali del Sud, via S. Sofia 62, Catania 95123, Italy
[2] Dipartimento di Fisica e Astronomia dell'Università, Viale Berti-Pichat 6/2, Bologna 40127, Italy
[3] INFN – Sezione di Bologna, Viale Berti-Pichat 6/2, Bologna 40127, Italy
[4] Friedrich-Alexander-Universität Erlangen-Nürnberg, Erlangen Centre for Astroparticle Physics, Erwin-Rommel-Strasse 1, Erlangen 91058, Germany

Abstract. The KM3NeT Collaboration has started the construction of a research infrastructure hosting a network of underwater neutrino detectors in the Mediterranean Sea. Two instruments based on the same technology are being built: KM3NeT/ORCA to measure the neutrino mass hierarchy and to study atmospheric neutrino oscillations and KM3NeT/ARCA to detect high-energy cosmic neutrinos both in diffuse and point source mode. The excellent angular resolution of the ARCA detector, with an instrumented volume of about one Gton, will allow for an unprecedented exploration of the neutrino sky searching for neutrinos coming from defined sources of sky regions, like the Galactic Plane and the Fermi Bubbles. It will also look for diffuse high energy neutrino fluxes following the indication provided by the IceCube signal.

This contribution will report on the sensitivity of the KM3NeT/ARCA telescope with particular attention to the region of the Galactic Plane. Comparisons with theoretical expectations are also discussed.

1. Introduction

The KM3NeT Collaboration has started the construction of the next generation neutrino detectors, being built in the Mediterranean Sea. A large volume neutrino telescope, called ARCA [1], will be devoted to the search for high energy cosmic neutrinos while a more dense and compact neutrino detector, ORCA [1], will study the neutrino mass hierarchy looking at low energy atmospheric neutrino oscillations. Cherenkov light emitted by charged particles produced by all flavour neutrino interaction will be detected to reconstruct the neutrino direction and energy.

Both detectors will be built following a modular approach. The basic detector element is the detection unit, a vertical slender string kept taut by a top buoy and holding 18 pressure resistant glass spheres, each housing 31 3-inch photomultipliers. 115 DUs will constitute a building block. ARCA will be made of two building blocks with an average spacing between DUs of 90 m and a vertical distance

[a] e-mail: lfusco@bo.infn.it

between DOMs on each DU of 36 m; the total instrumented volume will be of the order of 1 km^3. ORCA will comprise one building block having its DUs spaced horizontally by 20 m and an inter-DOM distance of 9 m for an instrumented volume of about 6 Mton.

The IceCube collaboration has observed cosmic neutrinos in various analyses [2–4], reporting a large excess of high energy neutrinos over the atmospheric flux expectations. The observed flux is compatible with isotropic origin and equipartition in the three neutrino flavours as expected from neutrino oscillations in vacuum [5]. Individual sources of this excess could not be identified, nor has the production mechanism for these high energy neutrinos been determined. Here, the one-flavour cosmic signal flux is parametrised as:

$$E_\nu^2 \frac{d\Phi(E_\nu)}{dE_\nu} = 1.2 \cdot 10^{-8} e^{-E/3\,\mathrm{PeV}} \mathrm{GeV\,cm^{-2}\,s^{-1}\,sr^{-1}} \qquad (1)$$

even though different fits are also possible. The results presented here do not depend strongly on the assumed signal flux.

2. Searches for diffuse neutrino fluxes

Two analyses have been developed to search for a diffuse neutrino flux with ARCA. The "track" analysis searches for through-going muon track events induced mainly by ν_μ CC interactions and the shower analysis that identifies all-flavour neutrino interactions which produce electromagnetic and hadronic showers at the interaction vertex. Both analyses have been optimised using the "model discovery potential" (MDP) [6] method for a 5σ discovery of the spectrum given in Eq. (1).

2.1 Track analysis

The first step in the event selection for the track channel analysis is a cut on the reconstructed track direction to get rid of most of the atmospheric muons. After optimisation a cut to keep events with reconstructed zenith angle $\theta_{zen} > 80°$ was chosen. This cut does not reject all the atmospheric muons: a significant fraction of down-going atmospheric muons is wrongly reconstructed as up-going. To reject the background due to mis-reconstructed atmospheric muons and to atmospheric neutrinos the MDP minimisation is performed selecting well-reconstructed tracks and removing less energetic events mainly due to atmospheric neutrinos. The two reconstruction parameters correlated with the reconstruction quality and the neutrino energy are the Λ and N_{hit} parameters, described in [7]. After one year of operation with the ARCA detector, a flux twice that of Eq. (1) can be discovered with 5σ significance. The sensitivity can be increased by using a statistical method based on the maximization of a likelihood function, rather than using a simple cut. This has been performed as described in [8], on a sample of preselected neutrino events with $\theta_{zen} > 80°$ and a Λ selection. The likelihood ratio function that has been maximised is given by:

$$LR = \sum_{i=1}^{n} \log \frac{\frac{n_s}{n} P_s(N_{hit}^i) + \left(1 - \frac{n_s}{n} P_b\left(N_{hit}^i\right)\right)}{P_b(N_{hit}^i)} \qquad (2)$$

where n_s is the number of signal events, n_b the number of background events, $n = n_s + n_b$ and P_s and P_b are the probability distribution functions (PDF) for a signal or background to be reconstructed with a final selected number of hits N_{hit}^i. This function is maximised with respect to the number of signal events n_s. The resulting significance is reported in Fig. 1 as a function of the number of observation years and reaches about 4σ after one year.

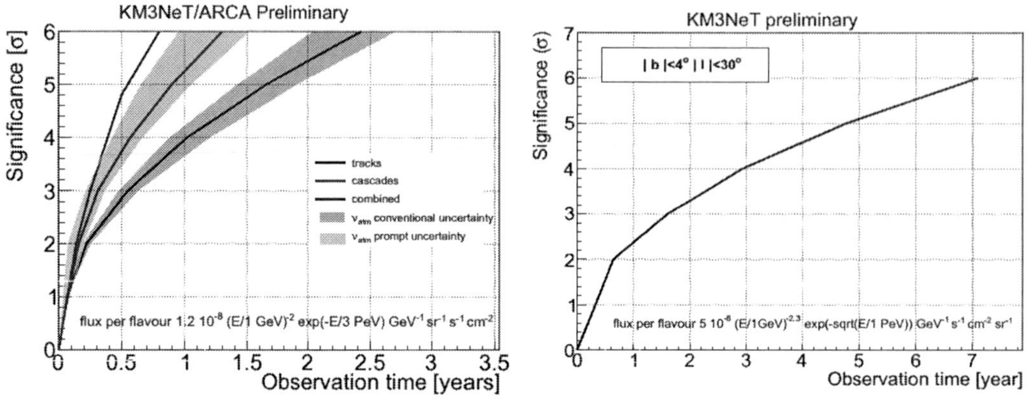

Figure 1. a) Discovery potential to the flux of Eq. (1) for ARCA analyses (see text for details). b) Performance of the track analysis in the search for a diffuse neutrino flux from the galactic plane.

2.2 Showers

An event selection procedure based on three consecutive cuts has been used in the shower analysis to isolate signal from background events. At the first selection level, only contained events are kept. Since most of the atmospheric muons enter the detector from above, their pseudo-vertices are reconstructed in the upper part of the detector. A large part of the background is rejected by cutting on coordinates of the reconstructed vertices with respect to the centre of the apparatus. A further rejection of atmospheric tracks, both from atmospheric muons and neutrinos, via an energy-related cut is obtained cutting on the total time over threshold (ToT) of hits selected by the shower reconstruction algorithm. The ToT is measured for each PMT and is a measure of the collected charge. Finally the signal is isolated using a machine-learning algorithm based on the boosted decision tree (BDT) from the ROOT TMVA package [9]. Quality parameters from the available shower and track reconstruction algorithms are used for the BDT classification. Its output is used together with the reconstructed shower energy E_{rec} to separate the signal from both the remaining atmospheric neutrino and muon backgrounds. In the shower channel a 5σ discovery of the assumed IceCube flux of can be achieved after 1.3 years of ARCA. A further improvement of this result is reached by applying the maximum likelihood method described above. The resulting significance as a function of the observation years is reported in Fig. 1.

2.3 Combined search

As both the track and shower analyses have a strong bias in efficiency towards one particular event topology, a further analysis strategy has been developed to incorporate the results from both the searches. The two analyses are indeed almost independent, as the overlap between the final event samples is limited to less then 2% of the surviving events. In order to develop an all-sky analysis with tracks and showers, the two hemispheres are treated separately. For downward-going events, where the atmospheric background is extremely high, only contained shower events are considered. The selection is then analogous to that of the usual shower analysis. For upgoing events, preliminary cuts are used to reject atmospheric muons that are wrongly reconstructed as upgoing, then a BDT method is used for a further rejection of background events. The BDT uses parameters coming from both the track and shower reconstruction algorithms and is optimised to reject atmospheric muons. Its output is used together with the shower energy estimation in the same maximum likelihood approach reported before. As a result a 4.8σ significance is obtained in 0.5 years of observation time, as shown in Fig. 1.

3. Diffuse neutrinos from the Galactic plane

One of the most promising regions of the sky from which a neutrino flux is expected is the Galactic Plane (GP) region. Neutrinos are expected to be produced in the interactions of cosmic rays with the dense galactic interstellar medium and radiation fields. The observation of γ-rays from the GP [10], part of which is expected to arise from the same hadronic processes that would produce high-energy neutrinos, strongly supports this hypothesis. Aiming at explaining the observed IceCube neutrino signal, some models evaluating the neutrino flux from regions in the GP near the Galactic center have been proposed. In [11] a CR propagation model with non uniform local transport properties has been adopted to explain the high energy γ-ray observation. To estimate the performance of the ARCA detector in searching for neutrinos from the GP, the resulting neutrino flux estimation has been adopted. In this framework the neutrino flux from a limited region of the GP near the galactic centre ($|l| < 30°$ and $|b| < 4°$), with b and l being the galactic latitude and longitude respectively, has been estimated and the performances of the ARCA detector have been evaluated following the same approach as in Sect. 2.1. The result of the maximum likelihood approach for the track channel is shown in Fig. 1b. After about 5 years of operation the ARCA detector will be able to discover such a signal at a significance level of 5σ. As a consequence its contribution to the IceCube diffuse flux would be estimated and constraints on CR propagation models in the inner part of our Galaxy could be placed.

4. Conclusions and outlook

Thanks to the purity of the event sample, a high-significance detection of a cosmic neutrino signal will be possible for both track-like and cascade-like events within one year of operation. The excellent angular and energy resolutions combined with the large effective mass, provide for a significant discovery potential to find neutrino sources in the Universe. A diffuse component from the Galactic plane could be identified in a few years of operation, bringing fundamental knowledge about the physics of primary cosmic rays in the Milky Way.

References

[1] S. Adrián Martínez et al., KM3NeT 2.0 Letter of Intent, arXiv:1601.07459 (2016)
[2] C. Kopper et al., proceedings of the 34th ICRC, ID 1081, The Hague, The Netherlands (2015)
[3] M.G. Aartsen et al., Phys. Rev. D **91**, 022001 (2015)
[4] M.G. Aartsen et al., Astrophys. J. **809**, 98 (2015)
[5] M.G. Aartsen et al., Phys. Rev. Lett. **114**, 171102 (2015)
[6] S. Adrián Martínez et al., Astr. Phys. **42**, 7–14 (2013)
[7] A. Trovato et al., proceedings of the 34th ICRC, ID 1279, The Hague, The Netherlands (2015)
[8] A. Trovato et al., proceedings of the 34th ICRC, ID 1282, The Hague, The Netherlands (2015)
[9] A. Hoecker et al., TMVA – Toolkit for Multivariate Data Analysis, PoS ACAT 040 (2007)
[10] M. Ackermann et al., Astrophys. J. **750**, 3 (2012)
[11] D. Gaggero et al., arXiv:1508.03681 (2015)

EPJ Web of Conferences **116**, 04009 (2016)
DOI: 10.1051/epjconf/201611604009

Interpretation of astrophysical neutrinos observed by IceCube experiment by setting Galactic and extra-Galactic spectral components

Antonio Marinelli[2,a], Daniele Gaggero[1], Dario Grasso[2], Alfredo Urbano[3], and Mauro Valli[4]

[1] GRAPPA Institute, University of Amsterdam, Science Park 904, 1090 GL Amsterdam, The Netherlands
[2] INFN and Dipartimento di Fisica "E. Fermi", Pisa University, Largo B. Pontecorvo 3, 56127 Pisa, Italy
[3] CERN, Theory division, 1211 Genève 23, Switzerland
[4] SISSA and INFN, via Bonomea 265, 34136 Trieste, Italy

Abstract. The last IceCube catalog of High Energy Starting Events (HESE) obtained with a livetime of 1347 days comprises 54 neutrino events equally-distributed between the three families with energies between 25 TeV and few PeVs. Considering the homogeneous flavors distribution (1:1:1) and the spectral features of these neutrinos the IceCube collaboration claims the astrophysical origin of these events with more than 5σ. The spatial distribution of cited events does not show a clear correlation with known astrophysical accelerators leaving opened both the Galactic and the extra-Galactic origin interpretations. Here, we compute the neutrino diffuse emission of our Galaxy on the basis of a recently proposed phenomenological model characterized by radially-dependent cosmic-ray (CR) transport properties. We show that the astrophysical spectrum measured by IceCube experiment can be well explained adding to the diffuse Galactic neutrino flux (obtained with this new model) a extra-Galactic component derived from the astrophysical muonic neutrinos reconstructed in the Northern hemisphere. A good agreement between the expected astrophysical neutrino flux and the IceCube data is found for the full sky as well as for the Galactic plane region.

1. Introduction

In the latest years the IceCube collaboration opened the era of neutrino astronomy detecting a astrophysical neutrino excess above the expected background flux. Recently, a preliminary analysis [1] based on four years experiment data-taking (1347 days of livetime), reported a total of 54 HESE between 28 TeV and 3 PeV. The astrophysical ν excess inferred by IceCube with HESE events collected in three years was fitted by a power law with index $\Gamma = -2.3 \pm 0.3$ and a best-fit per flavor flux of $E^2\Phi(E) = 1.5 \pm 0.7 \times 10^{-8}(E/100\,\mathrm{TeV})^{-0.3}\,\mathrm{GeV\,cm^{-2}\,s^{-1}\,sr^{-1}}$ above 60 TeV [2]. The four-year data favor a steeper spectrum: $\Gamma = -2.58 \pm 0.25$ and a best-fit flux $E^2\Phi(E) = 2.2 \pm 0.7 \times 10^{-8}(E/100\,\mathrm{TeV})^{-0.58}\,\mathrm{GeV\,cm^{-2}\,s^{-1}\,sr^{-1}}$ [1] for the same energy range. Although a statistically

[a] e-mail: antonio.marinelli@pi.infn.it

significant departure from isotropy cannot be claimed yet, recent analyses [3, 4] showed that the angular distribution of HESE events allows up to 50% of the full-sky astrophysical flux to have a Galactic origin. The Galaxy can be considered a guaranteed source of neutrinos up to a fraction of PeV energies at least; a sizable flux may either come from freshly accelerated CRs undergoing hadronic scattering with gas clumps, or from the hadronic interactions between the Galactic CR sea and the interstellar gas. The expected diffuse neutrino flux produced by these processes is strongly dependent on the CR transport scenario. In this work we present the expected Galactic diffuse neutrino flux, computed with a recently introduced scenario (KRA_γ) considering a radially-dependent diffusion coefficient [5], for different regions of the sky. Respect to a standard scenario, with a fixed diffusion coefficient for the whole galaxy, this model predict a diffuse Galactic neutrino flux two times higher for the full sky region and up to four times higher for the Galactic plane region. However, to explain the spectra obtained with IceCube measurements for these regions it is necessary to account also for a extra-Galactic (EG) component. Here we assume this EG component to be isotropic and use the astrophysical muonic neutrino IceCube measurements from the Northern hemisphere [6] to obtain an estimation of this flux. In particular we obtain the EG neutrino spectral features considering the best-fit analysis performed by the IceCube collaboration [6] without a particular assumptions for the EG neutrino emitters. We show that the measured full sky neutrino spectrum as well as the derived neutrino spectrum of the Galactic plane region (θ[1] $< 7.5°$) are well described by the sum of Galactic (KRA_γ) and EG (best-fit of ν_μ from Northern hemisphere) selected ν spectra.

2. Diffuse Galactic neutrino spectra obtained with KRA_γ model

As explained in details in [5], the KRA_γ model, used in this work to obtain the Galactic diffuse neutrino emission, adopt an index δ[2] for the rigidity dependence of the diffusion coefficient that increases with the Galactocentric radius R (implying spatially variable CR transport as originally suggested e.g. in [7]), and this turns into a hardening of CR propagated spectrum and ν and γ-ray emissivity above the TeV energies in the inner Galaxy region. The model is tuned to reproduce over the entire sky the diffuse γ-ray emitting spectrum of the Galaxy as measured by Fermi-LAT [8] as well as the local CR observables [9, 10]. The hadronic emission computed by considering the KRA$_\gamma$ setup is higher than the conventional model predictions, with important implications for Galactic diffuse neutrino expectations. We first compute the ν_e and ν_μ production spectra. For both flavors we use the emissivities provided in [11] (well tuned on accelerator and CR data) for projectile energies below ~ 500 TeV, while we adopt the ones provided in [12] above that energy range. Then we account for neutrino oscillations. Their effect is to almost equally redistribute the composition among the three flavors [13]. We only consider proton and helium CRs/gas – as for γ-rays – since heavier nuclear species give a negligible contribution in the energy range we cover in this work [14]. Two possible cut-offs have been selected for the interacting CR: $5PeV$ and $50PeV$; these choices match CREAM p and He data and roughly KASCADE [15] and KASCADE-Grande data [16].

3. Coupling a Galactic (KRA_γ ν) and a extra-Galactic (ν flux of the Northern hemisphere) component to explain IceCube data

In this work we present the interpretation of the astrophysical neutrino spectrum measured by IceCube adding the Galactic diffuse emission, obtained with KRA_γ scenario, to a diffuse EG neutrino component

---

[1] With θ we refer to the angular width of the selected region respect to the galactic plane line, in this case θ correspond to the galactic declination b.

[2] $D(\rho) \propto (\rho/\rho_0)^\delta$ with ρ representing the rigidity, and δ variable with R as $\delta(R) = a \times R + b$ where $a = 0.035$ Kpc^{-1} and $b = 0.21$.

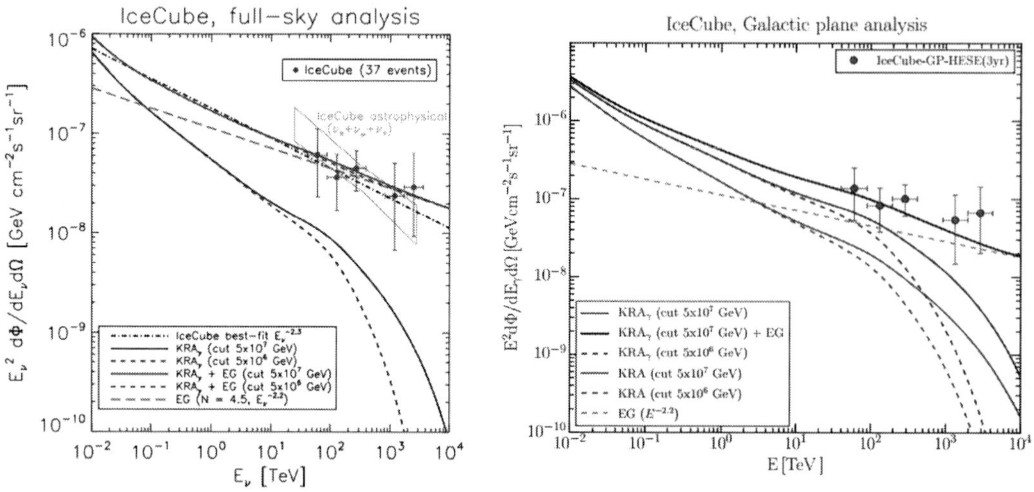

Figure 1. Comparison between IceCube spectral measurements and neutrino spectra obtained with the sum of a diffuse Galactic component (computed with the implementation of KRA_γ model inside the DRAGON framework [19]) with a isotropic EG ν spectrum derived from the ν_μ measurements of the Northern hemisphere. On the left plot the full sky analysis: with blue solid and dashed lines we report the KRA_γ ν spectra considering CR cut-offs respectively at 50 and 5 PeV. With the grey dashed line the isotropic EG component obtained from the Northern ν_μ measurements. The yellow region and the green points are referred to spectral analyses obtained with 3 years of HESE events. The black dash-dotted line represents the best-fit analysis of 3 years of HESE events while the red solid line represents the sum of KRA_γ with EG component obtained from ν_μ. On the right plot the Galactic plane analysis: with the blue and red solid and dashed thin lines we show the KRA_γ (new scenario for CR transport) and the KRA (conventional scenario for CR transport) ν spectra with the CR cut-offs at 50 and 5 PeV. The grey dashed line is the EG spectral component, while the blue solid thick line is the sum of KRA_γ neutrinos and EG spectrum. With the green points the derived spectral points for the Galactic plane region $\theta < 7.5°$.

derived from the best-fit analysis of muonic neutrinos coming from the Northern hemisphere [6]. Since the inner Galactic plane is located in the South hemisphere, we assume the muonic neutrino flux from the Northern hemisphere ($\Phi_{\nu_\mu}^{\rm North} = 1.7^{+0.6}_{-0.8} \times 10^{-18}$ $(E/100\ {\rm TeV})^{-2.2\pm0.2}$ $({\rm GeV\ cm^2\ sr\ s})^{-1}$) to be a good approximation of a isotropic EG neutrino component. The IceCube best-fit spectrum [6] was multiplied by three for taking into account the sum of ν_μ, ν_e and ν_τ. This analysis was done for the full-sky region as well as for the whole Galactic plane inside the latitude of $\theta < 7.5°$. For the full sky region we compare the sum of computed neutrino fluxes (KRA_γ + North $\nu_\mu \times 3$) with the IceCube spectral analyses corresponding to HESE events collected in three years [2, 17]. Instead for the Galactic plane we derived the spectral points following [18] guideline. In particular we renormalized the IceCube spectral measurements [2] considering the neutrino events reconstructed inside the Galactic plane window selected ($\theta < 7.5°$). As we can see in Fig. 1 adding to the KRA_γ the EG neutrino component (obtained from the ν_μ of Northern hemisphere) we are able to explain the IceCube spectral measurements. For the full sky analysis, where the Galactic contribution is expected to be subdominant, the computed ν spectrum with KRA_γ allows to reproduce the best-fit analysis of HESE events at low energies. This is due to the steeper and higher spectrum of KRA_γ ν at $E < 100\,{\rm TeV}$ respect to the considered EG spectral component. Considering the Galactic plane region the ν spectrum of KRA_γ is even more important for explaining the derived IceCube spectrum. The right plot in Fig. 1 shows that taking into account a standard CR transport scenario (with a fixed diffusion coefficient for the whole galaxy) for producing diffuse Galactic ν, it would be needed a excessive high EG spectral ν counterpart to reproduce the IceCube spectrum.

4. Conclusions

In the presented analysis we interpreted the astrophysical neutrino flux measured by IceCube as the sum of a computed diffuse Galactic ν component with an isotropic EG ν flux obtained directly from data. The diffuse Galactic ν's, produced by cosmic-ray interaction inside our Galaxy, are obtained following a new CR transport scenario, called KRA_γ, where the diffusion coefficient is assumed to be radially-dependent. Whereas the EG ν flux is obtained from a particular IceCube analysis. Since the inner Galactic plane is positioned in the South hemisphere, we took the ν_μ flux reconstructed in the Northern hemisphere as a good estimate of the isotropic EG neutrino component. The spectral features of the EG spectrum correspond to the best fit parameters given by the IceCube collaboration. The sum of these two astrophysical components well reproduce the IceCube spectral measurements for the full sky region as well as for the whole Galactic plane with $\theta < 7.5°$. For both regions we highlight the importance of using the KRA_γ model instead of a standard CR transport scenario to obtain the expected Galactic diffuse neutrino flux. In a near future the IceCube-ANTARES common analysis of the Galactic plane can confirm the features of neutrino spectrum predicted by the KRA_γ scenario and constrain the spectral characteristics of EG neutrino flux.

References

[1] M. Aartsen et al. (IceCube Collaboration), Proc. 34th ICRC, den Haag, Netherlands, PoS **id: 1081** (2015)
[2] M. Aartsen et al. (IceCube Collaboration), Phys. Rev. Lett. **113**, 101101 (2014), 1405.5303
[3] M. Ahlers, Y. Bai, V. Barger, R. Lu, ArXiv e-prints (2015), 1505.03156
[4] A. Neronov, D.V. Semikoz, Astropart. Phys. **75**, 60 (2016), 1509.03522
[5] D. Gaggero, D. Grasso, A. Marinelli, A. Urbano, M. Valli, ApJ Letters **815**, L25 (2015), 1504.00227
[6] M. Aartsen et al. (IceCube Collaboration), Phys. Rev. Lett. **115**, 081102 (2015), 1507.04005
[7] A.D. Erlykin, A.W. Wolfendale, Astropart. Phys. **42**, 70 (2013), 1212.2760
[8] M. Ackermann et al. (Fermi-LAT), Astrophys. J. **750**, 3 (2012), 1202.4039
[9] O. Adriani et al. (PAMELA Collaboration), Science **332**, 69 (2011), 1103.4055
[10] M. Aguilar et al. (AMS Collaboration), Phys. Rev. Letters **114**, 171103 (2015)
[11] T. Kamae, N. Karlsson, T. Mizuno, T. Abe, T. Koi, The Astrophysical Journal **647**, 692 (2006), astro-ph/0605581
[12] S. Kelner, F.A. Aharonian, V. Bugayov, Phys. Rev. **D74**, 034018 (2006), astro-ph/0606058
[13] V. Cavasinni, D. Grasso, L. Maccione, Astropart. Phys. **26**, 41 (2006), astro-ph/0604004
[14] M. Kachelriess, S. Ostapchenko, Phys. Rev. **D90**, 083002 (2014), 1405.3797
[15] T. Antoni et al., Astropart. Phys. **24**, 1 (2005), astro-ph/0505413
[16] W. Apel et al., Astropart. Phys. **47**, 54 (2013), 1306.6283
[17] M. Aartsen et al. (IceCube Collaboration), Astrophys. J. **809**, 98 (2015), 1507.03991
[18] M. Spurio, Phys. Rev. **D90**, 103004 (2014), 1409.4552
[19] DRAGON-web, *DRAGON code*, http://www.dragonproject.org/ (2015)

EPJ Web of Conferences **116**, 05001 (2016)
DOI: 10.1051/epjconf/201611605001

The Control Unit of KM3NeT data acquisition

Cristiano Bozza[a] for the KM3NeT Collaboration

University of Salerno, Dipartimento di Fisica, via Giovanni Paolo II 132, Fisciano 84084, Italy

Abstract. The KM3NeT Collaboration is building a new generation of neutrino telescopes in the Mediterranean Sea. With the telescopes, scientists will search for cosmic neutrinos to study highly energetic objects in the Universe, while one neutrino detector will be dedicated to measure the properties of the high-energy neutrino particles themselves. Control of the KM3NeT data acquisition processes is handled by the KM3NeT Control Unit, which has been designed to maximise the detector live time. The Control Unit features software programs with different roles, following the philosophy of having no single point of failure. While all programs are interconnected, each one can also work alone for most of the time in case other services are unavailable. All services run on the Common Language Runtime, which ensures portability, flexibility and automatic memory management. Each service has an embedded Web server, providing a user interface as well as programmatic access to data and functions. Data to and from detector components for monitoring and management purposes are transmitted using a custom designed protocol. The Control Unit is interfaced to one or more Message Dispatchers to control the data acquisition chain. A Data Base Interface provides fast and fault-tolerant connection to a remote Data Base.

1. Overview

The KM3NeT Collaboration [1] is installing water-Cherenkov neutrino telescopes [2] at multiple locations in the Mediterranean Sea. Data taking is expected to start by the end of 2015 when the first sensors will be installed. One telescope will be devoted to the search for astrophysical sources of neutrinos, whereas another will be used to study features of neutrinos themselves. Each detector is built of detection units, i.e. vertical lines supporting 18 Digital Optical Modules. Each optical module hosts 31 photomultipliers, monitoring instruments and piezo elements for the acoustic positioning of the module. A detector *building block* will be made of 115 detection units supporting in total 64,170 optical modules. The telescopes should be operational for at least 15 years. The detector and the Trigger and Data Acquisition System (TriDAS) [3] are controlled by a scalable software infrastructure called Control Unit that is already being used in test and validation benches to characterise and tune detector components.

Interaction events both of primary cosmic neutrinos and of secondary neutrinos from cosmic rays may occur at any time. Maximising the detector live time is one of the main priorities in the design of the Control Unit. It should have no single point of failure, and should be able to operate also in

[a] e-mail: cbozza@unisa.it

downgraded mode with one or more services stopped. Long time scales demand easy operation and configuration. The software is designed relying on widely adopted standards with a large development and user base, whereas the code of all non-standard components is completely under the control of the KM3NeT Collaboration. The Control Unit can run with relatively small hardware resources, which helps keeping the overall cost low.

All the controllable elements in the detector as well as in the TriDAS implement the same State Machine (SM), so the control logic is in principle the same for all of them. At any time, a target state is defined that each SM should reach, corresponding to an overall detector target:

Off: all systems are idle, photomultipliers are not powered and they do not generate data;
On: all systems active, photomultipliers are powered but they do not send data;
Run: detector taking data with all systems.

In principle, data acquisition should start when all subsystems are ready. However, after several years of operation, it is to be expected that some optical modules might be damaged or only partly functional; also in this case the acquisition must work without useless waits: the Control Unit software is tolerant with respect to failing hardware components. If software components fail in the TriDAS the acquisition cannot work because data cannot be processed; in that case, data acquisition becomes pointless.

The programs the Control Unit is composed of are called Control Unit Services. In the following, the shortened term *services* will often be used.

2. Networking protocols

Several networking protocols are used by the Control Unit, reflecting different criticality levels of the communication. For internal communications the following protocols are in place:

1) HTTP(S) for GUI interfaces to services mostly for human users but also for programmatic access.

2) Server Application with Web Interface (SAWI): remote calls among services are implemented on top of HTTP(S) on specific pages. SAWI transports data, method calls and exceptions in a transparent way as if everything were occurring in the same process. Since it runs on HTTP(S), it works on top of an existing HTTP(S) Web server library. Among the benefits of this approach, it is worth recalling that Web browsers can be used to mimic remote calls, so they can be used for debugging purposes. SAWI is mostly used for light messages, whereas for massive data transfer NFS is used (see below).

3) Simple Retransmission Protocol (SRP): this is a UDP-based protocol used to communicate with the Central Logic Boards (CLBs) [4] of the optical modules. Commands flow from the Detector Manager to CLBs, monitoring data go in the opposite direction.

4) One or more Message Dispatchers use stable TCP links to TriDAS programs to send control commands, receive reports, and distribute data streams.

5) Network File System (NFS): data to be transferred to the database are saved to a local shared directory so that they can be checked in case of needs. Because their storage relies on the file system, even in the unlikely case of a failure of either the sender or receiver service they are not lost.

For external communications, HTTP(S) is used for GUI interfaces, while SQLNet is used to connect the Data Base Interface (see 3.5) to the central database ([5], [6]) running Oracle Database Server [7].

3. Components

The modularity in the Control Unit design has two goals:

1) Resilience by avoiding single points of failure and explicitly foreseeing activity in downgraded conditions with one or more services unavailable;

2) Scalability by allowing the same service to work in multiple instances, sharing the workload.

3.1 Service structure

All services are designed to run on all major operating systems. Linux SLC6 is used for production, whereas Windows and MacOS are used for development. Neither recompilation nor cross-compilation is needed to change the operating system. This is possible thanks to the adoption of the Common Language Runtime (CLR) [8], supported by Mono [9] for Linux, Windows and MacOS, and by .NET on Windows. So far all programs have been written in C# [10], but any other language with CLR support may be used to add features. The core of all services provides four basic features:

1) Persistence/resilience tools tracking the current state of the service by a configuration file and a session file to allow restart with minimal or no human action in case of a power cut or sudden failure;

2) An HTTP server library (Web server) supporting a GUI interface; the SAWI protocol uses this HTTP server to expose methods for remote calls.

3) A Virtual Directory that allows for representing all the quantities and parameters that need to be exposed for monitoring in a convenient directory structure for access through the HTTP server;

4) Human-readable log generation.

Built on top of these features, each service has its own specific commands (from text-mode console), GUI, set of SAWI-exposed methods and control behaviour.

3.2 The Master Control Program

The Master Control Program (MCP) is the central reference point to define the current state, i.e. detector configuration, *runsetup* (set of operational parameters), run number and target (*Off/On/Run* as stated above). When the state changes, other services are notified (*push* message mode). They also periodically query the MCP to check the state (*pull* message mode). The MCP can also execute scheduled jobs with predefined start-end time and *runsetup*. State information is kept encrypted in the persistent files to preserve integrity.

3.3 The Detector Manager

The task of the Detector Manager (DM) is to control each Central Logic Board (CLB) of an optical module to fulfil the general detector state defined by the MCP. The SM of each CLB is driven according to the current target and operational parameters applied as specified in the current *runsetup*, defined in the KM3NeT database. Monitoring data about photomultiplier high voltage, temperature, humidity, compass, accelerometer, etc. are continuously collected in so-called DataLog files that are uploaded to the database. Communication with the CLBs is accomplished through SRP. A *swarm* of threads shares the load of controlling the CLBs optimising CPU occupancy. A single control thread manages reactions to external commands. The control logic and operational parameters can be fine-tuned for each single CLB if needed, which is useful to add test equipment to otherwise standard data-taking devices. The DM also features a special mode in which it merely routes and translates commands received by the HTTP server to CLBs. This is useful for automatic testing of large stocks of devices by running minimal installations of the Control Unit software driven by lightweight scripts.

3.4 The TriDAS Manager

The programs in the Data Acquisition Systems devoted to trigger and on-line data processing need to be driven and configured as a whole entity. This is the purpose of the TriDAS Manager (TM). In principle, there may be hundred instances of such programs running when the detector reaches the size of a *building block*. Because also TriDAS programs have the same SM as hardware components, the control logic is very similar. All communication from/to the TriDAS programs occurs using the Control

Host protocol. TM has notion of time and retry count to distinguish occasional response delays from stuck programs. If needed, it can start/restart/stop remotely each DAS program.

3.5 Database Interface and Database Stream Writer

Control Unit services do not connect directly to the KM3NeT database. Within each data-taking station, which can be a shore station of an underwater detector or a test-bench for product characterisation, all data exchange goes through HTTP (SAWI) calls and file exchange. The Data Base Interface (DBI) optimizes resource usage by hosting local data caches to avoid connections to remote databases that can be time-consuming. The list of users, detector definitions and *runsetups* are all automatically synchronized with the KM3NeT database. DataLog files generated by the DM are staged and buffered in the upload cache. The DBI runs data upload code that optimizes the physical arrangement of data on database disks. The Data Stream Writer (DSW), currently under development, has a similar approach to writing large data sets into the database. Calibration parameters and acoustic data are extracted from the serial flow of the Dispatcher and are written in batches to the database.

3.6 The Local Authentication Provider

In each data-taking station, a locally centralized authentication service provides users with access on the basis of their unique credentials, which are stored encrypted. With a similar logic, *service accounts* are used to provide a local directory service (i.e. maintaining the list of *which machine does what*) without having to specify other services' network addresses and port numbers in the configuration file of each member of the Control Unit. Only the Local Authentication Provider (LAP) address needs to be known to each service. Every time a user or service logs onto the Control Unit, it receives an access token. In addition, because all the GUIs are built using Web browsers, the LAP tracks all the tokens assigned, and allows handing out a token from one browser session to another so that users do not need to log in at each program window.

4. Performances

Services that come into action only during overall state changes raise no concerns about performance. Those that actively do work need some attention to correctly tune the hardware resources needed.

In tests run on the first detection units, each with 18 optical modules and a base controller, with DM controlling detector operation, the monitoring data polling frequency was set to 0.1 Hz, which was 10 times higher than envisaged for normal workload; 1.2 cores were used on average. This should scale to 13.8 cores for a full *building block* of detection units monitored at 10^{-2} Hz. However, this is expected to be a conservative guess, considering that in the aforementioned test up to 0.6 cores were used only to support the GUI.

The DBI and DSW will continuously feed the database with new data. While the expected data rate for a building block at 10^{-2} Hz is about 0.3 Mbps, tests of the SQLNet protocol with very simple tables, needing almost no ordering and indexing between two upload shots, showed it can use all the available network transfer speed if needed. The way data are stored in the database and their organization during the upload are known to affect the transfer speed by large factors, and, at the time of writing this report, work is focused on the optimization of this process, with promising preliminary results.

References

[1] `http://www.km3net.org`

[2] S. Adrián-Martínez et al., Eur. Phys. J. C **74**, 1 (2014)

[3] C. Pellegrino, T. Chiarusi, these Proceedings
[4] D. Real et al., these Proceedings
[5] A. Albert, C. Bozza, AIP Conf. Proc. **1631** 167 (2014)
[6] A. Albert, C. Bozza, these Proceedings
[7] `http://www.oracle.com`
[8] `http://www.ecma-international.org/publications/standards/Ecma-335.htm`
[9] `http://www.go-mono.com`
[10] `http://www.ecma-international.org/publications/standards/Ecma-334.htm`

EPJ Web of Conferences **116**, 05002 (2016)
DOI: 10.1051/epjconf/201611605002

KM3NeT Neutrino Telescope 1-ns Resolution Time To Digital Converters

David Calvo[a] and Diego Real

Instituto de Física Corpuscular, Catedrático Jose Beltrán, 2, 46980 Paterna, Spain

Abstract. The KM3NeT collaboration aims the construction of a multi-km3 high-energy neutrino telescope in the Mediterranean sea consisting of thousands of glass spheres, each of them containing 31 photomultiplier of small photocathode area. The main digitization system is composed by 31 Time to Digital Converter channels with 1-ns resolution embedded in a Field Programmable Gate Array. An architecture with low resource occupancy has been chosen allowing the implementation of other instrumentation, communication and synchronization systems on the same device. The 4-oversampling technique with two high frequency clocks working in opposed phases has been used together with an asymmetric FIFO memory. In the present article the architecture and the first results obtained with the Time to Digital Converters are presented.

1. The KM3NeT Project

The KM3NeT telescope is a deep-sea neutrino infrastructure to be deployed in the Mediterranean Sea (see Fig. 1, left). The telescope has been designed to detect extraterrestrial neutrinos with energies above 50 GeV by means of the Cherenkov photons induced by the passage of relativistic charged particles through the seawater [1]. When a neutrino interacts in the material surrounding the detector, it can produce a muon, which travels across the detector at a speed greater than the speed of light in water. Such a particle generates a faint blue luminescence called Cherenkov radiation. The arrival times of the photons collected by optical detectors disposed in a three dimensional array can be used to reconstruct the muon trajectory, and consequently that of the neutrino, which is strongly correlated. The main elements of a neutrino telescope are, therefore, the sensitive optical detectors which in the case of KM3NeT are small photocathode area photomultiplier tubes (PMTs) distributed around the glass sphere of the so called Digital Optical Module (DOM) [2] (see Fig. 1, right). Each DOM has 31 small photomultipliers that collect the Cherenkov light and convert it into electronic signals. Figure 1 shows the Digital Optical Module and the KM3NeT general overview.

In order to translate these signals into the arrival time of the photons, they are processed by Time to Digital Converters (TDCs) implemented on a Kintex-7 field-programmable gate array (FPGA). A TDC performs conversion of a time interval (TI) into a digital value. Many methods, both analogue

On behalf of the KM3NeT collaboration
[a] e-mail: dacaldia@ific.uv.es

Figure 1. Left: KM3NeT general overview. Right: Digital Optical Module.

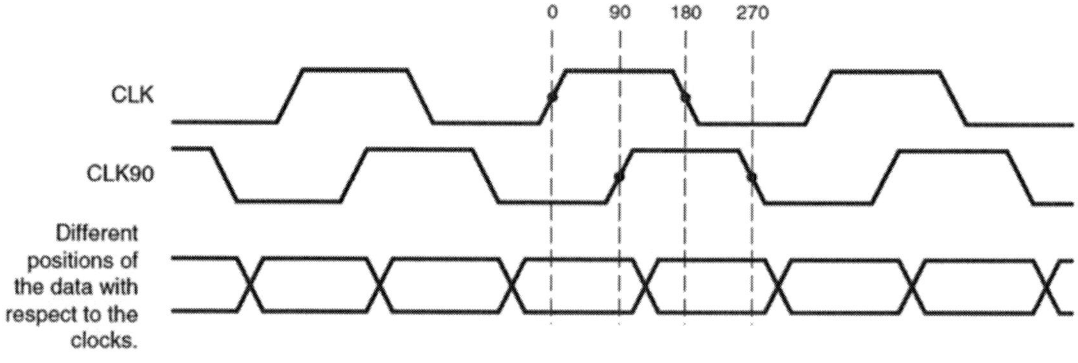

Figure 2. Oversampling technique using two phase-shifted clocks (see text).

and digital, can be used to achieve sub-nanosecond resolutions. These techniques can be designed both in application specific integrated circuit (ASIC) and FPGA devices. However, the design process of an ASIC device not only can be expensive, but also quite complex due to the long turn-around time and layout phase. Low cost and commercial availability are driving motivations for using general purpose FPGA to implement the TDC and the rest of the instrumentation control without using any external circuit in KM3NeT project.

The raw data from the PMTs consist of a continuous series of hits that should be digitized. Each hit corresponds to an analogue pulse of a signal that passed a preset threshold. The digital data correspond to the timestamp and length of the time-over-threshold signal that is produced by the active base of each PMT. The timestamp corresponds to the time when the leading edge of the signal crossovers the threshold, and the length is the time difference between the leading and trailing edges. Both, timestamp and length of pulse, require an accuracy of 1 ns.

2. Oversampling technique implementation

Oversampling method uses a sampling frequency significantly higher than twice the bandwidth (or highest frequency) of the signal being sampled. For the KM3NeT readout system, the "significantly higher" sampling frequency is obtained using different edges of multiple phase-shifted clocks. This method is called asynchronous oversampling because the clocks used to create the sampling frequency are nominally equal to the data stream accuracy 1 ns [3]. The function of the two extra clocks combination is shown in Fig. 2.

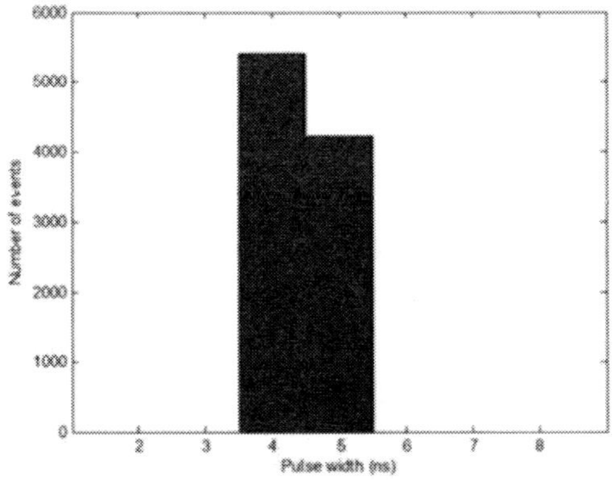

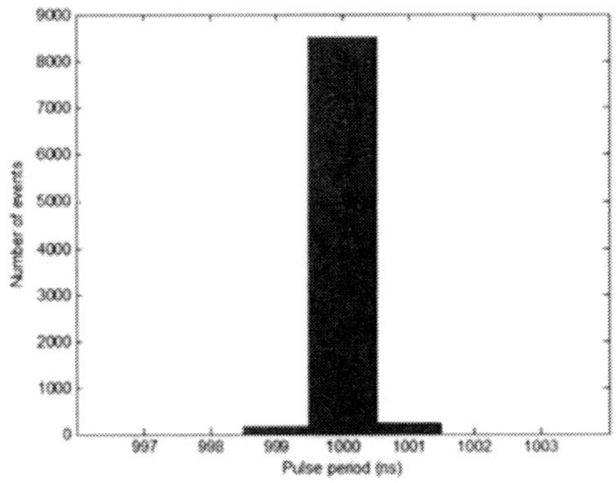

Figure 3. Left: histogram of pulse width measurements (mean = 4.4 ns, std = 0.5 ns). Right: histogram of pulse period measurements (mean = 1000.2 ns, std = 0.2 ns).

A 31 channel TDC has been designed in a Xilinx Kintex-7. Because of the programmable characteristic of an FPGA, a TDC readout implemented in this device has flexible characteristics. Here, a simplified TDC is designed to verify the idea of deserializing the raw data by means of dedicated input/output blocks of the FPGA. The CLB board provides small quartz of 25 MHz, the clock signal is first transferred from a clock pin to a buffer in the centre of the FPGA, and then fanned out to the inner PLLs to provide two high frequency clocks of 250 MHz and 90° phase shifted. 4-oversampling method increases the sampling frequency up to 1 GHz achieving the desired accuracy of 1 ns.

3. Results

The accuracy of the proposed TDC implementation was tested within the complete readout system. A high precision pulse generator developed on Virtex-6 evaluation board provides 5 ns input pulses including 500 ps rising and falling edge at 1 MHz. The test pattern consists of 10000 consecutive pulses. Pulse width and period measurements with the TDCs are reported in Fig. 3.

After validating the TDC accuracy, the first tests with PMTs were performed. The DOM was in a dark box for 19 hours. The high voltage in the PMT basis and the threshold were properly configured. The pulse widths generated by PMTs were computed by TDCs. Xilinx Kintex-7 FPGA is commercial and operates in a temperature range between 0°C and 85°C. TDCs were examined over the temperature range from 20°C to 30°C, the performance did not vary with temperature. The performance variations of the TDCs with the temperature and power supply are expected to be negligible.

4. Non linearities

A shift in the measurements means an offset error in the TDC results. In order to quantify this error we can use a density test with different input values. Ideally every bin will have the same number of counts (C_i) by the total number of samples (N_t). This value is compared with the expected fraction per bin for an ideal situation (F_i) where the code density is equal per bin. Finally minus one will give the final differential nonlinearity per bin because when the expected and actual counts are equal, a quantization error of 0 is needed. This can also be written as follows:

$$DNL_i = \frac{c_i/N_t}{F_i} - 1. \tag{1}$$

The Differential Non Linearities (DNL) obtained for the timestamp is below 0.03 LSB, that means, less than 30 ps of uncertainty. These results are shown in Fig. 4.

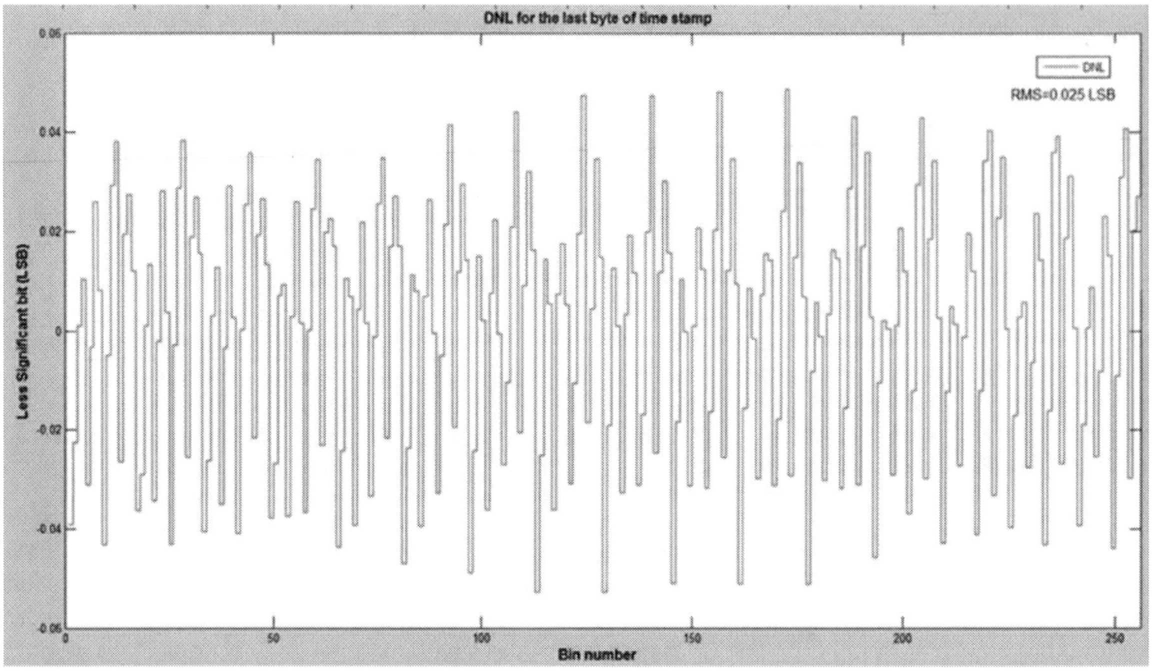

Figure 4. Differential Non Linearities analysis for the Time to Digital Converters.

5. Conclusions

The readout architecture for 31 Time to Digital Converter channels of high accuracy designed for KM3NeT has been presented. Deserializer primitives, within the input/output blocks of Kintex-7 FPGA, working in oversampling mode have been used. Four times oversampling technique using two high frequency clocks of 250 MHz has been implemented to achieve a resolution of 1 ns with DNL below 0.03 LSB. The accuracy of the proposed TDC readout system has been tested using an external pattern generator based on Virtex-6 FPGA and validated with the PMTs in the dark box.

The author acknowledges the financial support of the Spanish Ministerio de Ciencia e Innovación (MICINN), grants FPA2009-13983-C02-01, FPA2012-37528-C02-01, ACI2009-1020, Consolider MultiDark CSD2009-00064, YC-2012-10604, European Community's Sixth Framework Programme under contract n° 011937 and the Seventh Framework Programme under grant agreement n° 212525 and of the Generalitat Valenciana, Prometeo/2009/.

References

[1] E. de Wolf, Nucl. Instr. & Meth., *KM3NeT a new generation neutrino telescope*, **588**, pp 86–91 (2008)

[2] O. Kalekin, Nucl. Instr. & Meth., *Optical modules for the neutrino telescope KM3NeT*, **A623**, pp 312–315 (2015)

[3] J. Wu et al. IEEE Conf. Rec. NSS., *Firmware-only implementation of time-to-digital converters (TDC) in field programmable gate array (FPGA)*, **1**, pp 177–181 (2003)

EPJ Web of Conferences **116**, 05003 (2016)
DOI: 10.1051/epjconf/201611605003

Live monitoring and quasi-online event reconstruction for KM3NeT

Tamas Gal[a] for the KM3NeT Collaboration

ECAP, University of Erlangen-Nuremberg, Erwin-Rommel-Str. 1, 91058 Erlangen, Germany

Abstract. KM3NeT is a new generation neutrino telescope in the abyss of the Mediterranean Sea. It will instrument a volume of several cubic kilometres of sea water in its final configuration. Currently, the project is in its first phase with the aim of constructing and installing 31 detection units up to 700 m in height, each equipped with 18 digital optical modules. The optical modules are equipped with 31 3-inch photomultipliers to detect the Cherenkov light of charged secondary particles produced in high-energy neutrino interactions. This contribution describes a live detector monitoring system, which enables real-time parameter control and a reconstruction of events soon after the data acquisition. It also allows a rapid response to or provision of external alarms of multi-messenger campaigns. The data acquisition system of KM3NeT provides pre-filtered data in event form, as well as general detector status messages. The events will be processed almost in real-time – with a delay in the range of minutes – using fast reconstruction mechanisms. This allows for high-level monitoring of the detector status using derived distributions, such as time and charge distributions and event rates. The resulting data is displayed on a web page using a dedicated, flexible web service. The same service also displays low-level monitoring data such as trigger rates, PMT hit rates and the general status of the optical modules.

1. Introduction

KM3NeT is a new generation neutrino telescope in the abyss of the Mediterranean Sea. It will consist of several hundreds of digital optical modules (DOMs), each holding 31 photomultiplier tubes (PMTs). The DOMs send the acquired data along with calibration data through an optical network to the on-shore computer centre. In the following, a short description of software package Restless Oyster (ROy) is presented, which allows real-time monitoring of the detector by providing an easy to use web interface.

2. The concept of Restless Oyster – a software package for real-time monitoring

The main concept of the Restless Oyster[1] online monitoring package is to provide a lightweight, easy to setup monitoring system which can be smoothly integrated in any existing environment. It uses standard software for both the back-end and the front-end, which most modern operating systems already include.

---

[a] e-mail: `tamas.gal@km3net.de`

[1] `http://royweb.readthedocs.org`

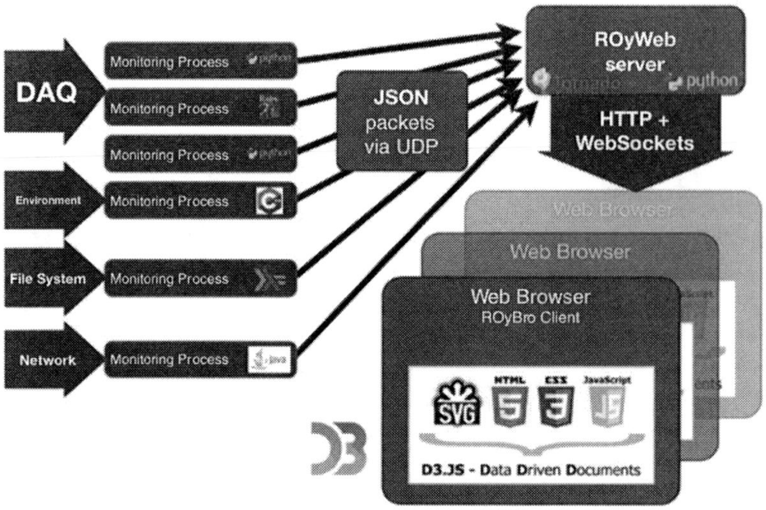

Figure 1. The data flow of the ROyWeb system. The blue boxes on the left represent the ROyCruncher, which are tiny daemons collecting and analysing online data and reading out sensors. The results are sent as JSON packets via UDP to the ROyWeb server. A web browser pointed to the ROyWeb server address is connected through HTTP to receive the web interface and through WebSockets to establish a permanent data stream which is then used to distribute the collected results in real-time.

The fact that ROy uses a central web server for data distribution allows to shift the workload of generating live graphs (like scatter plots or histograms) to the users workstation– namely the web browser –so the actual visualisation overhead is completely separated from the central monitoring server. In addition to that, each user is able to manage an own dashboard completely independent from others.

2.1 Components

ROy is a software package which consists of three main parts. ROyCruncher: small, standalone monitoring process or even a part of a bigger software package to read out sensors or calculate different parameters which need to be live monitored. ROyBro: the web interface which takes care about the user interaction, session management and data visualisations. ROyWeb: the main component of ROy, acting as a data dispatcher between ROyCruncher and ROyBro.

2.2 Data flow

Figure 1 shows the data flow of ROy. The ROyCrunchers collect all kinds of information like temperature, compass alignment, summary data of events, network usage etc. and send them as JSON packets through UDP to the ROyWeb web server. ROyWeb immediately stores the collected data in a local SQLite database and also redistributes them to all connected web clients via individual web socket connections between itself and the ROyBros.

2.3 Implementation

The core of ROyWeb is written in Python and is based on the Tornado[2] web server. The web client ROyBro is written in JavaScript and uses the D3.js[3] framework for data visualisation and HTML5/CSS

[2] `http://www.tornadoweb.org`
[3] `http://d3js.org`

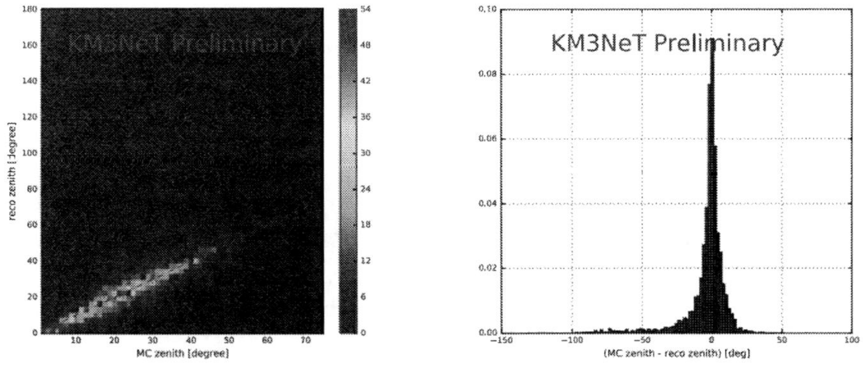

Figure 2. ROyFit – Fast muon reconstruction algorithm on 21857 MUPAGE events (atmospheric muon bundles) simulated for one detection unit of the KM3NeT neutrino telescope.

for the graphical user interface. Since operating systems like Linux and Mac OS X already include Python and a modern web browser like Safari, Firefox or Chrome, there is mostly no need to install any other dependencies in order to use ROy. The source code is MIT licensed and available on `https://github.com/tamasgal/royweb`.

3. ROy in the KM3NeT project

ROy is used both as a low-level monitoring tool for e.g. live display of PMT/DOM rates, trigger rates, time-over-threshold histograms, event size etc. as well as a high-level monitoring tool for quasi-online event reconstructions including fit parameters, reconstructed muon directions etc., which yield a nice overview of the detector performance.

Preliminary fit performance results of the fast muon reconstruction algorithm mainly based on BBFit[1] are shown in Fig. 2.

Reference

[1] ANTARES Collaboration (2011), *doi:10.1016/j.astropartphys.2011.01.003*

EPJ Web of Conferences **116**, 05004 (2016)
DOI: 10.1051/epjconf/201611605004

The data acquisition system for Baikal-GVD

A.D. Avrorin[1], A.V. Avrorin[1], V.M. Aynutdinov[1], R. Bannash[7], I.A. Belolaptikov[2], D.Yu. Bogorodsky[3], V.B. Brudanin[2], N.M. Budnev[3], I.A. Danilchenko[1], G.V. Domogatsky[1], A.A. Doroshenko[1], A.N. Dyachok[3], Zh.-A.M. Dzhilkibaev[1], S.V. Fialkovsky[5], A.R. Gafarov[3], O.N. Gaponenko[1], K.V. Golubkov[1], T.I. Gress[3], Z. Honz[2], K.G. Kebkal[7], O.G. Kebkal[7], K.V. Konischev[2], A.V. Korobchenko[2], A.P. Koshechkin[1], F.K. Koshel[1], A.V. Kozhin[4], V.F. Kulepov[5], D.A. Kuleshov[1], V.I. Ljashuk[1], M.B. Milenin[5], R.A. Mirgazov[3], E.R. Osipova[4], A.I. Panfilov[1], L.V. Pan'kov[3], E.N. Pliskovsky[2], M.I. Rozanov[6], E.V. Rjabov[3], B.A. Shaybonov[2], A.A. Sheifler[1], M.D. Shelepov[1], A.V. Skurihin[4], A.A. Smagina[2], O.V. Suvorova[1], V.A. Tabolenko[3], B.A. Tarashansky[3], S.A. Yakovlev[7], A.V. Zagorodnikov[3], and V.L. Zurbanov[3]

[1] Institute for Nuclear Research, 60th October Anniversary pr. 7A, Moscow 117312, Russia
[2] Joint Institute for Nuclear Research, Dubna 141980, Russia
[3] Irkutsk State University, Irkutsk 664003, Russia
[4] Skobeltsyn Institute of Nuclear Physics MSU, Moscow 119991, Russia
[5] Nizhni Novgorod State Technical University, Nizhni Novgorod 603950, Russia
[6] St. Petersburg State Marine University, St. Petersburg 190008, Russia
[7] EvoLogics GmbH, Berlin, Germany

Abstract. Baikal-GVD will be a neutrino telescope at the cubic-kilometer scale in Lake Baikal. The first out of 10-12 clusters of the first phase of GVD has been deployed and commissioned in April 2015. This paper describes design and implementation of the dataacquisition system of GVD.

1. Introduction

Baikal-GVD is a neutrino telescope of a new generation with an effective volume of $\sim 1\,\mathrm{km}^3$ [1]. It will consist of functionally independent *clusters*, each consisting of 8 strings each carrying 36 optical modules (OMs). The OMs consist of pressure-tight glass spheres housing 10″ Hamamatsu R7081-100 photomultiplier tubes (PMTs) and the control electronics [2].

The OMs on a string are grouped into *sections* – the lowest-level DAQ units, each of them serving 12 OMs. A section is a complete detection unit that forms a *section trigger* and provides data transmission from the OMs to the cluster DAQ center. Each string includes several sections, at present 2, in the future up to 4. The optimum number and configuration of the sections depend on the distance between the OMs and the instrumented length of the strings. Structuring into sections and clusters provides flexibility in adapting to the actual scientific requirements [3].

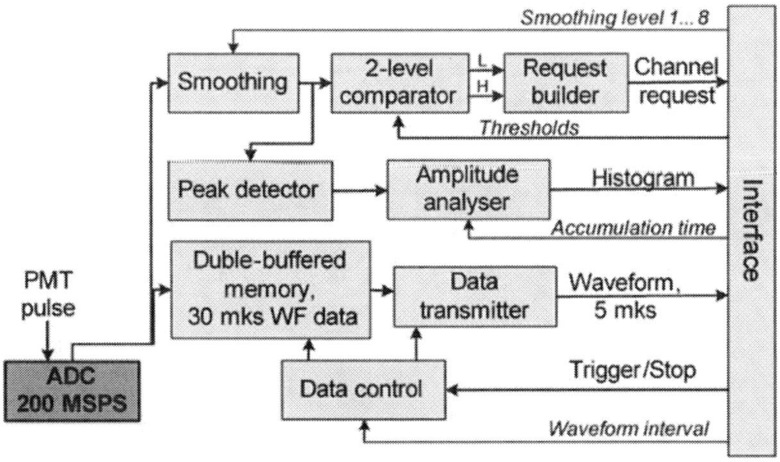

Figure 1. Functional diagram of one measuring channel of the ADC board.

2. Detection system and digitizing

The OMs house the PMT and electronics: a high-voltage power supply (SHV 12-2.0 K 1000 P, produced by TRACO Electronic AG), an OM controller (on the basis of a C8051F121 SiLabs micro-controller, designed and manufactured by SNIIP-AUNIS Ltd), a passive PMT divider (18 MΩ), a fast two-channel preamplifier and two calibration LEDs (about 470 nm wavelength). The analog signals from the PMTs pass through a two-channel amplifier and a dedicated deep-water cable 90 m long and then arrive at the center of the section, at the analog-to-digital converter (ADC) board for digitization and processing.

The analog signals are digitized in 12-channel ADC boards located at the center of the section. The functional diagram of one measuring channel of the ADC board is shown in Fig. 1. Each spectrometric channel is based on an AD9430 ADC with a resolution of 12 bit and a sampling frequency of 200 MHz. When digitized, the signal from the ADC is translated into the Xilinx Spartan-6 field-programmable gate array (FPGA). The FPGA software loaded over the data transmission channel ensures control and front-end processing of data arriving from the ADC. The 12 Kbyte FPGA memory buffer allows acquisition of data on the input signal waveform over a time frame as long as 5 μs.

A peak detector and a multichannel analyzer are connected to the measuring channel. They accumulate monitoring data (amplitude histograms) allowing control of the measuring channel performance. The trigger is based on the information formed in the channels of requests of ADC board. Each channel of requests includes a signal smoothing device (smoothing level, 1–8) and a two-level digital comparator with adjustable thresholds (low threshold L and high threshold H). Depending on the stated physical task, the values of the comparator thresholds may vary for different channels and, in normal operating mode of the detector, are 0.3 and 3.0 p.e. for L and H, respectively. The expected counting rate of an 8-string cluster at these thresholds is $\sim$100 Hz.

3. Triggering and data transmission

An effective suppression of noise events is achieved by using flexible andtunable trigger conditions in the requests analyzer. A Spartan-6 FPGA logic has been configured for generating a trigger, reading data from the ADC channels, and transmitting them over the Ethernet channel (the integrated network 100Base-TX interface) into the data acquisition center.

As a result of processing of L and H channel requests from all ADC channels of the section, the requests analyzer forms a request of the section. The embedded hardware–software Coincidence Matrix

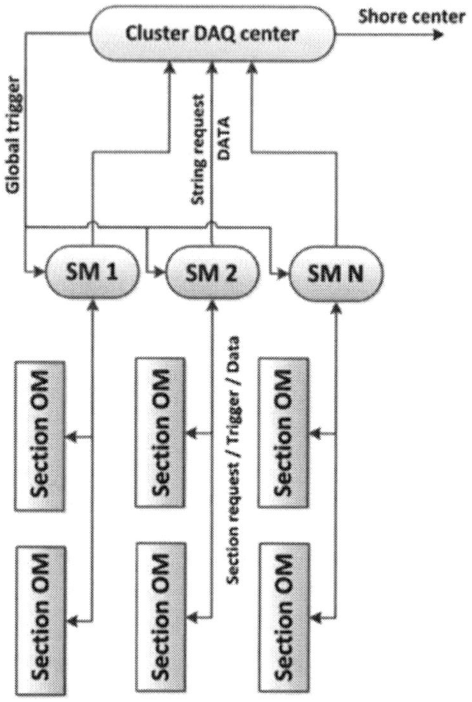

Figure 2. Diagram of the GVD cluster triggering and data transmission.

module (12H × 12L) is used to process requests of the channels. This module provides the simplest and fastest method for generating the section request. The trigger system is operative in two modes: majority coincidences of the L request signals in the selected time interval ("L trigger") and coincidences of the L and H requests in any neighboring optical modules (" L&H trigger").

The request signals of all sections are sent to the cluster center, where the acknowledgement signal being a common trigger for all strings is generated. The trigger signal initiates the generation of a common signal "Stop" for all measuring channels, read-out of the ADC data buffer, formation of the output buffer (the time frame) and its transmission to the DAQ center. The delay time of the common trigger is ~15 μs (the delay of the signal passage through 1.2-km line from the central module to the cluster center and backward, as well as the delay in the electronic units). For each channel, the beginning of the time frame may vary within a range of 0–30 μs relative to the moment of trigger arrival. The Spartan-6 FPGA logic also provides a real time data processing. As a result of this processing, only those parts of the time frame are transmitted to the data acquisition center where an excess over the threshold is observed. In this case, the data compression ratio is as large as 10^2.

The underwater local-area network (LAN) of the Baikal-GVD is a branched structure of star topology, consisting of a group of communication channels for data exchange between different detector units. Each ADC board with a Spartan-6 FPGA chip at the section center has its own IP address in the cluster LAN, which allows access to it and data readout from the registers according to the TCP/IP protocol. The functional diagram of the Baikal-GVD cluster triggering and data transmission system is shown in Fig. 2.

4. The first Baikal-GVD cluster

In April 2015 the first cluster of Baikal-GVD was deployed in Lake Baikal and commissioned [1]. The DAQ system of this first full-scale cluster is identical to the data acquisition system of the Gigaton

Volume Detector and comprises all assemblies of the future detector. During 2015 the DAQ system worked without critical failures. Some problems with string request transmission will be solved during the next deployment season in 2016.

This work was supported by the Russian Found for Basic Research (grants 13-02-12221, 14-02-00175, 14-02-00972).

References

[1] See the talk of Zhan Dzhilkibaev at this conference
[2] See the talk of Alexey Sheifler at this conference
[3] See the talk of Igor Belolaptikov at this conference

EPJ Web of Conferences **116**, 05005 (2016)
DOI: 10.1051/epjconf/201611605005

The Trigger and Data Acquisition System for the KM3NeT neutrino telescope

Carmelo Pellegrino[1,2,a] and Tommaso Chiarusi[1,b] on behalf of the KM3NeT Collaboration

[1] INFN Sezione di Bologna,V.le Berti Pichat 6/2, 40127 Bologna, Italy
[2] Dipartimento di Fisica e Astronomia Università di Bologna, V.le Berti Pichat 6/2, 40127 Bologna, Italy

Abstract. KM3NeT is a large research infrastructure in the Mediterranean Sea that includes a network of deep-sea neutrino telescopes. The telescopes consist of vertical detection units carrying optical modules, whose separation is optimised according to the different ranges of neutrino energy that shall be explored. Two building blocks, each one made of 115 detection units, will be deployed at the KM3NeT-IT site, about 80 km from Capo Passero, Italy, to search for high-energy neutrino sources (ARCA); another building block will be installed at the KM3NeT-Fr site, about 40 km from Toulon, France, to study the hierarchy of neutrino masses (ORCA). The modular design of the KM3NeT allows for a progressive implementation and data taking even with an incomplete detector. The same scalable design is used for the Trigger and Data Acquisition Systems (TriDAS). In order to reduce the complexity of the hardware inside the optical modules, the "all data to shore" concept is adopted. This implies that the throughput is dominated by the optical background due to the decay of $^{40}\mathbf{K}$ dissolved in the sea water and to the bursts of bioluminescence, about 3 orders of magnitude larger than the physics signal, ranging from 20 Gbps to several hundreds Gbps, according to the number of detection units. In addition, information from the acoustic positioning system of the detection units must be transmitted. As a consequence of the detector construction, the on-shore DAQ infrastructure must be expanded to handle an increasing data-rate and implement an efficient fast data filtering for both the optical and acoustic channels. In this contribution, the Trigger and Data Acquisition System designed for the Phase 1 of KM3NeT and its future expansion are presented. The network infrastructure, the shore computing resources and the developed applications for handling, filtering and monitoring the optical and acoustic data-streams are described.

1. Introduction

KM3NeT [1] deep-sea research infrastructure consisting of a network of neutrino telescopes will be installed in two sites in the Mediterranean Sea: off Toulon at a depth of about 2500 m, France, the ORCA (Oscillation Research with Cosmics in the Abyss) detector will study the neutrino mass hierarchy, using 115 string-like [2] Detection Units (DUs); off Portopalo di Capo Passero at a depth of 3500 m, Sicily,

[a] e-mail: `carmelo.pellegrino@bo.infn.it`
[b] e-mail: `tommaso.chiarusi@bo.infn.it`

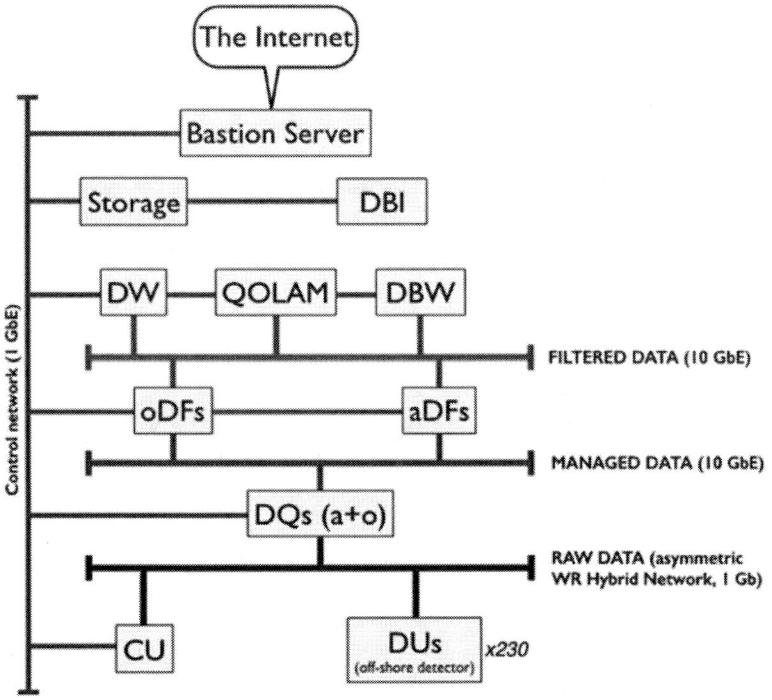

Figure 1. Sketch of the network topology of the KM3NeT detector and shore-station infrastructure. For a detailed description see the text.

the ARCA (Astronomy Research with Cosmics in the Abyss) detector will study ultra high energy neutrinos expected by astrophysical and cosmologic cosmic rays sources. ARCA will be made by about 230 string-like and 8 tower-like [3, 4] DUs arranged in a $\sim$ 100 meters-spaced grid. Each string carries 18 Digital Optical Modules (DOMs) [5, 6], which are vertically separated by 6 and 36 meters in ORCA and ARCA, respectively. Each DOM is a 17″ diameter glass sphere containing 31 3″ photomultiplier tubes. The detection principle is based on the collection of the Cherenkov light emitted along the path of charged particles produced by neutrino interactions in the detector vicinity.

Due to the high optical background caused by β decay of ^{40}K and bioluminescence phenomena in sea water, an on-line trigger is needed. In order to minimise the complexity of the underwater electronics, no hardware trigger is present and the all-data-to-shore approach is used: all detected optical signals are sent to shore via a computer network which comprehends both the shore-station and the detector. In the final configuration of the experiment, the expected data throughput from both sites is up to 500 Gbps.

2. Network topology

The DOMs, i.e. the off-shore detector, and the shore-station computing infrastructure are part of a computing network subdivided in 3 main subnetworks (see Fig. 1):

- Raw data network: aims at connecting the off-shore detector elements (DOMs and DU bases) with both the computers dedicated to the aggregation and distribution of data (DataQueue, see the next section) and the ControlUnit of the experiment;
- Managed data network: aims at connecting the DataQueues with the data filtering infrastructure;
- Filtered data network: is devoted to the transportation of post trigger data.

The Raw data network represents the main networking peculiarity of the experiment. It makes use of a White Rabbit [8] asymmetric network, which also provides for sub-nanosecond time synchronisation. The asymmetry comes from the physical topology of the optical links between the underwater detector elements and the on-shore station. Every single DOM has a unidirectional 1 Gb/s uplink to reach the on-shore station delivering the acquired data and the slow control (SC) replies. On the other side, the on-shore station has a unique unidirectional 1 Gb/s downlink to steer all the DOMs. This is called the *broadcast* and is shared by every single DOM, so that when a DOM receives a packet, all other DOMs also receive it. This topology highly reduces the communication resources cost (i.e. the number of optical fibres), but it requires a specific customisation of the communication elements as the switches and the DOMs.

The Managed and Filtered data networks are standard Ethernet networks that provide a data transfer speed of 10 Gbps.

A 1 GbE Control network allows the connection of all the shore-station computing elements with the ControlUnit and provides access to the Internet through a Bastion Server, which protects the internal networks.

3. The KM3NeT TriDAS

The KM3NeT Trigger and Data Acquisition System [7] (TriDAS) is the software and computing infrastructure deputed to the read-out, aggregation and filtering of all the detector data. All TriDAS components are connected through the network described in Sect. 2.

Each DOM carries 31 3″ PMTs and a piezo-electric acoustic sensor that produce the two main data streams from off-shore: the optical and acoustic data streams, respectively. The DOM digitises the signals from the PMTs as photon hits – defined by the integrated electric charge of the PMT signal over a certain threshold, the crossing time and the duration of the signal over the threshold – and collects them into optical data frames. Acoustic signals reaching the piezo-electric sensor are sampled at a constant frequency and with a proper resolution – both tunable from on-shore – and compose an acoustic data frame. Once a data frame is completed, the DOM sends it to shore using a UDP/IP protocol.

On-shore, the first data processing stage is represented by the DataQueue (DQ) farm which is in charge of the aggregation of unfiltered data and of their distribution to the computers dedicated to the on-line analysis and trigger, namely DataFilters (DFs). For each data stream a specific implementation of the DF exists, called AcousticDataFilter (aDF) and OpticalDataFilter (oDF) for acoustic and optical data streams, respectively. Each DQ is responsible for the collection of the data coming from a sector (i.e. a subset of DOMs) of the detector. Each oDF receives from all the DQs a bunch of data belonging to a precise time interval called Time Slice (TS). The duration of a TS is fixed, typically 100 ms. In this way, each oDF has a complete snapshot, limited to the TS duration, of the status of the whole detector. Triggered data coming out from the oDFs are sent to the DataWriter (DW) that collects them on permanent storage as ROOT [9] files. Each aDF, differently from the optical case, receives all data from a single DQ continuously, thus having the complete data time series from a subset of the acoustic sensors. The aim of the aDF is to provide on-line reconstruction of the Time Of Arrival (TOA) of the acoustic signals emitted by the acoustic beacons of the acoustic positioning system [10]. The output of all the aDFs is collected by the DataBaseWriter (DBW, see below) and written to a database.

The ControlUnit (CU) [11], which represents the user interface to the detector, aims at coordinating the TriDAS and operating the DOMs through a dedicated SlowControl (SC) protocol. It is composed of the following sub-systems:

- Master Control Program (MCP): is the main user interface and allows to schedule data taking of the experiment;
- TriDAS Manager (TM): steers the on-shore computing facility, setting up the TriDAS processes;

- Detector Manager (DM): controls and configures the off-shore detector;
- DataBaseInterface (DBI): provides local database caching and synchronisation with the remote DB;
- DataBaseWriter (DBW): collects slow-control data and TOAs from the aDFs and writes them to the DB.

The Quasi On-Line Analysis and Monitoring (QOLAM) [12] is a suite of programs that perform fast analysis on a subset of the collected post-trigger data, visualisation and monitoring. It is composed by:

- ROyWeb wich is a web-based program that allows to plot monitored parameters as time series, histograms or as synoptic plots;
- Rainbow Alga which is the 3D event display;
- ROyFit is the fast muon reconstruction software, based on an ANTARES algorithm [13].

4. Conclusions

The KM3NeT experiment will perform continuous data taking for at least 20 years in two sites, reaching a total throughput up to 500 Gbps per site. The construction and operation of the detector involves several challenging techniques, such as multi-PMT optical modules, asymmetric network and WhiteRabbit standard. The TriDAS is a custom designed system for data acquisition and triggering and has been designed to be modular and to scale with a growing up detector.

On December 3rd 2015 the first string of the ARCA detector has been deployed in the abyssal site of Capo Passero, in Sicily. Since then, the on-line Trigger and Data Acquisition System has been continuously running at the Portopalo control station, collecting data according to expected performances.

References

[1] KM3NeT Collaboration website, `http://www.km3net.de`
[2] Biagi S. for the KM3NeT Collaboration, Operation and results of the prototype KM3NeT Detection Unit, proceedings of the 5th Roma International Conference on Astroparticle Physics, Noto, Italy
[3] Aiello S. et al. (NEMO Collaboration), Measurement of the atmospheric muon depth intensity relation with the NEMO Phase-2 tower, Astropart. Phys. **66**, 2015, 1–7
[4] M. Favaro et al., *these proceedings*
[5] Circella M., The Digital Optical Module (DOM) for the KM3NeT Detector, proceedings of the 33nd ICRC, Rio de Janeiro, 2013
[6] D. Real et al., *these proceedings*
[7] Chiarusi T. et al. (NEMO Collaboration) KM3NeT: R&D and technical solutions for the next generation underwater neutrino telescope, proceedings of ICHEP 2014
[8] WhiteRabbit website, `http://www.ohwr.org/projects/white-rabbit`
[9] CERN `http://www.root.cern.ch`
[10] F. Simeone et al., *these proceedings*
[11] C. Bozza et al., *these proceedings*
[12] T. Gal et al., *these proceedings*
[13] J.A. Aguilar et al. (ANTARES Collaboration), A fast algorithm for muon track reconstruction and its application to the ANTARES neutrino telescope, Astropart. Phys. **34**, 2011, 652–662

EPJ Web of Conferences **116**, 05006 (2016)
DOI: 10.1051/epjconf/201611605006

Characterization of the electro-optical transceivers in the KM3NeT optical network

S. Pulvirenti[1,a], F. Ameli[2], A. D'Amico[3], G. Kieft[3], and J.-W. Schmelling[3] for KM3NeT Collaboration

[1] INFN Laboratori Nazionali del Sud, via S. Sofia 62, 95123 Catania, Italy
[2] INFN Sezione Roma, P.le A. Moro 2, 00185 Roma, Italy
[3] Nikhef, Science Park 105 1098 XG, Amsterdam, The Netherlands

Abstract. KM3NeT is a future research infrastructure hosting a network of neutrino telescopes in the abyss of the Mediterranean Sea. The whole data transport over an optical network is based on the Dense Wavelength Division Multiplexing technique with optical channels spaced 50 GHz apart and a bit rate of 1.25 Gbps. Over the telescope lifetime, precise temperature control of the laser is required to maintain stability of the central frequency, complying with the recommendations of the International Telecommunication Union. We will report on the characterisation of the tuneable features of the electro-optical transceivers required to compensate for the expected wavelength drift due to aging factors. Preliminary results will be showed.

1. Introduction

KM3NeT is a large second-generation detector in the abyss of the Mediterranean Sea. The first phase of the project (Phase-1) is funded mainly by Italian, Dutch and French contributions and foresees the construction and the deployment of 31 Detection Units (DU), 24 at the KM3NeT-It site (off-shore Portopalo di Capo Passero) and the remaining at the KM3NeT-Fr site (off-shore Toulon). The KM3NeT-It site will host also a set of 8 DUs constructed according to a previous design (NEMO tower [1]), which will be operated independently [2]. The data transfer over an optical network is based on the Dense Wavelength Division Multiplexing (DWDM) technique for both sites. The communication between on and off shore electronics is established over optical fibre using Small Form Factor Pluggable (SFP) transceiver. The one used in the DWDM network has optical channels spaced 50 GHz apart, a bit rate of 1.25 Gbps and up to 100 km span length. Over the telescope lifetime, precise temperature control of the laser is required to maintain stability of the central frequency, complying with the ITU-T G.694.1 [3] recommendations of the International Telecommunication Union.

[a] e-mail: spulvirenti@lns.infn.it

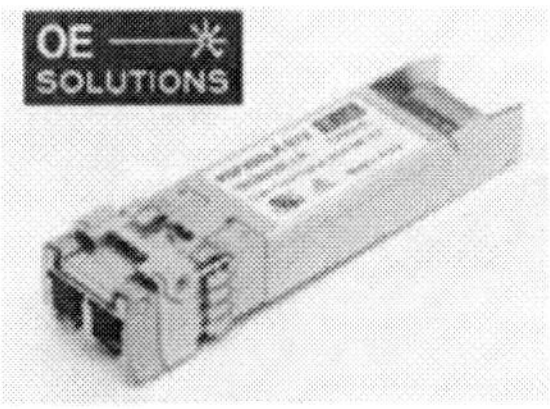

Requirements:	min	typ	max	
Wavelength range	1528		1564	nm
Extinction ratio	8			dB
Output power	0		5	dBm
Deviation From Central Frequency@EOL[1]	-6		+6	GHz
Bitrate transmitter[2]	1.25			Gbps
Link budget	24			dB
Lifetime	15			year
ITU channel spacing			50	GHz
Power consumption		1	1.2	Watt

Figure 1. Left: Picture of the OESolutions SFP transceiver used in the KM3NeT project. Right: main SFP requirements for the project optical network.

2. The Electro-optical transceivers in KM3NeT

The SFP transceivers used in KM3NeT optical network [4] are from OESolutions (see picture in Fig. 1) whose distributor for the project is Tallgrass [5]. Main KM3NeT SFP requirements are reported in Fig. 1.

OESolutions provides a digital diagnostic monitoring interface which allows real time access to the device operating parameters such as the transmitted and received optical power, the laser temperature, the supply voltage and the Thermo Electric Cooler (TEC) controller current.

3. Wavelength Tuning and fist test at Nikhef

The wavelength tuning feature of a SFP is a special aspect which is not required by the DWDM SFP Multiple Source Agreement (MSA) standards. The principle of the wavelength tuning is based on changing the temperature of the Transmission Optical SubAssembly (TOSA) or, to be more precise, changing the temperature of the laser cavity. There is a temperature control circuitry inside the SFP, which is controlled by a microcontroller. The actual change of temperature is done via a TEC controller; which is connected to the microcontroller where it gets its set-point from. The actual wavelength tuning performed by the user is done by changing the tuning value in the microcontroller memory, which is translated to the temperature setting of the TEC and thus the TOSA temperature. All the communication is done via the I2C bus of the SFP transceiver. As observed at Nikhef by means of a wavelength meter (model AQ6140) measuring the behaviour of various OESolutions SFPs, the tuning mechanism permits to shift the wavelength of approximately 100 pm/°C as expected. After a warm up time of the SFP of 15 minutes, the laser temperature was changed in order to observe a wavelength shift of ± 100 pm. We recorded several DAC factory settings before and after the wavelength tuning, verifying the precise control mechanism. The results show a maximum wavelength variation of 6% between different SFP transceivers whose wavelength was shifted by -100 pm and a 10% maximum variation for $+100$ pm wavelength shift of the same SFP under test. At Nikhef several checks of the SFP alarm controls were performed, the SFP transceivers have warning/alarm controls which shuts down the laser emission if the laser temperature drifts more than 1 °C from the setpoint. As shown in Fig. 2 the measurements were focused on the SFP wavelength response checking the biggest tuning step without raising any alarm. The results show a fine control of these devices without raising alarms.

4. Device aging and wavelength drift

Long term device reliability is characterized by the increase in threshold current and the decrease in optical power [6]. Aging of the laser is associated with loss of optical efficiency resulting in a drop of optical power and a shift in the transmitted wavelength whereas it is preferable to operate the device in

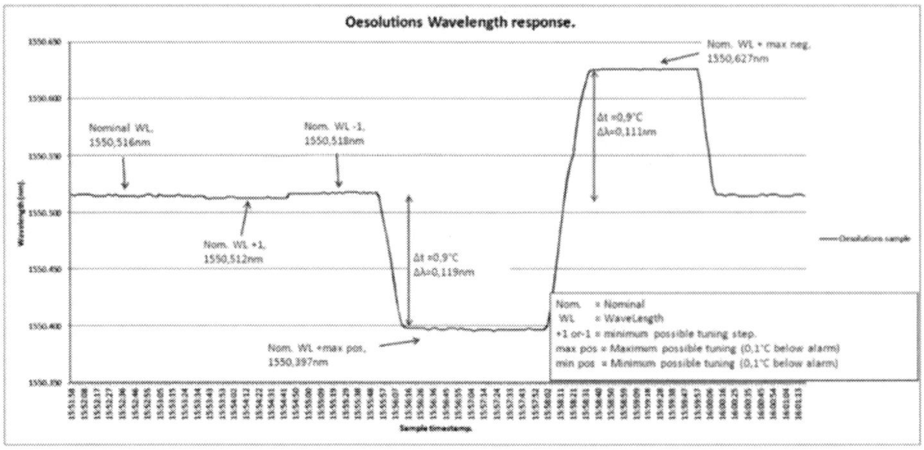

Figure 2. An example of SFP wavelength response changing the laser temperature (which time samples are reported on the x-axis).

constant output power mode. As the device undergoes aging the dissipated power increases resulting in a rise at the junction temperature to which the lasing wavelength is sensitive. The lasing wavelength of a Distributed FeedBack (DFB) laser diode is given by the simple equation:

$$\lambda = 2n\Lambda$$

where n is the effective refractive index and Λ the grating pitch. In order to obtain accurately the desired wavelength, both parameters have to be well controlled. Moreover, the effective refractive index can change with driving conditions such as laser temperature or laser current. Due to the aforementioned reasons, temperature control is often used to fine-tune the DFB wavelength precisely to an ITU-T grid [7]. Over the device lifetime, the laser diode will exhibit wavelength drift. Eventually the wavelength drift could drive the channel into the neighbouring ones and then disrupting the communication. For these reasons we have studied various SFP transceiver devices in order to "retune" the possible wavelength drift in the future.

5. Preliminary results

The measurements performed at LNS (Laboratori Nazionali del Sud) aimed at studying the behaviour of different SFP transceivers varying the laser temperature by means of the Floor Control Module (FCM). This board is the tower floor control module: it permits to collect digital streams of data from 6 PMTs and 2 hydrophones located at the tower floor and collect data from local sensors (Temperature, Relative Humidity, Currents and Voltage, etc.) installed on the same board. The test setup is composed of a SFP transceiver plugged in the FCM and whose optical output signal goes to an Optical Spectrum Analyzer (OSA) for monitoring. Dedicated software has been written in the FCM firmware in order to read and write the SFP registers. A host PC is connected through RS232 to the FCM board with the OEsolutions tuning procedure implemented for tuning. The tuning procedure first foresees a warm up time of 15 minutes in order to stabilize the laser temperature. After checking the DAC settings and storing the factory tuning value, we disable the transceiver alarms and start a temperature scan. There will be four acquisitions at the nominal wavelength at the beginning and at the end of the tuning procedure (in order to check the wavelength stability before and after the scan). The temperature will be changed for 8 samples following a slant trend of 2 °C steps (each temperature step corresponds to half a channel according to the ITU-T grid). The tuning scan starts from the lower point and reach the

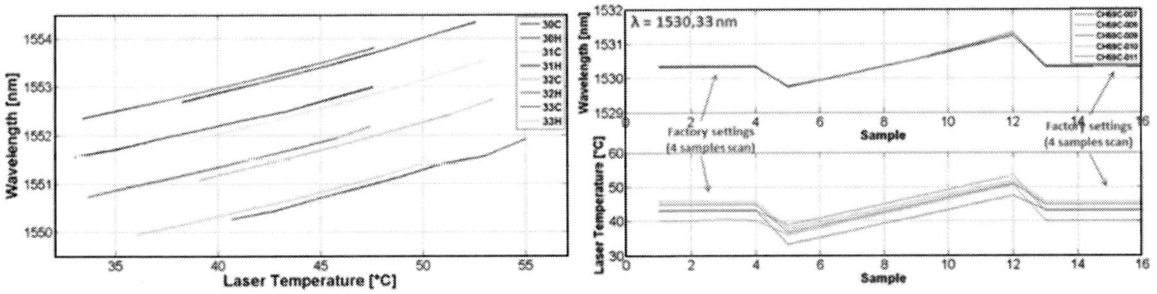

Figure 3. Left: various SFP transceiver trends of Wavelength [nm] vs. Laser Temperature [°C]. Right: Five transceivers CH 59C measured, on top the wavelength measured values and on the bottom the laser temperature.

upper point in about three minutes. Figure 3 shows on the left an example of 8 samples tuning of eight transceivers from channel 30C to channel 33 H. Changing the tuning value changes the laser temperature and consequently the wavelength. In the right plots five CH 59C transceivers were taken, i.e. devices with the same nominal frequency. It can be seen that when the wavelength ramps are kept the same (top figure on the right) the temperature trends are not the same (bottom figure on the right). The preliminary results show that the relation between wavelength and temperature is not linear. The laser cavity non linearity could affect this behaviour, so it seems not easy to find a common behaviour between different devices under test.

6. Conclusions

The KM3NeT telescope is under construction and tests on the optical communication link, based on SFP transceivers, are still on going. Next step will be to monitor the transceivers on a long period and check the wavelength stability at high temperature in order to predict their behaviour. Characterizing more devices in the whole C band within ± 100 pm would allow us to apply a wavelength correction if more than ± 50 pm drift occur.

References

[1] T. Chiarusi, S. Aiello, F. Ameli, M. Anghinolfi, G. Barbarino, E. Barbarito, F. Barbato, N. Beverini, S. Biagi, B. Bouhadef et al., Journal of Instrumentation **9**, C03045 (2014)

[2] A. Margiotta, Journal of Instrumentation **9**, C04020 (2014)

[3] `https://www.itu.int/rec/T-REC-G.694.1/en`

[4] `https://http://www.km3net.org/TDR/TDRKM3NeT.pdf`

[5] `http://www.tallgrass.nl`

[6] J.S. Huang, Applied Physics Research **4**, p15 (2012)

[7] M. Funabashi, H. Nasu, T. Mukaihara, T. Kimoto, T. Shinagawa, T. Kise, K. Takaki, T. Takagi, M. Oike, T. Nomura et al., Selected Topics in Quantum Electronics, IEEE Journal of **10**, 312 (2004)

EPJ Web of Conferences **116**, 05007 (2016)
DOI: 10.1051/epjconf/201611605007

KM3NeT Digital Optical Module electronics

Diego Real[a]

Instituto de Física Corpuscular (UV-CSIC), C/Catedrático José Beltrán 2, 46980 Paterna, Spain

Abstract. The KM3NeT collaboration is currently building of a neutrino telescope with a volume of several cubic kilometres at the bottom of the Mediterranean Sea. The telescope consists of a matrix of Digital Optical Modules that will detect the Cherenkov light originated by the interaction of the neutrinos in the proximity of the detector. This contribution describes the main components of the read-out electronics of the Digital Optical Module: the Power Board, which delivers all the power supply required by the Digital Optical Molule electronics; the Central Logic Board, the main core of the read-out system, hosting 31 Time to Digital Converters with 1 ns resolution and the White Rabbit protocol embedded in the Central Logic Board Field Programmable Gate Array; the Octopus boards, that transfer the Low Voltage Digital Signals from the PMT bases to the Central Logic Board and finally the PMT bases, in charge of converting the analogue signal produced in the 31 3" PMTs into a Low Voltage Digital Signal.

1. INTRODUCTION

KM3NeT is a European research facility being built in the Mediterranean Sea that, once finalised, will host, a neutrino telescope of cubic kilometer scale. Cherenkov light from neutrino-induced secondary particles will be detected by an array of Digital Optical Modules (DOMs) consisting of high pressure resistant glass spheres hosting photomultipliers inside. This vessel is composed of 31 small 3 inch photomultipliers (PMTs) distributed around the glass sphere, which collects the Cherenkov photons and transform them into electronic signals. The PMTs are suspended in a foam support structure: 19 in the lower hemisphere and 12 in the upper hemisphere. Each PMT has its own adjustable high voltage supply integrated in the PMT base. In addition to the PMTs the DOM contains a printed circuit board, the Central Logic Board (CLB), which converts the electronic signals from the PMTs into time, pulse duration and identification information in the Time To Digital (TDC) core embedded in the Field Programmable Gate Array (FPGA) of the CLB. The CLB integrates the White Rabbit Protocol (White Rabbit is a fully deterministic Ethernet-based network for general purpose data transfer and synchronization), which allows to synchronize all the KM3NeT DOMs within 1 ns resolution. The data provided by the PMT bases are collected and distributed to the CLB by means of two boards (one for each hemisphere), the so called Octopus Boards. It also contains the electronic and photonic components for an optical serial link to the shore. All necessary DC power is provided by the Power Board (PB).

[a] e-mail: `real@ific.uv.es`

Figure 1. Picture of DOM Power Board.

An aluminium dome called the *mushroom* provides heat conduction between the electronics inside and the exterior of the sphere. In the next section, all the components of the readout electronics of the DOM are described.

2. Power board

For an efficient transfer of the electrical power the voltage level V must be high and the current I low, because of the power loss depends on I2R of the cable. In addition, the different electronics of the DOM require many different voltage levels for their operation. Therefore, the power conversion board inside the DOM, shown in Fig. 1, derives all different client voltage levels from an input voltage of 12 V. Modern converters at high frequency are used to obtain a high efficiency power conversion. To protect the other electronics inside the DOM against possible high frequency noise interference, the converter board is located in a shielded part of the cooling mushroom.

3. CLB

The DOM Central Logic Board is the main electronics board in the readout chain of KM3NeT. The Low Differential Voltage Signals (LDVS) generated by the PMT bases and collected and distributed by the Octopus boards arrive to the CLB where they are discretized by means of 1 ns resolution TDCs. The oversampling method uses a sampling frequency significantly higher than twice the bandwidth (or highest frequency) of the signal being sampled. For the KM3NeT readout system, the "significantly higher" sampling frequency is obtained using different edges of multiple phase-shifted clocks. This method is called asynchronous oversampling because the clocks used to create the sampling frequency are nominally equal to the data stream accuracy. High-speed phase-shifted clocks are generated from a slow system clock provided by a local clock oscillator placed on the CLB. A Phase Lock Loop (PLL) inside the FPGA generates two clock phases (CLK0 and CLK90). These two phases are routed to a

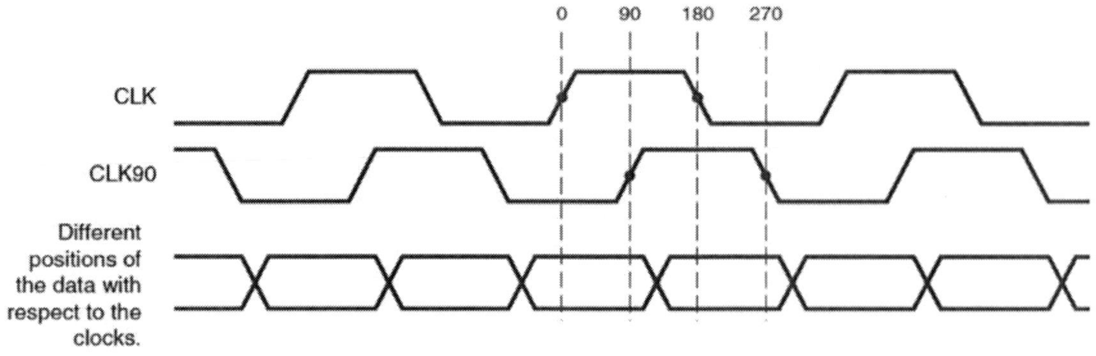

Figure 2. Oversampling technique using two phase-shifted clocks.

deserializer primitive, which is inside the input/output blocks of the FPGA. The generated CLK0 and CLK90 clocks make possible to oversample an incoming data stream on four edges, increasing four times the sampling frequency. The diagram of Fig. 2 shows how the distribution of clocks is performed in order to get 1 ns resolution in the TDCs. The TDC data are sent on-shore after being organized and timestamped at the CLB. The CLB takes care also of the read-out of several instruments, as it is the case of the compass, tiltmeter and temperature sensor, all of them integrated on the same CLB PCB, the piezo, the nanobeacon and the acoustic hydrophone. In order to synchronize the DOMs in KM3NeT, the CLB integrates the White Rabbit (WR) Protocol. It provides sub-nanosecond accuracy and picoseconds precision of synchronization for large distributed systems (more than 1000 nodes over optical fibre and copper lengths of up to 100 km). By using White Rabbit, we are able to achieve precision time-tag measured data and trigger data taking in large installations and, at the same time, the same network can be used for data transmission. The key technologies used are Synchronous Ethernet (SyncE) and Precision Time Protocol (PTP). The main component of the CLB is a Kintex FPGA. This device also allows the reconfiguration of the firmware of the CLB. It is feasible to store up to four FPGA images in a SPI memory, three of them reconfigurable. The non-reconfigurable image provides a safe start for the FPGA in case of corruption of the 3 reconfigurable images, being possible to choose to boot the FPGA with any of the four. The Kintex-7 FPGA is an FPGA family providing very low power consumption allowing a total CLB power consumption below 4 watts. The control of the CLB is achieved through the embedded software running in an LM32, an open source firmware microprocessor from Lattice. No operative system is used in order to reduce power consumption. The CLB contains two LM32 CPUs. One of them resides in the WR core which is dedicated to handle the PTP traffic and controlling the Phase Locked Loops (PLL) that are part of the timing system. The other CPU, the secondary LM32 takes care of the slow control communication with the shore station.

4. PMT base

The PMT base is in charge of discretizing the signal read by the PMT and to provide the High Voltage (HV) for the PMT. The PCB contains a pre-amplifier, a comparator (Time over Threshold) and a identifier. Every PMT must give the same output signal when it is hit by a single photon. The gain of a PMT depends on the supplied high voltage. The HV for each PMT is individually adjustable from 800 to 1400 V. Consequently each PMT gets its own HV circuit board. I2C protocol is used to be able to control the PMT base and to change the HV. The power consumption of each PMT base is around 4.5 mW. An additional function of the PMT base is the digitisation of the analogue output signal of the PMT. The output signal is converted from a charge signal to a voltage signal, followed by a conversion to a digital level by a comparator, resulting in a Time Over Threshold (TOT) signal. The comparator can

Figure 3. Half DOM open: View of the PMT bases (attached to the PMTs) and the Octopus boards (in vertical position).

be adjusted to the required TOT value using I2C protocol. The TOT signal is transferred to the DOM logic by a LVDS connection.

5. Octopus board

Within each hemisphere, the LDVS signals from the PMT bases are collected on a custom electronics board (Fig. 3), called the Octopus board, and transferred to the DOM central logic board. The boards also provide connection for the electrical power to the PMT bases and the I2C communication control. For each PMT, the electrical power can be switched on/off individually by the slow control and in case of overload of a PMT the power will be switched off automatically. This can be monitored by the Fault Flag (FFLG). A clock enabled signal to the PMTs for the I2C communication avoids digital interference.

References

[1] `http://www.km3net.org/`

[2] M. deJong, "KM3NeT:the next generation neutrino telescope", Proceedings to the 33nd ICRC, Rio de Janeiro (2013)

[3] M. Circella, "KM3NeT: A second-generation neutrino telescope in the Mediterranean Sea", Proceedings to the 33nd ICRC, Rio de Janeiro (2013)

[4] D. Real,"The electronics readout and data acquisition system of the KM3NeT Neutrino Telescope Node", Proceedings to the VLVnT, Stockholm (2013)

[5] P. Timmer, "Very low power, high voltage base for a Photo Multiplier Tube for the KM3NeT deep sea neutrino telescope", Proceeding to the TWEPP, Aachen (2010)

[6] D. Gajanana, "ASIC design in the KM3NeT detector", Proceeding to TIPP, Oxford (2012)

[7] A. Belias, "Design and development of the Power Supply Board within the Digital Optical Module in KM3NeT", Proceeding to TIPP, Oxford (2012)

[8] P. Musico, "The Central Logic Board for the optical module of the KM3NeT detector", Proceeding to TIPP, Amsterdam (2014)

[9] S. Biagi, "The data acquisition system of the KM3NeT detector", Proceedings to the 34nd ICRC, The Hague (2015)

[10] Maciej Lipinski, "White Rabbit: a PTP Application for Robust Sub-nanosecond Synchronization", Proceedings to the ISPCS, Munich (2011)

EPJ Web of Conferences **116**, 05008 (2016)
DOI: 10.1051/epjconf/201611605008

NaNet^3: The on-shore readout and slow-control board for the KM3NeT-Italia underwater neutrino telescope

R. Ammendola[2], A. Biagioni[1,a], O. Frezza[1], F. Lo Cicero[1], M. Martinelli[1], P.S. Paolucci[1], L. Pontisso[3], F. Simula[1], P. Vicini[1], F. Ameli[1], C.A. Nicolau[1], E. Pastorelli[1], F. Simeone[1], L. Tosoratto[1], and A. Lonardo[1]

[1] INFN Sezione di Roma, Sapienza, P.le Aldo Moro, 2, 00185 Roma, Italy
[2] INFN Sezione di Roma, Tor Vergata, via della Ricerca Scientifica, 1, 00133 Roma, Italy
[3] INFN Sezione di Pisa, via F. Buonarroti 2, 56127 Pisa, Italy

Abstract. The KM3NeT-Italia underwater neutrino detection unit, the tower, consists of 14 floors. Each floor supports 6 Optical Modules containing front-end electronics needed to digitize the PMT signal, format and transmit the data and 2 hydrophones that reconstruct in real-time the position of Optical Modules, for a maximum tower throughput of more than 600 MB/s. All floor data are collected by the Floor Control Module (FCM) board and transmitted by optical bidirectional virtual point-to-point connections to the on-shore laboratory, each FCM needing an on-shore counterpart as communication endpoint. In this contribution we present NaNet^3, an on-shore readout board based on Altera Stratix V GX FPGA able to manage multiple FCM data channels with a capability of 800 Mbps each. The design is a NaNet customization for the KM3NeT-Italia experiment, adding support in its I/O interface for a synchronous link protocol with deterministic latency at physical level and for a Time Division Multiplexing protocol at data level.

1. Introduction

The ever-increasing bandwidth requirements for High Energy Physics experiments are often so constrained as to require custom technological solutions to be satisfied. The KM3NeT-Italia experiment [1], being an evolution of the NEMO Phase-2 experiment [2], inherits the interface of the on-shore endpoints for the electronic boards that gather the data collected by the underwater detector. Compared to NEMO Phase-2, the desired ramp-up in number of channels that KM3NeT-Italia aims for would cause, if the same readout boards as NEMO Phase-2 were used as endpoints, an unmanageable encumbrance at the on-shore receiving installation. The need of more tightly aggregated channels calls for an improved endpoint but incurs the obvious costs of its redesign. This endpoint consolidation can instead be achieved with a modicum of implementation effort – mainly due to the enforcement of deterministic latency on the board transceivers for system-wide distribution of a common timing – by

[a] e-mail: andrea.biagioni@roma1.infn.it

a customization of the NaNet design [3], the NaNet^3 board, that allows for a cost-effective scaling in the receiving installation.

2. The KM3NeT-Italia DAQ and data transport architecture

KM3NeT-Italia [4] is an underwater experimental apparatus for the detection of high energy neutrinos in the TeV÷PeV range based on the Čerenkov technique. The detector consists of a tridimensional array of photomultiplier tubes (PMTs) exploiting the Čerenkov effect induced by superluminal charged particles in water. The hit is the response of a PMT at the passage of particles. The temporal and spatial distribution of the hits allows reconstructing the particles direction. Accurate knowledge of the position of the PMTs is mandatory to accomplish this task. The required spatial accuracy is 40 cm.

The final assessment of the experiment foresees the installation of 8 detection units. The detection unit is called *tower* and consists of 14 floors vertically spaced 20 meters apart. The floor consists of 6 optical modules (OMs) containing a 10 inch PMT each, the front-end electronics needed to digitize the signal, format and transmit the data and 2 hydrophones to reconstruct in real-time the OM position. A board called *Floor Control Module* (FCM) collects the data coming from the underwater devices and communicates with the on-shore laboratory through an optical fiber.

The main requirement for the endpoint board in the laboratory is the distribution of a common timing all over the system in order to correlate signals from different parts of the apparatus with the required $\sim$ 1 ns resolution. Every data frame needs to be labelled with a "time stamp" in order to reconstruct the tracks of charged particles. The described constraints hinted to the choice of a synchronous link protocol which embeds clock and data with deterministic latency; due to the distance between the apparatus and shoreland, the transmission medium is forced to be an optical fiber.

The second requirement is keeping the infrastructure cost-effective: an on-shore readout board supporting multiple channel allows for more efficient scaling of the PC farm size.

3. NaNet^3

The NaNet board family aims at providing a bridge between the readout of the detectors and the computing nodes of a PC farm by means of a real-time data transport mechanism.

NaNet^3 is a PCIe Gen2 Network adapter implemented on the Terasic DE5-net board, which is equipped with an Altera Stratix V FPGA and features four I/O channels with maximum capability of 800 Mbps each. The hardware follows the RDMA paradigm to manage the I/O towards both CPU and GPU – for this latter the protocol is more precisely the GPUDirect RDMA [5], – in this way avoiding the use of bounce buffers and minimizing the latency jitter of data transfers to and from application memory. The virtual-to-physical address translation is entrusted to a proprietary Translation Look-aside Buffer based on Content Addressable Memory [6]. In order to sustain the $\sim$ 320 MB/s aggregate bandwidth of the multi-channel system, incoming data are sent to the computing node by PCIe DMA write processes. The outbound flow consists instead of messages for slow control management of the underwater devices; since requested bandwidth does not exceed $\sim$ 240 kB/s, the task can be accomplished with PCIe target read processes.

The NaNet^3 Router module can support up to 6 data streams with a capability of 2.8 GB/s, multiplexing the incoming messages towards the receiving port and applying a deterministic routing policy.

Finally, the KM3link data transmission system is based on Time Division Multiplexing (TDM) protocol and guarantees deterministic latency of the communication.

GPS signals – i.e. clock and IRIG data – are received from the two SMA connectors on board.

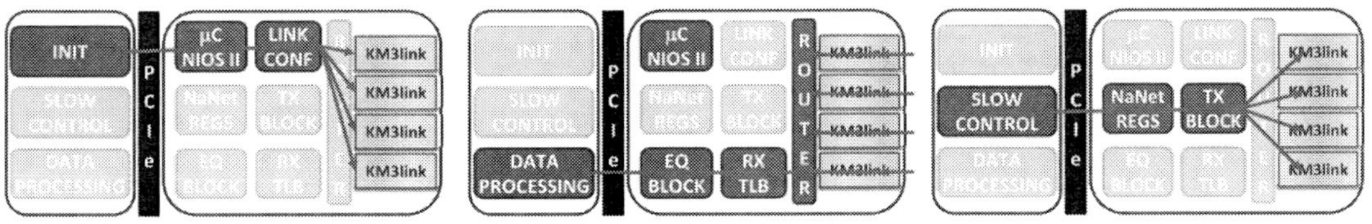

(a) (b) (c)

Figure 1. Basic functions paths within NaNet[3]. Software tasks are on the left side of the PCIe bus, the hardware tasks are on the right.

3.1 NaNet³ configuration and data flow

In this section, overall board operativity with a focus on major functionalities is provided.

During *Hardware/Software Initialization* (Fig. 1a), several hardware modules are enabled by the NaNet driver which initializes the board configuration registers. Incoming data are collected in circular lists of persistent buffers (CLOPs) on the target device memory. The driver registers the buffers and memory pages are pinned and locked. The Nios II microcontroller on board sets up the translation of their addresses from virtual memory space to the physical one the network adapter is able to handle. A custom hardware module takes care of setting up the CLOP parameters – i.e. number of buffers, buffer size and virtual address – for the KM3links. The result is a channel able to receive and transmit messages without further OS involvement – the network stack protocol is totally hardware-managed.

In the *Data Reception and Processing* stage (Fig. 1b), incoming TDM data is translated in a packet-based protocol optimizing PCIe communication. A custom module computes the destination virtual address to allocate the data into the CLOPs and patches it into the packet header. The four data stream, one for each KM3link, are dispatched to the receiving port through the *Router*. The receiving module manages the data flow and operates Virtual-to-Physical translations exploiting a TLB. Its entries are pre-registered in order to speed-up the hardware processing. A DMA write transaction moves data to the computing node memory and an event queue collects the completion messages of the writing process; an interrupt is raised to notify the system. The driver signals the reception of new frames to the application, kicking off the computing process.

Finally, *Slow Control Transmission* (Fig. 1c) is handled by the kernel driver. Data addressed to the underwater devices are stored in registers (one for each underwater device) through PCIe target mode, then a custom TX module dispatches the messages to the proper channel according to the register address. Data is allocated in a 10 kB sized frame according to TDM protocol. One data frame is transmitted every 125 μs, with a GPS tagging every frame with a 12.5 ns precision timestamp.

3.2 Deterministic latency

A channel featuring deterministic latency is the main requirement of the DAQ architecture in order to achieve system-wide common timing. The KM3link exploits the Altera Deterministic Latency PHY IP core enabling accurate delay measurements and known timing for the transmit (TX) and receive (RX) datapaths. The core Physical Coding Sublayer offers, among other things, 8B10B encoder/decoder, Word Aligner and TX bit slipper.

NaNet[3] is in charge of distributing the timing information and signal to the connected underwater devices enforcing the constraint of having a ns-precise offset between the wave-fronts of the system clocks. A GPS clock acts as reference and it is used for the optical link transmission from the on-shore board towards the underwater FCM. NaNet[3] acts as master of the communication managing the reference clock. The FCM recovers the clock from the incoming stream and employs it to send its payload from the apparatus back to NaNet[3]. Finally, the receiving module of KM3link recovers

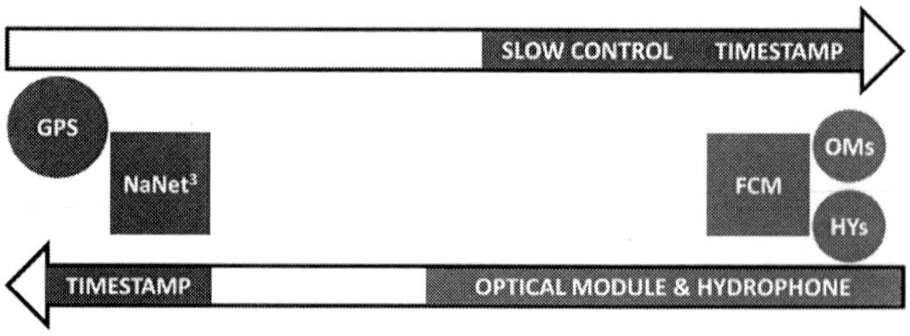

Figure 2. NaNet^3-FCM data flow; *timestamp* and *Slow control* segments are injected by NaNet^3 while *Optical module & Hydrophone* segment is injected by the FCM.

the clock by decoding this payload at loop end, therefore guaranteeing the establishment of a fully synchronous link. A scheme of the data flow is depicted in Fig. 2. Full control of the communication latency between NaNet^3 and FCM board, based on Xilinx FPGA, is mandatory to obtain the knowledge of timing of the overall system. The Altera Deterministic Latency PHY IP guarantees its functioning when sender and receiver are directly connected. In the present case, the off-shore device breaks the link making the TX bit slipper ineffective. The latency varies whenever alignment procedure of the communication channel is executed – i.e. the reboot of the host server or substitution of the cable. A custom module was developed to tackle the issue. The hardware logic allows for setting the desired bit slip value during the alignment procedure through a configuration register of NaNet^3. The procedure runs until the number of bit slip needed to acknowledge the alignment-word matches the required one.

4. Conclusion

In this paper we presented the NaNet^3 board, a customization of the NaNet design dedicated to the KM3NeT-Italia experiment. The NaNet -to-FCM slow control transmission was successfully tested for distances up to 100 Km. The multichannel reception system is being lab-tested and has reached 2 continuous weeks of error-free data collection.

References

[1] *KM3NeT ITALIA*, accessed: 26/Gen/2016, `https://web2.infn.it/KM3NeT-Italia/`

[2] S. Aiello, F. Ameli, M. Anghinolfi, G. Barbarino, E. Barbarito, F. Barbato, N. Beverini, S. Biagi, B. Bouhadef, C. Bozza et al., Astroparticle Physics **66**, 1 (2015), ISSN 0927-6505, `http://www.sciencedirect.com/science/article/pii/S0927650514001960`

[3] A. Lonardo, F. Ameli, R. Ammendola, A. Biagioni, A.C. Ramusino, M. Fiorini, O. Frezza, G. Lamanna, F. Lo Cicero, M. Martinelli et al., Journal of Instrumentation **10**(04), C04011 (2015), `http://stacks.iop.org/1748-0221/10/i=04/a=C04011`

[4] A. Margiotta, Journal of Instrumentation **9**(04), C04020 (2014), `http://stacks.iop.org/1748-0221/9/i=04/a=C04020`

[5] R. Ammendola et al., *GPU Peer-to-Peer Techniques Applied to a Cluster Interconnect*, in *Parallel and Distributed Processing Symposium Workshops PhD Forum (IPDPSW), 2013 IEEE 27th International* (2013), pp. 806–815

[6] R. Ammendola et al., *Virtual–to–Physical address translation for an FPGA–based interconnect with host and GPU remote DMA capabilities*, in *Field-Programmable Technology (FPT), 2013 International Conference* (2013), pp. 58–65

EPJ Web of Conferences **116**, 05009 (2016)
DOI: 10.1051/epjconf/201611605009

The Trigger and Data Acquisition System for the KM3NeT-Italia towers

M. Favaro[1,2,4,a], T. Chiarusi[1,b], F. Giacomini[2,c], M. Manzali[2,4,d], A. Margiotta[1,e], and C. Pellegrino[1,3,f]

[1] INFN BO, Bologna, Italy
[2] INFN CNAF, Bologna, Italy
[3] Dipartimento di Fisica e Astronomia, Università degli Studi di Bologna, Bologna, Italy
[4] Università degli Studi di Ferrara, Ferrara, Italy

Abstract. KM3NeT-Italia is an INFN project supported with Italian PON fundings for building the core of the Italian node of the KM3NeT neutrino telescope. The detector, made of 700 10″ Optical Modules (OMs) lodged along 8 vertical structures called *towers*, will be deployed starting from fall 2015 at the KM3NeT-Italy site, about 80 km off Capo Passero, Italy, 3500 m deep. The *all data to shore* approach is used to reduce the complexity of the submarine detector, demanding for an on-line trigger integrated in the data acquisition system running in the shore station, called *TriDAS*. Due to the large optical background in the sea from ^{40}K decays and bioluminescence, the throughput from the underwater detector can range up to 30 Gbps. This puts strong constraints on the design and performances of the TriDAS and of the related network infrastructure. In this contribution the technology behind the implementation of the TriDAS infrastructure is reviewed, focusing on the relationship between the various components and their performances. The modular design of the TriDAS, which allows for its scalability up to a larger detector than the 8-tower configuration is also discussed.

1. Introduction

The INFN's project KM3NeT-Italy [1], supported with Italian PON fundings, consists of 8 vertical structures, called Towers, instrumented with a total number of 700 Optical Modules (OMs) and will be deployed 3500 m deep in the Ionian Sea, at ∼ 80 km from the Sicilian coast [2, 3]. A Tower is made of 14 horizontal bars, piled up one by one with 90° heading difference. Each bar hosts 6 OMs. Each OM contains a 10″ PMT and the readout electronics. The detection principle exploits the measurement of the Cherenkov light from relativistic particles outgoing high-energy neutrino interaction within a fiducial volume around the telescope. In order to reduce the complexity of the underwater detector, the *all data to shore* approach is assumed, demanding for a Trigger and Data Acquisition System

[a] e-mail: matteo.favaro@cnaf.infn.it
[b] e-mail: tommaso.chiarusi@bo.infn.it
[c] e-mail: francesco.giacomini@cnaf.infn.it
[d] e-mail: matteo.manzali@cnaf.infn.it
[e] e-mail: annarita.margiotta@unibo.it
[f] e-mail: carmelo.pellegrino@bo.infn.it

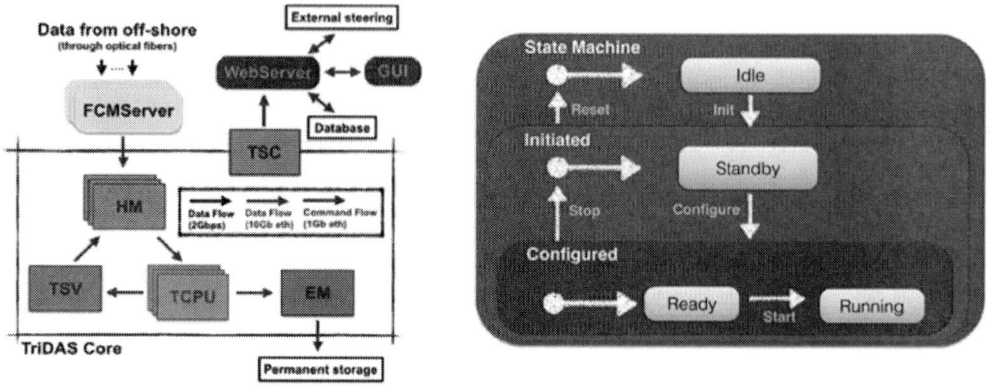

(a) (b)

Figure 1. The Fig. 1a (left) represents the TriDAS block diagram with the various managed data flows. The Fig. 1b (right) is the finite state machine adopted for steering the TriDAS components.

(TriDAS) [4] running at the shore station. The collected data stream from all the Towers is largely affected by the optical background in the sea [5], mainly due to the ^{40}K decays and bioluminescence bursts. Ranging up to 30 Gbps, such a large throughput puts strong constraints on the required TriDAS performances and the related networking architecture. In the following sections there are the description of the final implementation of the physics-data handling (TriDAS Core), the user and management interfaces (TriDAS Control) and the large-band network infrastructure.

2. The TriDAS software

2.1 TriDAS Core

The TriDAS Core [6] (Fig. 1a) is formed by the HitManager (HM), the Trigger CPU (TCPU), the TriDAS-SuperVisor (TSV) and the Event Manager (EM). The data to the TriDAS Core are provided from the FCMServer (FCMS) units that read the optical data from the detector and send them to the HitManager. The FCMServers are designed to be, onshore, the gate for all the kind of data streams (slow control, optical and acoustic), going to and coming from the offshore detector [7]. One single FCMServer can handle the optical data coming from 4 floors of a Tower. With 8 Towers, the total number of FCMServers is 32. The FCMServers forward the data coming from the OMs to the first layer of the TriDAS, the HitManagers. Every single HitManager process runs in a dedicated server and it is linked to a fixed number of FCMServers, which correspond to a portion of the detector, called Sector. Test proved that a HitManager process is capable to withstand the data from an entire Tower. All the HitManagers share the same time line, originated from a common timestamp, which is quantized in subsequent intervals of time of the same duration, called *TimeSlices*. In this way, a full set of optical data occurred during a particular TimeSlice are asynchronously managed by all the HitManagers, which organize their own fraction of data in a special data-structures called the "*Sector Time Slices*" (STSs). The role of the TriDAS-SuperVisor is then to steer all the HitManagers sending the STSs belonging to the same TimeSlice to the first available TriggerCPU, according to a free-token-scheduler mechanism. On its turn, one TCPU collects all the STSs of a TimeSlice into the so called "*Telescope Time Slice*" structure (TTS), then process it according with the trigger algorithms [4]. Many TriggerCPUs process different TTS from different TimeSlices at the same time. The fraction of data which fullfill the trigger selection criteria is sent to the EventManager, which records the filtered data in binary files on the local storage. Offline, the written post-trigger files are transferred from the Shore station infrastructure of

Portopalo to the storage facility at LNS via a dedicated 10 Gbps connection. The design for the TriDAS is modular and scalable with the number of deployed Towers. The required number of TriggerCPUs processes depends on the complexity of the trigger algorithms and increases with the number of OMs. It is currently under study the determination of the necessary computing resources as a function of the detector dimension.

2.2 TriDAS Control

The *TriDAS Control* (TSC) is the software component that orchestrates all the TriDAS processes running on the data acquisition farm. The TriDAS Control implements a simple hierarchical state machine with four states, as shown in Fig. 1b:

Idle. This is the initial state of the overall TriDAS state machine, where no processes are running. An *init* transition, which takes a *run setup* identifier as a parameter, executes an action that retrieves a run datacard corresponding to the given run setup. The datacard describes the geometry of the detector and the configuration of the TriDAS system (such as the role of each node) for this run. If the action is successful, the state machine moves into the *Initiated* sub-state machine.

Standby. This is the initial state of the *Initiated* sub-state machine. Here the TriDAS Control is aware of the configuration of the TriDAS system but no processes are running yet. A *configure* transition executes an action that retrieves the *run number* and starts the Trigger CPU, HitManager and Event Manager processes on the corresponding nodes. If all the processes start successfully the state machine moves into the *Configured* sub-state machine.

Ready. This is the initial state of the *Configured* sub-state machine. Here the Trigger CPUs, the HitManagers and the Event Manager are ready to acquire, filter and save physics data coming from the FCMServers. The *start* transition executes an action that computes the start time of the run and starts the TriDAS-SuperVisor. The TriDAS-SuperVisor's role is to schedule which Trigger CPU process will compute a given TTS. The scheduling follows a credit-based mechanism to balance the load among the Trigger CPUs. If the TriDAS-SuperVisor starts successfully the data acquisition starts and the state machine moves into the *Running* state.

Running. In this state the data acquisition is running.

Transitions exist to move the system back to the *Idle* and *Standby* states.

If any error occurs during a transition, the transition is aborted. Depending on the severity of the error, the system may stay in the current state or even shut-down completely.

The communication with the TriDAS Control, for example to trigger the transitions described above or to query the state of the system, is stateless and happens over a UNIX socket. Only one client can use the socket at any one time.

2.3 WebServer

The *WebServer* is the only entry point for controlling the DAQ. This component provides a set of RESTful API which allows to inquire the TriDAS Control. Therefore, it provides user authentication based on hierarchical configuration, which is implemented via different privileged groups. The privileged groups present are: administration, DAQ control and monitor. A user can have a combination of privileges, belonging to different groups. The TriDAS Control has a unique way of communication as described in 2.2. The TriDAS Control is a local-single-client program. The WebServer can be contacted from several different concurrent people at a time. The system allows controlling the DAQ via an escalation procedure only one user at a time, that permits users to acquire the privilege for controlling the TriDAS Control.

The use of WebSocket allows to implement a real time feedback system to the users. The WebServer can communicate instantly feedback and alarms during an acquisition phase or escalation.

3. Conclusion

The TriDAS has been improved to sustain the foreseen 8 towers detector. Its performances and scalability are under intense test, with long duration runs and varying the incoming throughput using either real FCMServer and simulation programs, using a test bench that reproduces the real farm in Portopalo. New trigger algorithms are under development serving different kinds of physics analysis, e.g. multi-messenger external alerts, high energy neutrino induced showers and astrophysical source detection.

The preliminary test phase demostrates that the system is stable and the users are able to control it properly. Moreover, in November 2015 the installation and functionality test of the farm in Portopalo has been completed. Extended tests of the TriDAS will be also realized in the Portopalo infrastructure, in advance with respect to the first deployment of the Towers. This will make possible tests in the real situation, exploiting the actual facility available in the shore station. In addition to that, increasing the computing resources with respect to what available in the test-bench, more realistic results will be achieved.

References

[1] *Km3net web site*, `http://www.km3net.org/home.php` (2015), [Online; accessed 11-Dec-2015]
[2] S. Aiello et al., Astroparticle Physics **66**, 1 (2015), ISSN 0927-6505, `http://www.sciencedirect.com/science/article/pii/S0927650514001960`
[3] M.S. T. Chiarusi, European Physical Journal **65**(649), 3 (2010)
[4] C. Pellegrino, et al. (Nemo), AIP Conference Proceedings **1630**(158), 7 (2014)
[5] Pellegriti, M. G., et al., AIP Conference Proceedings **1630**(1), 94 (2014)
[6] M. Manzali, T. Chiarusi, M. Favaro, F. Giacomini, A. Margiotta, C. Pellegrino, Nuclear Instruments and Methods in Physics Research Section A: Accelerators, Spectrometers, Detectors and Associated Equipment (2015), ISSN 0168-9002, `http://www.sciencedirect.com/science/article/pii/S0168900215014266`
[7] A.L. et al., Journal of Instrumentation **10**(04), C04011 (2015), `http://stacks.iop.org/1748-0221/10/i=04/a=C04011`

EPJ Web of Conferences **116**, 05010 (2016)
DOI: 10.1051/epjconf/201611605010

Digital and Analog Electronics for an autonomous, deep-sea, Gamma Ray Burst Neutrino prototype detector

K. Manolopoulos[a], A. Belias, C. Markou, P. Rapidis, and E. Kappos

Institute of Nuclear & Particle Physics, NCSR Demokritos, Athens, Greece

Abstract. GRBNeT is a Gamma Ray Burst Neutrino Telescope made of autonomously operated arrays of deep-sea light detectors, anchored to the sea-bed without any cabled connection to the shore. This paper presents the digital and analog electronics that we have designed and developed for the GRBNeT prototype. We describe the requirements for these electronics and present their design and functionality. We present low-power analog electronics for the PMTs utilized in the GRBNeT prototype and the FPGA based digital system for data selection and storage. We conclude with preliminary performance measurements of the electronics systems for the GRBNeT prototype.

1. Introduction

GRBNeT is a Gamma Ray Burst Neutrino Telescope that involves the development, construction and testing of a prototype for an autonomous array of deep-sea detectors, capable to operate continuously for a prolonged period of time, anchored at the sea bed without any cabled connection to the shore. A few tens of such arrays, separated by 300–400 meters from each other, will be able to cover a volume of $\sim 10\,\text{km}^3$ in the deep-sea, down to 5 km depth, thus providing a sufficiently large observational volume, capable of detecting high-energy neutrinos originating from Gamma Ray Bursts (GRBs).

In this paper we present the digital and analog electronics of the Data Acquisition (DAQ) and readout system for the *GRBNeT* prototype. This prototype consists of four Cherenkov light detectors, based on photomultiplier tubes (PMTs) inside a pressure resistant glass sphere. Immersed in the deep-sea these light detectors will be operated, and their analog signals will be transmitted to a fifth glass sphere, which houses the data acquisition and the readout system of the detector. One of the key features of *GRBNeT* is its ability to operate autonomously, without the need of a submarine cable to supply the necessary power from the shore. This autonomy is achieved by utilizing locally a cluster of batteries, also housed in glass spheres that provide the entire system with power. The main advantage of this feature is the cost reduction by omitting the submarine power cables to shore, while on the other hand, it creates a restriction to the detector's power consumption.

The design of the DAQ architecture demands low power analog electronics and it is well suited for a low power FPGA implementation. The Xilinx Spartan-6 FPGA LX16 evaluation board is targeted to implement and to validate the functionalities of the DAQ design. Moreover, a PIC microcontroller

[a] e-mail: kmanolo@inp.demokritos.gr

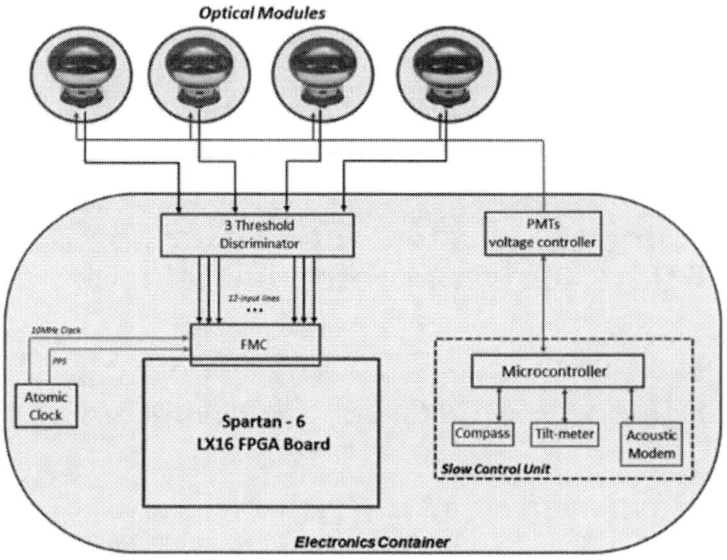

Figure 1. Block diagram of the detector unit. The prototype consists of two such units, operating independently of each other.

is used to handle the slow-control tasks, thus partitioning the system's tasks to achieve low power dissipation.

2. Design description

The main functions of the DAQ system and the readout electronics are the following: i) detect events of interest, based on a digital analysis of the light detectors' digitized signals, ii) measure the temporal and spatial characteristics of selected events (i.e. duration, intensity etc.), iii) use an atomic clock as an autonomous, free running, reference clock for synchronicity and time-stamping of each event, iv) store these results locally to permanent data storage, in order to be analyzed after retrieval, and v) collect and store operational data, as well as data from auxiliary devices such as orientation sensitive devices (i.e. compass, tiltmeter etc.). Figure 1 depicts the detector unit with the block diagram of the electronics container, which consists of both analog and digital electronics. The detection unit utilizes a set of 4 Optical Modules (OMs). Each OM consists of a 17-inch diameter Benthos glass-sphere and contains a single R8055 Hamamatsu PMT of 13-inch diameter.

2.1 Analog electronics

In order to maintain the power consumption for the PMT bases at a minimum, instead of a resistive divider design, we have developed a Cockroft-Walton (CW) voltage multiplier (Fig. 2b) and a High Voltage (HV) control unit, that utilizes ultra-low power microcontrollers. In each OM the HV-controller drives the Cockroft Walton (CW) voltage multiplier to produce the desired voltage for the specific PMT. When a PMT detects Cherenkov light it produces an analog pulse, whose characteristics (i.e. duration, height) depend on the intensity of the light detected. Each OM transmits the analog signal from its PMT to a 3 threshold-level discriminator. A single discriminator unit implements 4 independent input channels, one for each OM, where the analog signals are compared to three distinct threshold levels (low, medium and high). When a threshold level is crossed the discriminator outputs a leading edge, followed by a trailing edge that lasts for as long as the analog signal remains above that specific threshold. The

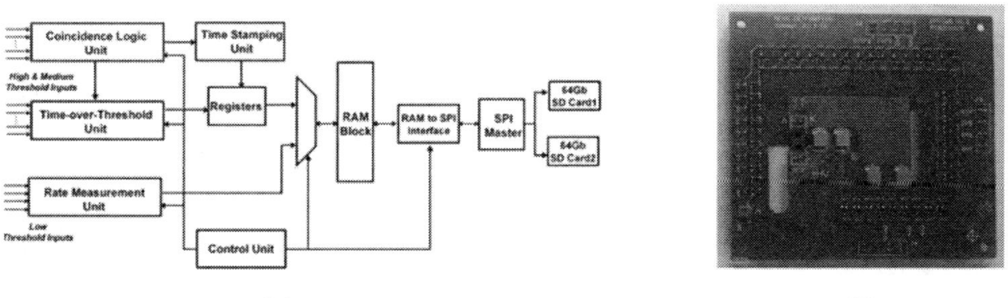

(a) (b)

Figure 2. (a) Block diagram of the FPGA design, (b) Cockcroft-Walton board.

discriminator's digitized outputs enter as input to the digital circuit implemented on the FPGA board. Since each PMT uses a dedicated 3 threshold-level discriminator, with 4 PMTs operated in a cluster, a total of 12 input lines are produced for the FPGA design.

2.2 FPGA architecture

The Data Acquisition (DAQ) hardware system, in this prototype version, of the *GRBNeT* module consists of the Spartan-6 LX16 Evaluation board [2] and the Microsemi SA.45 s Chip Scale Atomic Clock [3]. The atomic clock provides the system with a stable, low-jitter reference clock of 10 MHz and a Pulse-Per-Second (PPS) signal. This clock is free-running for the duration of the deployment and is used by the FPGA to derive the other clock domains that will be used by the design.

Figure 2a shows the block diagram of the FPGA architecture. The main task of the design implemented on the Spartan-6 FPGA is to detect and record events that could be related to high-energy neutrinos. The selection of such events is based on the inputs received by the 4 PMTs of the cluster and it is performed by the *Coincidence Logic Unit* (CLU). The *CLU* searches for PMT signals that, after passing a threshold, coincide within a fixed time frame. These coincidences are considered as events of interest and therefore, when detected by the *CLU* a trigger is generated in order to save all the corresponding PMT data to permanent local storage.

In this prototype, any PMT signal passing at least the programmable threshold level will be considered in the search for coincidences. Whenever this occurs a 200 ns time-window is generated. When a second PMT signal, that has passed the same threshold, is detected within the time-window a 2-fold coincidence occurs. When a 3-fold coincidence occurs it triggers the expansion of the time window for a subsequent period, a post-trigger window, of an extra 100 ns. At the end of the expanded 300 ns time-window, the data of all 4 PMTs are stored to the RAM block.

The data of interest for each detected event are the length of each PMT pulse and the timing information regarding when the event has been detected. To determine the length of a PMT pulse we measure the Time-over-Threshold (ToT), the time duration for which the pulse remains above the specific threshold. This is performed by the *ToT* Unit that is implemented in the Spartan-6 according to [4].

As we mentioned previously, there are 3 input lines per PMT that correspond to the three threshold levels. The medium and the high threshold are taken into account by the CLU in order to detect events of interest, while the low threshold level is used by the Rate Measurements Unit. The purpose of this unit is to measure the rate of the pulses generated by each PMT. The result of these measurements will also be recorded, regardless of coincidences, providing additional information during the off-line data analysis to determine the operational functionality of each PMT and to monitor the pulse rates of each PMT.

The data gathered from the previous units are temporarily stored to a RAM block, in order to be permanently stored to a flash memory. Two high speed 64 Gb SD memory cards are used for this purpose, communicating with the FPGA via the SPI protocol.

The *Control Unit* handles the synchronization and the control signals of the FPGA design. It provides all the parameters to the design's components, it is responsible for all data and command handling, and it periodically initiates the data storage from the RAM block to the SD cards.

2.3 Slow Control Unit

Throughout the deployment a number of different parameters are monitored, such as the PMTs voltage levels, the detector's orientation and the temperature inside the glass sphere that houses all the electronic components. These tasks are performed by the *Slow Control Unit* (SCU) by utilizing a PIC microcontroller that communicates via the I2C protocol with a compass and a tiltmeter in order to record changes in the detector's orientation. Moreover, the SCU is directly connected to the PMTs voltage controller in order to control and monitor the voltages. The slow control data gathered by the microcontroller are stored periodically and independently of the events detected by the FPGA design.

3. Implementation analysis and tests

The presented design was tested at a hardware level targeting the Spartan-6 LX16 evaluation board, utilizing initially a function generator to produce the incoming pulses and subsequently performing test with the actual PMTs. The maximum operating frequency of the FPGA is 200 MHz, but in order to reduce the power consumption slower clock domains are also used, depending on the operation of the various design components. To further minimize the power consumption the design has incorporated low power techniques, such as avoiding free rolling counters and implementing clock gating techniques to keep several components of the design inactive for the time period where there is no PMT activity. As a result, the total power consumption of the design was reduced to 375 mW for a typical operational scenario, of which 93 mW is the quiescent power, and the dynamic power equals to 282 mW.

4. Conclusions

This paper has presented the digital and analog electronics of the DAQ system for an autonomous deep-sea Gamma Ray Burst Neutrino Telescope prototype. The design features programmable functionality based on a scalable design, while maintaining a low power consumption with the ability to operate autonomously utilizing a local cluster of batteries. The FPGA implementation shows the benefits of the presented architecture and field experiments are expected to provide measurements for the design's parameter tuning to optimize its performance.

This study was performed for the GRBNeT Project, funded by the THALIS Program of the Greek Ministry of Education and Religious Affairs (MIS 360381).

References

[1] K. Pikounis et al., *GRBNeT - A Prototype Autonomous Neutrino Detector* (this conference)

[2] `http://www.em.avnet.com/en-us/design/drc/Pages/Xilinx-Spartan-6-FPGA-LX16-Evaluation-Kit.aspx`

[3] Quantum SA.45s Chip Scale Atomic Clock, `http://www.miscrosemi.com`

[4] A. Aloisio, P. Branchini et al., *FPGA implementation of a high-resolution time-to-digital converter*, 3rd ed. Nuclear Science Symposium Conference Record, 2007, Vol. 1, pp. 504–507

EPJ Web of Conferences **116**, 05011 (2016)
DOI: 10.1051/epjconf/201611605011

KM3NeT tower data acquisition and data transport electronics

C.A. Nicolau[1], F. Ameli[1], A. Biagioni[1], A. Capone[1,3], O. Frezza[1], A. Lonardo[1], R. Masullo[3], C.M. Mollo[2], A. Orlando[4], F. Simeone[1], and P. Vicini[1] on behalf of the KM3NeT-IT Collaboration

[1] INFN Sezione di Roma, Sapienza, P.le Aldo Moro, 2, 00185 Roma, Italy
[2] INFN Sezione di Napoli, Complesso di Monte S. Angelo Ed. 6, via Cintia, 80126 Napoli, Italy
[3] Dipartimento di Fisica, Università degli Studi di Roma "Sapienza", P.le Aldo Moro, 2, 00185 Roma, Italy
[4] INFN Laboratori Nazionali del Sud, via S.Sofia 62, 95123 Catania, Italy

Abstract. In the framework of the KM3Net European project, the production stage of a large volume underwater neutrino telescope has started. The forthcoming installation includes 8 towers and 24 strings, that will be installed 100 km off-shore Capo Passero (Italy) at 3500 m depth. The KM3NeT tower, whose layout is strongly based on the NEMO Phase-2 prototype tower deployed in March 2013, has been re-engineered and partially re-designed in order to optimize production costs, power consumption, and usability. This contribution gives a description of the main electronics, including front-end, data transport and clock distribution system, of the KM3NeT tower detection unit.

1. Introduction

The KM3NeT tower is a semi-rigid three-dimensional structure composed of 14 *floors* and a *base-module* (Fig. 1a). The floors are 8 m long structures that, after unfolding, are vertically spaced by 20 m, with the lowermost one located 100 m above the sea bottom, and are mechanically interlinked by a set of tensioning ropes kept taut by a system of buoys at the top of the tower. The base-module anchors the whole structure to the seabed. In the next sections, an overview of the layout of the KM3NeT tower and a brief description of the electronics devoted to data acquisition and data transport will be given.

2. Floor electronics

Each floor supports 6 Optical Modules (OMs), two digital hydrophones, and an optional oceanographic instrument (Fig. 1b). A *Center Floor Vessel* (CFV) is located at the center of each floor, and is connected to the aforementioned peripherals by dedicated electrical cables, one per each peripheral. Each CFV is in turn connected to the tower-base module by a dedicated electro-optical cable containing two optical fibers and two conductors.

[a] e-mail: carlo.nicolau@roma1.infn.it

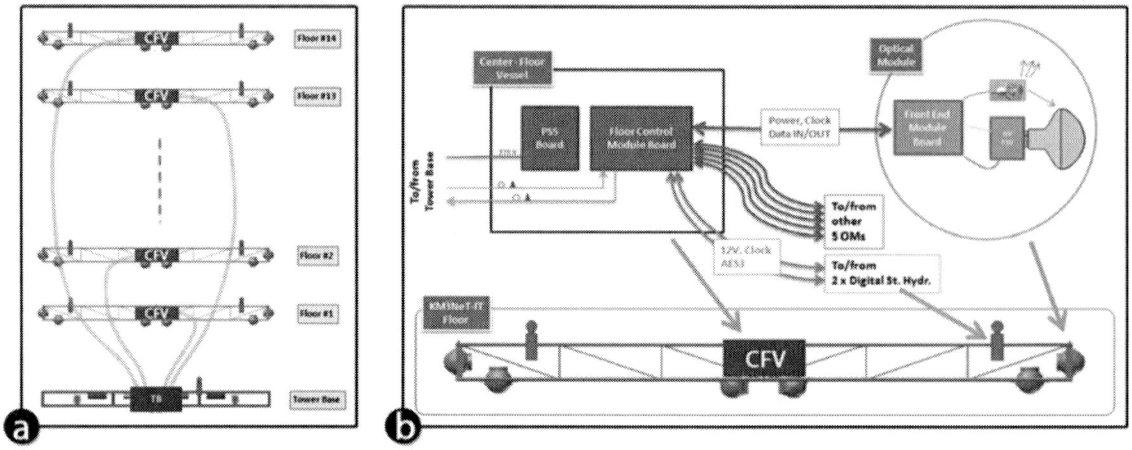

Figure 1. Layout of a KM3NeT tower (a) and of a floor (b).

From the standpoint of the data communication and connections topology, the tower is thus organized hierarchically, starting from the tower-base module, passing through the floors CFVs, up to the floor peripherals. Each floor is assigned a pair of wavelenghts in the Dense Wavelength Division Multiplexing (DWDM) spectrum to allow bidirectional communication; the wavelenghts are optically multiplexed/demultiplexed at the tower-base, where a pair of optical fibers provide the link to the shore station. A system of junction boxes provides the connectivity for the various detection units [1].

2.1 Optical module electronics

Each OM consists in a 13 "glass sphere containing a 10" photomultiplier (PMT) with its high voltage power supply and the Front-End Module (FEM) board. Two OMs per floor are equipped with a LED beacon, which can be used to calibrate the detection unit [2]. The FEM is a 9 cm by 8 cm board mainly dedicated to acquire the pulses generated by the PMT and transmit them to the CFV, and to perform slow-control tasks at OM level, including high voltage control, environmental monitoring, and LED beacon control.

The FEM board is built around a Xilinx Spartan 6 FPGA and an ADS4149 (a 14 bit differential FastADC by Texas Instruments) and is connected to the CFV by means of three differential twisted pairs using Low-Voltage Differential Signaling (LVDS) standard. One of the pairs is used to receive the clock signal generated by the CFV, one pair is used to receive slow-control commands and timing synchronization information, one pair is used to send the acquired data and the slow-control data back to the CFV. The common modes of two of the twisted pairs are used to feed the 5 VDC power.

The FEM board was designed with an "all-data-to-shore" philosophy: no pre-elaboration is performed and the most of the acquired information (in this case, the waveforms generated by the PMT) is transmitted to the shore-station as far as possible. In particular, the current pulses generated by the PMT are low-pass filtered and digitized at 200 Msps; in order to retain only the meaningful part of the acquired waveform the FPGA logic implements a zero-skipping mechanism with programmable threshold. The digitized pulses are locally time-stamped and put in a 2 ksample FIFO. Although time-stamping is performed at the ADC sampling rate (5 ns), hit times can be calculated with a precision of the order of 200 ps by implementing interpolation algorithms at the shore station. The acquired pulses are then 8b/10b encoded and sent to the CFV at a maximum rate of 20 Mbps. To further reduce the produced datarate (and therefore sustain higher hit-rates) a non-lossy compression is implemented on this data channel; in particular, low level signals such as those of single photo electrons (s.p.e.) are encoded using an 8 bit sample word size. Given the 50 kHz average s.p.e. rate at the selected submarine site, the average data-rate produced by each FEM is of the order of 6 Mbps and the maximum sustainable s.p.e. rate if of the order of 150 kHz.

The FPGA logic also embeds a microblaze-based system on chip that performs the slow-control duties, including control and monitoring of the analog front-end and the high voltage power supply, temperature and humidity sensor monitoring, and full remote re-programmability.

2.2 Center floor electronics

The CFV hosts two electronic boards: the *Floor Control Module* (FCM) and the *Power Supply System* (PSS).

The main tasks of the FCM consist in providing a communication channel for each of the floor peripherals (the 6 OMs, the two digital stereo hydrophones, and the optional oceanographic instrument), distributing clock and timing information, distributing, controlling and monitoring their power and performing monitoring at CFV level. The link toward the shore station consists in a DWDM standard electro-optical Small Form-factor Pluggable (SFP) transceiver that is attached to a couple of single-mode optical fibers contained in the back-bone cable connecting the CFV to the tower-base module. A custom self-synchronizing 800 Mbps communication protocol embedding clock, data and timing information is implemented in a constant-length frame-oriented fashion. Each peripheral in the floor (including the FCM itself) is assigned a sufficient and constant amount of space in each frame payload, following a time-division multiplexing scheme.

The heart of the FCM board is a Xilinx Spartan 6 FPGA which implements the main functionalities, including the fixed-latency transceivers, the PLLs used for generating local and distributed clocks, the communication interfaces to the 6 OMs, a couple of stereo AES3 standard receivers for the hydrophones, two RS232 ports (externally opto-isolated), and a microblaze-based system on chip. The latter is in charge of performing all the slow-control duties consisting in monitoring the power consumpion of all the floor peripherals (by means of dedicated onboard ADCs), controlling the power switches, the optical transceiver (including wavelength shift monitoring and tuning), and the on-board temperature and humidity sensor. Also, it implements full remote re-programmability.

The PSS board provides the various DC voltages needed by the floor peripherals, including sequenced 1.2 V and 3.3 V for the FCM itself, 5 V for the OMs, and 12 V for the digital hydrophones and the optional external oceanographic instrument. The PSS is built around a Vicor DC/DC converter which receives the 375 V DC voltage from the tower-base module through a couple of conductors in the back-bone and produces a 5 V voltage that is in turn converted to the other needed voltages by separate low power DC/DCs.

3. Tower-base electronics

The tower-base is primarily devoted to provide connectivity and power to all the tower devices; it is connected to the shore station, through the junction boxes, by a pair of optical fibers (one per data direction) each carrying 15 DWDM wavelengths (14 for the floors, one for the tower-base itself). It includes three pods – the *photonics pod*, the *electronics pod*, and the *power pod* – and supports a digital hydrophone and an optional laser beacon.

The photonics pod contains the optical components to multiplex/demultiplex the DWDM wavelengths from/to the CFVs and the electronics pod to/from the two optical fibers coming from the junction box.

The power pod is oil filled and contains the *Power Control System* (PCS) board that implements the set of switches working at 375 V DC voltage that independently control the power lines of the floors. Both voltage and current of each floor are monitored. A Microchip PIC microcontroller communicates with the electronics pod in order to actuate switch commands and provide monitoring information through a RS232 interface. In case of failure of the electronics pod, the microcontroller starts an autonomous procedure to sequentially switch the floors on at known times, while monitoring current absorption to prevent short-circuits.

The electronics pod contains the same electronics hosted by the CFV (a FCM and a PSS) dedicated to control and monitor the electronics pod and to provide connectivity to the photonics pod, the power pod, the digital hydrophone, and the laser beacon.

4. Data and timing link model

As already mentioned, the communication channel between the shore station and each of the FCM boards consists in a couple of wavelengths in the DWDM spectrum. In the shore station, an equivalent number of transceivers are hosted by a set of PCIe boards (the NaNet^3 boards [3]) that are devoted to communicate with the floors. Data collected from the floors are then sent to a CPU through the PCIe bus to be further elaborated. Each NaNet^3 board can host up to four SFP transceivers, allowing to communicate with up to four floors.

All the boards receive, from a single GPS receiver, a common clock together with an IRIG-B encoded timing reference signal, thus sharing the same time-base. The time information is then sent to the FCMs by means of a self-synchronous protocol embedding both data and timing in the same stream. The FCMs, in turn, propagate this timing information to all the floor peripherals, including the FEMs, which use it to time-stamp the PMT hits. Thanks to the fixed-latency implementation along all the relevant paths (including transceivers in the NaNet^3 boards, the FCM, and the FEM), the whole apparatus (i.e. the 8 towers and the relevant elements in the shore station) is fully synchronized and all devices are time-stamp aware with a sub-nanosecond accuracy.

5. Conclusions

In this paper the main electronics developed for the KM3NeT tower was outlined. The whole architecture was fully tested and characterized. Due to the necessity of extremely tight timings and the adoption of a custom self-synchronous transmission protocol, extensive tests where performed on the propagation of the timing information from the shore-station, passing through the whole transmission chain, up to the OMs; a jitter of the order of 50 ps was measured on the 200 MHz clock driving the FastADC. The presented electronics was mass-produced and is currently being integrated into the 8 towers that will be deployed in the next future.

References

[1] R. Papaleo, *Km3NeT Italia – Seafloor Network*, this conference

[2] G. De Bonis et Al., *The LED Beacon prototype system for the on-shore time calibration of the KM3NeT-IT Towers*, this conference

[3] A. Biagioni et Al., *NaNet^3: the on-shore readout and slow-control board for the KM3NeT-IT underwater neutrino telescope 2*, this conference

EPJ Web of Conferences **116**, 06001 (2016)
DOI: 10.1051/epjconf/201611606001

A Precision Optical Calibration Module (POCAM) for *IceCube-Gen2*

M. Jurkovič[a], K. Abraham, K. Holzapfel, K. Krings, E. Resconi, and J. Veenkamp for the IceCube-Gen2 Collaboration[b]

Physik-Department, Technische Universität München, James-Franck-Str. 1, 85748 Garching, Germany

Abstract. We present here a new concept of an *in-situ* self-calibrated isotropic light source for the future *IceCube-Gen2* neutrino detector called the Precision Optical Calibration Module (POCAM). *IceCube-Gen2* will be a matrix of light sensors buried deep in the ice at the geographic South Pole. The timing, the location, and the amount of Cherenkov light deposited by the secondary charged particles are used to reconstruct the properties of the incident neutrinos. The reconstruction relies on a detailed detector model that includes the response of optical modules to the Cherenkov light, as well as the optical properties of the detector medium – the natural Antarctic ice. To understand these properties, both natural, and artificial light sources are already used for calibration. New calibration devices are being developed in order to improve the precision of these measurements, and reduce systematic errors. The POCAM concept is based on the principle of an inverted integrating sphere. The main components are LEDs emitting light at several wavelengths and solid-state light sensors e.g. calibrated photodiode or silicon photomultipliers to monitor the emitted light intensity. We report on the current status of the POCAM R&D.

1. Motivation

Building on the success of the IceCube detector, a next generation detector is under design, the *IceCube-Gen2* facility [1, 2]. Two new detectors are under consideration: the Precision IceCube Next Generation Upgrade (PINGU) [3, 4] for determining the neutrino mass ordering and the *IceCube-Gen2* high-energy array [5] for high-statistics observation of astrophysical neutrino sources. Both in-ice detector concepts share a common Digital Optical Module (DOM) design including a calibration system similar to that of the current detector, where all DOMs include LED flashers. This system has been used in IceCube to measure ice properties, DOMs' sensitivity, timing, orientation, and position after deployment. It has provided a level of precision required for the present operation but it lacks an *in-situ* monitoring of the emitted light. LED-to-LED variations, individual DOM sensitivity variations and the optical properties of the refrozen hole ice are the main factors limiting the precision currently obtained in IceCube. An improved calibration system is needed to enable a better understanding of the optical properties of the deep ice and thus significantly reduce systematic effects. We propose here a new calibration device called the Precision Optical CAlibration Module (POCAM).

---

[a] e-mail: `martin.jurkovic@tum.de`

[b] `http://icecube.wisc.edu`

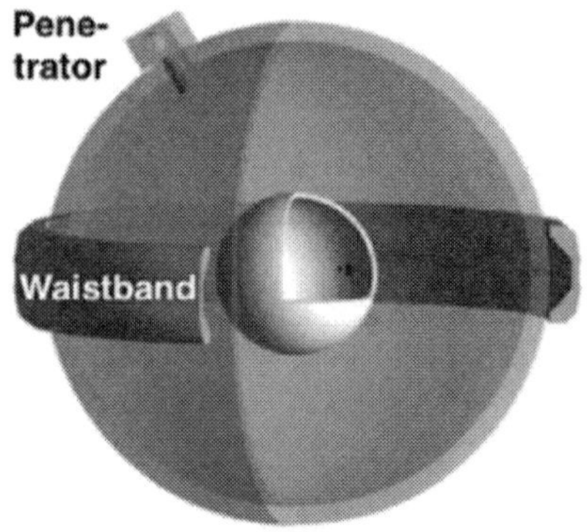

Figure 1. POCAM's baseline geometry consisting of a semi-transparent integrating sphere placed in the middle of the IceCube pressure sphere.

2. POCAM design and performance goals

The POCAM, shown schematically in Fig. 1, will be a self-calibrated homogeneous calibration light source. With the POCAM we aim to verify the energy scale, and measure the energy resolution at the few percent level. Here the energy scale verification means the comparison of the expected number of photoelectrons (NPEs) to the NPEs measured by all DOMs' PMTs. Additionally, an even more precise understanding of the optical ice properties and DOM sensitivity will be provided.

In both in-ice detector designs, the POCAM will be located on the communication/power cable alongside the DOMs. For the *IceCube-Gen2* high-energy array, POCAM's requirement on the homogeneity can be weakened as compared to PINGU because the inter-string distance on the order of 200–300 m is much larger than the effective scattering length of up to ~ 70 m in the clearest ice [6]. Thus, photons emitted by a bright source will undergo several scattering processes before reaching DOMs at nearby strings. For PINGU, the distance between strings will be 22 m only. This distance is comparable or even smaller than the effective scattering length in the ice below the dust layer [6]. Therefore, photons will propagate mostly directly from the source to the receiving DOMs at the closest strings. Given the detector symmetry (Fig. 2) in the azimuthal direction and the POCAM's random orientation along the string axes after deployment, we aim for a homogeneous ice illumination on the order of 1% in the azimuth direction. Because the PINGU baseline geometry has an up-down asymmetry due to the DOMs' asymmetric construction, with a single downward-facing PMT in the lower hemisphere, we allow an inhomogeneity in illumination in zenith direction on the order of 10%.

2.1 POCAM components

The POCAM (Fig. 1) is a homogeneous, intense, and fast light source incorporating an *in-situ* light monitoring system, placed in the IceCube pressure sphere, a housing that will allow POCAM's safe deployment and operation in the deep Antarctic ice. As a light source for the POCAM we consider a matrix of monochromatic fast-switched LEDs driven by the Kapustinsky circuit [7] covering a wavelength range $370 < \lambda/\text{nm} < 500$. Alternatively, the feasibility of using laser diodes will be investigated, which have become affordable in recent years. The light emitted by a discrete light source will be reflected on the inner wall of a diffusing sphere made from a highly reflective polytetrafluorethylene (PTFE) [8] to generate a homogeneous ice illumination. For the construction of the diffusing sphere we consider two options: (a) a multi-port configuration made of a PTFE with wall thickness > 1 cm having equidistantly distributed holes (ports) through which photons can escape into the ice and (b) a semi-transparent configuration having a wall thickness of ~ 2 mm allowing photons to

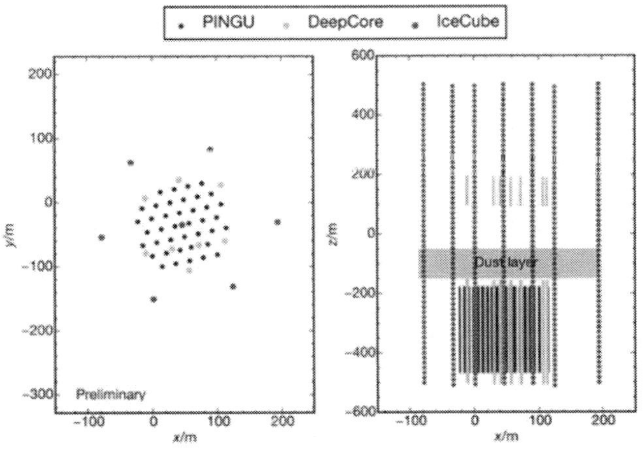

Figure 2. Top (left) and side (right) view of the PINGU baseline geometry showing the position of IceCube, DeepCore, and planned PINGU strings. The $z = 0$ m position coresponds to a depth of 1948.07 m below the surface.

Table 1. POCAM configurations: integrating sphere diameter D, number of ports n, port opening angle α, effective photon transparency T, photon absorption A, inhomogeneity for three different zenith angle zones and decay constant τ. Configurations *C3* and *C4* correspond to the semi-transparent setup.

Config.	D/cm	n	$\alpha/°$	T	A	Inhomogeneity			τ/ns
						$\theta < 60°$	$60° \leq \theta \leq 120°$	$\theta > 120°$	
C1	24	768	1	1.46%	61.0%	11.9%	18.5%	11.6%	26.4
C2	6	768	1	1.46%	74.4%	14.7%	56.4%	10.3%	8.1
C3	24	–	–	–	50.3%	12.4%	14.5%	12.4%	13.1
C4	6	–	–	–	50.7%	14.3%	61.3%	10.3%	3.1

escape into the ice through the wall with a probability of $\sim 2.5\%$. The results of the integrating sphere geometry optimization are summarized in the section 3, and more details can be found in [9]. For the continuous *in-situ* monitoring of the emitted light intensity with a precision of about 1% we are going to implement fast solid state detectors such as photodiodes or silicon photomultipliers. For the housing of the POCAM the well-tested IceCube pressure sphere is used. This choice reduces the homogeneity of the POCAM via the presence of the penetrator, the waistband/harness, and the cable shadowing. It enables an easy integration of the POCAM into the standard deployment operation.

3. Results

To optimize POCAM's geometry we implemented a dedicated Geant4 [10] simulation, for details see [9]. Here we report preliminary results on homogeneity and timing, comparing the multi-port and semi-transparent configuration of the integrating spheres with 6 and 24 cm diameters, see Table 1. Properties of all photons leaving POCAM were recorded in 20 cm distance from POCAM's centre. The average photon emission direction in azimuth direction for different zenith angle bands θ with $\Delta\theta = 1°$ is shown in Fig. 3 (left) together with the time profile of the photon emission (right). All configurations show pronounced decrease in photon density at zenith angles around $\theta = 30°$ and $\theta = 90°$ caused by the shadow of the penetrator and the waistband. Generally, these shadows get smaller but more pronounced for the configurations with smaller integrating spheres. In the regions without any shadow, e.g. for $\theta > 120°$, the overall inhomogeneity goes down to 10.3% for configurations *C2* and *C4*. In narrow zenith bands the semi-transparent configuration *C4* has the smallest inhomogeneity on the order of 1.8%. This configuration shows in our simulation not only the most homogeneous illumination, but also the fastest time response with a decay constant $\tau_4 = 3.1$ ns.

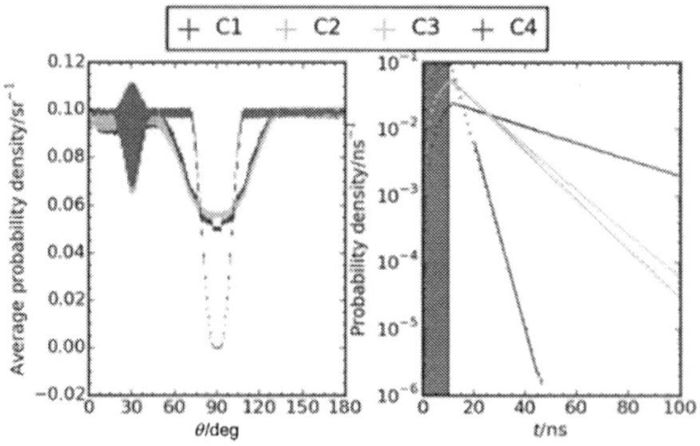

Figure 3. Left: average photon emission direction in azimuth direction for different zenith angle bands. The error bars show the standard deviation. Right: emission time profile with an exponential distribution fitted to each profile for $t > 20$ ns (solid line). The shaded area indicates the rectangular LED time profile.

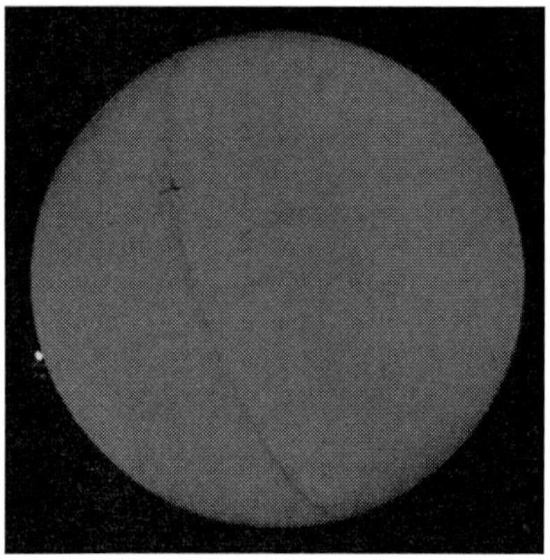

Figure 4. First prototype of the semi-transparent integrating sphere for the POCAM manufactured from commercially available PTFE with 2 mm wall thickness.

Based on our simulation results, the semi-transparent configuration with 6 cm diameter is our new baseline geometry for prototyping the integrating sphere for the POCAM. It is shown in Fig. 4 and consists of two hemispheres held together by thin strings. The hemispheres with a wall thickness of 2 mm were machined out of a massive block of commercially available PTFE. A white LED illuminates the interior of the prototype through a small hole at the pole of the left hemisphere (bright spot). Already by eye, several dark spots close to the equator are clearly visible. All hemispheres produced in this batch show the same pattern. These defects could be explained by the material itself or by the material stress during the machining. For the final prototype we are going to use an optical PTFE-based material from SphereOptics [8]. Nevertheless, this prototype can be used to test and optimize our setup for measuring the light emission from the integrating spheres and from the POCAM.

4. Conclusions and outlook

The Precision Optical Calibration Module (POCAM) is a new isotropic light source featuring *in-situ* light intensity monitoring being developed for the verification of the energy scale and resolution in *IceCube-Gen2*. With a dedicated Geant4 simulation we were able to demonstrate that a semi-transparent configuration of the integrating sphere for the POCAM is favored over a multi-port configuration allowing slightly more homogeneous ice illumination (1.8% inhomogeneity in a narrow zenith band)

and shorter light pulses with $\tau = 3.1$ ns. Following this result we build a first prototype of the integrating sphere from a commercially available PTFE with a wall thickness of 2 mm.

Finally, while the POCAM is being developed for *IceCube-Gen2*, it is a calibration device that has potential use more broadly in the neutrino astronomy community.

References

[1] K. Hanson (IceCube-Gen2), in *VLVnT Workshop, Rome* (2015)
[2] M. Aartsen et al. (IceCube-Gen2) (2014), `1412.5106`
[3] M. Aartsen et al. (IceCube-PINGU) (2014), `1401.2046`
[4] K. Clark (IceCube-Gen2), in *International Cosmic Ray Conference* (2015)
[5] E. Blaufuss (IceCube-Gen2), in *International Cosmic Ray Conference* (2015)
[6] D. Chirkin (IceCube), in *International Cosmic Ray Conference* (2013), `1309.7010`
[7] J. Kapustinsky et al., Nucl. Instrum. Meth. **A241**(2–3), 612 (1985)
[8] *Zenith Polymer® from SphereOptics (*`www.sphereoptics.de`*)*
[9] K. Krings (IceCube-Gen2), in *International Cosmic Ray Conference* (2015)
[10] S. Agostinelli et al. (GEANT4), Nucl. Instrum. Meth. **A506**, 250 (2003)

EPJ Web of Conferences **116**, 06002 (2016)
DOI: 10.1051/epjconf/201611606002

Aspects of the optical system relevant for the KM3NeT timing calibration

Gerard Kieft on behalf of the KM3NeT Collaboration

Nikhef, Science Park 105, 1098 XG Amsterdam, The Netherlands

Abstract. KM3NeT is a future research infrastructure in the Mediterranean Sea housing the large Cherenkov telescope arrays of optical modules for neutrino detection. The detector control and data transmission system is based on fibre optical technology. For timing calibration of the detector signals the optical system is used to send and fan-out an onshore clock signal, derived from a GPS receiver, to all optical modules in the deep sea. The optical modules use this clock signal to time stamp the light pulses detected by the photomultipliers inside the modules. The delay time between the GPS clock on shore and the clock in each optical module is measured with sub-nanosecond precision using a White Rabbit based timing calibration system. The aspects of the optical system relevant for the timing calibration and the quantification of their effect will be presented.

1. Introduction

The final configuration of the arrays of optical modules of the KM3NeT neutrino telescope being installed at the bottom of the Mediterranean Sea will comprise a total instrumented volume of several cubic kilometres to detect Cherenkov light from traversing neutrino-induced leptons [1]. During the first construction phase 31 detection units (DU) will be installed configured as long vertical lines each containing 18 Digital Optical Modules (DOM) and each DOM containing 31 photomultipliers [2]. These DUs will be deployed at two locations in the Mediterranean at depth of 2500 m (KM3NeT-Fr) and 3500 m (KM3NeT-It) and connected to seabed networks of electro-optical cables and junction boxes. Vertical electro-optical cables (VEOCs) run the full length of a DU. At the anchor of each DU a dedicated base-module is mounted for connection of the VEOC to the seabed infrastructure. Main electro-optical cables (MEOCs) connect the seabed infrastructures with the shore station at distances between 40 km (KM3NeT-Fr) and 100 km (KM3NeT-It). For communication between the shore station and each DOM the implemented optical system consists of a single optical broadcast from shore to all the DUs and a DWDM based optical paths over several fibres for transport of the data from the DUs to shore. For the broadcast one wavelength is used to distribute a GPS derived system clock and the slow control data to each DOM. To send data to shore every base-module and each DOM use a dedicated wavelength in a band of at maximum 72 wavelengths per fibre.

Temperature, pressure and wavelength have an effect on the propagation delay of optical pulses over the fibres. The KM3NeT timing calibration depends on the measurement of the delay between the system clock on shore and all the DU-base modules and DOMs. After an introduction on the principle

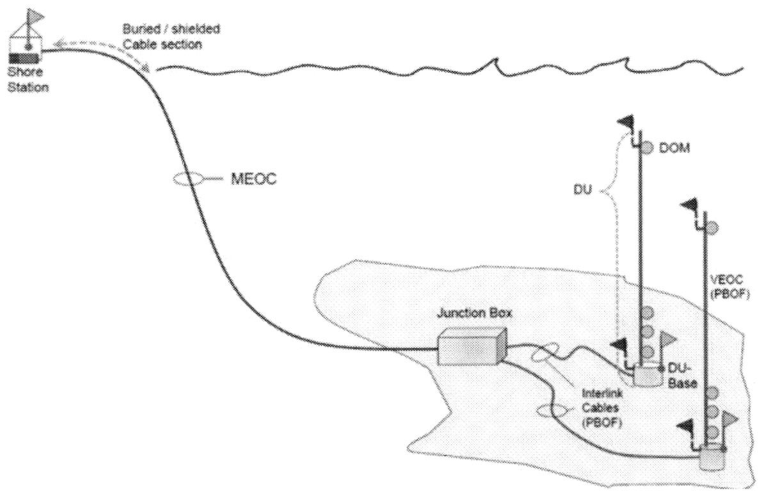

Figure 1. Schematic view of the KM3NeT optical scheme.

of timing calibration in Sect. 2, measurements of the wavelength-dependent propagation delays of the optical fibres in the MEOC are reported in Sect. 3. The effects of pressure and temperature on the propagation delays over the optical fibres in the VEOC of a DU (Fig. 1) are described in Sects. 4 and 5.

2. Principle of the timing calibration

The aim of the timing calibration system is to measure for each DOM the time delay between a GPS derived system clock on shore and the arrival of a timed laser pulse on a reference PMT in each DOM. This delay is measured in 2 steps:

1. White Rabbit (WR) protocol and hardware [3] is used to measure the RTT (Round Trip Time) between the system clock on shore and the clock on the CLB (Central Logic Board) in a DU-base module (between the yellow flags in Fig. 1). When the difference in the propagation time (asymmetry) from shore to DU-base and from DU-base to shore is known, the delay between the system clock on shore and the clock in the DU-base can be derived from the RTT.
2. The delay in the VEOC between the DU-base and each DOM in a DU is measured before deployment during a darkroom test on shore (between the red flags in Fig. 1).

3. Wavelength-dependent propagation delay

The propagation delay of optical pulses over a fibre is determined by the Group Refractive Index (GRI) which depends on the wavelength. For the broadcast signal from shore to a DU-base a different wavelength is used than for the Detector Control (DC) signal from a DU-base back to shore (Fig. 2). Due to the different propagation delay times, the measured RTT between the shore and a DU-base is asymmetric. To derive the propagation delay from shore to a DU-base from the RTT, the asymmetry must be known. Therefore, the wavelength dependencies of the propagation delay of the optical fibres in the MEOCs are measured. For an accurate calculation of the propagation delay of optical signals over fibres, the Group Velocity (GV) at the wavelength of interest must be known. The GV for optical signals in the MEOCs are measured for a set of 8 wavelengths.

Using WR the RTT in a stand-alone reference set-up is measured (Fig. 3a). This is followed by a RTT measurement of the same reference set-up but now including a MEOC fibre (Fig. 3b). By subtracting the

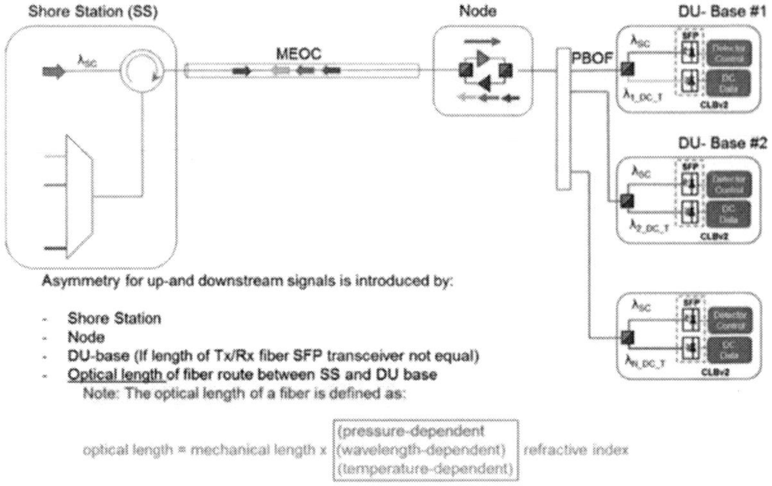

Figure 2. Wavelengths between shore station and DU-base.

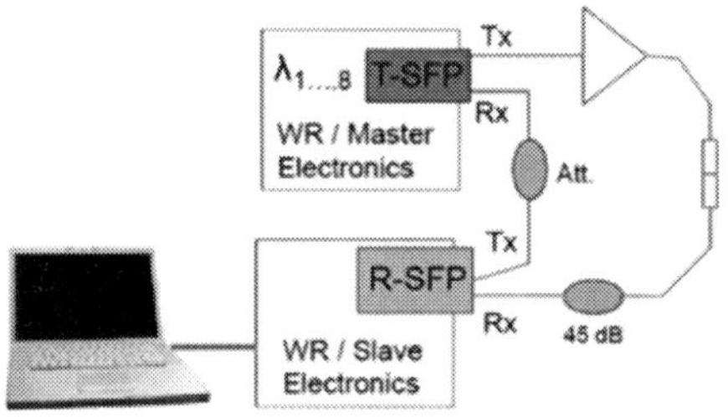

Figure 3a. Reference set-up.

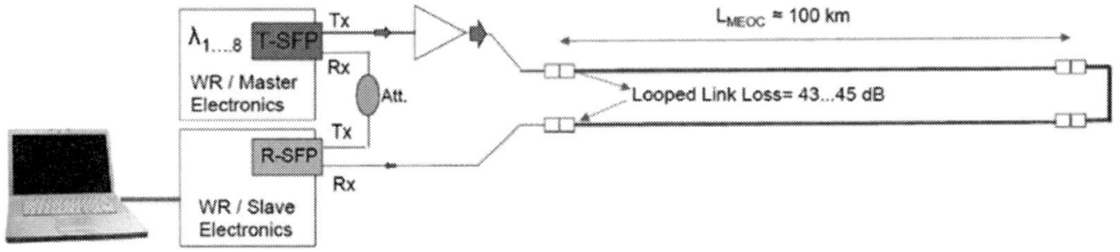

Figure 3b. Reference set-up including the MEOC fibre.

reference RTT from the second RTT, the propagation delay for the specific wavelength in the MEOC fibre can be calculated. For the MEOC at KM3NeT-Fr six fibres, to be used for the SC/DC optical signals, were measured prior to deployment of the MEOC. Since at KM3NeT-It the MEOC was already deployed, the measurement could only be performed on three sets of two fibres in series. Each set of fibres was looped-back in the offshore cable termination.

Figure 4 shows the Relative Group Delay (RGD, the difference in GV) measured for the MEOC of KM3NeT-It. The results represent the average values for 3 different loops of 2 fibres in series. For this MEOC, using a wavelength range from 1530–1544 nm, the RGD for a fibre of ~96.76 km length ranges between ca. 5.2 ns and 5.4 ns. Figure 5 shows the RGD measured for the MEOC of KM3NeT-Fr. The results represent the values for 6 single fibres. For this MEOC, using a wavelength range from 1530–1552 nm, the RGD for a fibre of ~62 km length ranges between ca. 7.2 ns and 8.0 ns.

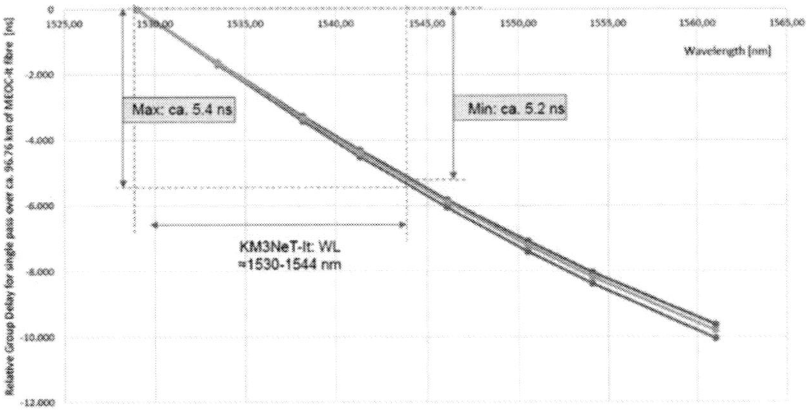

Figure 4. RGD measured for 3 fibre loops in MEOC-Italy.

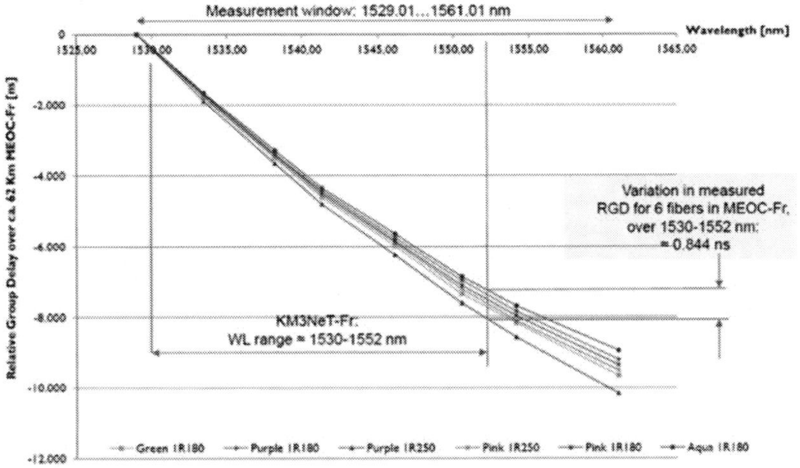

Figure 5. RGD measured for 6 fibres in MEOC-France.

For accurate calculation of the GV for specific wavelengths, all RTT measurements were fitted to a 2nd order (quadratic) Sellmeier function which appeared to give the best fit. The fit results showed regression coefficients ranging between $r^2 = 0.9998$ and $r^2 = 1.0$.

4. Effect of temperature on propagation delay

The timing in a VEOC of a DU is calibrated in a darkroom under "Room Temperature" conditions (e.g. 23 °C). After deployment, the BendBrightXS fibres in VEOC of the DU will be operated in sea water of ~13 °C. This temperature change will affect the propagation delay of optical pulses in the VEOC fibres. In literature, for standard SMF28 fibre with an UV-acrylate primary coating, an effect of ca. 35 till 38 ps/km.°C is reported [4]. In a climate chamber the effect of temperature on the propagation delay for three spools of 4.4 km BendBrightXS UV-acrylate primary coated fibre[1] was measured for different wavelengths. To avoid errors due to thermal expansion of the drum material, each fibre was strain-free spooled on a half-circular drum and thus partially suspended in air. The effect of the temperature on

[1] Provided for free by the company Prysmian, Eindhoven, The Netherlands.

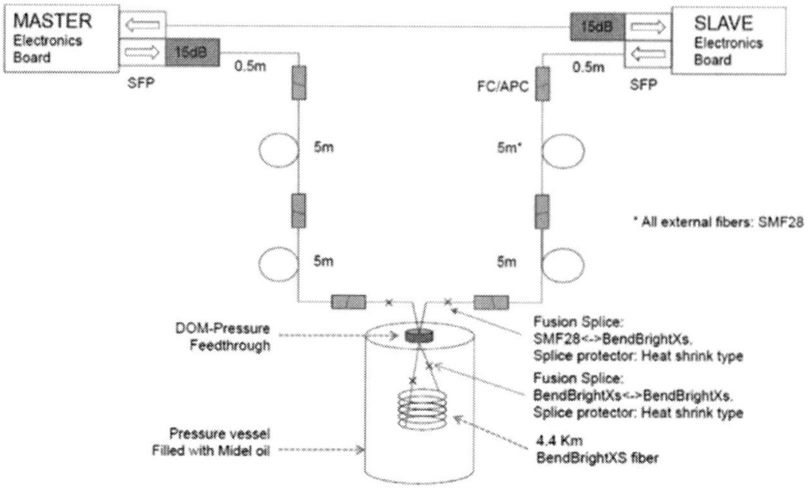

Figure 6. Set-up to measure RTT during pressure test.

the propagation delay for BendBrightXS fibre is measured as 36.7 ps/km.°C and does not significantly depend on the wavelength. Due to a temperature change of ca. $-10\,°\mathrm{C}$ for a DU after deployment, the change in propagation delay between the DU-base and the highest DOM (spanning $\sim$750 m) is: $-10 * 0.75 * 36.7 = -0.275$ ns.

5. Effect of pressure on propagation delay

The VEOC of the DU is calibrated under atmospheric pressure. After deployment of the DU, the BendBrightXS fibres in the VEOC will be operated at a pressure between 25 and 35 MPa (250 and 300 bar) depending on the deployment site. This pressure change will affect the propagation of optical pulses running over the VEOC fibres. The effect of pressure on the same three spools of 4.4 km BendBrightXS fibre was measured in a pressure vessel. The variation of the optical attenuation was measured over a pressure range between 0.1 and 60 MPa. Within an accuracy of 0.02 dB no changes in optical attenuation were observed.

Using the WR protocol to measure the RTT at 1530.33 nm, the pressure-dependent propagation delay of the three fibre spools is determined to be between -1.58 and -1.86 ps/bar.km. In literature for standard SMF28 fibre an effect of ca. -3.4 ps/bar.km is reported [5]. The effect of pressure on the propagation delay of BendBrightXS fibre is about half the value as reported for SMF28 fibre. This is assumed to be due to the different material properties of the BendBrightXS fibre. At a deployment depth of 3500 m the average pressure over the $\sim$750 m length of the VEOC fibre for the highest DOM is 31.25 MPa (312.5 bar). Due to the pressure change after deployment, the change in the propagation delay after deployment with respect to that measured on-shore during the DU calibration is: $-1.86 * 312.5 * 0.75 = -0.436$ ns.

6. Summary and conclusions

For fibres in the MEOCs of the KM3NeT deep sea infrastructures the dependency of the propagation delay as function of the wavelength has been measured for 8 wavelengths. The resulting data have been fitted to a 2nd order Sellmeier function. Using this function the asymmetry in the RTT between shore and offshore due to different wavelengths in the downstream and upstream optical paths, could be calculated. The effect of temperature and pressure changes on the propagation delay in the BendBrightXS fibres

Table 1. Condensed overview of optical effects on the timing calibration.

Component	λ	T	P	Other effects	Remarks
MEOC fibres (50-100 km)	●	●	●		* Temperature interval: between sea floor and shore. * Fibres in MEOC in stainless steel tube, with pressure seal.
Interlink fibres (≈1km)	●	●	●		* Fibres in (horizontal) PBOF cables (Interlink cables) are exposed to constant hydrostatic pressure.
VEOC - fibres	●	●	●	Strain-induced effects in VEOC-fibres	* Fibres in VEOC are exposed to variable hydrostatic pressure and (assumed to be minor) vertical temperature gradients of sea water. * VEOC delay times are calibrated in the lab. After deployment the VEOC is exposed to a lower temperature and higher pressure and assumed not to have a significant change of the fibre strain.
EDFA's in containers	●	●	●	EDFA Pump power settings	* Erbium Doped fibres show generally a higher (also pump power-dependent) Chromatic Dispersion than 1550nm-Dispersion Shifted fibres, but are only 10-40m in length. This effect will be negligible.
Passive components in containers	●	●	●		* Temperature inside containers will be higher than sea water temperature but propagation times are relatively very small.
Transceivers	●	●	●	Input power-dependent time delay	* Internal time delay of a PIN-based SFP receiver varies with ca. 225 ps for a 20 dB change in received optical power. Internal time delay of an APD-based SFP receiver varies < 25 ps over 20 dB range of received optical power.

Legend:
● Effective (known from fiber specifications or measured).
● Less effective /small effect due to relative short cable lengths and stable environment
● Negligible effect (calculated / literature)

of the VEOC has been measured. The resulting temperature and pressure coefficients will be used to calculate after the deployment of a DU in the deep sea, for each DOM in the DU the change of the propagation delay as measured on shore during the darkroom calibration. A condensed overview of the optical effects on the KM3Net timing calibration is shown in Table 1.

This work is part of the research programme of the 'Stichting voor Fundamenteel Onderzoek der Materie (FOM)', which is financially supported by the 'Nederlandse Organisatie voor Wetenschappelijk Onderzoek (NWO)'. The research leading to these results has received funding from the European Community's Sixth Framework Programme under contract n° 011937 and the Seventh Framework Programme under grant agreement n° 212525.

References

[1] KM3NeT, Technical Design Report for a Deep-Sea Research Infrastructure in the Mediterranean Sea Incorporating a Very Large Volume Neutrino Telescope, ISBN 978-90-6488-031-5 (April 2008)
[2] P. Kooijman, et al., Nuclear Instruments and Methods in Physics Research, Section A **626–627** (2011) S139
[3] White Rabbit Specification v2.0 (July 2011) `http://www.ohwr.org/documents/160`
[4] L.G. Cohen and J.W. Flemming: Effect of Temperature on Transmission in Lightguides, The Bell System Technical Journal, **58**, No.4, April 1979
[5] D. Monnier-Bourdin, Système de synchronization picoseconde et métrologie associée, Application au laser megajoule, Thèse de doctorat de l'Université Paris, 1998

EPJ Web of Conferences **116**, 06003 (2016)
DOI: 10.1051/epjconf/201611606003

Calibration methods and tools for KM3NeT

Vladimir Kulikovskiy[a] on behalf of the KM3NeT collaboration

INFN, Laboratori Nazionali del Sud (LNS), via S. Sofia 62, 95123 Catania, Italy

Abstract. The KM3NeT detectors, ARCA and ORCA, composed of several thousands digital optical modules, are in the process of their realization in the Mediterranean Sea. Each optical module contains 31 3-inch photomultipliers. Readout of the optical modules and other detector components is synchronized at the level of sub-nanoseconds. The position of the module is measured by acoustic piezo detectors inside the module and external acoustic emitters installed on the bottom of the sea. The orientation of the module is obtained with an internal attitude and heading reference system chip. Detector calibration, i.e. timing, positioning and sea-water properties, is overviewed in this talk and discussed in detail in this conference. Results of the procedure applied to the first detector unit ready for installation in the deep sea will be shown.

1. Introduction

The KM3NeT detectors [1] are large 3-dimensional arrays of optical modules. Digital optical modules (DOMs) are arranged by modulo eighteen on the flexible lines attached to the sea bottom. Each line has a base module close to the anchor which performs the line power control and the optical link signal amplification. For the neutrino events reconstruction with tracking precision better than 1°, the optical modules need to be synchronized with sub-nanosecond precision and their position determined with better than meter precision. Each DOM and the base have an identical Control Logic Boards which perform the signals processing and transfer, time synchronization and control of the instruments.

The detector is connected to the shore via the main electro-optical cable. The single fiber in the cable (about 40 km in Toulon, and 100 km in Capo Passero) is common for the base and the DOMs. For the base synchronisation a fully compliant White Rabbit[1] system was chosen. White Rabbit is developed to synchronise the clocks between the nodes with a bidirectional point to point optical fiber connection. The system continuously monitors the delay of the fiber connecting the nodes and keeps the node clocks synchronus. The current KM3NeT optical network architecture does not allow White Rabbit synchronisation for the DOMs. A custom protocol was developed for their synchronisation. Time and slow control broadcast from the shore master White Rabbit switch is done to all DOMs, while data return is performed to normal data switch. Synchronisation is achieved thanks to the fixed electronic latencies in the system. The most important delay in the system (main electro-optical cable) is measured continiously by the shore – line base synchronisation system [3]. Time synchronisation

[a] e-mail: `kulikovs@ge.infn.it`
[1] `http://www.ohwr.org/projects/white-rabbit/wiki/`

between different detector components is monitored in situ by light propagation time measurements between light emitters (LED nanobeacons, lasers) and PMTs.

Each DOM is equipped with an internal acoustic piezo detector for the acoustic positioning system [5]. Each line base is equipped with an external hydrophone which has better sensitivity compared to the piezo detectors to reference the line position on the sea bed. The acoustic signal transit time is measured periodically and for three or more emitters position of the receiver can be measured with trilateration. The acoustic emitters are installed in the calibration units [4] that will be placed at known positions on the sea bed. In addition, each calibration unit contains a laser beacon for interline time calibration and water optical properties monitoring, a long-baseline acoustic beacon for DOMs positioning, an hydrophone for their own positioning, and environmental monitoring instruments (conductivity, salinity, temperature, sound velocity, sea currents). The DOM orientation is measured thanks to the internal attitude and heading reference system board.

The key detector components are tested and calibrated at different production steps: PMTs are tested and calibrated in the DarkBox setup. Integrated DOMs are qualified with the GreenBox setups, integrated lines are calibrated with the dark room setups.

2. PMT characterization

For the whole infrastructure more than hundred thousand PMTs need to be installed. The Dark Box setup [2] is designed to provide a facility for a fast automatic pre-calibration of 62 PMTs simultaneously which corresponds to the amount of PMTs needed for 2 DOMs. The setup contains a picosecond laser emitter synchronized with the data acquisition system. The laser signal is optically split to reach each PMT with similar power and equal time delay. In KM3NeT a PMT signal is discriminated using a fixed threshold (about 1/3 of the photoelectron signal), and discriminated signal is digitized by time stamping the arrival time with respect to the detector master clock and measuring its duration (Time over Threshold, ToT). The measurement procedure consists of three main phases: HV tuning, darkening and PMT tests with laser.

The tuning is done with 25 V steps of HV and ToT is calculated by means of a Gaussian fit function to ToT distribution during 1 min of data taking for each step. The proper HV, corresponding to ToT 26.4 ns, is found from a parabolic fit to ToT vs HV plot. This ToT value corresponds to the average ToT value estimated on a subset of PMTs with properly calibrated gain ($3 \cdot 10^6$).

The optimized HV is set to each PMT and the dark rates are monitored during 9 hours. The mean dark count rate is computed by an average over the last 100 s.

The measurements with the laser provide PMT time offsets (mainly due to PMT transit times) and provide an estimation of spurious pulses such as delayed pulses, prepulses and after pulses.

3. DOM qualification

After integration DOMs are qualified using the GreenBox setup (Fig. 1).

Since the PMT dark rates can vary after integration due to the contact with structure, glass and gel, their dark rates are rechecked. Also the calibration of the PMTs is checked with a ToT measurement using the dark noise signals.

Radioactive potassium decay in the glass and the gel can be simultaneously seen by the neighbour PMTs in the DOM. This feature is used to verify the PMT mapping and to perform inter PMT time calibration. The example of differences between hit times on two PMTs is shown in Fig. 1. The gaussian peak due to ^{40}K is centered at 0 for properly intercalibrated PMTs. The area of the peak is proportional the rate of the decays seen simultaneously by two PMTs and it is dependent on the angular distance between them.

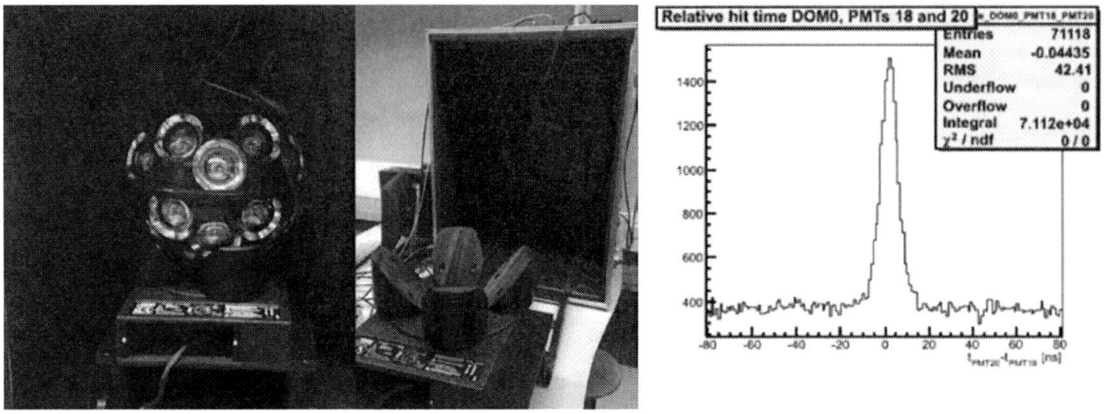

Figure 1. Left: GreenBox images. Right: Distribution of differences between hit times on two PMTs.

The GreenBox is equipped with a laser source which is split to four reference PMTs for the time offsets measurement. This measurement is used later for the time calibration of the line using the same reference PMTs. For the acoustic piezo qualification, the acoustic emitter is placed in the GreenBox. The piezo sensitivity to the signal that will be used for the acoustic positioning is checked.

To calibrate the attitude and heading reference system sensors, inside their working environment, the DOM is placed on a rotating table inside the GreenBox. Three-axis magnetometers can be calibrated by rotating the DOM in orthogonal planes to find the offsets due to magnetic components inside the DOM. The Earth magnetic field designs a sphere in the magnetometers coordinate system during the rotation of the DOM. The sphere is shifted in the presence of the internal magnetic field rotating with the DOM. This offset is found by fitting the rotation measurements with a sphere assumption.

Environmental conditions inside the DOM are checked with an analog pressure manometer. Temperature and humidity are monitored with digital instruments integrated to the slow control system. The integrated LED emitter (nanobeacon) is checked by PMT rate monitoring. DOMs passing the qualification are sent to the line integration sites.

4. Line qualification

All the integrated line components are qualified in a dark room setup. Dark room uses laser sources split into 18 equal length channel fibers arriving to the reference PMTs with the same position inside the DOM, one per each DOM. The delay of the laser signal is measured for each DOM to perform the line time calibration (Fig. 2). The delay of about 200 ns between DOMs corresponds to about 40 m of the optical fiber connecting the DOMs. The total delay for the light detection by reference PMTs of each DOM is calculated using known laser system propagation time and the cable length connecting the line to the test shore station. After the deployment, this delay will be increased by the known delay introduced by the underwater infrastructure.

The base is calibrated in the laboratory with two short 5 m cables connecting it to the shore station (time spread due to the different fiber lengths is less than 20 ps). Time delays in the electronics at the base and at the shore station are calculated and inserted in the White Rabbit calibration tables. These delays ensure that the system estimates properly the cable length and that the reference signals (pulse per second) of the CLB and the shore station are synchronised (Fig. 2). The same calibration is repeated with reference CLB which has a frozen firmware to monitor possible delay changes due to CLB firmware upgrades in the base or in the shore station.

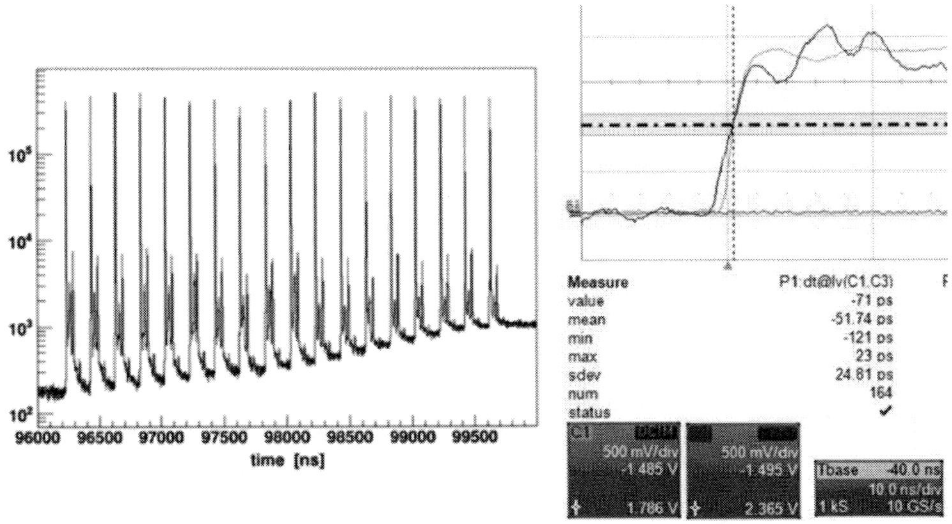

Figure 2. Left: laser photons detection time for the reference PMTs. Each peak corresponds to one DOM. Right: pulse per second signal for the shore station (yellow) and the line base (blue) synchronised at sub-ns level.

The dark room is equipped with several synchronized acoustic emitters which can be placed nearby the DOMs. The delay of the acoustic signal arrival is measured for DOM's piezos. For this purpose a correlation analysis is performed in which the emitted signal with a known shape is compared to the detected signal plus noise.

5. Calibration after deployment

Several instruments will be used to perform the verification and adjustment of the time calibration in-situ. ^{40}K decay in water will be used for PMT time delay calibration between PMTs in the same DOM. The nanobeacons installed in each DOM provide light detection from one DOM to the neighbor DOMs for their inter calibration. Vertical atmospheric muons provide alternative inter-DOM calibration. Laser beacons installed on the Calibration Units provide light detection by DOMs of different lines for the time calibration between the lines. Positioning of the DOMs will be performed thanks to the acoustic instruments installed in the DOMs, DU bases and calibration units.

6. Summary

The KM3NeT detector is under construction. Detector components are qualified and calibrated on different production stages: PMT delivery, DOM integration, line integration, deployment and in-situ operation. These procedures were tested and optimised during first DU line integration.

References

[1] R. Coniglione in this conference, D. Samtleben in this conference
[2] C. M. Mollo, this conference
[3] G. Kieft, this conference
[4] A. Kouchner, B. Baret, this conference
[5] F. Simeone, this conference

EPJ Web of Conferences **116**, 06004 (2016)
DOI: 10.1051/epjconf/201611606004

Moon shadow observation with ANTARES and KM3NeT neutrino telescope

Matteo Sanguineti[1,2,a] and Carla Distefano[3,b]

[1] INFN – Sezione di Genova, via Dodecaneso 33, 16146 Genova, Italy
[2] Dipartimento di Fisica dell'Università, via Dodecaneso 33, 16146 Genova, Italy
[3] INFN – Laboratori Nazionali del Sud (LNS), via S. Sofia 62, 95123 Catania, Italy

Abstract. The ANTARES detector is the largest neutrino telescope currently in operation in the Northern Hemisphere. The search for point-like neutrino sources is one of the main goals of ANTARES, so a reliable way to evaluate the detector pointing performance is needed. The Moon shadow measurement, i.e. the deficit in the atmospheric muon flux in the direction of the Moon induced by absorption of cosmic rays, allows the estimation of the angular resolution and the pointing accuracy. The 2007–2012 ANTARES data set shows a significance of Moon shadowing around 3σ. This is the first measurement of the ANTARES angular resolution and absolute pointing for atmospheric muons using a celestial calibration source. The same approach has been used to estimate also the expected Moon shadow significance with the KM3NeT-ARCA detector, the future cubic kilometre scale detector that will be installed in Sicily.

1. Introduction

Neutrino telescopes have opened a new era of observation of the Universe. Neutrinos are chargeless, weakly interacting particles that can cross dense matter or radiation fields without being absorbed for cosmological distances. Therefore they can provide unperturbed information on the far Universe and the interior of astrophysical sources.

The ANTARES neutrino telescope [1] is designed for the identification of point-like sources of high energy cosmic neutrinos, like starburst galaxies, GRBs, Supernova remnants and AGNs. Following the experience from the ANTARES experiment, the KM3NeT collaboration has started the construction of the next generation deep-sea neutrino telescope. It will be a cubic kilometre scale detector with several hundreds of detection units.

The pointing accuracy and the angular resolution of these detectors are critically important for the detection of point-like sources and a proper way to evaluate these performances is needed.

A deficit in the event density of the atmospheric muon flux corresponding to the direction of the Moon disk is expected due to the absorption of cosmic rays by the satellite. Several experiments, like

[a] e-mail: matteo.sanguineti@ge.infn.it
[b] e-mail: distefano_c@lns.infn.it

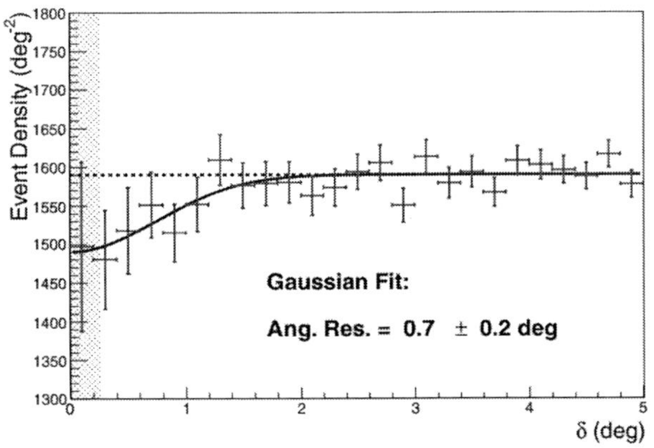

Figure 1. Event density of muons after selection cut versus the angular distance from the Moon centre.

CYGNUS [2], TIBET [3], CASA [4], MACRO [5], SOUDAN [6] , ARGO [7] and IceCube [8], used the so-called Moon shadow effect to test the pointing performance of the detector.

In this work we exploit this technique to measure the ANTARES angular resolution for atmospheric down-going muons and we show the expected performances of the KM3NeT-ARCA detector.

2. Detection of the Moon shadow with ANTARES

The MUPAGE [9] code has been used to simulate the atmospheric muon events, where geo-magnetic deflection is not taken into account. The correction of the muon trajectory is negligible at detector level because only low energy muons (absorbed before reaching the detector) are strongly deflected [10], so the geo-magnetic effect can be neglected in this analysis. The detector response to the Cherenkov light emitted by the muon tracks is simulated [11]. Finally a robust track fitting procedure based on a maximum likelihood method is used to reconstruct the direction of the atmospheric muon tracks [12]. The shadowing effect is simulated rejecting the muons generated within the Moon disk ($R_{Moon} = 0.259°$). The simulation live time is 2080 days in this data analysis (years 2007–2012).

The region of the sky around the Moon centre is divided in concentric rings with increasing radius (ring size = 0.2°). Event tracks detected when the Moon is above the Horizon and reconstructed as down-going are selected.

A test statistic function t is defined as:

$$t = \sum_{rings} \frac{(n_m - n_{exp,NO\_Moon})^2}{n_{exp,NO\_Moon}} - \frac{(n_m - n_{exp,Moon})^2}{n_{exp,Moon}}, \tag{1}$$

where the sum is over all the rings around the Moon centre; n_m is the number of events detected in a ring, $n_{exp,Moon}$ is the expected number of events in "Moon shadow" hypothesis and $n_{exp,NO_{Moon}}$ is the expected number of events in "no Moon shadow" hypothesis [13].

The significance of the Moon shadow deficit was estimated optimising the event selection using the test statistic t (Eq. (1)). The expected Moon shadow significance with optimized event selection is 2.9σ. The optimized selection criteria were applied to the data showing a 3.1σ significance for the rejection of "No Moon shadow" hypothesis. The event density for selected muons as a function of the angular distance from the Moon centre is shown in Fig. 1.

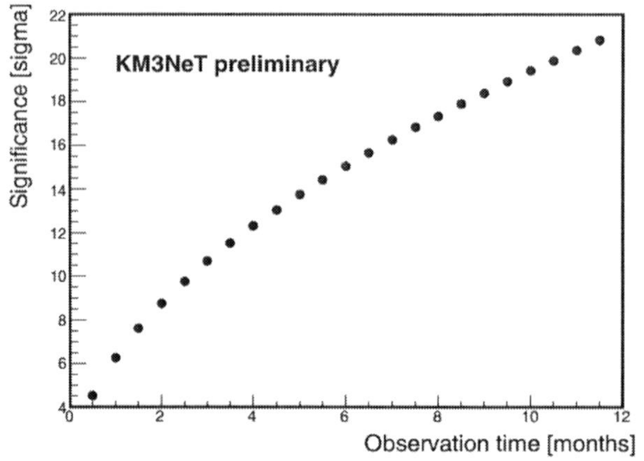

Figure 2. Moon shadow significance as a function of the measure time based on a simulation of the KM3NeT-ARCA detector with one building block.

The detector angular resolution can be evaluated fitting the event density with the formula:

$$\frac{dn}{d\delta^2} = k\left(1 - \frac{R^2_{Moon}}{2\varsigma^2} e^{-\frac{\delta^2}{2\varsigma^2}}\right), \quad (2)$$

where $R_{Moon} = 0.259°$ is the Moon angular radius and δ is the angular distance from the Moon centre. The fit free parameters k and ς are respectively the off-source density level and the detector angular resolution. We have assumed a Gaussian shape for the detector point spread function [6]. From the fit we can estimate the angular resolution: $\varsigma = 0.7° \pm 0.2°$.

3. Simulation of the Moon shadow with KM3NeT-ARCA

The same approach of Sect. 2 has been used to estimate the expected Moon shadow significance with the KM3NeT-ARCA detector.

A single building block of KM3NeT-ARCA has been considered. It is composed of 115 detection units and each unit has 18 digital optical modules (DOMs). The spacing between the DOMs is 36 m, whereas the spacing between the detection units is 90 m.

In Fig. 2 the Moon shadow significance is shown as a function of the measure time.

The expected significance reached in one year is around 21σ. The corresponding plot of event density for selected muons as a function of the angular distance from the Moon centre is shown in Fig. 3. The fitted angular resolution (Eq. (2)) is $\varsigma = 0.43° \pm 0.01°$.

4. Conclusions

The Moon shadow observation is a good way to estimate the pointing performance of neutrino telescopes. It has been observed with a 3.1σ significance in the ANTARES data sample 2007–2012. The Moon shadow profile has been fitted assuming a Gaussian shape for the detector point spread function, hence we derived the angular resolution for the atmospheric muon flux: $0.7° \pm 0.2°$. The simulation of

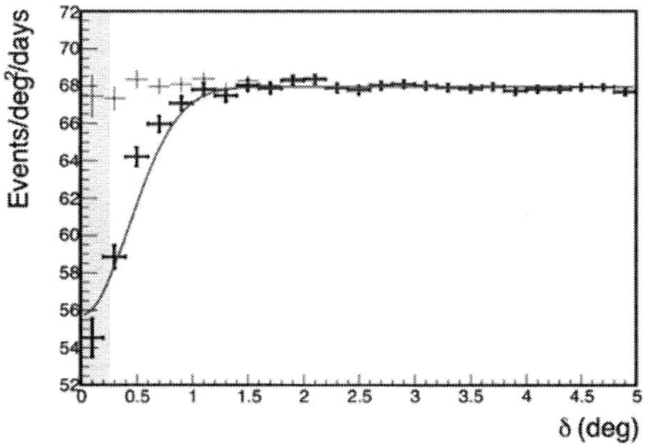

Figure 3. Expected event density of muons after selection cut versus the angular distance from the Moon centre.

the Moon shadow effect with the KM3NeT-ARCA detector predicts a 21σ significance in one year and the expected angular resolution for down-going muons is $0.43° \pm 0.01°$.

References

[1] M. Ageron et al. (ANTARES), Nucl. Instrum. Meth. **A656**, 11 (2011), 1104.1607
[2] D.E. Alexandreas et al. (CYGNUS), Phys. Rev. **D43**, 1735 (1991)
[3] M. Amenomori et al. (Tibet AS gamma), Phys. Rev. **D47**, 2675 (1993)
[4] A. Borione et al., Phys. Rev. **D49**, 1171 (1994)
[5] M. Ambrosio et al. (MACRO), Phys. Rev. **D59**, 012003 (1999), hep-ex/9807006
[6] J.H. Cobb et al. (Soudan 2), Phys. Rev. **D61**, 092002 (2000), hep-ex/9905036
[7] B. Bartoli et al., Phys. Rev. **D84**, 022003 (2011), 1107.4887
[8] M.G. Aartsen et al. (IceCube), Phys. Rev. **D89**, 102004 (2014), 1305.6811
[9] G. Carminati, A. Margiotta, M. Spurio, Comput. Phys. Commun. **179**, 915 (2008), 0802.0562
[10] C. Distefano (ANTARES), Nucl. Instrum. Meth. **A626-627**, S223 (2011)
[11] J. Brunner (ANTARES), *Antares simulation tools*, in *Technical aspects of a Very Large Volume Neutrino Telescope in the Mediterranean Sea. Proceedings, Workshop, VLVnuT Workshop, Amsterdam, Netherlands, October 5–8, 2003* (2003), pp. 109–113, http://www.vlvnt.nl/proceedings/
[12] A. Heijboer, Ph.D. thesis, Amsterdam U. (2004), http://www.nikhef.nl/t61/thesis/index.html
[13] M. Sanguineti, *Moon shadow observation with the ANTARES neutrino telescope*, in *Proceedings, 34th International Cosmic Ray Conference (ICRC 2015)* (2015), Vol. ICRC2015, http://pos.sissa.it/cgi-bin/reader/conf.cgi?confid=236

EPJ Web of Conferences **116**, 06005 (2016)
DOI: 10.1051/epjconf/201611606005

LED based calibration systems of the Baikal-GVD neutrino telescope

A.D. Avrorin[1], A.V. Avrorin[1], V.M. Aynutdinov[1], R. Bannash[7], I.A. Belolaptikov[2], D.Yu. Bogorodsky[3], V.B. Brudanin[2], N.M. Budnev[3], I.A. Danilchenko[1], G.V. Domogatsky[1], A.A. Doroshenko[1], A.N. Dyachok[3], Zh.-A.M. Dzhilkibaev[1], S.V. Fialkovsky[5], A.R. Gafarov[3], O.N. Gaponenko[1], K.V. Golubkov[1], T.I. Gress[3], Z. Honz[2], K.G. Kebkal[7], O.G. Kebkal[7], K.V. Konischev[2], A.V. Korobchenko[2], A.P. Koshechkin[1], F.K. Koshel[1], A.V. Kozhin[4], V.F. Kulepov[5], D.A. Kuleshov[1], V.I. Ljashuk[1], M.B. Milenin[5], R.A. Mirgazov[3], E.R. Osipova[4], A.I. Panfilov[1], L.V. Pan'kov[3], E.N. Pliskovsky[2], M.I. Rozanov[6], E.V. Rjabov[3], B.A. Shaybonov[2], A.A. Sheifler[1], M.D. Shelepov[1], A.V. Skurihin[4], A.A. Smagina[2], O.V. Suvorova[1], V.A. Tabolenko[3], B.A. Tarashansky[3], S.A. Yakovlev[7], A.V. Zagorodnikov[3], V.A. Zhukov[1], and V.L. Zurbanov[3]

[1] Institute for Nuclear Research, 60th October Anniversary pr. 7A, Moscow 117312, Russia
[2] Joint Institute for Nuclear Research, Dubna 141980, Russia
[3] Irkutsk State University, Irkutsk 664003, Russia
[4] Skobeltsyn Institute of Nuclear Physics MSU, Moscow 119991, Russia
[5] Nizhni Novgorod State Technical University, Nizhni Novgorod 603950, Russia
[6] St. Petersburg State Marine University, St. Petersburg 190008, Russia
[7] EvoLogics GmbH, Berlin, Germany

Abstract. Baikal-GVD is a cubic-kilometer scale neutrino telescope, which is currently under construction in Lake Baikal. GVD will consist of an array of optical modules arranged in clusters of strings. The first GVD-cluster has been deployed and put in operation in April 2015. We describe equipment and methods for the calibration of the first GVD-cluster and discuss the accuracy of the calibration procedures.

1. Introduction

Baikal-GVD will be a kilometer-scale high-energy neutrino observatory, which is currently under construction in the southern basin of Lake Baikal [1–5]. The main scientific goal of GVD is to map the high-energy neutrino sky in the Southern Hemisphere including the region of the Galactic center. The Baikal-GVD will have a modular structure: it will consist of functionally independent setups – clusters of vertical strings of optical modules (OM). Each optical module comprises a light sensor (PMT R7081-100), which detects Cherenkov radiation produced by the relativistic charged particle moving through water. The first GVD-cluster has been deployed and commissioned in April 2015. It comprises 192 OMs arranged along 8 strings. Each string includes a chain of 24 OMs spaced uniformly at depths 900–1250 m. Event reconstruction with the GVD-cluster is based on the measurements of time and charge of the pulses detected with the OMs. Except the light sensors located in the OM, measuring channels include 200 MHz ADCs that provide measurements of the pulse parameters: time and charge. The measuring channels have slightly different time and amplitude characteristics: time offsets and sensitivity. The precise measuring of these characteristics (channel time and charge calibration) is the necessary condition of reliable event reconstruction. In this article we describe equipment and methods for the calibration of the first GVD-cluster and discuss the precision of calibration procedures.

2. LED based calibration system

Methods of the GVD-cluster calibration are determined by the design of the data acquisition system (DAQ) of the installation [6–9]. The DAQ of the GVD-cluster is constructed from three basic building blocks: optical modules (OMs), sections of OMs (main detection units), and strings. OMs contain light sensors and pulse amplifiers. The section includes 12 OMs connected to the central 12-channel 200 MHz ADC unit by 90 meter coaxial cables. Each string consists of two sections of OMs. The section electronics produces the section trigger request that is transferred to the cluster DAQ centre. The electronics of the DAQ centre generates a global trigger for all sections of the GVD-cluster. The global trigger produces a stop signal simultaneously for all ADC channels and initiates the "section event". Each section event contains waveform data for 12 ADC channels, global trigger number and local time. The trigger number and local time of the section provide the possibility to combine all section events into a unified cluster event.

There are three basic calibration parameters for such a DAQ organisation: relative time offset of the channels, relative time offset of the sections and channel sensitivity. Correspondingly, the calibration of the array recording system consists of the following procedures: charge and time calibrations of the measuring channels, time calibration of the sections, and channel charge calibration. Channel calibration procedures are based on the usage of OM's internal calibration LEDs L-7113 (two LEDs inside each OM). The LED beacons are used for the time calibration of the sections. The dominant wavelength of LEDs is 470 nm, the LED pulses have a width of $\sim$5 ns (FWHM). LED light intensity can be regulated in the range from $\sim$1 up to about 10^8 photons. The cross talk between LED channels is less than $<1\%$.

2.1 Time calibration of the channels

During the time calibration of the channels the relative time offsets of signals recorded by the OMs are derived. The signal delay of a channel is formed by the internal PMT delay and the delay caused by signal passing through an about 90 m long cable connecting OM and ADC unit. Cable delays are measured once in the laboratory. They are the same during array operation. The PMT delay depends on the power voltage and thus it requires regular calibration during array operation. There is a specialized test pulse which is generated by the OM controller and is delivered to the point of signal creation in the PMT preamplifier. The test pulse initiation is synchronized with the start time of the LED. From the measured difference between arrival times of LED signal and test pulse the PMT delay is obtained (see Fig. 1).

The measured PMT delays for the GVD-strings #4 and #5 as a function of PMT power voltages and the distribution of the OM cable delays (RMS = 0.6 ns) are shown in Fig. 1. The sum of the PMT and cable delays forms the channel time offset: T_{TST}. A straightforward test of the obtained channel time offsets is to pulse the internal LED in an OM and measure the arrival time of photons at this OM and the OMs above. The distances between adjacent OMs are known with good accuracy, and the differences between expected and measured arrival times give the relative time offsets of two channels: dT_{LED}. A similar parameter dT_{TST} can be obtained using results of the measuring of the PMT and cable delays. The distribution of the channels with respect to the differences between dT_{LED} and dT_{TST} is shown in Fig. 2 for the GVD-strings #4 and #5 (mean = 0.6 ns, RMS = 1.8 ns). The relative time offsets measured with

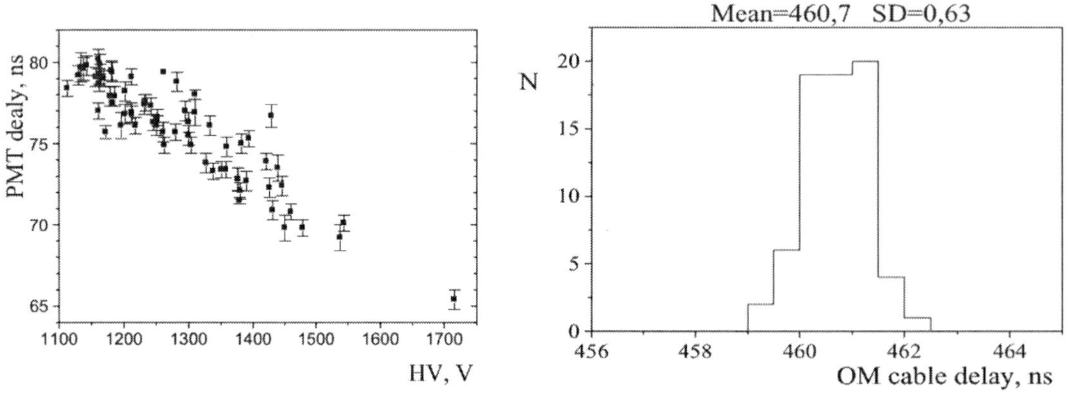

Figure 1. The PMT delays in dependence of the PMT voltages (left) and distribution of the OM cables on the delay (right).

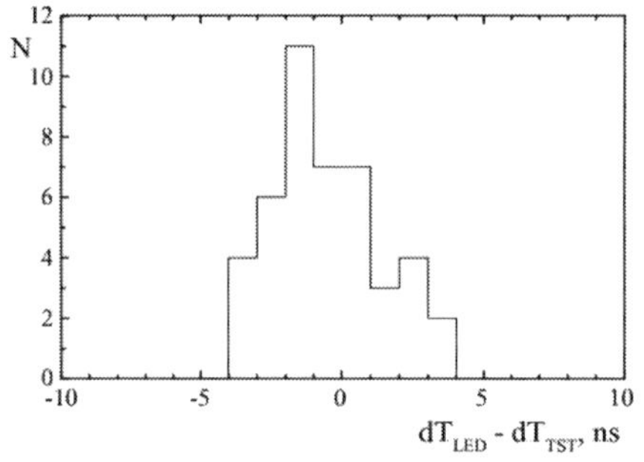

Figure 2. Distribution of channels with respect to the parameter $dT_{LED} - dT_{TST}$.

LEDs are consistent with the results of the channel time calibration. The uncertainties of the calibration coefficients are less than 2 ns.

2.2 Time calibration of the sections

OM LEDs are faced vertically and, therefore, cannot be used for the time calibration of sections located on different strings. For this purpose LED beacons were installed. A LED beacon contains 12 LEDs. Six of them point upward and six horizontally. The LED flashes can be detected simultaneously by OMs of different sections. The GVD acoustic positioning system provides information about the OM positions. The relative time offset of the sections is calculated as the time difference between expected time delay between OM signals dT_{EXP}, and the measured one, dT. An example of time calibration of the sections located on string #2 and string #8 is presented in Fig. 3. The calibration data contain information obtained with 9 pairs of OMs: the mean relative time offset is 142 ns, the RMS is 1.7 ns. The section calibration accuracy is about 2 ns including channel calibration uncertainties.

2.3 Charge calibration

We apply a standard procedure for the charge calibration of PMTs, based on an analysis of the single photoelectron spectrum (s.p.e.). In this calibration mode the pulses of two LEDs of an OM are used. The intensity of the first LED is fitted to provide a detection of s.p.e. signals with a detection probability of about 10%. These pulses are used to measure the s.p.e. distribution of channel signals. Pulses of the

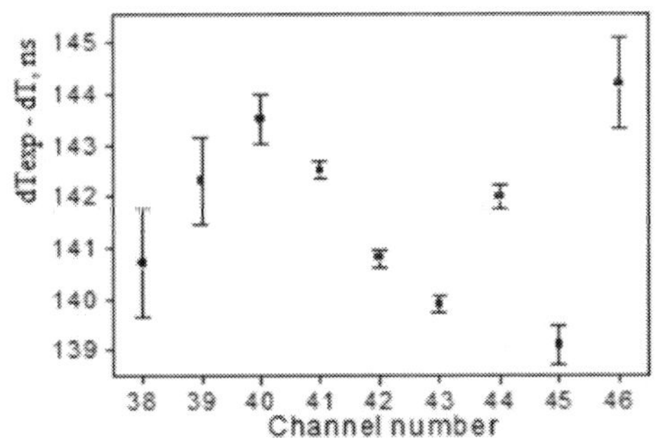

Figure 3. Results of time offset measurement for string #2 vs string #8.

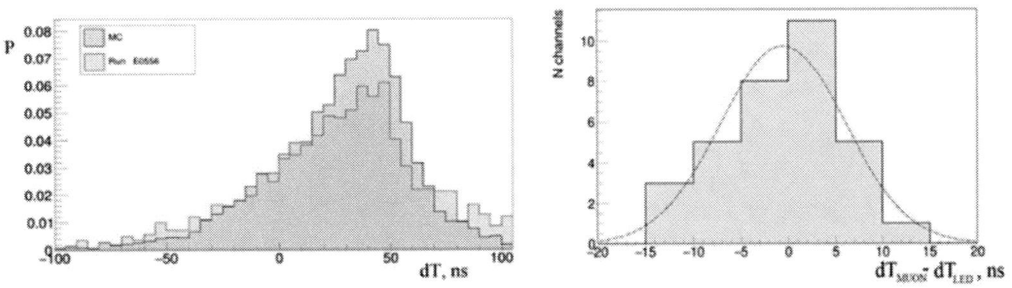

Figure 4. Experimental and modeled time difference distributions for adjacent channels obtained with muons (left), and distribution of the measuring channels (string #4 and #5) on differences of pairwise time offsets dT obtained with muons and LEDs (right).

second LED with intensities corresponding to about 50 p.e. PMT's signal are delayed by 500 ns and are used as a trigger to suppress background signals with small amplitudes initiated by PMT dark current, as well as light background of the lake's deep water.

3. Time calibration crosscheck with atmospheric muons

Atmospheric muons can be utilized to verify the data acquired via the LED calibration system. The relative time offset between two measuring channels can be obtained via the time difference of signals from down-going muons on neighboring OMs. In this case, pairwise channel delays are equal to the time shift between the experimental and modeled muon time difference distribution. An example of experimental and modeled time difference distributions after time shift optimization is presented in Fig. 4. Comparing pairwise channel delay differences acquired by LED flashers and by muons yields a near-normal distribution with an average value of 0.4 ± 1ns (see Fig. 4), that gives an estimation of the systematic uncertainties of the calibration procedure.

4. Conclusion

The Baikal-GVD calibration system comprises LED flashers installed in each optical module as well as LED beacons. The OM LED flashers provide time and amplitude calibration of the measuring channels, the LED beacons are used for time calibration of the GVD-sections. Comparison of two independent calibration methods gives a calibration accuracy of about 2 ns. A time calibration crosscheck with atmospheric muons does not show any systematic biases.

This work was supported by the Russian Found for Basic Research (grants 13-02-12221, 14-02-00175, 14-02-00972).

References

[1] A. Avrorin, et al., Astronomy Letters **37**, 7 (2011)
[2] A. Avrorin et al., Nucl. Instr. and Meth. **A626**, 6 (2011)
[3] A. Avrorin et al., Nucl. Instr. and Meth. **A630**, 4 (2011)
[4] A. Avrorin et al., Nucl. Instr. and Meth. **A639**, 3 (2011)
[5] A. Avrorin et al., Instruments and Experimental Techniques **54** (2011)
[6] A. Avrorin et al., Nucl. Instr. and Meth. **A692**, 7 (2012)
[7] A. Avrorin et al., Nucl. Instr. and Meth. **A693**, 10 (2012)
[8] A. Avrorin et al., Nucl. Instr. and Meth. **A725**, 4 (2013)
[9] A. Avrorin et al., Instruments and Experimental Techniques **3**, 12 (2014)

EPJ Web of Conferences **116**, 06006 (2016)
DOI: 10.1051/epjconf/201611606006

The Calibration Units of the KM3NeT neutrino telescope

B. Baret[1, a], P. Keller[2], and M. Lindsey Clark[1] for the KM3NeT Collaboration

[1] APC, Université Paris Diderot, CNRS/IN2P3, CEA/Irfu, Observatoire de Paris, Sorbonne Paris Cité, France
[2] CPPM, Aix-Marseille Université, CNRS/IN2P3, Marseille, France

Abstract. KM3NeT is a network of deep-sea neutrino telescopes to be deployed in the Mediterranean Sea that will perform neutrino astronomy and oscillation studies. It consists of three-dimensional arrays of thousands of optical modules that detect the Cherenkov light induced by charged particles resulting from the interaction of a neutrino with the surrounding medium. The performance of the neutrino telescope relies on the precise timing and positioning calibration of the detector elements. Other environmental conditions which may affect light and sound transmission, such as water temperature and salinity, must also be continuously monitored. This contribution describes the technical design of the first Calibration Unit, to be deployed on the French site as part of KM3NeT Phase 1.

1. Introduction

The KM3NeT Collaboration is currently constructing the first phase of a next-generation neutrino telescope on two sites in the Mediterranean Sea: KM3NeT-Fr near Toulon, France, and KM3NeT-It near Capo Passero in Sicily [1]. Each site will host a three-dimensional array of thousands of photosensors that will detect the Cherenkov light resulting from neutrino interactions in the vicinity of the detector. The French site will be mainly dedicated to the study of oscillation effects with ~GeV atmospheric neutrinos (ORCA) while the Italian one will focus on high-energy (TeV-PeV) neutrino astronomy (ARCA). The KM3NeT detector relies on a novel design for its Digital Optical Modules (DOMs) [2] which are distributed along vertical flexible strings anchored to the seabed. The anchor (or base) at the bottom of the string is the interface with the seabed infrastructure: it supports the interlink cable, equipped with a wet-mateable connector, and the base container, which houses a power converter and dedicated optical components. Each string with its anchor constitutes one detection unit (DU) of the KM3NeT detector.

The performance of a neutrino telescope crucially depends on the accurate time and energy calibration of the detector elements, as well as on the precise knowledge of their position, which is affected by sea currents. Measurements of water properties are also important for evaluating the detection efficiency. The water transparency affects the propagation of Cherenkov light in sea water, and

[a] e-mail: baret@in2p3.fr

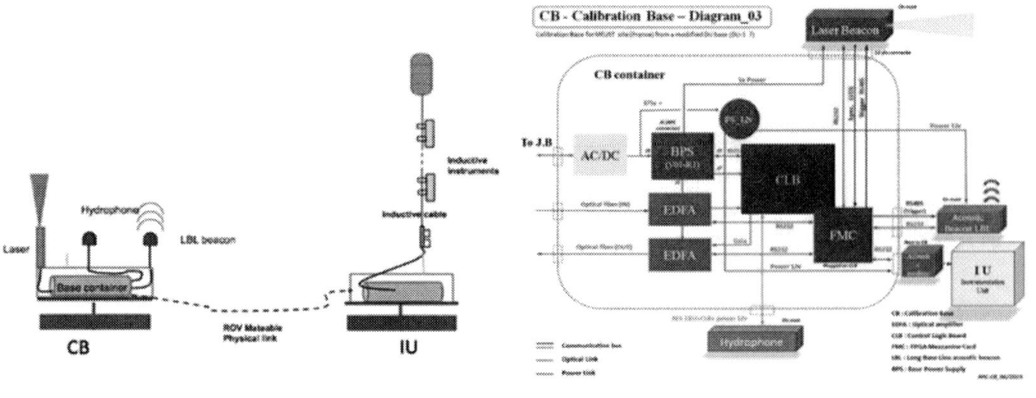

Figure 1. Left: General design of the KM3NeT Calibration Unit. Right: Preliminary electronics scheme for the Calibration Base, with the acoustic option for the communication with the IU.

hence the effective volume and angular resolution. The monitoring of the evolution of these parameters with time and of environmental conditions (temperature, sea currents, bioluminescence...) will be achieved using monitoring tools and probes installed on the detector elements (DOMs and anchors) and on dedicated Calibration Units (CUs) distributed among the detector lines.

The KM3NeT positioning system is based on a network of acoustic transmitters (beacons) and receivers (hydrophones) that will be distributed in the detector installation field, on the CUs (emitter and receiver), DU bases (receiver) and standalone tripods, forming a long-baseline reference system for the determination of the DOM positions by triangulation [3]. The timing calibration will be performed thanks to LED beacons mounted on the DOMs, and to a few laser beacons installed on the CUs [4]. This beacon network will be also used to monitor the water transparency. Other external parameters (sea currents, temperature, salinity...) will be continuously monitored by commercial probes installed on the CUs. Their study is also of great interest for the Earth and Sea Sciences, as already illustrated by joint publications using ANTARES data [5].

2. General description of the Calibration Unit

The general design of a KM3NeT Calibration Unit is shown in Fig. 1. It will comprise two sub-systems: a Calibration Base (CB) and an Instrumentation Unit (IU), powered through the CB. The CB will host the detector calibration devices, namely a laser beacon, a long-baseline acoustic beacon and a hydrophone. The IU consists of an instrumentation base (IB) incorporating an electronics interface board and an inductive modem to be connected to an inductive, semi-autonomous and recoverable instrumentation line (IL) equipped with environmental monitoring instruments distributed in clusters at different depths. The height of the IL will be adapted to the detector configuration (750 m for ARCA and 200–250 m for ORCA are currently being considered). The CB and the IB will be separated by approximately 80 m. The default option for the CB-IU connection is to use a ROV mateable electrical Ethernet connector, but the possibility to use an acoustic modem is also under study, as discussed in Sect. 4.2.

The full CUs equipped with a calibration base and an instrumentation unit will be placed in peripherical positions around the detector, in order to allow easy periodical recovery and redeployment of the IU. In a sparse detector configuration (ARCA), CBs alone can also be distributed more homogeneously inside the detector, thereby ensuring a complete coverage of the whole network of strings. The possibility to attach standard detection lines to the bare CBs is also being considered and taken into account in the CB mechanical design. The first KM3NeT CU to be deployed on the French site will be a full unit with a CB and an IU.

3. The Calibration Base

The CB will consist of a DU base modified so as to host the following instruments on its external frame:

- **a Laser Beacon** for time calibration and measurement of water optical properties. The laser chosen for the calibration of KM3NeT DOMs is the Nd-YAG model STG-03E-040 and its associated control board MLC-03A-BP1 from Teem PhotonicsIt emits at a wavelength of 532 nm (green) and provides high peak power (> 6 kW) pulses of duration < 0.5 ns; it can be triggered at a maximum repetition rate of 4 kHz. The laser beacon equipped with attenuator and diffuser will be mounted in an Aluminium (class 6) cylindrical container with protective coating. One of the sides of the container hosts the connector, while the other one supports the cylindrical quartz rod that serves as optical window.
- **an acoustic receiver (hydrophone)** providing the position of the CB on the seafloor
- **an acoustic long-baseline emitter** for DOM positioning made of a transducer and an Aluminum shielded container hosting the electronics board detailed in [3]
- possibly an **acoustic modem** for connection to the IU, if this solution is retained.

All the electronics, optical and power devices are enclosed in a Titanium cylindrical Base Container attached to the CB frame. This container will in particular incorporate

- **one KM3NeT Central Logic Board (CLB)** [6] **and its FPGA Mezzanine Card (FMC)**, for data acquisition and transmission
- **two optical amplifiers (EDFA)** to read and write from/to the optical fibers
- **a power conversion board** to power all the CB elements and the IU.

A preliminary synoptic view of the electronics components and their connections to the various CB instruments is shown in Fig. 1.

4. The Instrumentation Unit

The Instrumentation Unit is composed of one Instrumentation Base container (IB) laid on the seabed and of an inductive Instrumentation Line (IL), maintained vertical by a buoy, and supporting various autonomous instruments powered on internal batteries. The IU is recoverable in order to allow the change of batteries and the periodical recalibration required by some of the instruments. The design is directly derived from the ALBATROSS project (*Autonomous Line with a Broad Acoustic Transmission for Research in Oceanography and Sea Sciences*) which was recently deployed close to the KM3NeT-Fr site. It is totally autonomous and will communicate its data through an acoustic modem to an Instrumented Interface Module (IIM) to be connected to the KM3NeT-Fr node, as part of the European Multidisciplinary Seafloor and water-column Observatory network.

4.1 The Instrumentation Base

The instrumentation Base consists of a Titanium cylindrical container attached to a dead weight, that will contain

- **an Inductive Modem (IM)** foreseen IM to be the Seabird IMM, with a RS232 data interface
- **the power system** for the IMM (either provided by the CB or by an internal battery)
- **the communication interface with the CB.**

Based on the experience of the ALBATROSS project, the possibility to use an acoustic modem instead of an Ethernet cable is currently under study, with the potential advantage of reducing the number of wet mateable connectors and hence the risk of failure at (de)connection.

4.2 The Instrumentation Line

The inductive mooring line is made of a steel wire rope covered by a plastic jacket except on its ends, which are grounded in seawater, allowing a current loop through the cable. Inductive Modem devices couple to this loop inductively along the insulated part of the cable, without direct electrical connection. The signal induced on the line by the instruments is retrieved by a dedicated Seabird inductive cable coupler (ICC). The configuration of the IL will allow the instrumentation of the full water column of the KM3NeT detector. It will comprise clusters of identical instruments, distributed at different depths depending on the detector configuration (ORCA or ARCA):

- **one CTD (Conductivity, Temperature, Depth) probe.** These measurements are used to calculate the sound velocity. The foreseen instrument is the SeaBird SBE 371IMP-0D0-1b.
- **one sound velocimeter** which performs direct measurements of the celerity by measuring the time of flight of an acoustic signal over a known fixed distance. The foreseen velocimeter is the Valeport Midas Mini SVS which will be powered and interfaced to the mooring line via the RS232-inductive coupler SBE44 from Seabird.
- **one Acoustic Doppler Current Profiler (ADCP)** that measures the current profile over a water column whose length depends on the acoustic signal frequency (from ~ 20 m at 1000 kHz to more than 500 m at 75 kHz). This characteristic strongly influences the cost, weight and dimensions of the instrument and different options are still under study.

The measurement periodicity of all these instruments is typically of 20 minutes.

5. Deployment, calibration and maintenance of the Calibration Units

Specific mechanical tools must be designed for the KM3NeT CUs, in particular to ensure safe deployment and recovery of the IU. Base on the ALBATROSS experience, it is foreseen that the IU will be laid on the seabed from the surface boat using the winch. The deadweight of the IU will support acoustic releases that will be activated in order to recover the IL. The IU instruments are in principle provided already calibrated. The foreseen periodicity for recalibration is every one to three years, to be adjusted taking into account the necessity of battery replacement. The IL recovery will be planned during one of the sea operations for DU deployment in order to minimize the costs. It is foreseen that the first CU will be deployed on the KM3NeT-Fr site by 2017, to allow a validation of the calibration procedure with the first lines of the ORCA demonstrator array.

References

[1] KM3NeT Technical Design Report ISBN-978-90-6488-033-9

[2] R. Bruijn and D. Van Eijk [KM3NeT Collaboration], Proc. of the 34rd ICRC conference, ID **1157** (2015)

[3] S. Viola [KM3NeT Collaboration] Proc. of the 34rd ICRC conference, ID **1169** (2015)

[4] M. Bouwhuis [KM3NeT Collaboration], Proc. of the 34rd ICRC2015 conference, ID **1170** (2015)

[5] C. Tamburini et al. [ANTARES Collaboration]; H. van Haren et al. [ANTARES Collaboration], Ocean Dynamics **64** (2014) 507–517

[6] S. Biagi and A. Orzelli [KM3NeT Collaboration], JINST **9** (2014) C12033

EPJ Web of Conferences **116**, 06007 (2016)
DOI: 10.1051/epjconf/201611606007

Characterization benches for neutrino telescope Optical Modules at the APC laboratory

Avgitas Theodore[a], Creusot Alexandre[b], and Kouchner Antoine for the KM3NeT collaboration

Laboratoire APC – Université Paris Diderot – Paris VII, France

Abstract. As has been demonstrated by the first generation of neutrino telescopes Antares and IceCube, precise knowledge of the photon detection efficiency of optical modules is of fundamental importance for the understanding of the instrument and accurate event reconstruction. Dedicated test benches have been developed to measure all related quantities for the Digital Optical Modules of the KM3NeT neutrino telescope being currently deployed in the Mediterranean sea. The first bench is a black box with robotic arms equipped with a calibrated single photon source or laser which enable a precise mapping of the detection efficiency at arbitrary incident angles as well as precise measurements of the time delays induced by the photodetection chain. These measurement can be incorporated and compared to full GEANT MonteCarlo simulations of the optical modules. The second bench is a 2 m × 2 m × 2 m water tank equipped with muon hodoscopes on top and bottom. It enables to study and measure the angular dependence of the DOM's detection efficiency of the Cherenkov light produced in water by relativistic muons, thus reproducing in situ detection conditions. We describe these two benches and present their first results and status.

1. Introduction

Two major neutrino projects take place in the depths of the Mediterranean Sea. The Antares neutrino telescope already provides the scientific community with data while the next generation detector KM3NeT would be a major step towards the understanding of fundamental neutrino properties and will open a new window for the study of astrophysical objects through neutrino detection.

The basic observation units of these telescopes are the Optical Module (Antares) and the Digital Optical Module (KM3NeT). The knowledge of their operational characteristics is of major importance for the efficiency of these telescopes. In this context the APC Laboratory has already constructed a fully functional test bench, implementing a black box with a wide variety of light sources, while the development of a second test bench has just finished, combining a water tank with a hodoscope for muon Cherenkov detection. With this two test benches the APC provides a full calibration system for OMs and DOMs.

[a] e-mail: avgitas@apc.in2p3.fr
[b] e-mail: creusot@apc.in2p3.fr

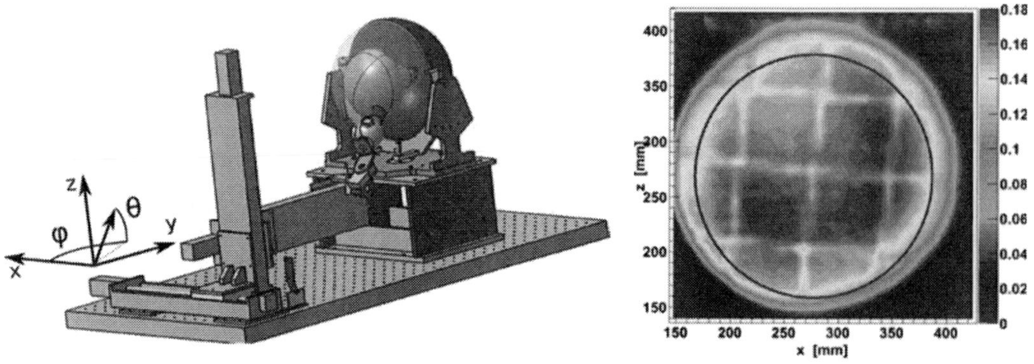

Figure 1. In the left, scheme of the content of the black box. The origin of the coordinate system stands at the corner of the table. The emission direction of the photons is completely defined by the 2 angles θ (zenith) and ϕ (azimuth). The origin (θ and $\phi = 0°$) corresponds to normal incidence when the light source is facing the PMT in its center. In the right, scan of the photocathode surface of an Antares OM. 11449 points of measurement have been taken. The points are positioned regularly on the optical module surface with a separation of 0.75°. For each point, 10^6 triggers were saved (statistical error of 1%).

2. Black box

2.1 Test bench set up

The black box is a parallelepiped 2 m long, 1 m large and 1 m high. It is equipped with an Antares Optical Module (OM [1]), a light source, 5 motor arms and their respective structures (see Fig. 1). All the holding structures are kept fixed on an optical table. The communications with the components within the box are allowed by several junction panels.

The 5 motors installed in the box allow choosing the position (x, y, z) of the light source and the direction of the emission (v_x, v_y, v_z) of the photons. The motor step size for the position and the direction is smaller than 0.01 mm and 1 arcsecond, respectively. The motors are driven via the *LabVIEW* software installed in a computer.

The light source, an integrating sphere, has been developed at APC [2]. It consists of a sphere fitted with a matrix of *Light Emitting Diode* (LED) centered on $\lambda = 378$ nm. A small fraction of the light of the diodes can escape by a small hole made in the sphere, allowing to have a really weak and calibrated light emission. An additional collimator selects the direction of emission of the photons within a cone of aperture half-angle 5.4°. This device allows reproducing with good accuracy the experimental environment, i.e. a majority (95%) of one photon pulses. One of the main characteristics of the sphere is that the power escaping through a hole is proportional to the area of the hole allowing an absolute calibration of the device.

The command voltage of the OM is set so that the PMT gain is nominal, i.e. the signal from a detected photon has an amplitude of about 45 mV. The anode signal is sent to the acquisition system [3, 4]. Above a threshold of 1/3 pe, the signal charge is integrated over 80 ns. Knowing the number of photons exiting the sphere for each LED trigger, it is straightforward to estimate the absolute efficiency of detection of the OM.

2.2 Results

For the scan presented in this article, the direction of emission of the photons is kept colinear to y (see left plot of Fig. 1) and the position of the light source follows the OM surface (steps of 0.75°) in order to simulate a parallel beam of light. The results are shown in the right plot of Fig. 1. The color scale represents the absolute detection efficiency of the OM photocathode. An average detection efficiency

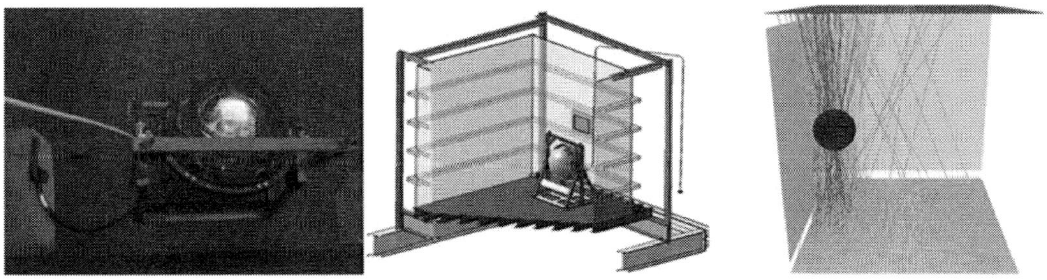

Figure 2. The OM attached with clamps on its base (left). An artistic representation of the OM inside the tank (center). Visualization of the muon tracks (right). The grids on top and bottom represent the hodoscope. The tracks crossing outside the OM are plotted because only time correlation criteria are applied thus extracting multiple track candidates per OM event.

value of 15.8% has been found within the red circle representing a photocathode of 500 cm^2. The mu-metal grid can clearly be seen in the scan as yellow strips of lower efficiency.

3. APC Water Tank

For the characterization of the OMs a Water Tank with a Hodoscope, originally developed for LAGUNA/MEMPHYNO [5], is maintained at the facilities of the APC.

3.1 Test bench Set up

The photo proof water tank is of cubic shape with edge size of 2 m. It is made out of High Density Polyethylene (*HDPE*). The inner walls are of black color. Inside the tank the OM is fixed on its base via multiple clamps (see left plot of Fig. 2). The OM support also allows different Zenith angles for the OM that are manually adjusted through a rotor (see center plot of Fig. 2).

Hodoscope

It consists of four identical scintillation detectors originated from the Track Tracker of the OPERA experiment [6], properly customized to fit the needs of the present set up. Each detector is 2 m in length, 1.7 m in width and approximately 1 cm thick. The scintillator (polystyrene, 2% p-Terphenyl, 0.02% POPOP) is arranged in independent strips of 1.6 m length, 2.6 cm width and height of 1cm. The strips are coated with TiO_2 as reflective material. An optical fiber collects and guides the photons from the scintillator to the photodetector (an 8×8 multianode PMT).

Two detectors placed perpendicular to each other form a Matrix of 64×64 channels, providing an accuracy of 2.6 cm $\times$ 2.6 cm for the (x, y) positions. One such pair is placed on top of the tank while another one resides below it. A commercial Hamamatsu *H*8808 PMT is used for the detection of the scintillator signals [7]. A CAEN multichannel high voltage supply is used for the operation of all the PMTs of the Hodoscope.

Hodoscope Read Out

The 64 PMT anode signals are sent to a MAROC3 chipset [8] equipped with an FPGA card for signal treatment and triggering. The MAROC3 has been designed to amplify individually each channel and to produce, below a common threshold, a trigger. Each trigger is then time stamped and sent to an acquisition computer together with the strip and hodoscope identities. The four cards are adjusted so that the channels have similar trigger rates at a threshold level equivalent to 1 pe. They are also synchronized though a 40 MHz GPS and a 100 Hz PPS.

Antares OM set up

The OM was placed on its base and inside the empty photo proof tank facing upwards, $\theta = 0°$. The communication with the OM is established through its control box, which serves two purposes. It supplies the OM with high voltage and provides an output for the signal. The signal is then delivered, using a 16 ns RG-58 cable, to a MAROC3 card signal input (channel 0). The threshold of the channel is set at ~ 0.02 mV. For what concerns the OM, the output of the card is the measured charge in ADC units (instead of the strip identity) and the time of the event. The time of this card is also synchronized with that of the hodoscope through the same GPS and PPS. The OM was operated at its nominal voltage.

3.2 Results

The acquired data have been used to run a visualization program indicating those particle tracks detected together by the 4 scintillators and the OM in a time window of 100 ns. This result is presented in the right plot of Fig. 2. Some tracks are obviously not crossing the OM, this could be due to the fact that the program uses time correlation for the events and it happens (rarely) to find two candidate tracks for the same OM event and then plots both.

References

[1] P. Amram et al. [Antares Collaboration], Nucl. Instrum. Meth. **A484** (2002) 369–383
[2] P. Gorodetzky, Sphères intégratrices Brevet France N° 06/09088
[3] D. Breton, E. Delagnes, J. Maalmi, Picosecond time measurement using ultra fast analog memories, in proceedings of TWEPP-09, Topical Workshop on Electronics for Particle Physics, Paris, September 2009
[4] D. Breton, E. Delagnes, Fast analog sampler with great memory depth, FR2824177 and US No.6859375 patents
[5] A. Tonazzo et al. [LAGUNA-LBNO Collaboration], Nucl. Part. Phys. Proc. **265–266** (2015) 192–194
[6] T. Adam and others, The OPERA experiment Target Tracker, Nucl. Instrum. Meth. **A577** (2007) 523–539
[7] M. Marafini, Physics studies and R&D towards the MEMPHYS experiment: A water Cherenkov Detector in Europe, Université Paris Diderot, École Doctoral 517, 2011
[8] S. Blin et al. 2010 JINST 5 C12007

EPJ Web of Conferences **116**, 06008 (2016)
DOI: 10.1051/epjconf/201611606008

The LED Beacon prototype system for the on-shore time calibration of the KM3NeT-IT Towers

Giulia De Bonis[1,a], Fabrizio Ameli[2], Carlo A. Nicolau[2], and Francesco Simeone[2] on behalf of the KM3NeT-IT Collaboration

[1] Dipartimento di Fisica, Università degli Studi di Roma "Sapienza", Italy
[2] Istituto Nazionale di Fisica Nucleare, Sezione di Roma, Italy

Abstract. The first implementation of the KM3NeT-IT neutrino telescope consists in the installation of 24 Strings and 8 Towers. Focusing on the Towers, the idea behind this work is to exploit the LED sources mounted in the OMs to develop a complementary system, on shore and before the deployment, for the determination of time delays, aiming at the characterizations of the time response of the different elements of the detector. During the assembling of the first tower, a set of measurements has been carried out; the test set-up and the measurement procedure are described, together with preliminary results of the calibration system. Lesson learnt is quite encouraging: uncertainties of the order of 400 ps are reached with very few cautions taken during the short calibration session, and with large room for improvement, making this system feasible and effective for the KM3NeT-IT experiment.

1. Introduction

KM3NeT [1] is a deep-sea research infrastructure consisting of a network of neutrino telescopes in the Mediterranean Sea. The KM3NeT Collaboration gathers the decennial experiences of ANTARES [2] NEMO and NESTOR in the design and implementation of sub-marine telescopes. The main detection principle is based on the Cherenkov technique; the Cherenkov light is collected via photomultiplier tubes (PMTs), arranged in three-dimensional structures (detection units) deployed underwater. The design of the detection units, the dimensions of the structures, the distances between the different elements and the geometry of the installation, all these features determine the scientific target of the detector and the energy threshold of the observations. The present project of KM3NeT is "multi-site" and "multi-purpose", and consists of two detectors to be installed in two separate underwater locations in the Mediterranean Sea: KM3NeT-FR, in the proximity of the ANTARES submarine site, with the "ORCA" component of the experiment [3] focused on the low energies (neutrino oscillations and neutrino mass hierarchy), and KM3NeT-IT, in the submarine site selected by the NEMO Collaboration (about 100 km offshore Capo Passero, Sicily, at 3500 m depth), with the "ARCA" [4] component, focused on the high energies and on the astrophysical targets of the experiment. KM3NeT-IT will host two detecting

[a] e-mail: giulia.debonis@roma1.infn.it

structures, the Strings and the Towers; the first phase of the experiment will include 24 Strings and 8 Towers.

The layout of the Towers reflects the design of the NEMO Tower [5] and consists in a three-dimensional flexible structure made up of 14 floors, for a total height of about 400 m; each floor is a 8 m long frame holding 6 Optical Modules (OMs), 2 at each end and 2 at the centre. The OM is a 13-inch pressure-resistant glass sphere, housing a 10-inch PMT and electronics boards for power supply and data acquisition [6]. Concerning the electronics for data acquisition and data transport of the Tower [7], the FEM board is the Front End Module for each OM and the FCM board is for the Floor Control Module; the backbone ensures a point-to-point electro-optical connection between each floor and the tower base. The on-shore station in Capo Passero is where the collected data are received, the power is supplied and the clock distribution is provided; each FCM has an on-shore counterpart which receives data from the floor (OMs, hydrophones, slow-control instruments) and sends the clock signal and the slow control commands by means of a synchronous point-to-point optical link. All the off-shore devices are thus synchronous to a common clock generated on-shore: as a consequence, a synchronization signal generated by the on-shore electronics can be used in a calibration set-up aimed at determining the latency of the different elements of the detector.

2. The LED Beacon Test

Two of the six optical modules in the floor (the ones located at the edge of the floor and downward-looking) host a 470 nm LED source, whose emission is controlled by the FEM, generating trigger pulses in a fixed phase relation with respect to the global timing of the detector. Taking into account the features of the data transport electronics of the tower, the idea behind this work is to exploit the LED sources to develop a complementary system for the determination of time delays, aiming at the characterizations of the time response of the different elements of the detector. The calibration procedure, hereafter called LED Beacon Test, is intended to be carried out during the integration phases of the tower and exploits the facilities of the integration site, in particular the rack system that implements the on-shore electronics in the experimental hall. For each tower, two tests are planned: the Floor Test and the Tower Test. The first is executed for each OM that holds a LED, and thus consists of two separate sets of measurement for each floor. It aims at estimating the floor latency, considering for each LED the system constituted by the FCM, the FEM and the connecting cables. The Floor Test gives indications on the operation, performance and behaviour of each LED and of the floor electronics. Once that the Floor Test has produced an estimate of the floor latency, the information can be combined with the results of the Tower Test, executed on one of the LED of the floor, to compute the latency of the backbone for each floor. Due to the large number of elements to be tested (and the severe time-restrictions imposed during the integration phase), constraints on the time required for a complete set of measurement the possibility to perform the test in parallel with other activities during the integration phase. A fast data analysis is a further item in the wish list, for a fast check of the tested elements and for permitting to take action before the deployment, in case some elements do not to match the specifications. The prototype system here presented fulfils the above listed requirements; it has been employed on October 2014 during the integration of Tower 8 in INFN – Laboratori Nazionali del Sud (LNS, Catania, Sicily) and is a promising procedure to evaluate the time response of the apparatus.

2.1 Description of the time calibration system

The LED sources

The LEDs are placed inside the OM, at the North Pole of the glass sphere. The distance between the LED emitting point and the internal surface of the sphere is less than 0.5 mm; the thickness of the sphere is 12 mm.

The light sensor

The instrument used to detect the LED light is the semiconductor device Hamamatsu C11208-350, of the MPPC series (Multi-Pixel Photon Counter). 3600 pixels cover an effective photosensitive area of $3 \times 3\,\mathrm{mm}^2$; each pixel is an APD (Avalanche Photo Diode) operating in Geiger mode. The advantages of the device reside in its excellent photon counting capability, the insensitivity to magnetic fields, the low-voltage operation (provided via USB connection) and its pocket-size. The product used in the prototype system is a starter kit and produces both a digital and an analog output; the peak sensitivity wavelength is at 450 nm.

The optical fibre

An optical fibre is used to collect the light emitted by the LED and to transmit it to the sensor; FC/PC connectors are mounted at each end of the fibre. The coupling of the fibre with the light source and the light sensor is an important mechanical issue to be faced for the accomplishment of the LED Beacon Test. While the task concerning the coupling with the light source is still an open point, the coupling of the fibre with the light sensor has been solved with a PVC "cap" (designed and realized by INFN-Roma), hosting an adaptor for a FC/PC connector and placed on the side of the sensor facing the detection window.

Time difference measurement

Time differences are measured with the TDC Agilent 53230A: the instrument receives two signals, S_A and S_B, at times t_A and t_B, and computes the time difference $\Delta t = t_B - t_A$. The reference signal (S_A) is the synchronization signal produced by the on-shore electronics; signal S_B is the digital output produced by the MPPC sensor when light is detected. The distribution of measured time differences is the observable of the LED Beacon test.

3. The test on Tower 8 (LNS, October 2014)

3.1 Experimental set-up

During the tests performed on Tower 8, a multi-mode 50/125 fibre, 40 m long, was employed; a temporary solution was adopted for positioning the termination of the fibre in front of the LED. 20×10^3 readings were acquired via the TDC for each LED. The analog and the digital output of the MPPC and the reference signal from the on-shore electronics were also visualized on the oscilloscope.

Because of the shortage of time, the Floor Test was performed on Floor 01 only, and no Tower Test was done. The optical modules were not covered nor obscured during the data taking, causing fluctuations in the amount of collected light and of ambient light entering the sensor.

3.2 Results

Because of the severe limitations of the experimental set-up, only preliminary considerations and no conclusive results can be extrapolated from the outcome of the prototype calibration system applied to Tower 8. Nevertheless, what is obtained, shown in Fig. 1, is fully encouraging about the potential of the measurement to investigate on the latency of the system and to help in understanding the different components of time delay. In particular, it is notable that the mean value of the distribution is not the same for the two OMs of the same floor, with a difference of about 1 ns that can reasonably be due to systematic effects, for instance the different emission time of the LEDs or the actual length of the cables for the two modules in symmetric position along the floor.

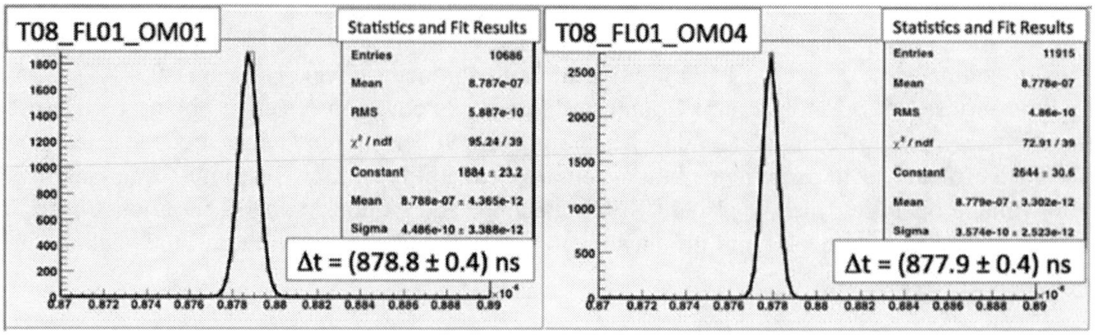

Figure 1. Distribution of the time differences (in seconds) between CH2 of the TDC (the MPPC digital output) and CH1 (the reference signal). Both the measurements have been done setting the same value for the number of readings acquired with the TDC, and the statistics is comparable for the 2 OMs. The result of a Guassian fit is superimposed, showing comparable uncertainties (400 ps) for the two set of measurements.

4. Conclusions

The lesson learnt, valid for the future experiences, is that the LED Beacon Test is already feasible and quite accurate with the instrumentation available. Only few minor improvements may be introduced, for instance the usage of 62.5/125 multi-mode fibres (larger core, for a larger amount of collected light); the use of collimators for optimizing the opto-mechanical interface between the LED and the fibre; the definition of a simple mechanical support to hold the fibre termination in a fixed position in front of the LED source. The test is feasible also with no cover or obscuration of the optical modules: the presence of background light does not affect significantly the measurement. With 20×10^3 readings of the TDC, about half an hour is required for completing the acquisition for a floor having the same statistics collected for T08/FL01; about one working day is taken for completing the Floor Test on the full tower and less than one working day is needed for the Tower Test on the integrated tower. Improved systems for light collection can reduce the number of TDC readings, and thus the time required for the LED Beacon Test.

References

[1] *KM3NeT*, `http://www.km3net.org/`

[2] J. Aguilar et al. (ANTARES), Nucl. Instr. and Meth.A **656**, 11 (2011), `1104.1607v1`

[3] D. Samtleben (KM3NeT), *KM3NeT/ORCA: status and plans*, this Conference

[4] D. Coniglione (KM3NeT), *KM3NeT/ARCA: status and plans*, this Conference

[5] S. Aiello et al. (NEMO), Astropart. Phys. **66**, 1 (2015), `1412.0849`

[6] E. Leonora et al. (KM3NeT), *Design and Mass Production of the Optical Modules for KM3NeT-Italia project*, this Conference

[7] C.A. Nicolau et al. (KM3NeT), *The KM3NeT-IT Tower data acquisition and data transport electronics*, this Conference

EPJ Web of Conferences **116**, 06009 (2016)
DOI: 10.1051/epjconf/201611606009

Measurement of light scattering in deep sea

N. Maragos, K. Balasi, T. Domvoglou, I. Kiskiras, D. Lenis, M. Maniatis, and G. Stavropoulos

NCSR DEMOKRITOS, Institute for Nuclear and Particle Physics, Agia Paraskevi 153 42, Greece

Abstract. The deep-sea neutrino telescope in the Mediterranean Sea, being prepared by the KM3NET collaboration, will contain thousands of optical sensors to readout. The accurate knowledge of the optical properties of deep-sea water is of great importance for the neutrino event reconstruction process. In this study we describe our progress in designing an experimental setup and studying a method to measure the parameters describing the absorption and scattering characteristics of deep-sea water. Three PMTs will be used to measure in situ the scattered light emitted from six laser diodes in three different wavelengths covering the Cherenkov radiation spectrum. The technique for the evaluation of the parameters is based on Monte Carlo simulations and our results show that we are able to determine these parameters with satisfying precision.

1. Introduction

The sensors of the deep-sea neutrino telescope KM3NET will detect the Cherenkov photons, emitted by the charged secondary particles created by neutrino interactions in the detector medium. The study presented below is motivated by the need of achieving an accurate description of the formation of the sensors signal, which the simulation and reconstruction techniques mainly rely on. This means that the characteristics of the absorption and scattering of the Cherenkov light, have to be determined as accurate as possible.

An experimental setup to measure in situ the optical properties of the detector medium is being studied and work concerning the design and the data analysis methods to be used, is presented below. In Sect. 2 a description of the experimental setup is made, including the specifications of the detector, the light sources and details about the geometry and the mechanical structure to be used. Section 3 is dedicated to the description of the Monte Carlo simulation background theory, its implementation details and the description of the method used for the parameters estimation.

2. The experimental setup

2.1 Overview

The measuring system will consist of light sources and light detectors. Photons emitted from the light sources will be scattered by sea water and detected by light detectors at fixed distances giving information about the optical properties of the medium. Sources and Detector will be placed inside of

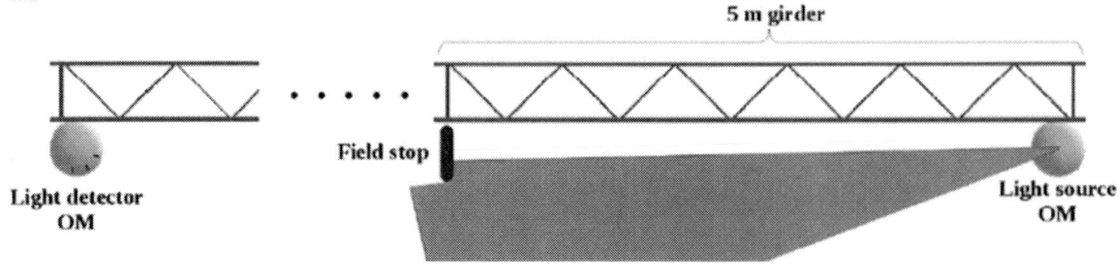

Figure 1. The experimental setup.

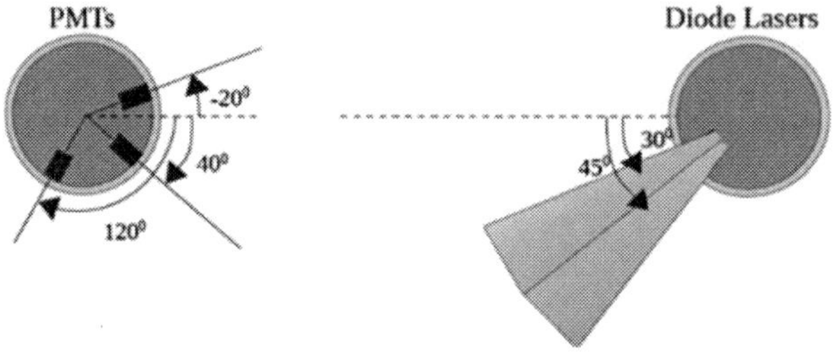

Figure 2. left: the detector Optical Module with three PMTs viewing at directions −20, 40, 120 deg with respect to the axis joining the centers of the Optical Modules. Right: the light source Optical Module with one pare of lasers emitting in 30 and 45 deg.

pressure resistant 1.5 cm thick glass spheres of outer diameter 43.2 cm (optical modules, OMs). The mechanical frame supporting both spheres will consist of 5 m long titanium girders attached to each other to form a long linear structure. The light source OM will be attached at the outer side and in the front end of the first girder (Fig. 1). The detector OM will be attached opposite to the light source OM and in the rear end of the last girder. A metal circular disk will be placed in front of the light source OM at the rear end of the first girder, preventing unscattered light from being detected. Three different separation distances will be used inserting zero, one and two girders in-between the first and the last (∼10, 15 and 20 m).

2.2 Light source OM

The light source will consist of three pairs of laser diodes emitting in three different wavelengths (405 ± 2 nm, 450 ± 2 nm, 520 ± 2 nm) covering the Cherenkov spectrum. The orientation of the two laser diodes in each pair will be 30° and 45° with respect to the axis formed by the centers of the OMs and in direction opposite to the mechanical frame (Fig. 2). The laser diodes will emit light pulses with pulse width 10 ns, pulse energy 2 pJ and repetition rate around 1 MHz. Their beam divergence will be ∼4 degrees.

2.3 Detector OM

The detector will consist of three HAMAMATSU H10682-210 PMTs modules operating at photon counting mode with their field of view centered at directions −20°, 40° and 120° with respect to the axis formed by the centers of the OMs (Fig. 2). These PMTs will measure the detected photon arrival times inside 1 μs "time windows" triggered together with the light pulses emitted from the laser diodes. The measured time accuracy will be 2.5 ns.

2.4 Data acquisition

The MAIN COUNTER works at 400 MHz, giving a time resolution of 2.5 ns. Every 1280 ns the counter restarts and at the same time a signal is send to the light souces triggering a light pulse. Each time a PMT gives a hit signal a buffer holds the MAIN COUNTER. The number of emitted laser pulses are counted and there is a record every 1024 pulses along with a mask indicating the active laser. Just before the MAIN COUTER restarts the PMT registers are guided to a FIFO along with a mask indicating which PMT gave the signal. A micro-controller is responsible for choosing witch laser will be active and for how long. It is also responsible for transferring the data to an SD card.

3. Simulation and parameter estimation technique

3.1 Theoretical model used for the simulations

The scattering model used for this study is a combination of Mie and Rayleigh scattering [1]. Its scattering angle distribution is expressed as a weighted sum of molecular and particulate scattering:

$$\frac{\mathrm{dP}}{\mathrm{d}\Omega_{\mathrm{s}}} = \mathrm{F}\left(\cos\theta_{\mathrm{s}}; \mathrm{p}, \mathrm{a}_{\mathrm{Rayl}}, \mathrm{a}_{\mathrm{Mie}}\right) = \mathrm{p} \times \mathrm{g}\left(\mathrm{a}_{\mathrm{Rayl}}, \cos\theta_{\mathrm{s}}\right) + (1-\mathrm{p}) \times \mathrm{f}(\mathrm{a}_{\mathrm{Mie}}, \cos\theta_{\mathrm{s}}) \tag{1}$$

where θ_s is the scattering angle, p the Rayleigh contribution, a_{Mie} is the average cosine of the scattering angle for Mie scattering and a_{Rayl} defines the shape of the Rayleigh scattering function. $f(a_{Mie}, \cos\theta_s)$ is a Henyey-Greenstein function given by:

$$f\left(\mathrm{a}_{\mathrm{Mie}}, \cos\theta_{\mathrm{s}}\right) = \frac{1}{4\pi} \frac{\left(1 - \mathrm{a}_{\mathrm{Mie}}^2\right)}{\left(1 + \mathrm{a}_{\mathrm{Mie}}^2 - 2\mathrm{a}_{\mathrm{Mie}}\cos\theta_{\mathrm{s}}\right)^{\frac{3}{2}}}. \tag{2}$$

The phase function $g(a_{Rayl}, \cos\theta_s)$, which is describing the Rayleigh scattering from sea water molecules, has been parametrized as:

$$\mathrm{g}(\mathrm{a}_{\mathrm{Rayl}}, \cos\theta_{\mathrm{s}}) = \frac{(1 + \mathrm{a}_{\mathrm{Rayl}}\cos^2\theta_{\mathrm{s}})}{4\pi\left(1 + \frac{1}{3}\mathrm{a}_{\mathrm{Rayl}}\right)}. \tag{3}$$

3.2 Method of parameter estimation

The experimental setup is simulated with the use of GEANT4 packages. Monte Carlo photon events are generated with their propagation defined by the described theoretical model. In a real life experiment, the 3×6 arrival time distributions from the measured data would be compared with the corresponding Monde Carlo (MC) distributions (Fig. 3). In this study we produced an independent MC sample of events which is considered as the data set (hereafter pseudo-data).

Since in the real experiment we won't know the real values of the parameters, we conduct another MC simulation with parameters different from that of the pseudo-data and we try to fit the corresponding arrival time histograms to the pseudo-data. The technique we use for the χ^2 fit relies on re-weighting the photon events according to their track probability [2]:

$$\mathrm{P}\left(\mathrm{L_a}, \mathrm{L_s}, \mathrm{p}, \mathrm{a}_{\mathrm{Rayl}}, \mathrm{a}_{\mathrm{Mie}}\right) \propto \mathrm{e}^{-\mathrm{L}/\mathrm{L_a}} \times \left[\prod_{\mathrm{i}=1}^{\mathrm{N}-1} \frac{1}{\mathrm{L_s}} \mathrm{F}(\cos\theta_{\mathrm{i}}; \mathrm{p}, \mathrm{a}_{\mathrm{Rayl}}, \mathrm{a}_{\mathrm{Mie}})\right] \times \mathrm{e}^{-\mathrm{l_N}/\mathrm{L_s}}, \tag{4}$$

where $L = \sum_{i=1}^{N} l_i$ is the total photon track length, L_a is the absorption length and L_s is the scattering length. The right part of (4) consists of three terms. e^{-L/L_a} is the probability for a photon not to be absorbed after traveling length L, the middle term is proportional to the probability for a photon to be

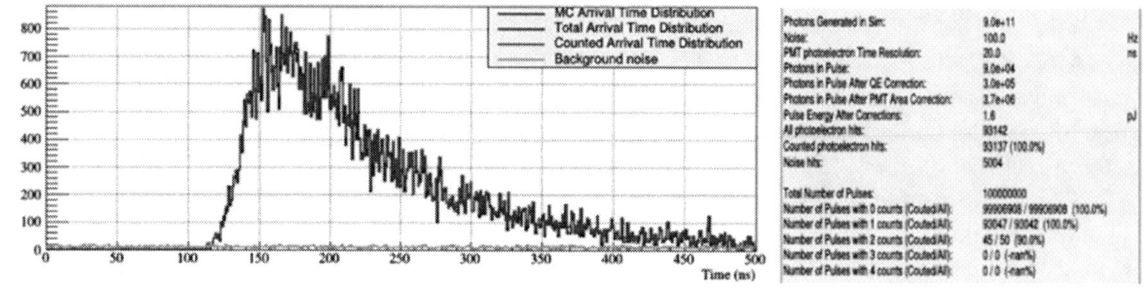

Figure 3. PMT hit arrival time distribution derived from simulation.

scattered with an angle θ_i after having crossed a distance l_i, and e^{-l_N/L_s} gives the probability for the photon not to be scattered at the last segment.

Using the estimated parameters from the fit we produce an additional set of MC events and we make a new fit to estimate again the parameters. We repeat this process until simulation and estimated parameters converge and their difference does not exceed 1σ.

To test the reliability of the fit method we make an application of it using the following two sets of parameters for pseudo-data and simulation respectively:

$$L_a = 73\,\text{m},\quad L_s = 56.4\,\text{m},\quad p = 0.21,\quad a_{Mie} = 0.75,\quad a_{Rayl} = 0.853 \quad \text{pseudo-data}$$

$$L_a = 65\,\text{m},\quad L_s = 48.4\,\text{m},\quad p = 0.17,\quad a_{Mie} = 0.924,\quad a_{Rayl} = 0.853 \quad \text{simulation data.}$$

The minimization process is made in the five dimensional parameter space. We set $a_{Rayl} = 0.853$ fixed, since it is a well known factor, attributable to the anisotropy of water molecules [3], and additionally its variations have insignificant impact in the resulting arrival time distributions. We also insert a normalization parameter c that can be related with the total number of generated photons. Using the ROOT MINUIT package we get the parameter estimations and the correlation matrices. A convergence is reached after five repeats of the fit process. The "real" parameters are well estimated from the final fit with the following uncertainties:

$$\delta \mathrm{p} = 0.0019(0.9\%),\quad \delta L_s = 1.34\,\mathrm{m}(2.4\%),\quad \delta \mathrm{L_a} = 2.14\,\mathrm{m}(2.9\%)\ \text{and}\ \delta \mathrm{a_{Mie}} = 0.0023(0.3\%).$$

4. Conclusions

Motivated by the need of accurate knowledge of the parameters determining the behavior of Cherenkov light in the deep-sea water medium for the KM3NeT we studied the design of an experimental setup using diode lasers and PMT sensors inside pressure resistant Optical Modules dedicated for in situ measurements. We investigated important parameters of the setup and we studied and tested a method to estimate the absorption and scattering parameters by fitting model based re-weighted Monte Carlo simulations to the expected PMT photon arrival time distributions. Our test results show that the accuracy of the parameter estimation is $\sim$2% for absorption and scattering lengths and less than 1% for the Rayleigh contribution and Mie angular distribution factor. Guided from the present design study we plan to realize the experiment.

Financial support through the European Fund for Regional Development, NSRF 2007–2013 action "Development of Research Centers – ΚΡΗΠΙΣ" is acknowledged.

References

[1] D.J.L. Bailey, Wolfson College, Oxford, thesis (2002)
[2] A. Papaoikonomou, A Leisos, A. Tsirigotis and S. Tzamarias, Workshop on very large volume neutrino telescopes, Stockholm, Sweden (2013)
[3] J.A. Aguilar et al., Astroparticle Physics, v23 (2005)

EPJ Web of Conferences **116**, 06010 (2016)
DOI: 10.1051/epjconf/201611606010

Development and performances of a high statistics PMT test facility

Carlos Maximiliano Mollo[a] on behalf of the KM3NeT Collaboration

INFN Napoli, Complesso Universitario di Monte Sant'Angelo, via Cintia, 80126 Napoli, Italy

Abstract. Since almost a century photomultipliers have been the main sensors for photon detection in nuclear and astro-particle physics experiments. In recent years the search for cosmic neutrinos gave birth to enormous size experiments (Antares, Kamiokande, Super-Kamiokande, etc.) and even kilometric scale experiments as ICECUBE and the future KM3NeT. A very large volume neutrino telescope like KM3NeT requires several hundreds of thousands photomultipliers. The performance of the telescope strictly depends on the performance of each PMT. For this reason, it is mandatory to measure the characteristics of each single sensor. The characterization of a PMT normally requires more than 8 hours mostly due to the darkening step. This means that it is not feasible to measure the parameters of each PMT of a neutrino telescope without a system able to test more than one PMT simultaneously. For this application, we have designed, developed and realized a system able to measure the main characteristics of 62 3-inch photomultipliers simultaneously. Two measurement sessions per day are possible. In this work, we describe the design constraints and how they have been satisfied. Finally, we show the performance of the system and the first results coming from the test of few thousand tested PMTs.

1. Introduction

The KM3NeT Collaboration [1] is aimed at realize a telescope for the detection of high energy astrophysical neutrinos by reconstructing the trajectory of the muon generated by the neutrino interaction with the Earth. Digital optical modules (DOM) will detect the Cerenkov light generated by muons in the seawater, where the telescope will be placed. Each DOM is equipped with 31 3-inch PMTs manufactured by Hamamatsu [2]. Approximately 115 strings or Detector Units (DUs) will constitute the first so-called building block of the telescope (3 building blocks will constitute the complete telescope), each DU has 18 DOMs then 64170 PMTs will be needed to complete the first building block. The knowledge of all principal parameters of each PMT is essential in order to have a good calibration of the telescope. In addition, it is necessary to determine the optimal high voltage of each PMT in order to have almost the same gain for all photodetectors. The measurement of the characteristics of 64170 PMTs could be a bottleneck in the telescope realization. Therefore, a system capable to measure several PMTs in parallel is mandatory. The realization of such a system has several challenges: timing synchronization and

[a] e-mail: maximil@na.infn.it

Figure 1. Left: PMTs on a tray ready to be tested; right: dark cylinders and cable extension connectors inside the Dark Box.

uniformity of a laser light needed to illuminate the PMTs for time calibration measurements. For the test facility we decided to use the same DAQ electronics of the DOMs. As said before each DOM electronics is able to handle the signals of 31 PMTs. Using two complete DOM electronics and synchronizing them by using a White Rabbit Switch we are able to test 62 PMTs in parallel.

2. Mechanical and optical setup

In order to test 62 PMTs in parallel we need to optically insulate a PMT with respect the others using dark cylinders (Fig. 1 right), in this way timing measurements like TTS (Transit Time Spread) or spurious pulses can't be corrupted by light coming from the near or even distant PMT test docks. At the base of each cylinder there is an opal diffusing glass disc that ensures a near uniform distribution of the light illuminating each photocathode uniformly even at single photon-electron condition. This is very important for TTS measurements.

In order to perform timing measurements like TTS, TT and spurious pulses, we have used one single light source, a picosecond LASER source from PiLas with excellent timing performances. The wavelength is 470 nm and the jitter between the trigger signal and the light pulse is less than 40 ps. To illuminate all 62 PMTs concurrently we have adopted an optical splitter 1 to 70. It shows a good power output distribution as shown in Fig. 2.

The PMTs are placed in a tray of 57 cm by 66 cm. Each tray has 4 pins that ensure the correct distance from the table when positioning the PMTs and also are used to guide the tray insertion into the Dark Box.

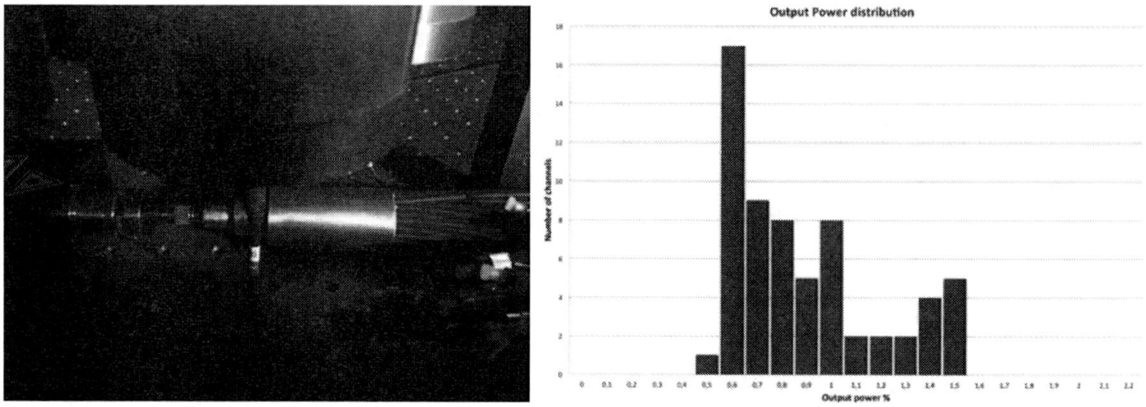

Figure 2. The optical splitter inside the Dark Box (left). The splitter output power distribution (right), all power outputs are within 0.5% and 1.5% of the input power.

3. Electronics

The DAQ electronics is based on the KM3NeT Central Logic Board (CLB). The CLB is the main logic board of the KM3NeT DOM. It is directly connected with a Power Conversion Board which provides all the needed rails for the board (1.0 V, 1.8 V, 2.5 V, 3.3 V, 5.0 V). The purpose of the CLB is to manage all the instrumentation present inside the DOM, acquire the data from the 31 PMTs and from the acoustic device, compress the data into User Datagram Protocol (UDP) packets and send them through the optical line. A sub-ns time synchronization level between all the DOMs is achieved by using White Rabbit switchs. The CLB is based on a Xilinx Kintex 7 FPGA, which is connected to all the rest of the board, including in particular:

- 1 Small Form-factor Pluggable (SFP) connector for laser transceivers for the optical communication;
- 2 custom Octopus connectors for the PMTs signals;
- Tunable oscillators (White Rabbit compliant);
- Embedded instrumentations (temperature and humidity sensor, tilt & compass);
- External connectors (for Nanobeacon LED, acoustic devices, expansion boards).

For the purposes of the PMT test facility we only use the White Rabbit gear and acquire the data from PMTs. In order to record the data from 62 PMTs simultaneously we used two CLBs synchronized using a White Rabbit switch. The laser pulse emission is synchronized and syntonized to the CLB common clock and absolute time allowing for the measurement of PMT transit time (TT) [4].

4. DAQ software

The most part of the software tools were taken or re-adapted from the KM3NeT DAQ system, which was still in development phase (`www.km3net.org/TDR/TDRKM3NeT.pdf`). The software chain, together with providing fast analysis of the PMTs characteristics, has to be interfaced with the KM3NeT Database System. This step is necessary to: retrieve in real time information on the PMT assembly (e.g. serial number, vendor's HV setting, test and integration history); store the results of the calibration and make them available to the DOM integration teams for cross checks and further analysis.

Two data types are analysed: monitoring and optical data. The first one contains the number of pulses from each channel in the time-slice; the second contains arrival time (timestamp) and ToT (Time over Threshold) of each hit [5].

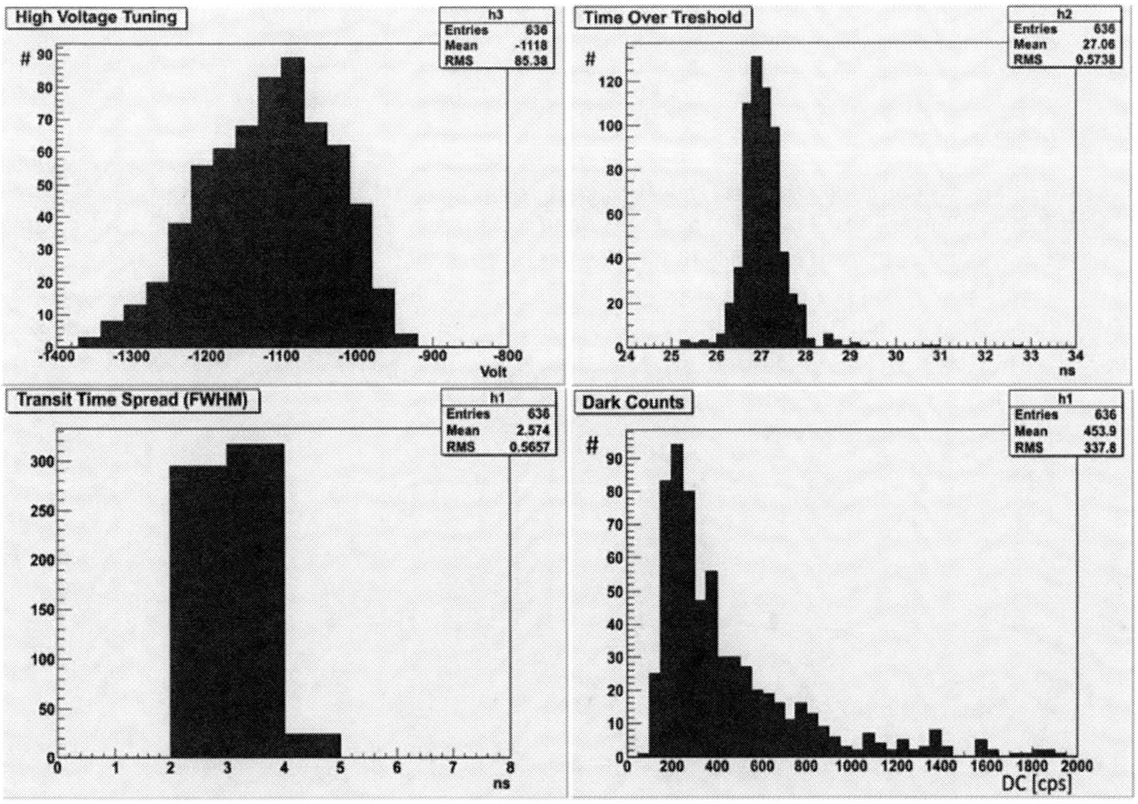

Figure 3. Distributions of the results of the test of 636 PMTs. High voltage tuning distribution (top-left), ToT distribution (top-right), Transit time spread distribution (bottom-left) and Dark counts distribution (bottom-right).

The data analysis code is as a C++ program interfaced with ROOT libraries to create ROOT files and text output with the final plots and values. XML file creator is a ROOT script which creates a file with the PMT test results to be uploaded to the database. Also it generates a separate PDF test sheet for each PMT with the test results.

5. Measurement results and conclusions

The full test and calibration procedure for 62 PMTs takes about 12 hours including the analysis of all the data acquired. During the test almost all the PMT characteristics are measured. At the end of a test session all raw data and processed data are recorded into the KM3NeT Data Base in order to be available during the DOM production. The distributions of Dark Counts, ToT, TTS, and Tuned HV for about 600 PMTs are shown in Fig. 3.

The Dark Box system allows to test up to 124 PMTs per day, with usual methods the test of such number of PMTs would require weeks or months. The performance in terms of reliability were tested by comparing the test results obtained with the Dark Box with data from Hamamatsu. To speed up the sorting of PMTs to the integration sites it was adopted a triage (RED, YELLOW, GREEN) that allows a quick selection of those PMTs ready to be integrated (within the specifications) "GREEN". The PMTs that present characteristics out of specifications (for example: dark counts over 3000 cps, prepulses more than 1%, etc.) are labelled "RED". The PMTs with characteristics slightly above the specifications are labelled "YELLOW" (for example: dark counts within 2000–3000 cps). In conclusion, the system is in all aspects (amount of PMTs tested daily, reliability, safety, etc.) ready for the mass production of the KM3NeT neutrino telescope.

References

[1] KM3NeT collaboration – `http://www.km3net.org/collaboration.php`
[2] KM3NeT consortium, O. Kalekin et al. Photomultipliers for the KM3NeT optical modules. Nuclear Instruments and Methods in Physics Research A **695** (2012) 313–316
[3] LASER PiLas – `http://www.alsgmbh.com/`
[4] White Rabbit Switch – `http://www.ohwr.org/projects/white-rabbit/wiki/Switch`
[5] V. Kulikovskiy, this conference

EPJ Web of Conferences **116**, 06011 (2016)
DOI: 10.1051/epjconf/201611606011

Measuring the optical properties of IceCube drill holes

Martin Rongen[a] for the IceCube collaboration[b]

III. Physikalisches Institut B, RWTH Aachen University, 52074 Aachen, Germany

Abstract. The IceCube Neutrino Observatory consists of 5160 digital optical modules (DOMs) in a cubic kilometer of deep ice below the South Pole. The DOMs record the Cherenkov light from charged particles interacting in the ice. A good understanding of the optical properties of the ice is crucial to the quality of the event reconstruction. While the optical properties of the undisturbed ice are well understood, the properties of the refrozen drill holes still pose a challenge. A new data-acquisition and analysis approach using light originating from LEDs within one DOM detected by the photomultiplier of the same DOM will be described. This method allows us to explore the scattering length in the immediate vicinity of the considered DOMs.

1. Motivation

The IceCube Neutrino Observatory [1] is situated in the deep glacial ice below the geographic South Pole. The IceCube sensors are 5160 digital optical modules (DOMs) which detect Cherenkov light produced by relativistic charged particles traversing the detector. Each DOM incorporates a 10 inch downward facing photomultiplier as well as several LED light sources for calibration [2] (see Fig. 1). The DOMs were deployed in a depth between 1450 m and 2450 m in 86 holes where the ice had been melted with a hot water drill [3].

The deep Antarctic ice has long been known to have exceptional optical properties, as trapped air bubbles shrink and eventually vanish under large pressure forming an air clathrate hydrate [4]. For accurate reconstruction of the properties of traversing relativistic particles, a precise knowledge of the absorption and scattering lengths of the ice traversed by the Cherenkov photons is crucial. The vertical structure of the undisturbed glacial ice or "bulk ice" was originally measured by so called dust-loggers [5], which were lowered into the drill holes. A small fraction of the light from a horizontal laser beam was redirected by scattering centers in the ice, mainly dust particles, to a downward facing PMT optically separated from the light source. In addition, data from flashing LEDs of known brightness situated in each DOM has been used to derive an accurate model of the optical absorption and scattering in the bulk ice [6].

---

[a] e-mail: `martin.rongen@icecube.wisc.edu`

[b] `https://icecube.wisc.edu/collaboration/authors/2015/09`

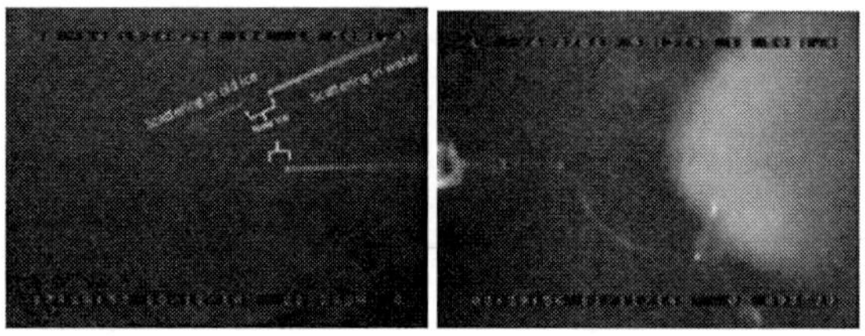

(a) Camera looking sideways into the bulk ice (b) Camera looking downward into the hole ice

Figure 1. The hole ice as seen by the camera.

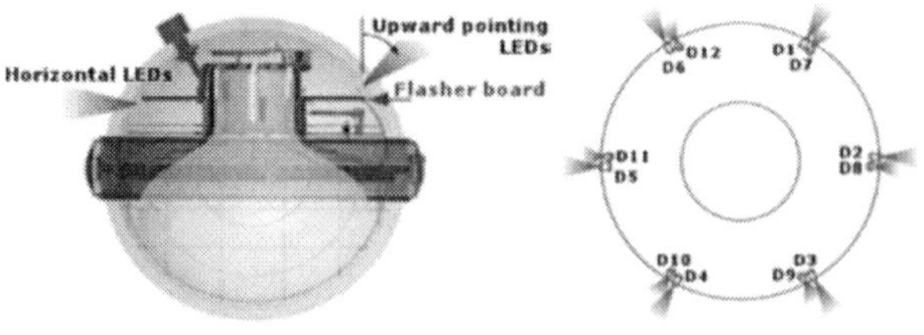

(a) Location of flasher LEDs (b) LED numbering

Figure 2. Schematic overview of an IceCube DOM.

While the dominant part of the propagation of Cherenkov photons happens in the bulk ice, each photon detected by a DOM also has to propagate through the refrozen water in the drill holes, called "hole ice". The optical properties of the hole ice are less understood than the bulk ice and present one of the largest uncertainties to neutrino oscillation measurements in IceCube [7]. This paper presents a new method to determine the scattering length in the hole ice.

1.1 Previous hole ice measurements

The hole ice was directly imaged by a camera installed at the bottom of a drill hole in the center of IceCube [8]. It suggests two hole ice components: A clear outer region (Fig. 1(a)) and a central column of $\sim$8 cm diameter with a very small scattering length (Fig. 1(b)). This observation is consistent with cylindrical freezing, where impurities or air bubbles are pushed along the freezing boundaries until they merge in the center. A quantitative evaluation of the the hole ice was first conducted with IceCube's predecessor AMANDA [9] by measuring the arrival time distribution of photons from flasher LEDs with optical modules (OMs) in adjacent holes [10]. The distributions were compared to simulations assuming scattering on spherical air bubbles, resulting in an effective scattering length between 100 cm and 350 cm. The effective scattering length l_{eff} is related to the mean geometric scattering length l_{geo}, the mean distance between two scattering processes, by the mean of the scattering function:

$$l_{\mathrm{eff}} = l_{\mathrm{geom}}/(1 - \langle \cos\theta \rangle). \qquad (1)$$

This result was incorporated into the AMANDA detector simulation as a modification of the angular acceptance function of the OMs with respect to the lab measurement (see Fig. 5) and is also used by IceCube as the default hole ice model [6]. In this model the acceptance in the forward direction of the PMT is reduced with respect to the lab measurements as the hole ice scatters away incident photons, while the acceptance in the backward direction is increased as photons have a chance to be scattered to the front of the DOM. Since then, multiple AMANDA and IceCube studies using inter-string LED data [11, 12] or muon data [13, 14] have qualitatively broadened our understanding of the hole ice, however arriving at partially conflicting angular acceptance functions.

2. Same-DOM flasher measurements

For the determination of hole ice properties it is unfavorable to measure the light emitted by one DOM with a another DOM, as the bulk ice properties will dominate the propagation. Each DOM possesses twelve 405 nm LEDs situated on a dedicated board above the DOM mainboard (Fig. 1). The LEDs are organized in six pairs 60° apart. One LED of each pair shines into the ice horizontally and one LED is angled $\sim$45° upward. The LEDs can be controlled individually. The dimmest stable pulse contains $\sim 257 \cdot 10^6$ photons in a $\sim$30 ns long pulse. By turning on the LEDs sequentially and measuring the

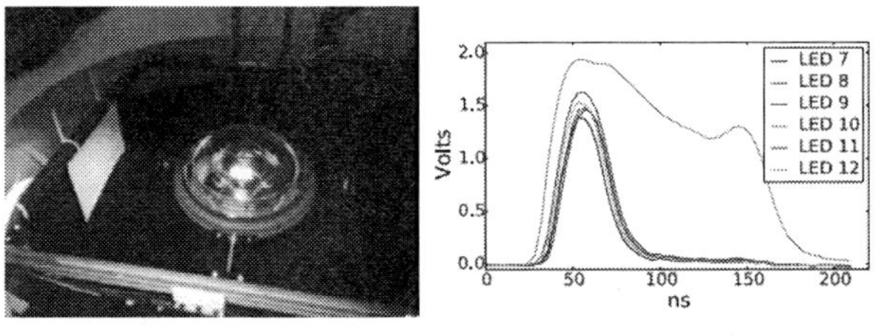

(a) DOM in water tank (b) PMT waveforms

Figure 3. Lab demonstration of the measurement principle.

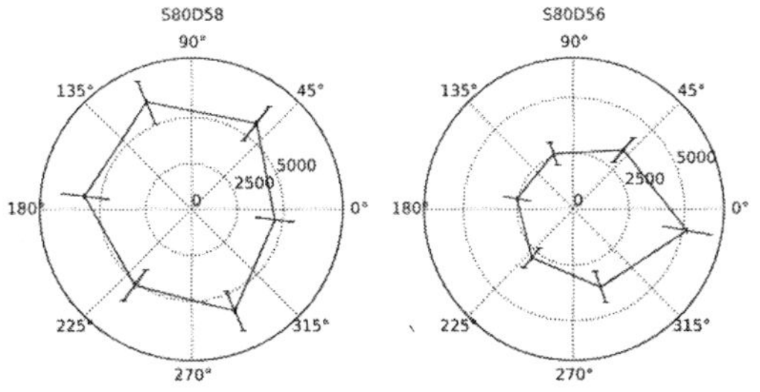

Figure 4. Typical azimuthal patterns of detected PEs on two in-situ DOMs, ranging from ~2000 to ~7000 PE.

intensity of light returning to the PMT of the same DOM a measurement similar to the dust logger can be performed, where the measured intensity is proportional to the scattering coefficient in the direction of the LED. This method was demonstrated by conducting the described procedure with a DOM in a water tank with an absorptive black lining, which contained a reflective sheet on one side (Fig. 3(a)). Figure 3(b) shows the average recorded PMT waveforms for each tilted LED. The LED pointing in the direction of the reflective sheet (LED 10) yields the strongest PMT signal.

2.1 DARD: Data acquisition for a flasheR DOM

The DOM firmware used for standard data acquisition (DAQ) [15] disables the high voltage to the PMT while the flasher board is enabled. Thus, a custom DAQ based on a debugging firmware had to be developed. The resulting system controls a single IceCube string, while the rest of the detector remains in normal operation. After setting up the high voltage to a gain of $\sim 10^5$, 100 waveforms are recorded for each LED configuration. This procedure has been applied to a total of 20 DOMs in two runs.

2.2 Photon counting

The raw waveforms are averaged, corrected for the digitizer pedestals and then calibrated [1]. The integral over the resulting waveform voltage bins (V_i) is directly proportional to the number of recorded photoelectrons (PEs), where Z is the frontend impedance and f the sampling frequency of the digitizers:

$$\mathrm{PE}_{\mathrm{exp}} = \frac{1}{Z(\Omega)} \cdot \frac{1}{e} \cdot \frac{1}{\mathrm{f(Hz)}} \cdot \sum_i V_i \cdot \frac{1}{\mathrm{gain}}. \tag{2}$$

Two examples of the number of measured PE as a function of the azimuthal LED direction can be seen in Fig. 3. Note, that light output varies by about 20% between different LEDs.

For this study only information from the tilted LEDs is used, as the horizontal LEDs showed non-negligible saturation effects and are possibly subject to internal light contamination. Where a small fraction of the photons can reach the PMT before leaving the DOM.

3. Simulation

A detailed Geant4 model of a DOM has been placed in the center of a 60 cm diameter hole ice column surrounded by a 600 m bulk ice volume. For both media the Henyey-Greenstein scattering function [16], which is a good approximation of Mie scattering and which is also used in the analysis of the bulk ice, is assumed with an average deflection angle $\langle\cos(\theta)\rangle = 0.9$. The different assumption regarding the scattering process compared to [10] implies that impurities in the hole ice dominate the light scattering

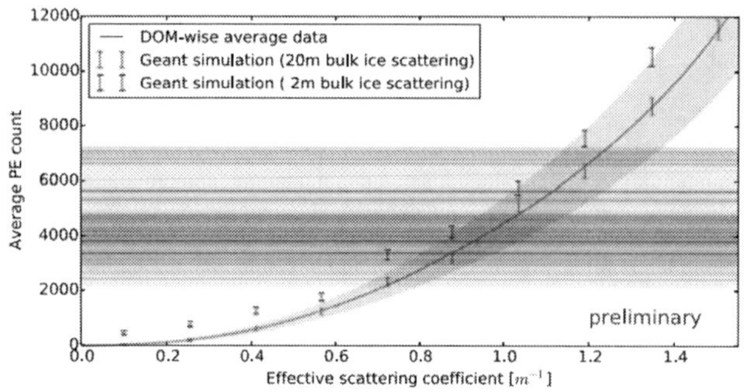

Figure 5. Comparison of the experimental and simulated light intensity arriving at the PMT of the emitting DOM.

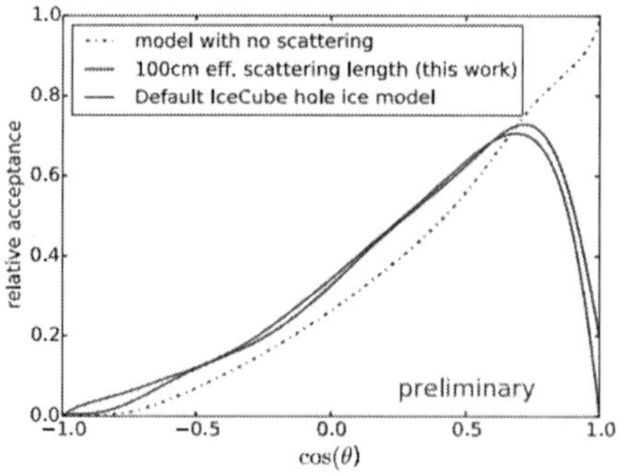

Figure 6. Comparison of angular acceptance models derived from lab measurements without scattering and modified for the hole ice properties.

instead of air bubbles. The number of PEs detected by the DOM (N_{hits}) is then given by the amount of photons arriving at the photocathode times the quantum efficiency and mean charge:

$$PE_{sim} = N_{hits} \cdot 0.25 \cdot 0.85 \, [PE]. \tag{3}$$

The primary systematic uncertainty involved in this simulation is a 20% uncertainty on the absolute photon output of the flasher LEDs. Currently not considered systematics include absorption in the hole ice, the collection efficiency of the PMT at low voltages, shadowing by the DOM cable as well as positional uncertainties of the LEDs. The simulation has been performed for two extreme values of the geometric scattering length of the bulk ice of 20 m and 2 m and a wide range of effective scattering coefficients of the hole ice column as shown in Fig. 5. The results stated below are for a bulk ice of 20 m scattering length that applies to most of the selected DOMs.

4. Extracting ice parameters

The measured number of returning photons does not show a strong dependence on the azimuthal flasher direction for any DOM. A strong inhomogeneity of the hole ice, such as small column that encompasses some but not all of the flashers, is not supported by the current data. In order to quantify the average hole ice properties, the number of PEs observed from the different LEDs have been averaged for each DOM. The distribution of measured PEs compared to the simulated expectation for different scattering coefficients is shown in Fig. 5. The scattering coefficient at which the simulated number of PEs crosses the measured PE count describes the mean properties of the hole ice around the specific DOM (for the assumptions used in the simulation). The systematic uncertainty of the simulation is denoted by the green shaded area. The average effective scattering length for all 20 DOMs is 100 ± 15 cm, while all DOMs are covered by effective scattering lengths ranging between 75 cm and 150 cm. This result agrees to the effective scattering length found by the original AMANDA analysis (100 cm to 350 cm) within the stated uncertainties.

5. Modeling the angular acceptance

One can employ the same Geant4 simulation described in Sect. 3 to calculate an effective modification to the angular acceptance due to the hole ice. This can be simulated by placing a uniform beam some distance away from the DOM. As the beam is rotated around the DOM the relative photon counts between simulations with no hole ice and a given hole ice column give a relative correction factor for each simulated angle. Applying this correction factor to the angular acceptance measured in the lab yields the average DOM angular acceptance in the ice, assuming a homogeneous 60 cm diameter hole ice column. The simulation has been performed for a variety of sizes and distances of the uniform beam, as different size regions of the hole ice column are probed. The resulting correction curve was found to be stable for beams larger than 1 m in radius placed more than $\sim$2.5 m away from the DOM. Figure 5 depicts the obtained angular acceptance for the best fit effective scattering length of 100 cm compared to the lab measurement and the currently used default IceCube hole ice model [6]. The default IceCube hole ice model and the acceptance obtained from the presented method agree well, especially considering the very different experimental approaches and simulation assumptions.

6. Summary and outlook

A new experimental approach to determine the scattering properties within the refrozen IceCube drill holes has been presented. Light originating from LEDs within one DOM detected by the photomultiplier of the same DOM has been used for this measurement. The measured light intensities have been compared to a simulation assuming Henyey-Greenstein scattering in a 60 cm homogeneous hole ice column with the DOM situated in its center. The best fit scattering length has been used to simulate an effective modification to the angular acceptance of the IceCube DOMs, which agrees well with the currently used model [6, 10]. In the future more complicated models, representing the structure as seen by the cameras, will be tested. These should result in a higher relative acceptance for head-on illumination, which would draw a consistent picture with recent inter-string flasher analysis.

References

[1] A. Achterberg et al., Astropart. Phys. **26**, 155 (2006)
[2] R. Abbasi et al., Nucl. Instr. Meth. Phys. Res. **618**, 139 (2010)
[3] T. Benson et al., Annals of Glaciology **55**, 105 (2014)
[4] H. Shoji, C.C. Langway, Nature **298**, 548 (1982)
[5] N.E. Bramall et al., Geophys. Res. Lett. **32** (2005)
[6] M. Aartsen et al., Nucl. Instr. Meth. Phys. Res. **711**, 73 (2013)
[7] M.G. Aartsen et al., Physical Review D **91** (2015)
[8] P. Hulth (VLVNT, 2013)
[9] E. Andres et al., Astropart. Phys. **13**, 1 (2000)
[10] A. Karle, K. Woschnagg (AMANDA internal, 1998)
[11] P. Niessen (AMANDA internal, 1998)
[12] D. Chirkin (IceCube internal, 2015)
[13] P. Olbrechts et al. (AMANDA internal, 2001)
[14] L. Gerhardt (IceCube internal, 2007)
[15] R. Abbasi et al., Nucl. Instr. Meth. Phys. Res. **601**, 294 (2009)
[16] L.C. Henyey, J.L. Greenstein, ApJ **93**, 70 (1941)

EPJ Web of Conferences **116**, 07001 (2016)
DOI: 10.1051/epjconf/201611607001

Using IKAROS as a data transfer and management utility within the KM3NeT computing model

Christos Filippidis[1,a], Yiannis Cotronis[2], and Christos Markou[1]

[1] NCSR Demokritos, Greece
[2] University of Athens, Greece

Abstract. KM3NeT is a future European deep-sea research infrastructure hosting a new generation neutrino detectors that – located at the bottom of the Mediterranean Sea – will open a new window on the universe and answer fundamental questions both in particle physics and astrophysics. IKAROS is a framework that enables creating scalable storage formations on-demand and helps addressing several limitations that the current file systems face when dealing with very large scale infrastructures. It enables creating ad-hoc nearby storage formations and can use a huge number of I/O nodes in order to increase the available bandwidth (I/O and network). IKAROS unifies remote and local access in the overall data flow, by permitting direct access to each I/O node. In this way we can handle the overall data flow at the network layer, limiting the interaction with the operating system. This approach allows virtually connecting, at the users level, the several different computing facilities used (Grids, Clouds, HPCs, Data Centers, Local computing Clusters and personal storage devices), on-demand, based on the needs, by using well known standards and protocols, like HTTP.

1. Introduction

Large-scale scientific computations tend to stretch the limits of computational power and parallel computing is generally recognized as the only viable solution to high performance computing problems. I/O has become a bottleneck in application performance as processor speed increases, leaving storage hardware and software struggling to keep up. Parallel file systems have be developed in order to allow applications to make optimum use of available processor parallelism. The most important factors affecting performance are the number of parallel processes participating in the transfers, the size of the individual transfers and of course the access patterns. The I/O access patterns are generally divided into the following subgroups [1]:

(1) Compulsory;
(2) Checkpoint/restart;
(3) Regular snapshots of the computation's progress;

[a] e-mail: cfjs@outlook.com

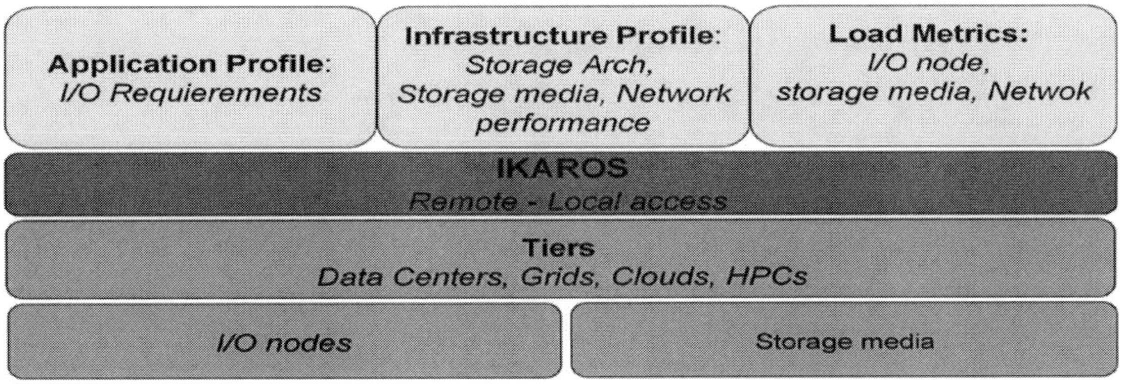

Figure 1. IKAROS Framework.

(4) Out-of-core read/writes for problems exceeding available memory;
(5) Continuous output of data for visualization and other post-processing.

The factors affecting performance are increasing if we consider the overall data flow (remote-local access) within an international collaborative scientific experiment, like KM3NeT. Such experiments will generate datasets that increase exponentially in both complexity and volume, making their analysis, archival, and sharing one of the grand challenges of the 21st century. These experiments, in their majority, adopt computing models consisting of Tiers (each Tier is made up of several computing centres and provides a specific set of services) and for the different steps of data processing (simulation, filtering, calibration, reconstruction and analysis) several software packages are utilized.

In order to confront the aforementioned problems and limitations we introduced IKAROS as a framework that enables us to create ad-hoc nearby storage formations and is able to use a huge number of I/O nodes in order to increase the available bandwidth (I/O and network) [3]. It unifies remote and local access in the overall data flow by permitting direct access to each I/O node, regardless of the tier. In this way we can handle the overall data flow at the network layer, limit the interaction with the operating system and minimize disk and network contention. This approach enables us to virtually connect, at user level, the several different computing facilities used (Grids, Clouds, HPCs, Data Centers, Local computing Clusters and personal storage devices), on-demand, based on the needs, by using well known standards and protocols, like HTTP.

2. IKAROS Framework

IKAROS [3] provides a dynamically coordinated I/O architecture for I/O accesses according to infrastructure topology/profile, load metrics, and the I/O demands of each application. At Fig. 1 we show an overview of the IKAROS framework: the input parameters that we feed at IKAROS (application I/O requirements, load metrics, infrastructure profile/topology) and the resources that IKAROS manages (I/O nodes and storage media) in all the tiers of the computing model. Currently we feed the appropriate input parameters to IKAROS manually. IKAROS allows data in a file to be striped across multiple disk volumes on multiple heterogeneous nodes and provides the utility for the storage system to access and transfer a data file in parts and in parallel mode, without a specific order according to a client request.

2.1 IKAROS overall data flow scenario (remote-local access)

In this scenario we analyse a data transfer from a remote storage server to the local parallel file system, which is actually a typical use case in a tiered system computing model, e.g: LHC and KM3NeT

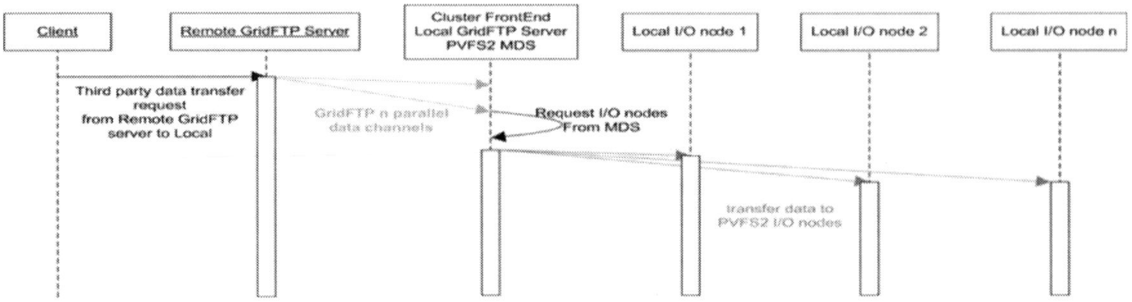

Figure 2. Overall data flow, GridFTP + PVFS2.

experiments. In Fig. 2 we show the implementation of this action by combining GridFTP and PVFS2. The client initiates a third party data transfer in order to transfer the data file from the remote GridFTP server to the local parallel file system in this case we are using PVFS2. We implement the local GridFTP server and the PVFS2 MDS at the frontend machine of the local computing cluster. Due to the client request, the remote GridFTP server starts sending the data file to the local GridFTP server by using N parallel data channels. The local GridFTP server moves the data chunks to the PVFS2 I/O nodes. The combined use of GridFTP and PVFS2, in order to implement the overall data flow, forces us to initiate many independent transfers incurring much overhead to set up and release connections. This approach can significantly impact performance due to the unnecessary network and disk contention.

The network and disk contention mainly appears due to the lack of proper coordination between the two systems, GridFTP/PVFS2. There is no guarantee that the remote access protocol and the local parallel file system, in this case GridFTP and PVFS2, have the same stripe size and the same stripe mapping.

A solution could be to use the GridFTP striped server technique, which is not exactly the case we analysed in Fig. 2. In the striped server scenario the data file will be striped at several remote GridFTP servers and the local I/O nodes will have to act both as PVFS2 I/O nodes and GridFTP servers. In this scenario, there is no guarantee that the GridFTP servers will stripe the data across the data nodes in the same sequence as PVFS2 does across the I/O servers.

In Fig. 3 we analyse, again, a data transfer from a remote storage server to the local parallel file system, but now we are using IKAROS for the overall data flow in order to avoid the unnecessary network and disk contention. Techniques like the reverse read implementation of the write request, introduced by IKAROS [3], combined with reverse HTTP tunneling techniques can help us provide proper coordination between remote and local access and achieve better performance.

This approach permits us to mainly route the data at the network level and minimize the usage of the operating system. In Fig. 3 the IKAROS client requests from the MDS system the available I/O nodes and decides the file partition distribution schema, based on the input parameters. Using this information, the client triggers each local I/O node, which participates in the transfer. Then each I/O node, on the basis of the previous trigger, requests from the remote storage server the corresponding data chunk. In this way we actually apply only coordinated parallel data transfers in contrast with the GridFTP/PVFS2 case in which we must manually synchronize the stripe size and the stripe mapping between them. In [5] we present measurements which shows that IKAROS approach outperforms the combined GridFTP+PVFS2 system.

3. KM3NeT–EGI flowchart (first use-case)

The KM3NeT computing model is based on the LHC computing model. The general concept consists of a hierarchical data processing system, commonly referred to as Tier structure and consists of several

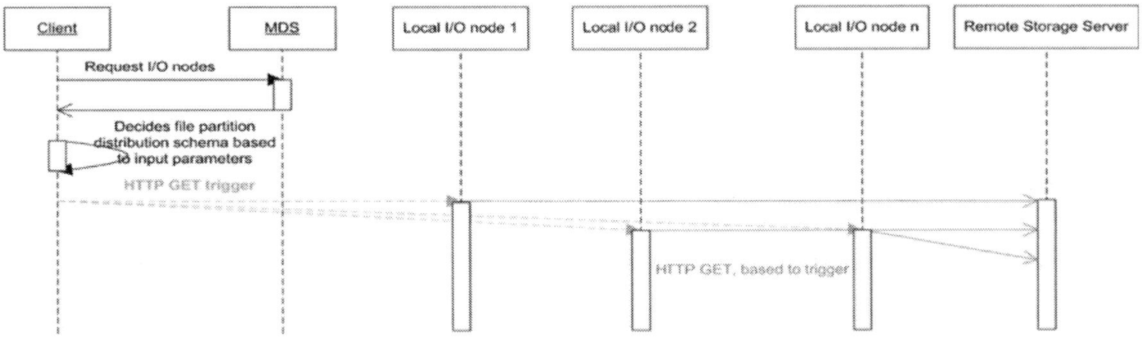

Figure 3. IKAROS overall data flow.

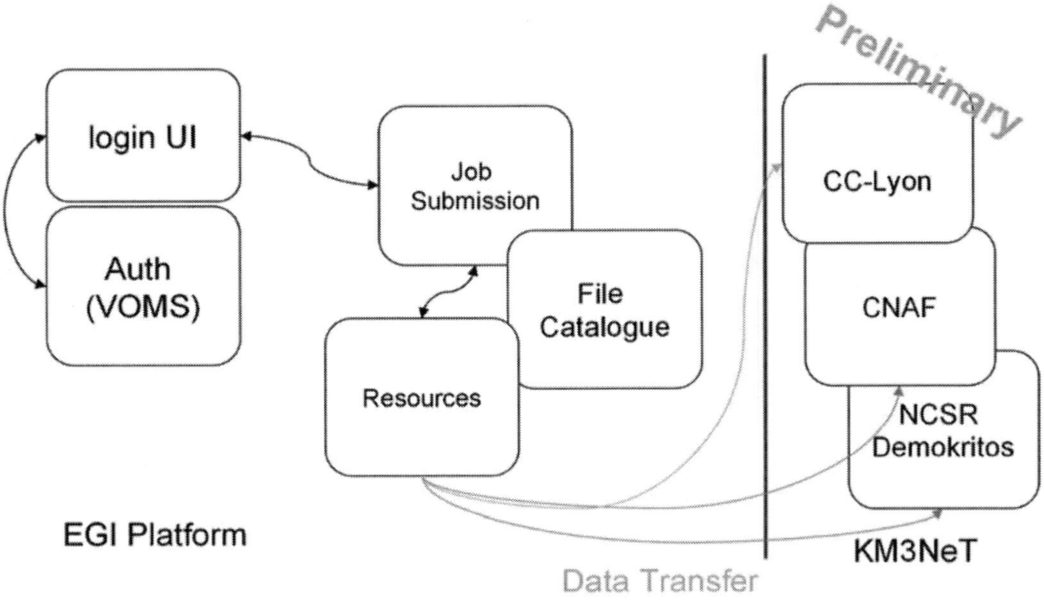

Figure 4. KM3NeT-EGI flowchart, first use-case.

Data Centers and Grid Computing infrastructures, like EGI. One of the main tasks is the efficient distribution of the data between the different computing centres – CC-IN2P3, CNAF will act as central storage, i.e. the resulting data of each processing step is transferred to those centres. The data storage at the centres is mirrored. In Fig. 3 we show the preliminary flowchart between KM3NeT and the resources hosted by EGI [4]. The data output for each process is transferred from the Grid worker node to the Grid storage element at the Napoli ReCaS Grid site and then to CC- IN2P3 at Lyon by using the GridFTP protocol or directly to user destination (laptop, local computer cluster, CC-Lyon) by using the eT-IKAROS [3] utility and the concept we demonstrated at Sect. 2.

References

[1] Schmuck, F., Haskin, R.: GPFS: a shared-disk file system for large computing clusters. In: Proceedings of the FAST 2002 Conference on File and Storage Technologies. IBM Almaden Research Center, San Jose, CA (2002)

[2] Dongarra, J., Beckman, P. et al. The International Exascale Software Roadmap," , Volume 25, Number 1, 2011, International Journal of High Performance Computer Applications, ISSN 1094-3420. Exascale Nearby Storage, Cray Position paper

[3] Filippidis, C., Cotronis, Y., & Markou, C. IKAROS: an HTTP-based distributed File System, for low consumption & low specification devices. Journal of Grid Computing, Springer, **11**, Issue 4 , pp 681–698 (2013)

[4] EGI Web site, `http://egi.eu` (2015)

[5] Filippidis, C., Cotronis, Y., & Markou, C. IKAROS: The IKAROS Metadata service as a Utility. 2014 IEEE/ACM 7th International Conference on Utility and Cloud Computing (UCC), `http://doi.ieeecomputersociety.org/10.1109/UCC.2014.75`

EPJ Web of Conferences **116**, 07002 (2016)
DOI: 10.1051/epjconf/201611607002

Enabling Grid Computing resources within the KM3NeT computing model

Christos Filippidis[a] for the KM3NeT collaboration

NCSR Demokritos

Abstract. KM3NeT is a future European deep-sea research infrastructure hosting a new generation neutrino detectors that – located at the bottom of the Mediterranean Sea – will open a new window on the universe and answer fundamental questions both in particle physics and astrophysics. International collaborative scientific experiments, like KM3NeT, are generating datasets which are increasing exponentially in both complexity and volume, making their analysis, archival, and sharing one of the grand challenges of the 21st century. These experiments, in their majority, adopt computing models consisting of different Tiers with several computing centres and providing a specific set of services for the different steps of data processing such as detector calibration, simulation and data filtering, reconstruction and analysis. The computing requirements are extremely demanding and, usually, span from serial to multi-parallel or GPU-optimized jobs. The collaborative nature of these experiments demands very frequent WAN data transfers and data sharing among individuals and groups. In order to support the aforementioned demanding computing requirements we enabled Grid Computing resources, operated by EGI, within the KM3NeT computing model. In this study we describe our first advances in this field and the method for the KM3NeT users to utilize the EGI computing resources in a simulation-driven use-case.

1. Introduction

KM3NeT is a future European deep-sea research infrastructure hosting a new generation neutrino telescope with a volume of several cubic kilometres located at the bottom of the Mediterranean Sea. The KM3NeT research infrastructure will comprise a deep-sea neutrino telescope at different sites (ARCA), a neutrino-mass-hierarchy detector (ORCA) and nodes for instrumentation for measurements of earth and sea science (ESS) communities. ARCA/ORCA, the node(s) for ESS instrumentation, the cable(s) to shore and the shore infrastructure will be constructed and operated by the KM3NeT collaboration.

International collaborative scientific experiments, like KM3NeT, are generating datasets which are increasing exponentially in both complexity and volume, making their analysis, archival, and sharing one of the grand challenges of the 21st century. These experiments, in their majority, adopt computing models consisting of different Tiers with several computing centres and providing a specific set of services for the different steps of data processing such as detector calibration, simulation and data filtering, reconstruction and analysis. The computing requirements are extremely demanding and,

[a] e-mail:cfjs@outlook.com

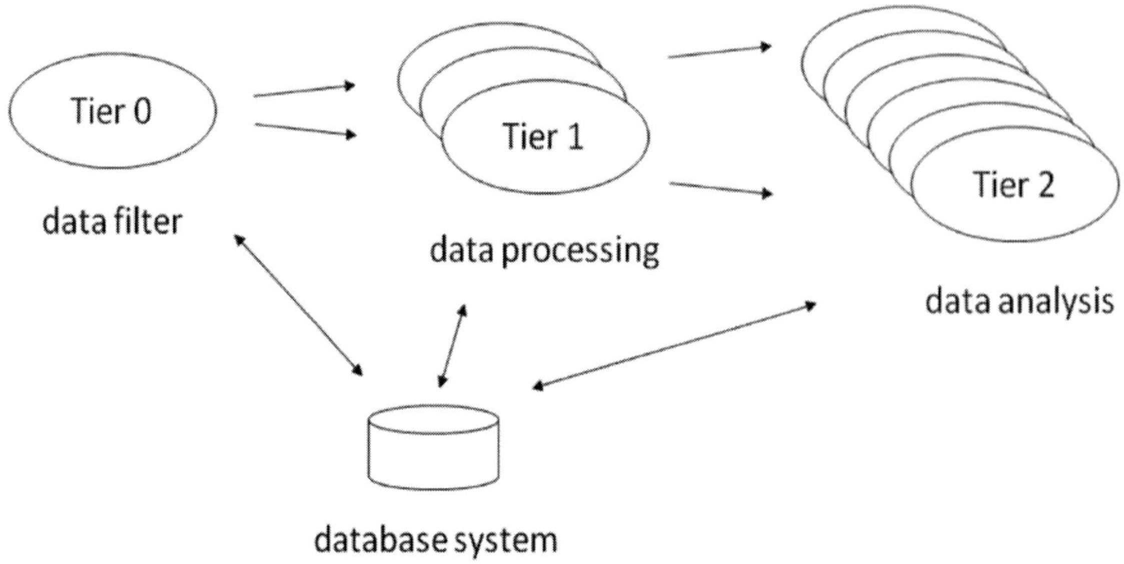

Figure 1. Data Distribution Model.

usually, span from serial to multi-parallel or GPU-optimized jobs. The collaborative nature of these experiments demands very frequent WAN data transfers and data sharing among individuals and groups. Typical, such a computing model utilizes several different computing infrastructures like: Grids, Clouds, HPCs, Data Centers and Local computing Clusters.

2. KM3NET computing model

The KM3NeT computing model (data distribution and data processing system) is based on the LHC computing model. The general concept consists of a hierarchical data processing system, commonly referred to as Tier structure (see Fig. 1).

At the first layer (Tier 0), a farm of computers filters the full raw data from the neutrino detectors (and other instruments) in real-time (all-data-to-shore strategy). This computer farm is located in the shore station at each detector site. The main requirement is the reduction of the overall data rate by about three orders of magnitudes (from 5 GB/s to 5 MB/s for one building block). The output data are temporarily stored on a persistent medium and distributed with fixed latency (typically less than few hours) to various computers centres. For detector monitoring and alert handling, selected events are reconstructed using fast algorithms with a low latency of the order of minutes (quasi-online reconstruction).

In the second layer (Tier 1), the filtered data are processed. Various models for hit patterns are fitted to the data (usually referred to as reconstruction) and the results of the fits are stored for further analysis. The overall rate of reconstructed data amounts to less than 1 MB/s.

Calibration data are used to determine the time offsets and positions of the photo-multiplier tubes (PMTs). The results of the calibration are stored in a central database system. The typical update frequencies of the PMT positions and time offsets are 10^{-2} Hz and 10^{-5} Hz, respectively. The corresponding data rate amounts to less than 1 MB/s. These data are necessary input to the reconstruction to achieve an optimal pointing resolution of the detectors. An automated calibration procedure is needed to process the filtered data in a timely manner. Ultimately, the calibration and the reconstruction procedures should operate with fixed and limited latency (few hours). The results from the reconstruction are made available to all scientists for detailed analyses and dissemination of scientific results.

Table 1. KM3NeT Computing Resources.

Tier	Computing Facility	Main Task	Access
Tier-0	at detector site	online processing	direct access, direct processing
Tier-1	CC-IN2P3	general offline processing and central data storage	direct access, batch processing and grid access
	CNAF	central data storage	grid access
	ReCaS	general offline processing processing, interim data storage	grid access
	HellasGRID	reconstruction of data	grid access
	HOU computing cluster	simulation processing	direct access, batch processing
Tier-2	local computing clusters	simulation and analysis	varying

To assess the detector efficiency and systematics, dedicated Monte-Carlo-simulations are needed. Due to the changing running conditions of the detector in the deep-sea environment, time (condition)-dependent simulation data sets (run-by-run simulations) are considered – the data taking is split in sufficiently small time intervals with stable conditions (referred to as runs) and the detector response is simulated individually for these periods, allowing for a direct comparison of experiment and simulation. Due to the construction of the detector in independent, nearly identical building blocks, simulations can be performed for the fixed geometry of two types of building blocks (ARCA and ORCA) – the overall detector performance can be directly deduced from that.

The analyses of the reconstructed data are primarily processed at the local computing clusters of the partner institutes (Tier 2). The following computing centres and pools provide resources for the KM3NeT experiment:

3. KM3NET- EGI flowchart (first use-case)

For the steps of job submission and user authentication-authorization the default EGI [1] procedure is being used. For these purposes the collaboration has set up the "KM3NeT.org" Virtual Organization (VO). At Table 2 we show the preliminary KM3NeT.org central services, within the EGI platform.

One of the main tasks is the efficient distribution of the data between the different computing centres – CC-IN2P3, CNAF will act as central storage, i.e. the resulting data of each processing step are transferred to those centres. The data storage at the centres is mirrored. In Fig. 3 the preliminary flowchart between KM3NeT and the resources hosted by EGI is shown. The data output for each process is being transferred from the Grid worker node to the Grid storage element at the Napoli ReCaS Grid site and then to CC- IN2P3 at Lyon by using the gridFTP protocol or directly to user destination (laptop, local computer cluster, CC-Lyon) by using the eT-IKAROS [2] utility.

Our biggest challenge for the near future is to finalize the VO policy and develop data management tools and utilities that will create a smooth connection between the EGI platform and the KM3NeT, in order to fulfill the KM3NeT computing model requirements.

Table 2. KM3NeT.org VO central services.

Service	Grid Site
Authentication/authorization system (VOMS)	RECAS-NAPOLI
User Interface (UI)	RECAS-NAPOLI, HG-08-Okeanos
Logical File Catalogue (associate a logical file name to a physical file)	RECAS-NAPOLI
Job submission and management system (WMS)	HG-03-AUTH

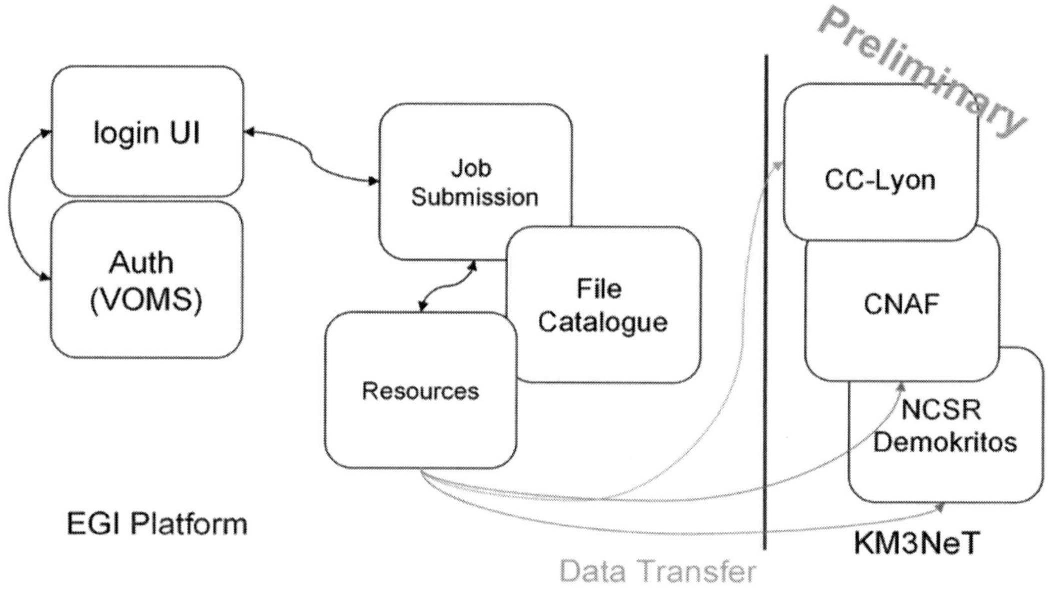

Figure 2. KM3NeT-EGI flowchart, first use-case.

References

[1] EGI web site, `http://www.egi.eu/` (2015)

[2] Filippidis, C., Cotronis, Y., & Markou, C. IKAROS: an HTTP-based distributed File System, for low consumption & low specification devices. Journal of Grid Computing, Springer, **11**, Issue 4, pp 681–698 (2013)

EPJ Web of Conferences **116**, 07003 (2016)
DOI: 10.1051/epjconf/201611607003

IceCube simulation production and the transition to IceProd2

David Schultz[a]

Wisconsin IceCube Particle Astrophysics Center, University of Wisconsin-Madison, 222 W Washington Ave, Suite 500, Madison, WI 53703, USA

Abstract. IceCube's simulation production relies largely on dynamic, heterogeneous resources spread around the world. Datasets consist of many thousands of job workflow subsets running in parallel as directed acyclic graphs (DAGs) and using varying resources. IceProd is a set of Python daemons which process job workflow and maintain configuration and status information on jobs before, during, and after processing. IceProd manages a complex workflow of DAGs to distribute jobs across all computing grids and optimize resource usage. IceProd2 is a new version of IceProd with substantial increases in security, reliability, scalability, and ease of use. It is undergoing testing and will be deployed this fall.

1. Introduction

IceProd has served the IceCube collaboration well over the last 8 years, processing Monte Carlo simulations, detector data, and analysis levels [1]. It has run thousands of CPU-core years and stored over 2PB of data. But it has experienced its share of issues and the initial software is running into fundamental design limitations. This article describes that history and the development of a second version of IceProd. The new version is a complete rewrite of the code base, fixing long-standing issues and preparing for the increased load of the next few years.

2. IceCube and computing

The IceCube collaboration uses two main data centers, one in North America (University of Wisconsin-Madison) and one in Europe (DESY Zeuthen). Other institutions generally only have limited access to shared campus resources or smaller compute clusters. Additionally, there is significant availability of opportunistic resources via providers like the Open Science Grid [2]. Currently, over 50% of all CPU resources are opportunistic. One of the highlights of the IceCube simulation chain is the direct propagation of photons using GPU accelerators [3]. Over the last 3 years IceCube has deployed more than 500 GPUs to help run this part of the simulation chain.

The IceCube simulation and data processing jobs have a large variability in resource requirements. Although most simulation jobs need between 2–4 GB of memory, several types of jobs require 10+ GB of memory. Certain simulation generation jobs require significant scratch disk, on the order of

[a] e-mail: david.schultz@icecube.wisc.edu

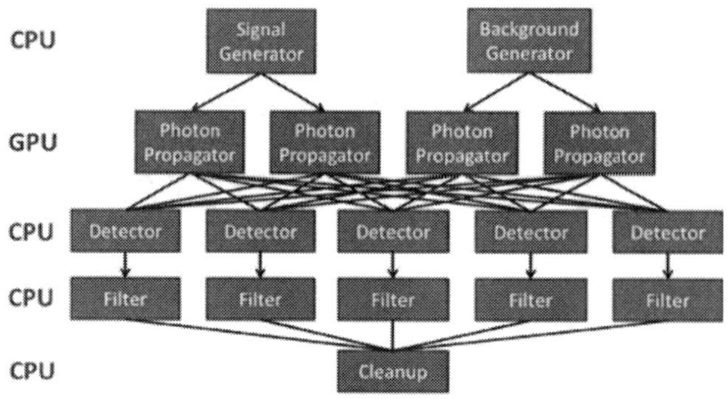

Figure 1. Simulation job streams contain CPU and GPU jobs running at different sites. This DAG has many interdependent jobs making up one "stream."

100 GB, to manage temporary files. Finally, one of the problems with Monte Carlo simulations is that rare events, such as extremely high energy events, can take an order of magnitude more resources, especially walltime. Handling these randomly occurring events is an ongoing issue.

Based on this information, requirements have become apparent [4]. Because the resources aren't owned, anything that requires root permission is unlikely to work. The mix in resources leads to multiple OS types and schedulers that need to be supported. Many of the smaller sites cannot handle glideinWMS, an easy way to join multiple sites into a global pool [5]. This points to a light-weight and flexible approach, with needs not met by other workflow systems.

2.1 IceProd job workflow

The core software framework of IceCube, called IceTray, is stream-based with modules that act on events in the stream [6]. Simulation jobs run similarly, with a series of modular jobs forming a stream (or multiple streams) of data that is incrementally processed to further levels. A series of these streams can run in parallel with different random seeds, and forms a dataset. It is typical to have up to a few hundred datasets running concurrently.

The technical term for the individual job stream is a directed acyclic graph (DAG). In Fig. 1 a more complex DAG can be seen, where two simulation generators run on CPUs, feeding four GPU jobs, which in turn feed their output to five parallel CPU jobs. These complex DAGs are used in real production, but they currently create significant scheduling overhead with extra database lookups to check dependency status; thus, they are only used when really needed. Simpler DAGs are preferred.

3. IceProd v1

The initial version of IceProd is a set of user-space site-local Python daemons backed by a central database [7]. All knowledge of state is stored in the database, with the Python daemons interacting with the queueing system or the jobs directly. One daemon submits new jobs and keeps track of the queue. Another daemon listens for XML-RPC from running jobs so their status can be updated as the job runs through its lifecycle. A third daemon is responsible for any cleanup tasks that are required, and a final daemon responds to user commands such as dataset submission, suspension, or deletion. A web interface was developed in PHP that interfaced with the database and displayed its contents in an easy to read format, along with forwarding user commands to the IceProd daemon to control datasets. Plugins have been designed for each batch system or other job submission interface. This turns a multitude of heterogeneous systems into a homogeneous API that IceProd can interact with. It also allows for new systems to be incorporated over time with minimal effort.

One of the major additions to IceProd came with the use of GPUs, making it necessary to break up jobs into CPU and GPU parts. Initially, with GPUs in a testing phase, this was accomplished locally at a few sites with DAGMan for HTCondor [8]. This worked fairly reliably, but did not extend to other job submission systems and did not allow jobs from one site to utilize GPUs at another site. The DAG feature was integrated directly into IceProd two years ago and solved cross-site linking.

When IceProd was first designed 8 years ago computing clusters were generally small, only a few hundred nodes at most, and with jobs taking hours to complete the queue rate was fairly slow. Today shorter jobs and clusters of thousands of nodes are showing bottlenecks in multiple places of IceProd v1. With jobs being individually queued and typically taking a second or more to queue, we are reaching the situation where constantly queueing can't saturate a cluster. The move to DAGs of jobs has led to a substantial rise in the number of queries on the database. At times the sheer number of queries per second is enough to slow things down, and we have to be very careful of the new queries we add. We are also starting to hit connection limits because of simultaneous updates of job statuses.

Several design decisions are hampering new ideas and current activity. There are better ways to do some things, but they are almost impossible to integrate into the current framework. On top of this is the code cruft accumulated over 8 years of patches without much cleanup.

4. IceProd v2

The second version of IceProd is a complete rewrite of the codebase, focusing on current and future usage and eliminating old usage patterns. Instead of being add-ons, features like global DAG are now standard, with a more efficient implementation that was not possible before. There is also a strong focus on unit tests, code coverage and documentation, which were lacking in IceProd v1. Python 3 and other future-looking software is fully supported.

There are several major improvements, but one of the overarching ones is to be able to run on stock Python with no extra packages installed. While performance is degraded, it is sufficient for individual users or testing purposes. It also allows a single-site IceProd instance to be downloaded, set up, and started very quickly. This would allow, for instance, individual analyzers to set up their own grid infrastructure [4].

4.1 Database

The database is the single largest change to IceProd v2. It has moved from one central database to a local database at each site. This allows standalone single-site instances as well as continued running when the master/global queue is down. It also has a side effect of spreading out the load, since most requests from jobs or users only need to talk to the local database instead of all hitting the master.

The primary database type is SQLite, which comes integrated with Python. Support is also available for MySQL for higher performance; MongoDB support is planned to handle the large amount of statistics accumulated at the master node.

4.2 Scalability

All external communications with IceProd v2 flow through the web server, so it must be fast and scalable. Not only does it handle user-facing websites, but also job-to-site and site-to-site communications. It is built on non-blocking principles, offloading longer tasks to worker threads or other processes. It can easily handle 10 thousand connections per minute, even with over a thousand sustained connections.

4.3 Job optimization

One long-desired feature for IceProd has been a pilot infrastructure. Especially in the last few years, shorter jobs under an hour have become normal. Clusters and queueing systems generally don't like short jobs, so the obvious solution is to run multiple jobs one after the other inside the "single" job the queueing system is running. The pilot infrastructure takes this a step further by only asking for jobs that match the node's resources where the pilot is started. This allows jobs to run where they are best matched and reduces failures.

4.4 Software distribution

While not technically part of IceProd, software distribution is strongly related to it. Jobs need software to run, and for IceProd v1 this meant tarballs shipped with each job and extracted on the worker node. While this will still be supported for legacy operations in IceProd v2, the primary mode of software distribution is moving to CernVM-FS [9]. For sites that have CernVM-FS installed, it is used natively. Otherwise, an I/O interposition agent called Parrot is used to make the filesystem appear to exist, dynamically fetching necessary files from the network and presenting them to the application in the correct location [10].

4.5 JSON API

IceProd v2 will feature a full JSON API as the primary interface to controlling IceProd. This allows great flexibility because of JSON's ease of use and ubiquity. This gives a single, shared interface that both the website and command line clients can use as a backend. It also allows easy creation of new frontends in the future.

5. Conclusions

IceCube simulation production has heterogeneous, significantly opportunistic resources and diverse job requirements. IceProd is an easy to install user-level middleware designed to easily manage and monitor large sets of distributed jobs. With IceProd v1 straining under current usage, IceProd v2 is a complete rewrite designed to easily handle current production and more easily cope with future needs. It can scale from a single user at one site up to sites around the world and many millions of jobs. IceProd v2 is currently in beta testing for production deployment, which is scheduled for late 2015. Though it was originally developed for managing IceCube simulation production, IceProd is general enough for many types of grid applications and is generally available to the scientific community.

The following entities supported this work: U.S. National Science Foundation; Center for High Throughput Computing at the University of Wisconsin–Madison; Open Science Grid and the U.S. Department of Energy; National Energy Research Scientific Computing Center; WestGrid and Compute/Calcul Canada; Deutsches Elektronen-Synchrotron Grid Centre.

References

[1] M.G. Aartsen et al., J. Parallel Distrib. Comput. *The IceProd framework: Distributed data processing for the IceCube neutrino observatory* **75**, 198–211 (2015)

[2] R. Pordes et al., J. Phys. Conf. Ser. **78**, 012057 (2007)

[3] D. Chirkin, Nucl. Instrum. And Methods Phys. Res. Section A: Accelerators, Spectrometers, Detectors Associated Equipment *Photon tracking with GPUs in IceCube* **725**, 141–143 (2013)

[4] D. Schultz, J. Phys. Conf. Ser. **664**, 062056 (2015)
[5] I. Sfiligoi, J. Phys. Conf. Ser. **119**, 062044 (2008)
[6] T. DeYoung, Int. Conf. on Comp. in High-Energy Phys. And Nucl. Phys. (CHEP 2004) *IceTray: a software framework for IceCube* 463–466 (2005)
[7] J. C. Díaz-Vélez, J. Phys. Conf. Ser. **513**, 032026 (2014)
[8] P. Couvares, T. Kosar, A. Roy, J. Weber, K. Wenger, Workflows for e-Science *Workflow in Condor* (2007)
[9] J. Blomer, P. Buncic, T. Fuhrmann, Proc of the 1st int. workshop on Network-aware data management (NDM'11) 49–56 (2011)
[10] D. Thain, M. Livny, Scalable Computing: Practice Experience **6**, 9 (2005)

EPJ Web of Conferences **116**, 07004 (2016)
DOI: 10.1051/epjconf/201611607004

The relational database system of KM3NeT

Arnauld Albert[1] and Cristiano Bozza[2,a] for the KM3NeT Collaboration

[1] Université de Strasbourg, Université de Haute Alsace, GRPHE, 34 rue du Grillenbreit, Colmar, 68008 France

[2] University of Salerno, Dipartimento di Fisica, via Giovanni Paolo II 132, Fisciano 84084, Italy

Abstract. The KM3NeT Collaboration is building a new generation of neutrino telescopes in the Mediterranean Sea. For these telescopes, a relational database is designed and implemented for several purposes, such as the centralised management of accounts, the storage of all documentation about components and the status of the detector and information about slow control and calibration data. It also contains information useful during the construction and the data acquisition phases. Highlights in the database schema, storage and management are discussed along with design choices that have impact on performances. In most cases, the database is not accessed directly by applications, but via a custom designed Web application server.

1. Overview

The KM3NeT Collaboration [1] is installing water-Cherenkov neutrino telescopes [2] in several sites in the Mediterranean Sea. First data are expected by the end of 2015. The detectors of the telescopes comprise many detection units, i.e. vertical lines supporting 18 regularly spaced digital optical modules; each optical module hosts 31 photomultipliers, monitoring instruments and piezo elements for acoustic positioning of the modules. The detectors are subdivided into detector *building blocks* each comprising 115 detection units with 64,170 photomultipliers in total. The complexity of tasks of the detector and the inherently multi-site nature of the challenge call for suitable data management tools. A relational database system [3] has been set up and is already providing operational functionality while it is still partially in development.

Many tables and concepts in the design of the database are inherited from the database of ANTARES [4], but substantial evolution was needed scaling performances up to meet the KM3NeT needs. This work has already started a few years ago [5]. The foundation of the system is Oracle Database Server [6]. Reflecting the geographical distribution of the telescopes, a multi-master database network is in place: a cluster of 3 servers is hosted and managed by the CC-IN2P3 centre in Lyon; another database server is hosted at the University of Napoli (UNINA), under the management of the ReCaS consortium. Two web servers, also hosted at CC-IN2P3 and at UNINA/ReCaS, are used to provide access to the databases. Both servers have interchangeable links, i.e. each one can connect to either master database.

---

[a] e-mail: cbozza@unisa.it

Table 1. Groups of tables in the KM3NeT database schema.

Group	Description
KM3NeT accounts	Members and external personnel, affiliation, roles, security credentials
Locations	Locations relevant to the KM3NeT research infrastructure
Institutions	Institutions and entities that have membership or other relationships with KM3NeT Collaboration
Operations	Simple or complex actions (possibly hierarchically described) that contain space and time information, e.g. construction, qualification tests, detector operation, etc.
Products	Hardware and software components of each detector and the related data acquisition and post-processing chain
Parameters	Definition of quantities that can be documented
Values	Values of parameters from datasheets, calibrations, tests, detector setup
Data streams	Streams of calibration data from quasi-on-line post-processing and from acoustic sensors (mostly used to determine the position and orientation of the optical modules and the shape of the detection units)

The database system is conceived as a living system that directly supports and takes part in several tasks, e.g. Collaboration management, detector construction and operation, data reconstruction and analysis and simulations. High performances both in writing and reading data are crucial for the whole system to be useful. System reliability and availability are very important as well for involving the database in real time data production and data collection; without these features the database would be just an offline container.

At the time of writing, the full schema of the database contains 92 tables which will not be described in detail. Instead, groups of tables that are logically linked are sketched as shown in Table 1.

2. Security

In order to ensure data security and reliable database operation, a layered security model has been adopted to provide user and batch process access. At the lowest layer, Oracle Database roles define consistent sets of privileges and permissions for access to tables, views, functions and procedures. Oracle Database accounts are provided to services with specific roles, such as the Web server and the shore station database interface. All duty accesses use HTTP(S) services, e.g. both the Web server and the shore station database interface behave as HTTP(S) servers. Physical users have KM3NeT accounts with irreversibly encrypted passwords that do not correspond to any Oracle account; and their privileges are defined in terms of KM3NeT responsibilities, e.g. membership of governing bodies, construction management or data-taking management. All KM3NeT IT services, such as the Wiki, SVN, TRAC and Google Drive access the database via the Web server as the central provider of authentication credentials.

3. Data flows

All relevant information is described in terms of *parameters*. The scalar type for each parameter is defined, and, when applicable, also the international standard unit is declared. Datasheets and tests for qualification, characterisation and acceptance are all described in terms of the set of parameters they yield. In addition, the database contains the information needed to prepare commonly used plots and graphs from the data of each dataset. This enables the Web server to automatically prepare reports in a user-friendly fashion.

The most relevant data flows in the KM3NeT database are related to detector construction, data-taking and offline event-reconstruction/simulation. The construction data flow starts with the description of purchased components; next, each object that is the result of an integration of smaller components is described in terms of a template and the real components that were used to assemble the product; cabling information is also added when needed. All results of tests (acceptance/qualification/characterisation) on objects are documented in the database. An object that is the result of an integration operation may on its turn be part of a more complex product, e.g. several optical modules are used to build a detection unit. This implies that the database is part of a loop in the information flow. A single photomultiplier test can involve up to about 10^5 measurements, corresponding to about 10 MB; this leads to the estimate of about 1 TB/*building block* for test results.

The information collected about a complex product is later used to drive data acquisition tasks, which need the full description of the detector. In addition, for several operational parameters of the detector defaults are determined automatically from test information. For instance, the supply voltage of a photomultiplier is extracted from the characterisation data provided by the vendor; then it is first optimized on a test bench; later it is re-optimized once the detection unit is fully assembled. This last value is then used during normal data acquisition tasks to set the working point of the photomultiplier. A *runsetup* is a full set of tuneable parameters for an acquisition task. It is possible to have several *runsetups*, according to different tasks and working conditions of the detector: for example, in times of high bioluminescence one might have a different tuning of supply voltage, threshold and trigger to reduce the trigger rate. The database hosts all available *runsetups* and the timeline of all *runsetups* used during the detector lifetime, so that data analysis and event reconstruction can be tuned accordingly. The output parameters from detector monitoring and slow control are read continuously (10^{-2} Hz) and fed into the database, with an estimated data rate of a few TB/y for a detector *building block*. The detector is calibrated and its shape is recomputed continuously. The expected data rate for calibration information and for acoustic data and shape parameters also tops a few TB/year.

The KM3NeT detectors will be built incrementally: each detection unit can start taking data independently of the availability of the others, as soon as it is installed on the seabed and connected to the seabed infrastructure. Hence, the telescope as a whole has a history that requires bookkeeping.

Finally, also for the purpose of simulating the expected output of the detector, detailed knowledge of detector configurations and working *runsetups* are needed. The database is thus accessed also by simulation programs to extract such information.

4. Storage

Tables in KM3NeT database belong to either of two categories: *critical* or *non-critical*, depending on the expected cardinality in 10 years: the criticality threshold is set at about 10^6 in 10 years. Non-critical objects get a database-wide unique ID from a single character sequence. Their storage and retrieval does not require special care. The category of critical object is worth some discussion. Critical tables all contain time series of parameters or event logs. Usual heap-organized tables with several fields per record would pose several problems:

1) With 10^{12} records, adding or removing a single field is hardly feasible because the tables would be locked for more than one week.
2) Reading a single field requires reading the full row of the table, with extra disk activity and reduced access speed.
3) Accessing a time interval of values would require index scans of the primary key in addition to the table, implying sub-optimal performance.

The solution is to use a *columnar* database storage scheme; more properly, it is quasi-columnar because each record also includes the timestamp. Data tables differ only by the data type: byte, 16-bit word, 32-bit double-word, 64-bit quad-word, single-precision floating point, double-precision floating point, string or event. The primary key includes the identifier of the parameter, its source (e.g. which photomultiplier or optical module it is related to) and the timestamp; a single value field contains the timestamped value of the parameter or the event code. Data are physically stored as if they were a stream, not repeating the unchanged key fields at every record, but storing only the timestamp and related value. This is possible by using index-organized tables with key compression. Repeated fields are skipped by physically storing the table data into the primary key. This resembles a normal data file, which contains a small header section followed by time series of data. Obvious advantages are:

1) Because identical data are not repeated, it is common to reduce the total storage size by a factor 4 or more;
2) Record access speed improves by the same factor.

5. Access via Web services

An HTTP(S) server has been written as a custom C# [7] library that can be added to any Common Language Runtime (CLR) [8] executable, so its source code is completely under KM3NeT control. It is normally run with the Mono [9] CLR on Linux, but it works with all major operating systems as well without recompilation. The Web server that provides the main access point to the database is based on the HTTP(S) server, but does much more than just providing static or dynamic Web pages for graphical user interface: it has indeed been conceived rather as an Application Web server with data management and distribution tasks. For instance, if a large (GB-size) dataset is sent for insertion into the database, a deferred insertion job is started and the result (including possible errors) is mailed later to the user that posted the dataset. Commonly accessed datasets might be automatically cached as files on disk, ready to be served to incoming requests. The cache management algorithm may consider both data size and request frequency to maximize the throughput in a proactive way. Several IT services access the Web server via *server pages* (i.e. pages not intended for human usage) to read and write data from the database. Data can be exchanged in XML or JSON format. SQL queries are replaced by HTTP GET requests with query parameters in the URL. Simple scripting tools such as *wget* and *cURL* can also be used, as well as any Web client libraries ready for inclusion in programs that need access to database information.

In the shore station, access to the database goes via the Database Interface, providing access to specific datasets through HTTP GET requests and returning data in XML format. This service also takes care of uploading detector monitoring files and calibration/acoustic data to the database. Programs that need to write to the database just have to produce intermediate files in a proper format, but they don't need to store database passwords or to depend on SQL clients to work, because the Database Interface deals with all the technicalities in a single place.

References

[1] `http://www.km3net.org`
[2] S. Adrián-Martínez et al., Eur. Phys. J. C **74**, 1 (2014)
[3] E. F. Codd, Comm. ACM. **13**(6), 377–387 (1970)
[4] A. Albert et al., Nucl. Inst. Meth. A **626**, S240–S242 (2011)
[5] A. Albert, C. Bozza, AIP Conf. Proc. **1631**, 167 (2014)
[6] `http://www.oracle.com`
[7] `http://www.ecma-international.org/publications/standards/Ecma-334.htm`
[8] `http://www.ecma-international.org/publications/standards/Ecma-335.htm`
[9] `http://www.go-mono.com`

EPJ Web of Conferences **116**, 07005 (2016)
DOI: 10.1051/epjconf/201611607005

A study on implementing a multithreaded version of the SIRENE detector simulation software for high energy neutrinos

Petros Giannakopoulos[2], Michail Gkoumas[2], Ioannis Diplas[2], Georgios Voularinos[2], Theofanis Vlachos[1], Konstantia Balasi[1], Ekaterini Tzamariudaki[1], Christos Filippidis[1,2], Yiannis Cotronis[2], and Christos Markou[1]

[1] NCSR Demokritos, Greece
[2] National and Kapodistrian University of Athens, Greece

Abstract. The primary objective of SIRENE is to simulate the response to neutrino events of any type of high energy neutrino telescope. Additionally, it implements different geometries for a neutrino detector and different configurations and characteristics of photo-multiplier tubes (PMTs) inside the optical modules of the detector through a library of C++ classes. This could be considered a massive statistical analysis of photo-electrons. Aim of this work is the development of a multithreaded version of the SIRENE detector simulation software for high energy neutrinos. This approach allows utilization of multiple CPU cores leading to a potentially significant decrease in the required execution time compared to the sequential code. We are making use of the OpenMP framework for the production of multithreaded code running on the CPU. Finally, we analyze the feasibility of a GPU-accelerated implementation.

1. Introduction

SIRENE [1] receives as input an event file, ROOT [2] formatted, with the information from the generator level and uses a separate file for the description of the detector geometry. The Probability Density Functions of the arrival time of light (PDFs) [4] are read from 4 files, each containing the data of the Cumulative Density Functions (CDFs) of direct light from a muon, scattered light from a muon, direct light from an electromagnetic shower and scattered light from an electromagnetic shower, respectively. The processing steps followed by SIRENE are outlined as follows: 1) parse Event from input file, 2) propagate muon(s) by simulating energy loss and EM-showers and generating hits originating from direct and single scattered light, 3) process shower particles from primary vertex and generate hits from direct and single scattered light, 4) merge hits to speed-up TriggerEfficiency, 5) write computed event to ROOT-formatted output file.

2. Overview of SIRENE

SIRENE implements a number of nested "for" loops for the computation of hits on OMs (Optical Modules) and on the PMTs within each OM: 1) Event "for" loop: consists of a number of particles described as Tracks.

Events are independent of each other. 2) Tracks "for" loop: Describes the path of each particle, starting with an initial energy E_0 and initial position on z-axis z_0. Energy loss and propagation is calculated in steps on each iteration. Hits from both direct/scattered light from muons/EM showers are computed. Hits registered on PMTs along the particle's path are computed via two procedures: a) integration of the CDF via polynomial interpolation approximation for calculating photon emissions, b) Poisson Distribution applied on arrival times of photo-electrons for calculating the number of photo-electrons on an OM and included PMTs.

2.1 Time requirements

The most time-consuming functions of SIRENE, in descending order of share of the total execution time, are: 1) calculation of hit probabilities on OMs and their PMTs for each Track and 2) particle propagation that includes the coordinate system transformations for the Tracks and PMTs. Decomposing, 1) accounts for $\sim 90\%$ of total execution time, while 2) and various secondary functions, such as I/O, amount to the rest 10%. It is evident that the calculations of hit probabilities are by far the heaviest part of SIRENE, so speeding up its execution could have a significant positive effect on the total running time.

3. Multithreaded SIRENE

Events are independent of each other, do not exchange any data and a dataset for an analysis requires the amount of approximately one million events or higher. Therefore they can be spread over multiple compute nodes by sending a number of events for computation to each node. This can be trivially done via the use of batch jobs. Each node will produce an output for the events assigned to it. When all nodes are finished we can collect and sort the outputs to piece together the complete simulation. The main focus of our implementation is how to take advantage of the multiple cores available in the CPU(s) within each node in order to accelerate the computations pertaining to each event. By examining SIRENE's most time-consuming segments of the original sequential code, the next step is to identify the most suitable methods of parallelization for each of these segments:

a. Hit detection

For each particle track all the calculations for probabilities of a hit are performed on each OM of the detector. For each OM, the probabilities of this track reaching the OM and registering hit(s) on the PMT(s) inside are calculated. There are typically hundreds of OMs in a detector providing enough work that can be shared between many CPU cores. A team of OMs can be assigned to each core allowing a potentially significant speed-up over having all OMs assigned to a single core. The size of the team can vary from 1 to N, where N is the total number of OMs in the detector, depending on the available CPU cores. This is visualized in Fig. 1.

b. Coordinate system rotations

Energy loss and position are calculated in discrete steps along the particle path. To simplify the simulation, the coordinate system of a particle is rotated such as the path lies only on the z-axis. The same transformation is applied to each OM of the detector so that the enclosed PMTs point along the z-axis. Since this rotation is applied to every OM of the detector it can also be executed in parallel by assigning a team of OMs to each thread.

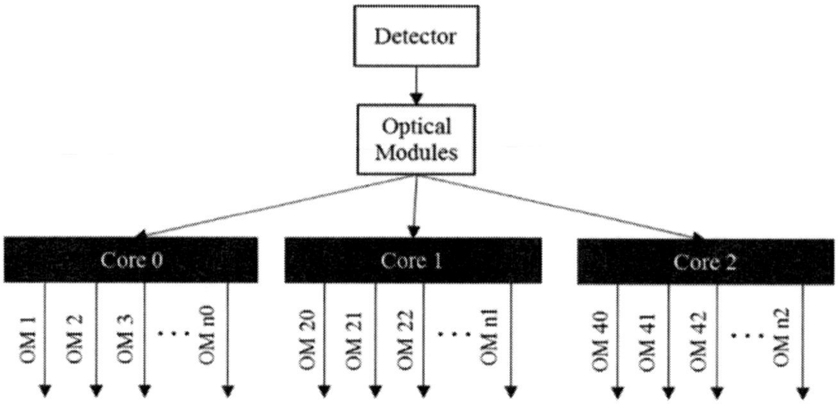

Figure 1. OMs spread over multiple CPU cores.

c. PDFs to CDFs conversion
Conversion of PDFs to CDFs doesn't require a significant amount of time, but the construction of the CDF tables can also be parallelized for additional speed-up.

3.1 Implementation

Analyzing SIRENE, as shown in Hit Detection, emerged the need for the use of more CPU cores. Another key point is the fact that the output is saved in a shared object, "Event", which is ROOT-formatted. As a result, one "machine" should be involved, otherwise multiple "machines" would require a great deal of overhead. OpenMP [4] is able to operate under these circumstances. It is implemented by enlisting one thread per OM. Each thread is capable of creating sub-threads as needed, according to the inner loops and the available system resources. The hit calculations of muon and EM showers are totally independent of each other, in different consecutive blocks inside different OM blocks. For the implementation, the appropriate OpenMP constructs are employed to share the work among several threads and for conflict avoidance when shared variables are accessed from multiple threads. Alterations to the original code structure are made wherever needed in order to facilitate more efficient parallel execution by optimizing order of execution and variable accesses for performance.

3.2 Performance results

Comparisons between the sequential and the parallel implementations of SIRENE were performed to gauge the benefits in performance offered by parallelization (Fig. 2). On our test system, the part of SIRENE responsible for detection of photon hits on PMTs required $\sim$10 ms per event for the sequential version and $\sim$4 ms for the multithreaded version across all particle energies. We can observe that with 4 cores (= 4 threads) available to the multithreaded version, hit detection on PMTs needs $\sim$2.5$\times$ less time to complete compared to the sequential version running on only 1 core (= 1 thread). Total execution time measures $\sim$11 ms for the sequential version and $\sim$5.5 ms for the multithreaded version, a $\sim$2.1$\times$ decrease in running time. Total running time includes non-parallelizable functions, such as I/O, which sets a limit to the maximum speed-up that can be achieved via parallelization. These functions make up only a small part of SIRENE so their negative impact on total speed-up remains small yet measurable.

By dividing the whole energy spectrum (*All-E*) into three slices (*Low-E* – 100GeV to 10 TeV, *Mid-E* – 10 TeV to 1 PeV, *High-E* – higher than 1 PeV), we can observe the contribution of each. In this case, most particles have *Low-E*, some *Mid-E* and few have *High-E*. The higher the energy, the more photons

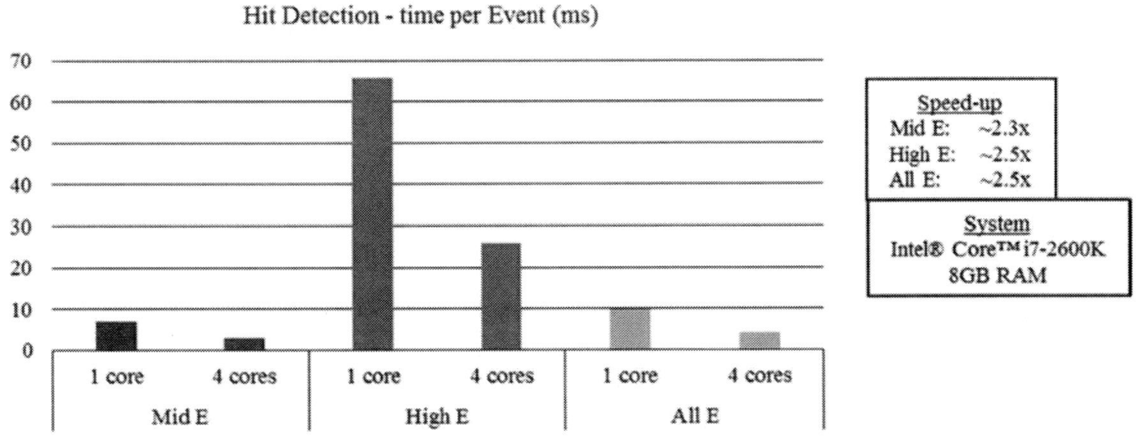

Figure 2. Running times of sequential and multi-threaded versions of SIRENE.

are produced and the higher the computational load, which is evident by the running times of *Mid-E* and *High-E* events in Fig. 2. *Low-E* events are not included because of their very short running time, yet, due to their large number, they have the highest contribution to the average running time across all energies. We can also observe a bit higher speed-up for the *High-E* events (~2.5×) compared to *Mid-E* events (~2.3×) and *Low-E* events (~2×) which can be accounted to the increasing computational load.

In order to test the scaling of the implementation with the number of threads, we also measured running using 2 cores. We observed ~1.7× speed-up with 2 threads compared to ~2.5× with 4 threads. This points to factors that limit the scaling of the implementation with the available threads. The most limiting factor is thread idling created by the computational load imbalance between different loop iterations on account of varying particle energies. Measures have been taken to mitigate the impact of load imbalance, such as using dynamic thread scheduling; however these methods can incur additional overhead.

4. Conclusion

SIRENE multithreaded implementation has delivered a positive result in terms of time saving by taking advantage of the multiple cores present in every modern CPU. However, it is possible that even higher gains can be achieved by utilising the compute capabilities of GPU. We have already implemented a GPU version of the coordinate systems transformations [5] by employing the CUDA framework. The relatively small computational workload provided by 1 event constrains the speed-up to ~4× and GPU is under-utilized. In order to get the maximum of the capabilities of GPU a parallel algorithm is needed to address more input data within the available system memory. Proof is the performance result of the GPU implementation. However, this can change when the hit detection computations, which are significantly heavier, are also offloaded to the GPU. In order to achieve this, significant functions which SIRENE uses repeatedly, such as Random Number Generation and Polynomial or Spline Interpolation, need to be implemented using parallel versions of these functions, suitable to run on a GPU. For example, the NVIDIA CUDA Random Number Generation library – cuRAND [6] can be used for fast random number generation on the GPU and the Cubic B-Spline interpolation algorithm for CUDA [7] can replace the CPU-oriented algorithm.

References

[1] C.M.M. Colnard, *Ultra-high energy neutrino simulations* (2009)

[2] R. Brun and F. Rademakers, *ROOT – An Object Oriented Data Analysis Framework*, Proceedings of AIHENP'96 Workshop, Lausanne (Sep 1996), Nucl. In st. & Meth. in Phys. Rev. A **389** (1997) 81-86. Also: `http://root.cern.ch/`

[4] M. de Jong, *The probability density function of the arrival time of light* (March 13, 2012)

[4] OpenMP Architecture Review Board, *OpenMP Application Program Interface - Version 4.0* (July 2013)

[5] I.D. Diplas, *Parallelization of calculations and creating tables for the scientific program "SIRENE"* (2015)

[6] NVidia, *CURAND LIBRARY*, PG-02829-001_v7.5 (September 2015)

[7] D. Ruijters and P. Thévenaz, *GPU Prefilter for Accurate Cubic B-Spline Interpolation*, The Computer Journal **55**, no. 1, pp. 15-20 (January 2012)

EPJ Web of Conferences **116**, 07006 (2016)
DOI: 10.1051/epjconf/201611607006

Application of data mining techniques in atmospheric neutrino analyses with IceCube

T. Ruhe[a] for the IceCube collaboration[b]

Department of Physics, TU Dortmund, Germany

Abstract. The selection of event candidates by machine learning algorithms has become an important analysis tool. Data mining, however, goes beyond the simple training and application of a learning algorithm. It also incorporates finding a good representation of data in fewer dimensions without losing relevant information, as well as a thorough validation of the results throughout the entire analysis. A data mining-based event selection chain has been developed for the measurement of the atmospheric ν_μ spectrum with IceCube in the 59-string configuration. It yielded a high statistics and high purity sample ($99.59 \pm 0.37\%$) of ν_μ, while allowing only $1.0 \times 10^{-4}\%$ of the incoming background muons to pass. In this paper the setup of the analysis chain is presented and the results are discussed in the context of atmospheric ν_μ analyses.

1. Introduction

In the context of machine learning and data mining the techniques used by the IceCube neutrino telescope [1] are interesting for several reasons. IceCube is taking data on the order of 1 TB/day on a daily basis. These data have to be stored and processed, which poses a general computational challenge. Moreover, ν-induced μ are hidden in a dominant background of atmospheric muons, which makes the efficient rejection of background a key task in the IceCube analysis chain. From the machine learning point of view the rejection of background corresponds to a two-class seperation task, with ν-induced muons being one class and atmospheric muons being the other one. In atmospheric ν_μ analyses the signal to background ratio is roughly 10^{-3}. A challenge therefore arises from the highly skewed distribution of classes.

In general, data mining starts with a set of annotated events, which in neutrino astronomy are often derived from Monte Carlo simulation. These annotated events are then fed into a learning algorithm, which delivers a model that can be applied to unseen data and will return a label for these. For neutrino telescopes, often a relatively large number of input variables is available (several hundred). It is preferrable to reduce the number of input variables in order to reduce the dimensionality of the problem and to limit the required resources – CPU time and memory – to a reasonable amount. The selection of

---

[a]e-mail: `tim.ruhe@tu-dortmund.de`
[b] `http://icecube.wisc.edu`

the variables as well as the performance of the learning algorithm has to be carefully validated before the analysis chain is applied to real data. In total 120,000 simulated ν_μ events and 4.96×10^6 simulated background events were available for verifying the analysis chain. This corresponds to a detector lifetime of $\approx$ 15 days.

The analysis chain used in atmospheric ν_μ analyses for the IceCube detector in the 59-string configuration consists of three consecutive steps. The first one is the application of simple straight cuts, intended to reduce obvious background events and the required CPU resources. These cuts were applied at the zenith angle ($\theta_{\text{Zenith}} > 88°$ and the estimated velocity of the lepton $v_{\text{Lepton}} > 0.19\,\text{c}$). Their application yielded a background rejection of 91.4% at a signal efficiency of 57.1%. The remaining background solely consists of falsely reconstructed atmospheric μ, which are removed by using machine learning algorithms.

In a second step the input variables for the learning algorithm are selected in a partially automated selection procedure. This selection procedure is carried out on a limited number of simulated events. Therefore, its stability with respect to statistical fluctuations in these sets needs to be ensured. In a third step a Random Forest [2] is trained and tested using the input parameters selected in step two. All machine learning algorithms were used in the data mining environment RapidMiner [3].

2. Variable selection

The selection of the input variables is carried out using the Maximum Relevance Minimum Redundancy (MRMR) algorithm [4]. MRMR is an iterative procedure, in which one variable is added to the subset of variables per iteration according to its relevance V_F, as well as its redundancy W_c. The redundancy is computed with respect to the variables selected in previous iterations [4, 5].

Compared to Forward Selection or Background Elimination, MRMR is independent of a specific learning algorithm. Also the resource requirements in terms of CPU time and memory are reduced by approximately a factor of 10.

Since the variable selection is carried out on a limited number of events, statistical fluctuations need to be taken into account by monitoring the stability of the variable selection as a function of the number of variables. Two indices, the Jaccard index and Kuncheva's index [6] were used to judge the stability of the variable selection.

The Jaccard index for two subsets of variables A and B is computed via:

$$J = \frac{|A \cap B|}{|A \cup B|} \tag{1}$$

whereas Kuncheva's index is calculated via [6]:

$$I_C(A, B) = \frac{rn - k^2}{k(n - k)}. \tag{2}$$

In Eq. (2) k represents the size of the subset, whereas $r = |A \cap B|$ represents the cardinality of the intersection. The total number of variables available is denoted by n. For a fixed number of variables n_{var} 10 subsets of attributes are selected using MRMR. The stability indices are then computed for every pair of subsets and divided by the number of possible combinations [5]. The stability of the MRMR variable selection is shown in Fig. 1. In this analysis 25 variables were selected as input to the learning algorithm. For $n_{\text{var}} = 25$ both indices were found to be larger than 0.7.

3. Training and testing of a Random Forest

A Random Forest [2] uses an ensemble of n_{tree} simple decision trees to obtain a classification of events. A given event j is assigned a label s_{ij} by tree i. This label is $s_{ij} = 1$ in case the event is recognized as

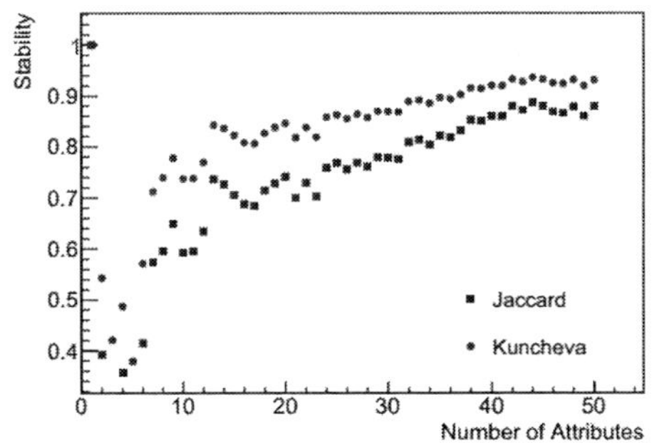

Figure 1. Stability of the Minimum Redundancy Maximum Relevance (MRMR) variable selection as a function of the number of variables selected.

signal and $s_{ij} = 0$ in case the event is recognized as background. The output returned by the forest s_j is an average over the output of the individual trees:

$$s_j = \frac{1}{n_{\text{trees}}} \sum_{i=1}^{n_{\text{trees}}} s_{ij}. \tag{3}$$

For the IC59 atmospheric ν_μ analysis, the Random Forest was trained setting the number of trees to $n_{\text{trees}} = 500$. The performance of the classification was validated in a five-fold cross validation using 750,000 simulated μ events and 70,000 simulated ν_μ events. Following the best practices for cross validating results, the events were divided in 5 statistically independent subsets. The training of the forest is then iterated 5 times using 4 of the 5 subsets, while holding out the 5th for testing. This leaves 650,000 simulated μ events and 56,000 simulated ν_μ events available for training per iteration. Studies showed that better results in terms of signal efficiency and background rejection were achieved using 27,000 simulated events of each class for training. Thus, in every iteration of the cross validation 27,000 events per class were sampled from the total number of events available for training. The use of a cross validation enables the analyzer to compute uncertainties on the performance of the learning algorithm, from which faulty behavior, such as overtraining, can be detected.

Background events were simulated with CORSIKA [7] according to the poly-gonato model [8]. Signal events were generated with NuGen an IceCube event generator based on ANIS [9], according to an E^{-2} spectrum. Although the spectrum of atmospheric ν_μ is roughly proportional to $E^{-3.7}$, events generated with a different spectral index can be used in the learning process. Tree-based algorithms classify events on an event-by-event basis, rather than on a statistical one. Furthermore, using events generated with flatter spectrum provides the learner with more events at high energies. The distribution of the random forest output variable s is shown in Fig. 2.

We find that due to the large number of background muons an additional cut needs to be placed on s. This cut was placed at $s = 1.0$, thus requiring that all trees in the forest classified an event as signal. The cut is rather strict due the overall analysis goal – an unfolding of an atmospheric ν_μ energy spectrum. In order to prevent a contamination of high energy bins with background muons, an event sample with a purity $\geq 99.5\%$ is required.

4. Overall performance of the event selection chain

The application of the event selection chain on data taken with IceCube-59 yielded 27,771 ν_μ candidates in a livetime of 346 days – roughly 80 ν_μ candidates per day. The number of remaining μ events was estimated to be 114 ± 103 for the entire lifetime. Comparing the number of detected events to 29,884 ν_μ

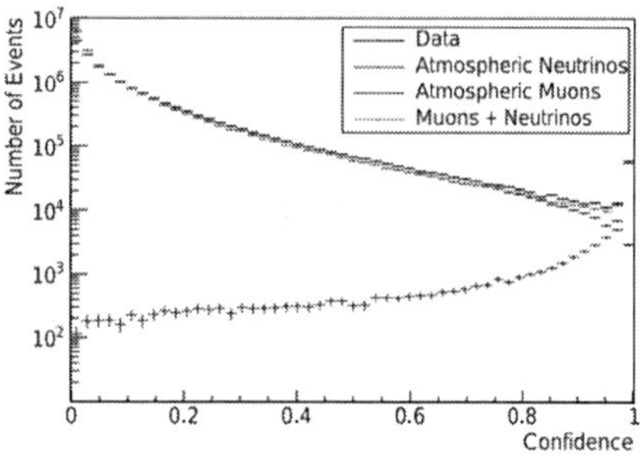

(a) Distribution of the Random Forest output s over the entire range.

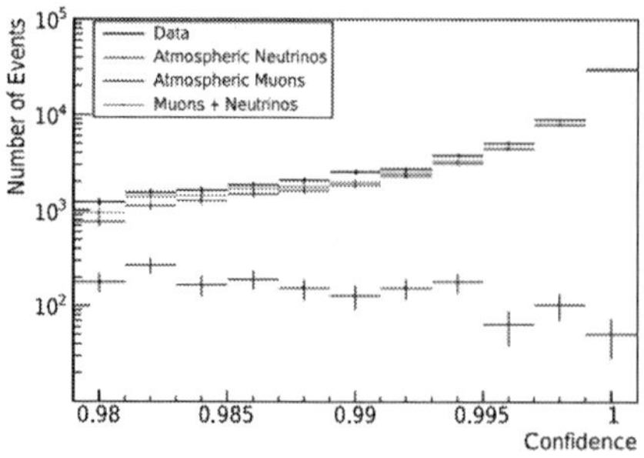

(b) Distribution of the Random Forest output in the region from $s = 0.98$ and $s = 1.0$.

Figure 2. Distribution of the Random Forest output for experimental data (blue), simulated atm. ν_μ (magenta), simulated atm. μ (red) and the sum of simulated μ and simulated ν_μ (green).

candidates expected from simulation, we find that these numbers disagree by approximately 7%. This deficit, however, is found to be well within the systematic uncertainty of the event selection.

Thus, a purity of $(99.59 \pm 0.37)\%$ was achieved while retaining $1,0 \times 10^{-4}$ of the incident background muons, computed with respect to the starting level of the analysis (before the application of straight cuts). Comparing these results to the event selection chain of IceCube in the 40-string configuration we find that the event rate is increased by $\approx 8\%$ at a comparable purity. The event selection chain presented was found to be robust and reliable and has been applied to other detector configurations with only minor changes.

References

[1] A. Achterberg, et al., Astroparticle Physics **26**, 155 (2006), astro-ph/0604450

[2] L. Breiman, Machine Learning **45**, 5 (2001)

[3] S. Fischer, et al., Tech. Rep. CI-136/02, Collaborative Research Center 531, University of Dortmund, Dortmund, Germany (2002)

[4] C. Ding, H. Peng, J. of Bioinformatics and Computational Biology **3** (2005)

[5] B. Schowe, *Feature Selection for high-dimensional data in RapidMiner*, in *Proceedings of the 2nd RapidMiner Community Meeting And Conference (RCOMM 2011)*, edited by S. Fischer, I. Mierswa (Shaker Verlag, Aachen, 2011)

[6] L. Kuncheva, *A Stability Index for Feature Selection*, in *Artificial intelligence and applications* (2007), pp. 421–427

[7] D. Heck, et al., *CORSIKA: A Monte Carlo code to simulate extensive air showers*, Vol. 6019 (FZKA, 1998)

[8] J.R. Hoerandel, N.N. Kalmykov, A.I. Pavlov, *The Knee in the Energy Spectrum of Cosmic Rays in the Framework of the Poly-Gonato and Diffusion Models*, in *Proceedings of International Cosmic Ray Conference 2003* (2003), Vol. 1, p. 243

[9] A. Gazizov, M. Kowalski, Computer Physics Communications **172**, 203 (2005), astro-ph/0406439

EPJ Web of Conferences **116**, 08001 (2016)
DOI: 10.1051/epjconf/201611608001

gSeaGen: A GENIE-based code for neutrino telescopes

Carla Distefano[a] for the ANTARES and KM3NeT Collaborations

INFN Laboratori Nazionali del Sud, via S.Sofia 62, 95123 Catania, Italy

Abstract. The gSeaGen code is a GENIE based application to generate neutrino-induced events in an underwater neutrino detector. The gSeaGen code is able to generate events induced by all neutrino flavours, taking into account topological differences between track-type and shower-like events. The neutrino interaction is simulated taking into account the density and the composition of the media surrounding the detector. The main features of gSeaGen will be presented together with some examples of its application within ANTARES and KM3NeT.

1. Introduction

Monte Carlo simulations play an important role in the data analysis of neutrino telescopes. Simulations are used to design and optimise trigger, reconstruction and event selection algorithms. A reliable simulator of neutrino events at the detector is therefore required.

GENIE [1] is a neutrino event generator for the experimental neutrino physics community. It has a focus on low energies and is currently used by a large number of experiments working in the neutrino oscillation field. The goal of the project is the development of a "canonical" Monte Carlo simulating the physics of neutrino interactions whose validity extends to all nuclear targets and neutrino flavours from MeV to PeV scales. At the moment GENIE is validated up to 5 TeV [2].

The gSeaGen code is a neutrino event generator for neutrino telescopes. It simulates particles created in all-flavour neutrino interactions, which may produce detectable Cherenkov light. gSeaGen uses GENIE to simulate the neutrino interactions. The kinematics of the generated particles is used as input to the codes simulating the detector response [3]. The code is written in C++. The main features are described in the following.

2. The neutrino interaction volume

The gSeaGen simulation code depends only on the detector size. The detector is defined inside the code by the so-called *can*. It represents the detector horizon, i.e. the volume sensitive to light. Within this volume the Cherenkov light is generated in the next steps of simulation to study the detector response. The *can* is a cylinder exceeding the instrumented volume by three light absorption lengths, L_a, bounded by the sea bed from which the light can not emerge (see Fig. 1).

[a]e-mail: distefano_c@lns.infn.it

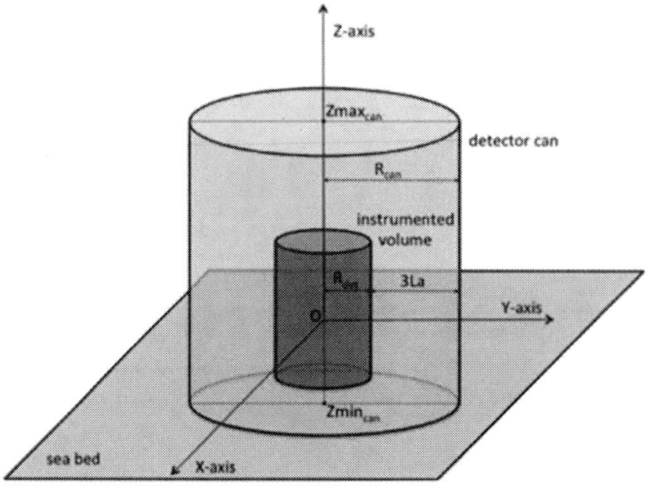

Figure 1. Definition of the detector can.

The *interaction volume* is the volume where a neutrino interaction can produce detectable particles. In case of electron neutrinos and neutral current interactions of muon or tau neutrinos, resulting particles may be detected only if they are generated inside the light sensitive volume (shower-like events) and the interaction volume is defined as a cylinder coincident with the *can* and entirely made by seawater. If muon or tau neutrino charged current interactions are simulated, secondary muons may be detected also if the interaction vertex is outside the can (track-type events). In this case the interaction volume is a cylinder made by a layer of rock and a layer of seawater surrounding the *can*. Its size is dimensioned according to the muon maximum range in water and in rock, evaluated at the highest energy of simulated neutrinos. The maximum muon range is input by the user. In the simulations reported in this work, the maximum muon range was evaluated through a Monte Carlo simulation using the MUSIC code [4].

3. The target media

Four different target media are defined: SeaWater, Rock (used to define the interaction volume), Mantle and Core (entering in the calculation of the transmission probability through the Earth). The composition of all media is set by the user, providing the possibility to study the systematics due to medium compositions and also to simulate under-ice detectors. The density profile of inner layers of the Earth (Mantle and Core media) is described according to the Preliminary Reference Earth Model (PREM) [5].

4. Simulation algorithm

The energy range considered for event generation is binned in equal divisions of log-energy. For each bin, the interaction probability is scaled up to reduce the number of trials. The probability scale is the maximum interaction probability (i.e. probability at maximum energy and for the maximum possible path length) summed over initial states and it is calculated by GENIE [1].

If track-type events may be generated, the interaction volume and the number of events are also scaled. The neutrino energy is drawn according to a power-law energy spectrum and its direction is randomly extracted according to a flat distribution in the solid angle. The track vertex is drawn on a circular surface (outside the interaction volume and covering its projection onto a plane perpendicular to each neutrino's direction).

Once the neutrino is generated, its interaction is simulated using the GENIE event generation driver class GMCJDriver [1]. The neutrino-induced particles generated inside the can are stored in the output file. Muons generated outside the can are propagated with the MUSIC code [4] and stored if they reach the can surface.

5. Calculation of the event weight

During the simulation, the code assigns a weight W_{evt} to each event in order to normalise the generation to a real neutrino flux. The event weight is the product of the real neutrino spectrum $\phi(E_\nu, \cos\theta_\nu)$, at the generated neutrino energy E_ν and direction θ_ν, and the generation weight W_{gen}:

$$W_{evt} = W_{gen} \cdot \phi(E_\nu, \cos\theta_\nu). \tag{1}$$

The generation weight is defined as the inverse of the simulated neutrino spectrum:

$$W_{gen} = \frac{I_E \cdot I_\theta \cdot T_{gen} \cdot A_{gen} \cdot N_\nu \cdot E_\nu^X \cdot P_{scale} \cdot P_{Earth}(E_\nu, \cos\theta_\nu)}{N_{Tot}}, \tag{2}$$

where I_E and I_θ are the energy and angular phase space factors, T_{gen} is the simulated time interval, A_{gen} is the generation area, N_ν is the number of simulated neutrino types and N_{Tot} is the total number of simulated neutrinos. P_{scale} is the GENIE interaction probability scale (see Sect. 4). The neutrino transmission probability through the Earth is calculated as:

$$P_{Earth}(E_\nu, \cos\theta_\nu) = e^{-N_A \cdot \sigma(E_\nu) \cdot \rho_l(\theta_\nu)}, \tag{3}$$

where $\sigma(E_\nu)$ is the total cross section per nucleon (taking into account the different layer compositions); $\rho_l(\theta_\nu)$ is the amount of material encountered by a neutrino in its passage through the Earth. The latter is computed with the line integral $\rho_l(\theta_\nu) = \int_L \rho_{Earth(r)} dl$, being L the neutrino path at the angle θ_ν and $\rho_{Earth(r)}$ is the PREM Earth density profile [5].

5.1 Calculation of the systematic weights

The accuracy of the input simulation parameters is known and the uncertainties related to nu interactions can be propagated with GENIE. For each input physics quantity P, a systematic parameter x_P is introduced. Tweaking this systematic parameter modifies the corresponding physics parameter P as follows:

$$P \rightarrow P' = P(1 + x_P \cdot \delta P / P), \tag{4}$$

where δP is the estimated standard deviation of P. The calculation of the systematic errors in GENIE is based on an event reweighting strategy. A description of the full reweighting scheme is reported in [6].

The evaluation of the systematics has been implemented in gSeaGen, using the GENIE class GReWeight [6]. The implementation accepts single parameters or a list of them as input. In the latter case, the code treats all parameters at the same time and calculates the global systematic weight. If the calculation is activated, the systematic weights w_{sys} are written in the output file. The modified distributions are obtained by multiplying the event weights by w_{sys}.

6. Applications

The development of gSeaGen code started within ANTARES Collaboration [7], providing the possibility to use modern and maintained neutrino interaction codes/libraries. Currently, the code is used as a cross-check for GenHen [8], the standard generator code written in FORTRAN, limiting the comparison within the present GENIE validity range. gSeaGen can generate high energy events in ANTARES and

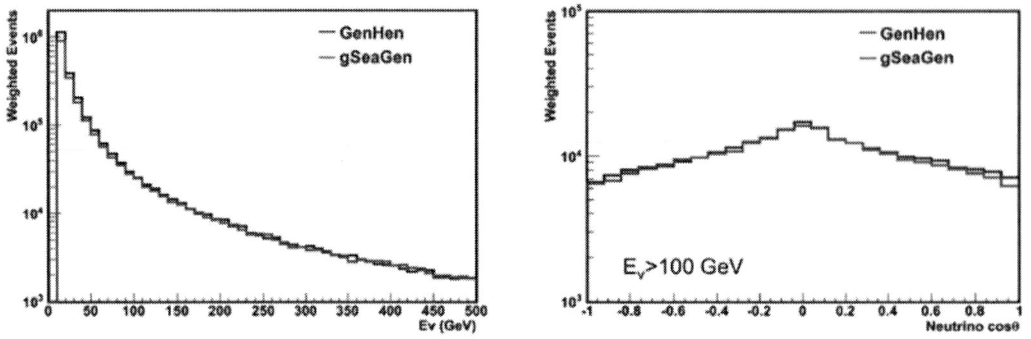

Figure 2. Energy spectrum and angular distribution of atmospheric muon neutrinos and anti-neutrinos producing detectable events at the ANTARES detector can. Generated events are weighted according to the Bartol atmospheric muon flux [11]. Results from the standard neutrino event generator GenHen are shown for comparison [8].

KM3NeT-ARCA [9], when the GENIE extension at the PeV scale will be available. At present, gSeaGen is the reference code for simulation of the KM3NeT-ORCA detector [10].

As an example of application of gSeaGen, a flux of muon neutrinos and anti-neutrinos have been generated for the ANTARES detector *can*. The spectrum of generated events have been shaped according to the Bartol atmospheric muon flux [11]. The results are reported in Fig. 2 in terms of the energy spectrum and angular distributions of neutrinos producing detectable events (i.e. inside or reaching the detector *can*).

References

[1] C. Andreopoulos et al., Nucl. Instrum. Meth. **A614**, 87 (2010), 0905.2517

[2] *GENIE web page*, http://www.genie-mc.org/

[3] A. Margiotta (ANTARES), Nucl. Instrum. Meth. **A725**, 98 (2013)

[4] P. Antonioli et al., Astropart. Phys. **7**, 357 (1997), hep-ph/9705408

[5] A.M. Dziewonski, D.L. Anderson, Physics of the Earth and Planetary Interiors **25**, 297 (1981)

[6] C. Andreopoulos et al. (2015), arXiv:1510.05494

[7] M. Ageron et al. (ANTARES), Nucl. Instrum. Meth. **A656**, 11 (2011), 1104.1607

[8] D.J. Bailey, Ph.D. thesis, University of Oxford, United Kingdom (2002), http://antares.in2p3.fr/

[9] P. Piatteli (KM3NeT), *All-flavour high-energy neutrino astronomy with KM3NeT/ARCA*, in *Proceedings, 34th International Cosmic Ray Conference (ICRC 2015)* (2015), Vol. ICRC2015, http://pos.sissa.it/cgi-bin/reader/conf.cgi?confid=236

[10] J. Brunner (KM3NeT), *KM3NeT – ORCA: Measuring neutrino oscillations and the mass hierarchy in the Mediterranean*, in *Proceedings, 34th International Cosmic Ray Conference (ICRC 2015)* (2015), Vol. ICRC2015, http://pos.sissa.it/cgi-bin/reader/conf.cgi?confid=236

[11] G.D. Barr et al., Phys. Rev. **D70**, 023006 (2004), astro-ph/0403630

EPJ Web of Conferences **116**, 08002 (2016)
DOI: 10.1051/epjconf/201611608002

Present theoretical uncertainties on charm hadroproduction in QCD and prompt neutrino fluxes

M.V. Garzelli[a], S. Moch, and G. Sigl

University of Hamburg, II Institute for Theoretical Physics, 22761 Hamburg, Germany

Abstract. Prompt neutrino fluxes are basic backgrounds in the search of high-energy neutrinos of astrophysical origin, performed by means of full-size neutrino telescopes located at Earth, under ice or under water. Predictions for these fluxes are provided on the basis of up-to-date theoretical results for charm hadroproduction in perturbative QCD, together with a comprehensive discussion of the various sources of theoretical uncertainty affecting their computation, and a quantitative estimate of each uncertainty contribution.

1. Introduction

Very large volume neutrino telescopes (VLVνTs) offer the opportunity of observing high-energy neutrinos, through their interactions deep in ice or water volumes where the detector modules are installed. The IceCube collaboration, in particular, has reported evidence for the existence of a high-energy lepton flux with increasing statistics over the years [1]. At present, the best-fit to the data looks quite compatible or even suggests an astrophysical interpretation of the observed signal, but no specific correlation with particular galactic or extragalactic sources has been identified, at least so far, thus the origin of the IceCube events still remains a mystery. Many tentative hypotheses have been formulated to explain this signal [2], taking into account that neutrinos produced in a number of astrophysical environments may actually reach us after travelling long distances undeflected by cosmic magnetic fields. Possibly interesting production sites range from galactic or extragalactic sources and their neighborhood, to dark matter (DM) populated regions where heavy DM decay or DM-DM annihilation could occur. In all cases, neutrinos produced by the interaction of cosmic rays (CRs) with the Earth's atmosphere represent a background that one has to subtract in order to disentangle the signal of truly astrophysical origin. Actually, taking into account that the CR spectrum at the top of the Earth's atmosphere is peaked in the region $E_{CR} \sim$0.2–1 GeV per nucleon for all ions, the bulk of atmospheric neutrinos, i.e., of neutrinos produced in the atmosphere, has energies $E_\nu \sim$0.05–0.4 GeV. These neutrinos are mainly the results of hadronic interactions between the impinging CRs and the light atmospheric nuclei (mostly nitrogen and oxygen), coming from intermediate charged pions and kaons, which subsequently decay leptonically. However, at increasing CR primary energy, the decay lengths of these light mesons increase, up to and exceeding the transverse size of the atmosphere. In these conditions, $\pi^\pm$ and $K^\pm$ decay probabilities are suppressed with respect to those of their

[a]e-mail: `maria.vittoria.garzelli@desy.de`

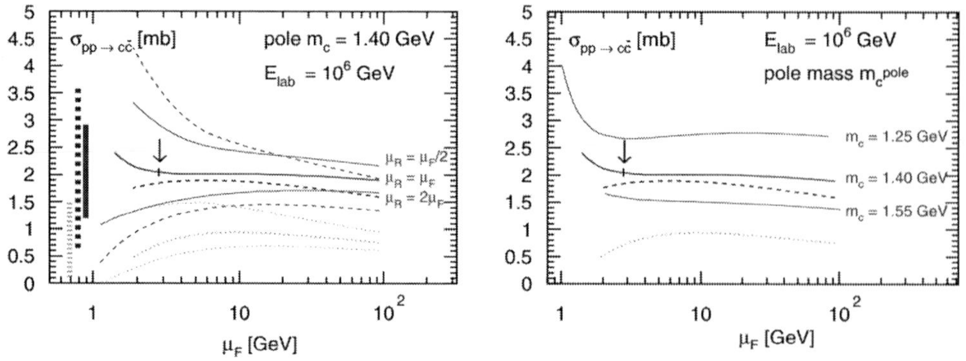

Figure 1. Left panel: cross-section for $c\bar{c}$ hadroproduction as a function of μ_F for different values of μ_R. Charm mass is fixed to m_{charm} = 1.4 GeV in the OS renormalization scheme. The predictions in red, blue and green refer to NNLO, NLO and LO approximations, respectively. At each fixed order, the central line refers to predictions with $\mu_R = \mu_F = m_{charm}$, whereas the upper and lower lines denotes predictions for $\mu_R = \mu_F/2$ and $\mu_R = 2\mu_F$, respectively. Right panel: cross-section for $c\bar{c}$ hadroproduction as a function of μ_F. μ_R is fixed to $\mu_R = \mu_F$. Varying the charm mass value in the range $m_{charm} = 1.4 \pm 0.15$ GeV gives rise to the three NNLO predictions denoted by red lines, whereas blue and green lines corresponds to NLO and LO predictions for the central m_{charm} value.

reinteraction, and ν's are mainly generated by other mechanisms, especially the atmospheric production of charmed mesons and baryons, followed by their prompt decay. Charmed hadrons in fact have decay lengths smaller than $\pi^{\pm}$ and $K^{\pm}$: the prompt neutrino flux, initiated by charm, becomes larger than the conventional one, initiated by light mesons, at an energy called "transition energy". The exact value of this energy is still subject to sizable uncertainties, mainly reflecting those affecting highly energetic hadronic interactions in the atmosphere. We have recently re-evaluated charm hadroproduction in the atmosphere [3], in the light of recent QCD theoretical and experimental progresses, triggered by the flourishing activities at hadron colliders, in particular the Large Hadron Collider (LHC): in this contribution we mainly review and further discuss the results presented more extensively in that paper.

2. Charm hadroproduction in QCD and prompt neutrino fluxes

In QCD the collinear factorization formalism allows the total cross-section for charm hadroproduction to be written as a convolution of a non-perturbative component, involving the parton distribution functions (PDFs) and the fragmentation functions (FF), with a perturbative part given by the partonic cross-sections for the production of charm quarks from initial state quarks and gluons. In the following, we assume that the collision of a CR with an atmospheric nucleus can be simply described as a superposition of Nucleon-Nucleon (NN) interactions. Recent progress in pQCD gives nowadays the possibility to calculate the total cross-section for $NN \to c\bar{c}$ pair hadroproduction including radiative corrections up to next-to-next-to-leading order (NNLO), by properly combining NNLO PDFs and NNLO partonic amplitudes. In practice, we obtained this cross-section as a function of the NN collision energy by extending the HATHOR event generator [4], originally developed for $t\bar{t}$ hadroproduction. The truncation of the QCD perturbative series at a fixed order makes the theoretical predictions for the cross-sections dependent of two unphysical scales: the renormalization scale μ_R and the factorization scale μ_F. Unfortunately, there is no unique unambiguous way derived from first principles to properly fix these scales. We fix μ_R and μ_F by looking for minimal sensitivity, i.e. in such a way to minimize the impact of radiative corrections when comparing next-to-leading order (NLO) to NNLO predictions. The point of minimal sensitivity approximately corresponds to the choice μ_F=μ_R=μ_0=$2m_{charm}$, denoted by an arrow in Fig. 1a. Besides the total cross-section, we are interested in differential ones. Thus we translate

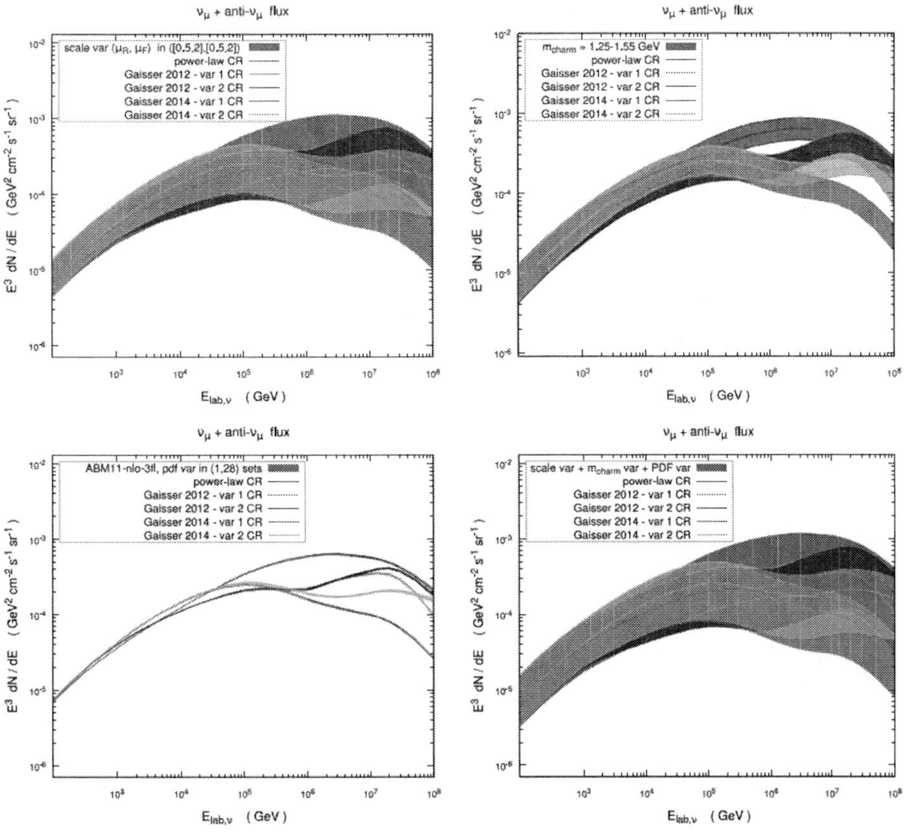

Figure 2. Prompt (ν_μ + $\bar{\nu}_\mu$) fluxes as a function of neutrino energies: QCD uncertainties due to scale, mass, PDF variation and their combination in quadrature are shown in panels a, b, c, d, respectively. For each panel, predictions related to the use of different primary CR all-nucleon spectra are shown by different colors: the broken power-law spectrum is in red, whereas four more modern spectra (see Ref. [13] and references therein) are shown with other colors.

this static scale into a dynamical one $\mu_0 = \sqrt{p_{T,charm}^2 + m_{charm}^2}$, to better catch the kinematical aspects of charm hadroproduction. We take this as our central scale choice for the generation of differential distributions, and then we calculate scale uncertainties by considering the independent variation of μ_R and μ_F in the range [0.5 , 2] μ_0, keeping out the two extreme combinations (μ_R, μ_F) = (2, 0.5) and (0.5,2) μ_0, as prescribed in Ref. [5].

Furthermore, the $c\bar{c}$ hadroproduction cross-sections are sensitive to the precise value of charm mass, as shown in Fig. 1. Particular care has to be taken in the choice of this parameter especially when adopting the on-shell (OS) scheme to renormalize heavy-quark masses, because this scheme shows a poor perturbative convergence, actually worse than the modified minimal subtraction ($\overline{MS}$) scheme [6]. We fix the charm mass value to $m_{charm}^{OS} = 1.40 \pm 0.15$ GeV. This way the cross-section in the OS scheme approximately reproduces that in the $\overline{MS}$ scheme. We notice that the accuracy on the pole mass is also limited by the renormalon ambiguity.

Further QCD uncertainties are related to the PDF choice. At present, many uncertainties still exist at low Bjorken x ($x < 10^{-4}$), due to lack of data. However this region is progressively important when going to higher collision energies, which is a crucial issue for astrophysics because CRs in the tail of the CR energy spectrum are even more energetic than the beams at colliders. Recently, LHCb data on charm and bottom hadroproduction in mid-peripheral collisions have been used to perform PDF fits down to lower values of x ($x \sim 10^{-6}$). The first fit in this direction was proposed by the PROSA collaboration in Ref. [7], followed by Ref. [8] applying the same concept to the NNPDF PDFs. In this paper we use the ABM11 PDFs [9]. We note that, although these PDFs do not take into account any LHCb data, their extrapolation in the range $10^{-6} < x < 10^{-4}$ turns out to lie within the PROSA PDF uncertainty band. We evaluate PDF uncertainties by considering the 28 ABM11 variations, accompanying the central fit available in the LHAPDF 6.1.5 interface.

Neutrino fluxes were computed by solving a system of coupled differential equations regulating particle evolution in the atmosphere, by using the Z-moment approach [10]. Inputs to the Z-moments are differential cross-sections for charmed hadron hadroproduction. Unfortunately, these distributions are not yet available at NNLO. We thus computed them through a NLO QCD + parton shower + hadronization approach, as provided by `POWHEGBOX` [11] + `PYTHIA` [12]. Charm mass, PDF and scale input of the computation and their intervals of variation were fixed on the basis of NNLO information, as explained above. The effect of uncertainties due to scale, charm mass and PDF variation on $(\nu_\mu + \bar{\nu}_\mu)$ prompt fluxes is shown in Fig. 2a, b, c, respectively. In the last panel, Fig. 2d, the combination of those QCD effects, summed in quadrature, is also shown. The total contribution of additional uncertainties related to hadronization and multiple particle interactions amounts to several ten percent. We observe that, among all QCD uncertainties, the one due to scale variation dominates. We also notice that at energies above the PeV scale, the contribution of astrophysical uncertainties, related to the unknown composition of the CR spectrum at primary energies above the knee, becomes progressively larger and ends up in dominating over the contribution from QCD uncertainties at ultra-high neutrino energies. The region around and above the PeV will be better probed by upgraded VLVνTs, like the IceCube-Gen2 project [14], providing an increase of the instrumented volume from ~ 1 km^3 to ~ 10 km^3. By the time they will start data taking, it would be advisable to reduce all uncertainties (both those from QCD and those from astrophysics) affecting the computation of prompt neutrino fluxes at high energies. Our most up-to-date predictions are available at `http://promptfluxes.desy.de`.

References

[1] IceCube Collaboration (M. G. Aartsen, et al.) Science **342**, 1242856 (2013) [arXiv:1311.5238]
[2] C.S. Fong, H. Minakata, B. Panes, JHEP **1502**, 189 (2015) [arXiv:1411.5318]
[3] M.V. Garzelli, S. Moch, G. Sigl, JHEP **1510**, 115 (2015) [arXiv:1507.01570]
[4] M. Aliev, H. Lacker, U. Langenfeld *et al.* Comput. Phys. Commun. **182**, 1034 (2011) [arXiv:1007.1327]
[5] M. Cacciari, S. Frixione, N. Houdeau *et al.* JHEP **1210**, 137 (2012) [arXiv:1205.6344]
[6] S. Narison, Phys. Lett. B **197**, 405 (1987)
[7] PROSA collaboration (A. Zenaiev et al.), EPJC **75**, 396 (2015) [arXiv:1503.04581]
[8] R. Gauld, J. Rojo, L. Rottoli, J. Talbert, JHEP **1511**, 009 (2015) [arXiv:1506.08025]
[9] S. Alekhin, J. Blümlein, S. Moch, Phys. Rev. D **86**, 054009 (2012) [arXiv:1202.2281]
[10] P. Lipari, Astropart. Phys. **1**, 195 (1993)
[11] S. Alioli, P. Nason, C. Oleari, E. Re, JHEP **1006**, 043 (2010) [arXiv:1002.2581]
[12] T. Sjostrand, S. Mrenna, P.Z. Skands, JHEP **0605**, 026 (2006) [hep-ph/0603175]
[13] T. Gaisser, T. Stanev, S. Tilav, Front. Phys. China **8**, 748 (2013) [arXiv:1303.3565]
[14] IceCube Collaboration (M. G. Aartsen, et al.) [arXiv:1412.5106]

EPJ Web of Conferences **116**, 08003 (2016)
DOI: 10.1051/epjconf/201611608003

Neutrino-nucleon cross sections at energies of Megaton-scale detectors

A. Gazizov[1,2,a], M. Kowalski[1], K.S. Kuzmin[3,4], V.A. Naumov[3], and Ch. Spiering[1]

[1] DESY-Zeuthen, Platanenallee 6, 15738 Zeuthen, Germany
[2] Gran Sasso Science Institute, viale Francesco Crispi, 7, 67100 L'Aquila, Italy
[3] Joint Institute for Nuclear Research, RU-141980, Dubna, Moscow region, Russia
[4] Institute for Theoretical and Experimental Physics, RU-117218, Moscow, Russia

Abstract. An updated set of (anti)neutrino-nucleon charged and neutral current cross sections at $3\,\text{GeV} \lesssim E_\nu \lesssim 100\,\text{GeV}$ is presented. These cross sections are of particular interest for the detector optimization and data processing and interpretation in the future Megaton-scale experiments like PINGU, ORCA, and Hyper-Kamiokande. Finite masses of charged leptons and target mass corrections in exclusive and deep inelastic $(\overline{\nu})\nu N$ interactions are taken into account. A new set of QCD NNLO parton density functions, ABMP15, is used for calculation of the DIS cross sections. The sensitivity of the cross sections to phenomenological parameters and to extrapolations of the nucleon structure functions to small x and Q^2 is studied. An agreement within the uncertainties of our calculations with experimental data is demonstrated.

1. Introduction

Future Megaton-scale neutrino detectors, such as PINGU, ORCA, and Hyper-Kamiokande can be used for determination of the neutrino mass hierarchy (NMH) with atmospheric neutrinos. Depending on the NMH, ν-oscillations produce a few percent differences in the event rate, appearing in the few-GeV energy region, since the matter effects act oppositely for ν's and $\overline{\nu}$'s. The NMH signature relies on the fact that both fluxes and cross sections are different for ν's and $\overline{\nu}$'s, so that the combined event rates show the remaining NMH dependence. For the proper interpretation of the data, the accurate predictions of the atmospheric neutrino fluxes have to be paired with the reliable knowledge of the (anti)neutrino-nucleon cross sections. We focus on the latter with an emphasis on the energy range $E_\nu \gtrsim 3$ GeV and present a self-consistent set of theoretical models and phenomenological parameters allowing us to describe the major contributions to the total νN and $\overline{\nu}N$ cross sections at these energies. The results of the present study may be included in future upgrade of the MC generator ANIS [1] in order to extend its validity to lower energies and improve the predictions for high and ultra-high energies.

[a]e-mail: askhat.gazizov@desy.de

2. Models and parameters

All νN-interactions, both proceeding via charged current (CC) and neutral current (NC),

$$(\bar{\nu}_\ell)\nu_\ell + N \to (\ell^+)\ell^- + X \quad \text{(CC)} \quad \text{and} \quad (\bar{\nu}_\ell)\nu_\ell + N \to (\bar{\nu}_\ell)\nu_\ell + X \quad \text{(NC)} \tag{1}$$

($\ell = e, \mu, \tau$; $N = p, n$) may be classified according to the number of mesons, pions, kaons, *etc.*, appearing in the final hadron state X. The total cross sections are combinations of the contributions from the channels *(i)* with no pions, – the (quasi)elastic scattering (ES or QES), *(ii)* with one pion, – the resonance single pion production (RES), and *(iii)* inclusive or deep inelastic scattering (DIS) with > 2 hadrons in the final state X: $\sigma_{\nu N}^{\text{tot}} = \sigma_{\nu N}^{\text{(Q)ES}} \oplus \sigma_{\nu N}^{1\pi} \oplus \sigma_{\nu N}^{\text{DIS}}$ (see, e.g., Ref. [2]).

For calculations of the (Q)ES contributions we use the standard approach with the nucleon axial mass parameter $M_A^{\text{QES}} = 1.02$ GeV extracted from available ν_μD, $\bar{\nu}_\mu$H, and $\pi^\pm$ electroproduction data (for details, see Refs. [3–5] and references therein).

For obtaining of RES contributions the extended Rein–Seghal model [6–8] is adopted with account for the pion-pole contribution to the hadronic axial current [9]. The value of the axial mass parameter $M_A^{\text{RES}} = 1.12$ GeV has been derived in Ref. [10] by fitting to the data available at that time. All known nucleon resonances with masses below ≈ 2 GeV are included and the interference of their amplitudes is properly taken into account according to Ref. [11].

For the DIS cross sections we use the approach of Ref. [12] with all 5 structure functions (SFs) being taken into account. In the case of CC processes this allows to account for the finite lepton mass, especially important for ν_τ. The functions $F_{1,2,3}(x, Q^2)$ are available by the OPENQCDRAD-2.0 code [13]; we use it with the new NNLO (both for light and charm production) parton distribution functions of ABMP15 [14]. Transitions from $F_{1,2,3}(x, Q^2)$ to target mass corrected (TMC) SFs, $F_i^{\text{TMC}}(x, Q^2)$, are performed according to Ref. [15, 16]. Below we do not discuss the complicated problem of the nuclear effects in the DIS SFs and focus only on purely (anti)neutrino-nucleon reactions.

To avoid a double counting, the phase spaces of RES and DIS are to be separated according to the mass of the final hadron state in Eq. (1), $W_{\text{cut}} = m_X \gtrsim m_N + 2m_\pi$. We found that $W_{\text{cut}} = 1.4$ GeV provides a good compatibility of these contributions: a variation of W_{cut} around 1.4 GeV leads to comparatively small variations of the sum $\sigma^{\text{RES}}(E) + \sigma^{\text{DIS}}(E)$ at all energies.

At intermediate energies significant uncertainties in calculations of the DIS cross sections arise from necessary extrapolations of $F_i^{\text{TMC}}(x, Q^2)$ to small Q^2 where the perturbative QCD fails. In accordance with the suggestion of Ref. [17], we smoothly switch all SFs off for a given x as $Q^2 \to 0$, assuming a power-law dependence on Q^2 (or on $\log(Q^2)$ in the case of $F_3^{\text{TMC}}(x, Q^2)$); $F_i^{\text{TMC}}(x, Q^2) = F_i^{\text{TMC}}(x, Q^2_{\text{min}}) \times [Q^2/(Q^2 + b)]^{\alpha(1-x)}$. The parameter values $Q^2_{\text{min}} = 1.2$ GeV^2, $\alpha = 0.5$ and $b = 0.645$ GeV^2 provide a reasonable agreement with the data.

Cross section for ν_μ and $\bar{\nu}_\mu$ scatterings off an isoscalar nucleon calculated using our set of parameters are shown in Fig. 1. Analogous plots for the $\nu_\tau N$ and $\bar{\nu}_\tau N$ cross sections are depicted in Fig. 2. All kinematic effects are taken into account.

In Figs. 1 and 2 the uncertainties for the QES and RES contributions are estimated by varying the QES and RES axial mass parameters within the ranges 0.9 GeV $\leq M_A^{\text{QES}} \leq$ 1.1 GeV and 1.1 GeV $\leq M_A^{\text{RES}} \leq$ 1.3 GeV, respectively. The upper bounds of the bands in Fig. 1 correspond to the higher values of the axial masses. In order to estimate the DIS uncertainties, we vary the parameter α within the range 0.3 to 0.7.

The solid curves in Fig. 1 show similar predictions of the Monte Carlo generator GENIE [18]. The reasons of our discrepancies with the GENIE predictions at low energies are mainly due to details of the implementation of the Rein–Sehgal model [11] for the resonance single-pion production and different cuts in W used in the calculations. This point will be discussed at length in a forthcoming paper. Here we only mention the effect of finite lepton mass to the leptonic currents [6], including the pion-pole contribution into the weak hadronic current [9], the effect of interference between the

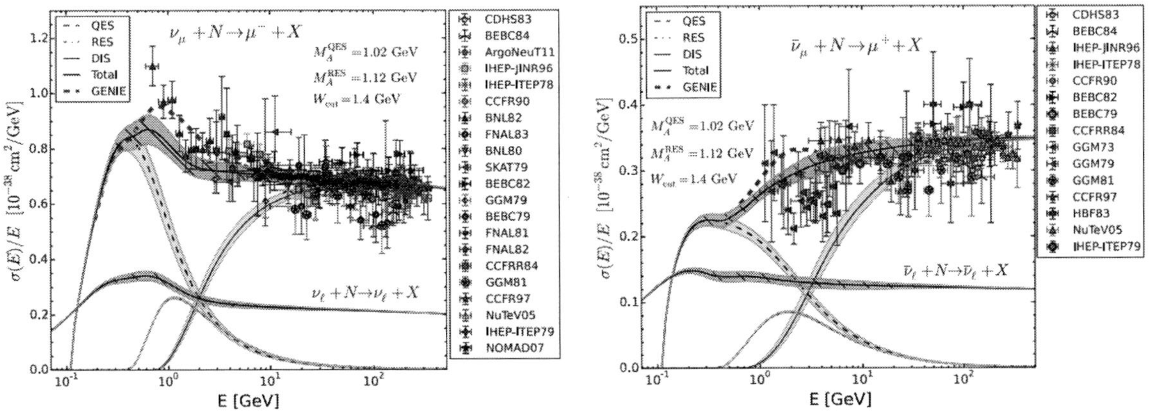

Figure 1. Left panel: Cross sections over energy for the CC $\nu_\mu N$ and NC scatterings off an isoscalar nucleon. Dashed, dash-dotted and thin solid lines correspond to contributions from QES, RES and DIS, respectively. Shaded bands show the uncertainties. Thick solid curve stands for the total cross section; the shaded hatched regions demonstrates the overall uncertainty. The thick dashed curve is the prediction of the GENIE neutrino MC generator [18]. Right panel: The same curves as in the left panel but for CC $\bar{\nu}_\mu N$ and NC scattering off the same target. The references to the data points can be found in Ref. [2].

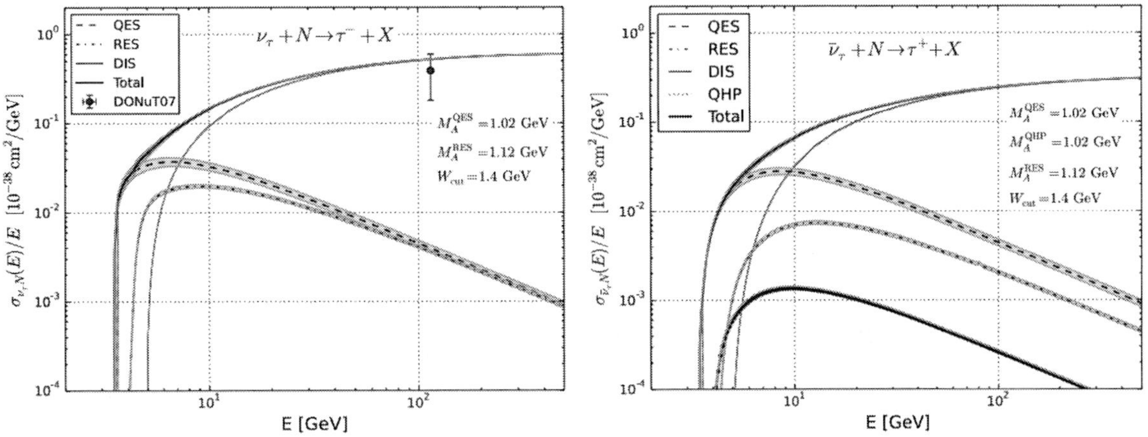

Figure 2. The same curves as in the left panel of Fig. 1 but for CC $\nu_\tau N$ and $\bar{\nu}_\tau N$ scatterings. Additional dotted QHP line in the right panel shows a contribution from hypercharge violating QES production of lightest hyperons [19]. The DONUT data point is from Ref. [20].

resonances having the same spin and orbital angular momentum of the final $N\pi$ state (both these effects are neglected in the cited version of GENIE), and also different "recipes" used for normalization of the Breit-Wigner (BW) factors and treatment of the unphysical BW "tails". The disagreements with GENIE at high energies (most notable for $\bar{\nu}$) are mainly due to the PDF models involved in the GENIE code and in our calculations, and due to distinct methods for extrapolating the DIS SFs to small Q^2.

3. Conclusions

We presented a comprehensive set of the total cross sections for the neutrino and antineutrino scattering off protons and neutrons at energies most important for the future experiments with Mton-scale neutrino detectors. Contributions from elastic and quasi-elastic scatterings (with the hypercharge change $\Delta Y = 0$ and 1), single-pion production, and deep-inelastic scattering were taken into account. For the ES, QES, and single-pion production contributions we rely on well-known methods and models. With the reasonable choice of parameters, our calculations are in reasonable agreement with the data within the experimental uncertainties. Calculations of the DIS cross sections were performed by using OPENQCDRAD-2.0 with the new set of NNLO PDFs of ABMP15 with account for the finite lepton masses and target mass corrections. We find that the uncertainties due the choice of the SFs extrapolations to small Q^2 are very significant. This partially explains the disagreements between our model and the calculations using the GENIE packet at high energies.

Present calculations will be used as a basis for the future upgrade of the Monte Carlo neutrino generator ANIS [1].

We are grateful to S. Alekhin, J. Blümlein and S.-O. Moch for making us available the new OPENQCDRAD and ABMP15 PDFs and to I. Kakorin for a valuable contribution. A.G. thanks the Helmholtz Alliance for Astroparticle Physics, HAP for support. The work of K.K. and V.N. was partially supported by the Russian Foundation for Basic Research under Grant No. 14-22-03090.

References

[1] A. Gazizov, M.P. Kowalski, Comput. Phys. Commun. **172**, 203 (2005), astro-ph/0406439
[2] K.S. Kuzmin, V.V. Lyubushkin, V.A. Naumov, Phys. Atom. Nucl. **69**, 1857 (2006)
[3] V. Bernard, L. Elouadrhiri, U.G. Meißner, J. Phys. **G28**, R1 (2002), hep-ph/0107088
[4] A. Bodek, S. Avvakumov, R. Bradford, H.S. Budd, Eur. Phys. J. **C53**, 349 (2008), 0708.1946
[5] K.S. Kuzmin, V.V. Lyubushkin, V.A. Naumov, Eur. Phys. J. **C54**, 517 (2008), 0712.4384
[6] K.S. Kuzmin, V.V. Lyubushkin, V.A. Naumov, Mod. Phys. Lett. **A19**, 2815 (2004)
[7] K.S. Kuzmin, V.V. Lyubushkin, V.A. Naumov, Nucl. Phys. B (Proc. Suppl.) **139**, 158 (2005)
[8] D. Rein, L. Sehgal, Phys. Lett. **B657**, 207 (2007), hep-ph/0606185
[9] C. Berger, L. Sehgal, Phys. Rev. **D76**, 113004 (2007); Erratum: *ibid.*, **D77**, 059901(E) (2008)
[10] K.S. Kuzmin, V.V. Lyubushkin, V.A. Naumov, Acta Phys. Polon. **B37**, 2337 (2006)
[11] D. Rein, L.M. Sehgal, Annals Phys. **133**, 79 (1981)
[12] C.H. Albright, C. Jarlskog, Nucl. Phys. **B84**, 467 (1975)
[13] https://www-zeuthen.desy.de/~alekhin/OPENQCDRAD/ (2012)
[14] S. Alekhin, J. Blümlein, S. Moch, R. Plačakytė (2015), arXiv:1508.07923 [hep-ph]
[15] S. Kretzer, M. Reno, Phys. Rev. **D66**, 113007 (2002), hep-ph/0208187
[16] S. Kretzer, M. Reno, Phys. Rev. **D69**, 034002 (2004), hep-ph/0307023
[17] A. Capella, A. Kaidalov, C. Merino, J. Tran Thanh Van, Phys. Lett. **B337**, 358 (1994)
[18] C. Andreopoulos *et al.*, Nucl. Instrum. Meth. **A614**, 87 (2010), 0905.2517
[19] K.S. Kuzmin, V.A. Naumov, Phys. Atom. Nucl. **72**, 1501 (2009)
[20] E.O. Maher (FERMILAB-THESIS-2005-90, UMI-31-92032), 2008

EPJ Web of Conferences **116**, 08004 (2016)
DOI: 10.1051/epjconf/201611608004

Measurement of the atmospheric muon neutrino energy spectrum with IceCube in the 79- and 86-String configuration

T. Ruhe[a], F. Scheriau, and M. Schmitz for the IceCube Collaboration[b]

TU Dortmund University, Dortmund, Germany

Abstract. IceCube is a neutrino telescope with an instrumented volume of one cubic kilometer. A total of 5160 Digital Optical Modules (DOMs) is deployed on 86 strings forming a three dimensional detector array. Although primarily designed for the detection of neutrinos from astrophysical sources, the detector can be used for spectral measurements of atmospheric neutrinos. These spectral measurements are hindered by a dominant background of atmospheric muons. State-of-the-art techniques from Machine Learning and Data Mining are required to select a high-purity sample of atmospheric neutrino candidates. The energy spectrum of muon neutrinos is obtained from energy-dependent input variables by utilizing regularized unfolding. The results obtained using IceCube in the 79- and 86-string configuration are presented in this paper.

1. Introduction

IceCube is a neutrino detector array located at the geographic South Pole. Its 5160 Digital Optical Modules (DOMs) are deployed on 86 strings, at depths between 1450 m and 2450 m forming an instrumented volume of 1 km^3 [1]. The DOMs are used to detect neutrinos via Cherenkov radiation initiated by their interaction products. Event properties, such as energy and direction are reconstructed from geometrical, timing and charge information obtained from the DOMs.

IceCube can be used to study neutrino spectra. The ν_μ energy spectrum is believed to consist of three major components. The first one is conventional atmospheric neutrinos originating from the decay of pions and kaons. The second component are prompt atmospheric neutrinos, which originate from the decay of charmed mesons. Both components are expected to follow a power law of the form $E^{-\gamma}$ and can be distinguished from their spectral indices [2]. Although the atm. ν_μ spectrum has been measured by various experiments, including Super-Kamiokande [3], ANTARES [4] and IceCube [5, 6], it is still subject to large uncertainties, especially at high energies [7]. Neutrinos from astrophysical sources [8] contribute to the spectrum as a third component. This component is expected to cause a flattening of the ν_μ energy spectrum at high energies.

This paper presents a model-independent measurement of ν_μ energy spectrum with IceCube in the 79- and 86-string configuration. The measurement determines the sum of all possible contributions to

---

[a]e-mail: `tim.ruhe@tu-dortmund.de`
[b]`http://icecube.wisc.edu`

Table 1. Cuts placed on different variables prior to training and testing the Random Forest. Only events to which the conditions of the cuts in this table apply were kept in the sample for further analysis.

IceCube-79		IceCube-86	
Cut Variable	Cut Value	Cut Variable	Cut Value
θ_{Zenith}	$\geq 86°$	θ_{Zenith}	$\geq 86°$
v_{Lepton}	$\geq 0.1\,\text{c}$	l_{Track}	$\geq 200\,\text{m}$
—	—	Empty Hits	$\leq 400\,\text{m}$

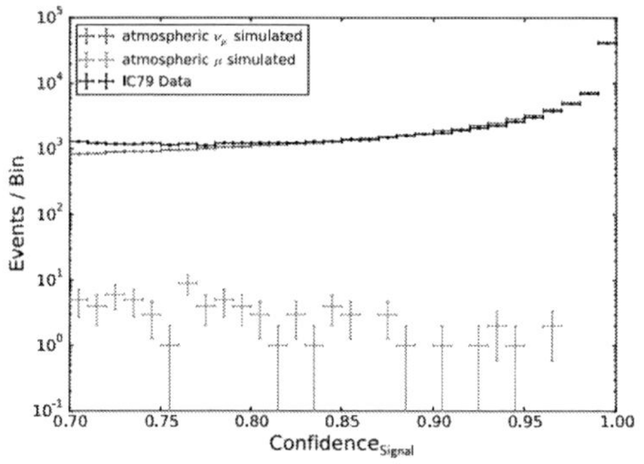

Figure 1. Confidence c for IceCube-79 in the region of the final selection. A good agreement between data and simulation is observed.

the spectrum. It is organized as follows: the selection of neutrino candidates is described in Sect. 2, whereas the unfolding of the ν_μ energy spectrum is addressed in Sect. 3. Section 4 concludes the paper with a discussion of the results.

2. Event selection

In IceCube muon neutrinos are detected via their interaction products – neutrino-induced muons. The background consists of atmospheric muons, entering the detector from above. At the starting level of this analysis the signal-to-background ratio is 10^{-3}. Therefore, a dedicated event selection is required in order to select neutrino candidates. The event selection is based on Machine Learning and carried out in a stepwise procedure.

In the first step we apply straight cuts to achieve a rejection of background events and a reduction of the required CPU-resources. The cuts placed on the different variables are summarized in Table 1. The cut on the zenith angle θ_{Zenith} removes a large portion of the muon background. The remaining muon events are falsely reconstructed as upward going and significantly harder to reject.

A Random Forest [9] is used to also reject those falsely reconstructed atmospheric muons. Simulated events generated with CORSIKA [10] and simulated neutrino events generated with NuGen [11] were used for training and testing. The testing was carried out in a five-fold cross validation. Input variables to the forest were selected using the Maximum Relevance Minimum Redundancy (MRMR) algorithm [12]. The stability of the variable selection was validated in a cross validation-like procedure [6, 13]. A Random Forest utilizes an ensemble of decision trees. The final output is generally referred to as confidence c and obtained as an average over the classification c_i of the individual trees:

$$c = \frac{1}{n_{\text{trees}}} \sum_{i=1}^{n_{\text{trees}}} c_i. \tag{1}$$

Figure 1 exemplarily shows the output of the Random Forest for the data sample obtained with the 79-string configuration of IceCube. Simulated atmospheric muons are shown in orange, simulated

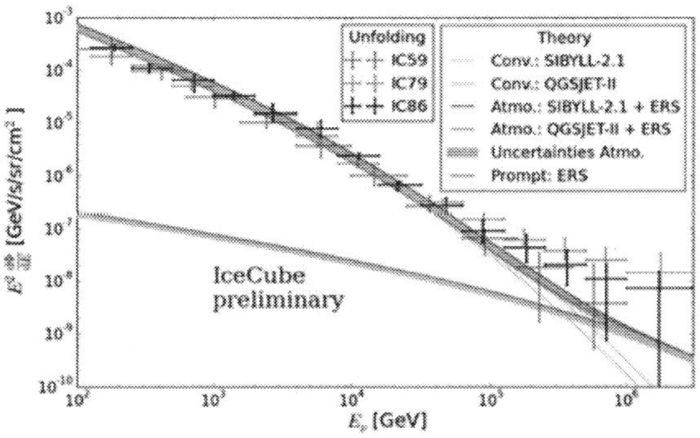

Figure 2. Atmospheric neutrino spectra obtained for the 79- and 86-string configuration of IceCube, compared to theoretical predictions obtained from two hadronic interaction model (SYBILL-2.1 [15] and QGSJET-II [16]) and the prompt flux prediction by Enberg et al. (ERS) [17].

atmospheric ν_μ in blue. Experimental data is depicted in black. A good agreement between experimental data and simulation is observed.

As a final selection criterion, an additional cut is placed on the output of the Random Forest. A cut at a confidence of $c \geq 0.92$ was used for the 79-string configuration. To increase the statistics of the event sample a two-dimensional cut is used for IceCube-86. This 2D-cut is simultaneously applied on the Random Forest output c and the estimated energy of the secondary muon. For $E_\mu \geq 10^4$ GeV events with $c < 0.7$ were rejected, whereas events with $c < 0.87$ were rejected for $E_\mu < 10^4$ GeV.

The event selection yielded a total of 66,885 neutrino candidates in 319.6 days of detector lifetime at an estimated background contamination of 330 ± 200 atmospheric muons for IceCube-79. For IceCube-86 92,060 neutrino candidates were obtained in 320.2 days of detector lifetime at an estimated background contribution of 410 ± 220 background muons.

3. Spectrum unfolding

The energy of the incident neutrino is not directly experimentally accessible. Therefore, it has to be inferred from energy-dependent variables and energy estimators. Furthermore, the production of secondary muons from neutrinos is governed by stochastical processes. Thus, an unfolding is required to reconstruct the ν_μ energy spectrum. Mathematically this is described by the Fredholm integral equation of first kind:

$$g(y) = \int_a^b A(x, y) f(x) \mathrm{d}x. \tag{2}$$

In Eq. (2) the energy distribution of the incident neutrino is denoted by $f(x)$ and $g(y)$ is the distribution of an energy-dependent observable. $A(x, y)$ represents the response function, which includes physics of muon production as well as muon propagation in ice and the limited resolution of the detector.

In this work Eq. (2) is solved using the software TRUEE [14]. TRUEE allows for the use of up to three input variables and the ν_μ spectrum is obtained in a maximum likelihood fit. In this work the estimated length of the muon's track, the number of direct photons and the estimated energy of the neutrino-induced muons were used as input variables for both detector configurations. A high correlation with energy was observed for all three input variables. For details on the unfolding of atmospheric $\nu_\mu$energy spectra and unfolding in general, we refer to [6] and [14].

4. Discussion

The atmospheric ν_μ energy spectra obtained with IceCube in the 79- and 86-string configuration are shown in Fig. 2. The theoretical flux predictions obtained for two hadronic interaction models (SYBILL-2.1 [15] and QGSJET-II [16]) are depicted together with the expected prompt flux according to the model by Enberg et al. (ERS) [17]. The spectrum obtained using IceCube in the 59-string configuration (brown) is shown for comparison [6]. The red shaded area visualizes the uncertainty of the theoretical flux predictions [7].

A flattening of the overall spectrum is observed for neutrino energies $E_\nu \geq 10^{5.3}$ GeV. This flattening is a clear deviation from the assumption of a solely atmospheric contribution to the overall ν_μ energy spectrum. It is attributed to a flux of high energy neutrinos from astrophysical sources. The observed excess is fully compatible with the recent observation of high energy astrophysical neutrinos by IceCube [8].

References

[1] A. Achterberg, et al., Astroparticle Physics **26**, 155 (2006), astro-ph/0604450

[2] J.G. Learned, K. Mannheim, Annual Review of Nuclear and Particle Science **50**, 679 (2000)

[3] C. Gonzalez-Garcia, M. Maltoni, J. Rojo, Journal of High Energy Physics **10**, 075 (2006), hep-ph/0607324

[4] S. Adrián-Martínez, et al., European Physical Journal C **73**, 2606 (2013)

[5] R. Abbasi, et al., Physical Review D **83**, 012001 (2011), 1010.3980

[6] M.G. Aartsen, et al., European Physical Journal C **75**, 116 (2015), 1409.4535

[7] A. Fedynitch, J. Becker Tjus, P. Desiati, Physical Review D **86**, 114024 (2012)

[8] M.G. Aartsen, et al., Physical Review Letters **113**, 101101 (2014), 1405.5303

[9] L. Breiman, Machine Learning **45**, 5 (2001)

[10] D. Heck, et al., *CORSIKA: A Monte Carlo code to simulate extensive air showers*, Vol. 6019 (FZKA, 1998)

[11] A. Olivas, et al., *Icecube simulation documentation*, http://wiki.icecube.wisc.edu/index.php/Simulation_Documentation_Wiki

[12] C. Ding, H. Peng, J. of Bioinformatics and Computational Biology **3** (2005)

[13] B. Schowe, *Feature Selection for high-dimensional data in RapidMiner*, in *Proceedings of the 2nd RapidMiner Community Meeting And Conference (RCOMM 2011)*, edited by S. Fischer, I. Mierswa (Shaker Verlag, Aachen, 2011)

[14] N. Milke, et al., Nuclear Instruments and Methods in Physics Research A **697**, 133 (2013), 1209.3218

[15] E.J. Ahn, et al., Physical Review D **80**, 094003 (2009), 0906.4113

[16] S. Ostapchenko, Nuclear Physics B Proceedings Supplements **151**, 143 (2006), hep-ph/0412332

[17] R. Enberg, M.H. Reno, I. Sarcevic, Physical Review D **78**, 043005 (2008), 0806.0418

EPJ Web of Conferences **116**, 08005 (2016)
DOI: 10.1051/epjconf/201611608005

Effect of atmospheric flux uncertainties on the determination of the neutrino mass hierarchy

Joakim Sandroos[1,a], Thomas Erhardt[1], Tim Arlen[2], and Sebastian Böser[1] for the IceCube-Gen2 Collaboration[b]

[1] Institute of Physics, University of Mainz, Staudinger Weg 7, 55099 Mainz, Germany
[2] Dept. of Physics, Pennsylvania State University, University Park, PA 16802, USA

Abstract. The next generation of large-volume neutrino telescopes will include low-energy subarrays which will be able to measure neutrinos with energies of a few GeV. In this energy range the primary signal below the horizon is neutrinos created by cosmic ray interactions in the atmosphere. The measured event rate will depend on the neutrino mass hierarchy, allowing determination of this quantity to a significance level of about 3.5 sigma within a 5-year period, mostly limited by systematic uncertainties. We present here the impact of the uncertainties on the atmospheric neutrino flux normalization on the determination of the neutrino mass hierarchy. We suggest constraining the systematic uncertainties by including the downgoing neutrino sample, which will increase the significance. This work was performed using simulation data from the low-energy extension to the IceCube detector located at the geographic south pole, PINGU, and is relevant to a wide range of other experiments.

1. Introduction

PINGU, the proposed low-energy extension to IceCube, located at the geographic south pole, will be able to measure neutrinos with an energy of a few GeV [1]. The experiment will be sensitive to the neutrino mass hierarchy (NMH) due to neutrino oscillations enhanced by the MSW-effect of neutrinos traveling through the Earth. These kinds of measurements usually only include upgoing neutrinos, as the downgoing neutrinos will not have had time to oscillate and as such carry no information regarding the NMH. In this contribution we suggest including the downgoing neutrinos in order to constrain the uncertainties. The neutrino energies considered fall in the range of 1 to 80 GeV.

2. Systematics

The uncertainties usually considered in relation to PINGU analyses are: the ratios $\nu/\bar{\nu}$, ν_e/ν_μ, the oscillation parameters θ_{13}, θ_{12}, Δm^2_{31}, and the effective area and the energy scale [1]. These however do

[a] e-mail: `jsandroo@uni-mainz.de`
[b] `http://icecube.wisc.edu`

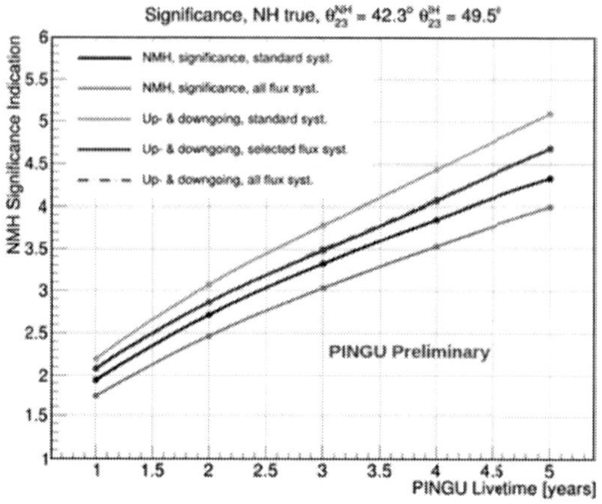

Figure 1. Significance vs. detector exposure for the neutrino mass hierarchy using PINGU. The black curve represents the standard set of 7 systematics, while the red includes the 5 most relevant atmospheric flux uncertainties. The green curve illustrates the significance when the downgoing neutrinos are included using the 7 standard systematics, while the red dashed curve includes the downgoing neutrinos and all systematics.

not touch upon the systematics stemming from the total normalization of the atmospheric neutrino flux. Using calculations from Barr et al. [2], we now incorporate the uncertainty on the atmospheric neutrino flux. This is considered from two sources: the primary cosmic ray flux and the hadronic interactions in the atmosphere. Barr et al. parametrizes the primary flux based on the GSHL model in the following way [2]:

$$\Phi(E_p) = a\left[E_p + b \times exp\left(c\sqrt{E_p}\right)\right]^{-d} \tag{1}$$

where d is the atmospheric spectral index, E_p is the energy per nucleon of the incoming cosmic ray, and a, b and c are normalization constants. The authors in [2] also consider the uncertainty in the hadronic interactions in the atmosphere, and divide the energy space into approximately independent regions as function of incoming cosmic ray energy and fraction of energy going into the decay products. The uncertainty of the atmospheric neutrino flux is thus different for each of these regions as well as for each of the four parameters on the primary CR flux, and are given individually in [2] as a function of energy $F_i(E)$. We take each function as a separate uncertainty input and associate each with a nuisance parameter P_i to scale each function. Here the index i refers to each of the independent uncertainties considered in [2] (Fig. 10). We can then vary the flux with the uncertainty as follows:

$$\Phi(E) \rightarrow \Phi'(E, P_i) = \Phi(E)\left[1 + \sum_i P_i F_i(E)\right]. \tag{2}$$

We use a Gaussian prior on the uncertainty parameters of $\pm 1\,P_i$. We now get a contribution to the overall flux which is linear in the parameter P_i by construction.

2.1 Method

The PINGU Simulation and Analysis software tool (PISA) is used for generating expected measurements from PINGU and analyzing those simulated data. It is not a traditional Monte Carlo generator: PISA takes as input external data such as a flux model and modifies it based on certain physical parameters to provide a model of the detector response under various assumptions of the nuisance parameters. In determining the neutrino mass hierarchy, a distinction between two possible physics scenarios is sought, and thus we compare simulated event rates in energy E and arrival direction $\cos(\theta_z)$ for each scenario and quantify their difference. This gives us the sensitivity to the NMH. Systematic uncertainties are implemented as individual parameters. In the following we investigate the impact of including the atmospheric neutrino flux uncertainties and the full sky on the neutrino mass hierarchy measurement. The flux is taken directly from MC flux tables in energy and $\cos(\theta_z)$, provided by Honda et al. [3] and is averaged over the azimuth angle. We use the flux at solar minimum. PISA incorporates several analysis methods; some based on a log likelihood ratio approach and other based on a $\Delta\chi^2$ approach under the Asimov assumption [1, 5, 6]. The $\Delta\chi^2$ method is computationally fast, but assumes a Gaussian likelihood space for all parameters. While this holds for the flux parameters, it is not true for all other parameters, especially the atmospheric mixing angle θ_{23}. We perform the $\Delta\chi^2$ analysis for 10 values of θ_{23} over a suitable range around its $\Delta\chi^2$ minimum, and calculate the pulls for all other pa- rameters. Even if this will not find the smallest $\Delta\chi^2$ the method can be used to give a quick comparison between models with different systematic parameters. Some other parameters included are also non-linear to some extent, and while it is in principle possible to scan the parameter space, the computation time will grow geometrically with each parameter. The main point is that all our calculations contain the same error, which is not critical insofar as we are interested in the relative difference between our results. The impact calculated this way of including the flux systematics on the neutrino mass hierarchy is shown as the red curve in Fig. 1.

3. Downgoing sample

In the following we test the idea of also including the downgoing neutrino flux in the NMH analysis. Doing this, the uncertainty parameters can be constrained, since the oscillations have little to no impact on the downgoing flux, leaving only the systematics impact in that part of the parameter space. This in turn should lead to a rise in the significance. We make two assumptions:Because of the by IceCube veto around PINGU, we chose to ignore any additional background, and assume the same effective area for the downgoing flux as for the upgoing. These are both overly positive estimates and needs to be adjusted for future studies. We realize these are rough assumptions, however for this first feasability study they were assumed for speed reasons. we use the same methods of analysis as in the previous section. The results are included in Fig. 1, as the green and blue curves, where we see a clear boost in NMH significance compared to only the upgoing sample.

3.1 Three samples

We proceed to investigate the source of this significance increase, using three data samples: The pure upgoing region, the pure downgoing region and a combined region. The "pure" samples are simulated independently, while the combined sample is simulated over all angles of $\cos(\theta_z)$. Three effects which could additionally contribute to the significance are investigated: Oscillation in the downgoing region near the horizon, mis-reconstruction of events from one region to the other and the constraint on the uncertainties in the downgoing region. The NMH significance in the pure downgoing sample is found to be negligible as illustrated in Fig. 2. Using the three distinct samples it is possible to estimate the amount of contamination from mis-reconstructed events each pure region will experience from the other. This is done by comparing the pure samples to the corresponding regions of the combined sample respectively. The effect of each region on the other is found to be asymmetric significance wise: the upgoing region significance gets washed out, while the downgoing region significance gets boosted. Without including any uncertainties this effect is summarized in Table 1. The effect of constraining the uncertainties is estimated by running our analysis twice: first using the pure downgoing sample, in order to get a constraint on the uncertainties – then again using those constraints as priors on the uncertainty

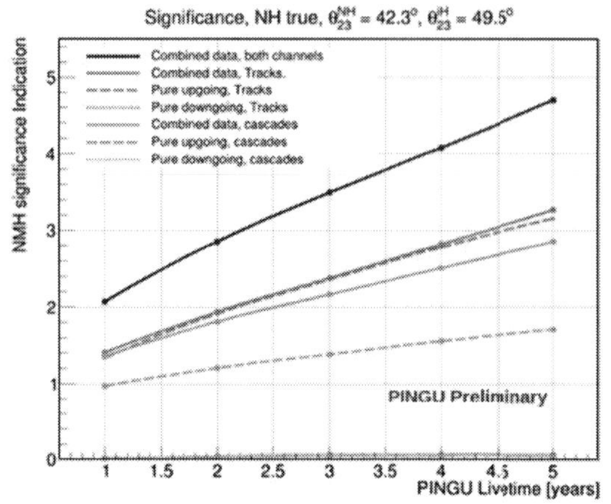

Figure 2. NMH Significance by sample and for track and cascade channels, as a function of time. The blue curve displays the combined sample and thus matches the blue curve in Fig. 1. The orange and green curves are from the track and cascade channels respectively. The dashed lines give the significance for the pure upgoing sample, while the fine dashed curves at the bottom represent the pure downgoing region.

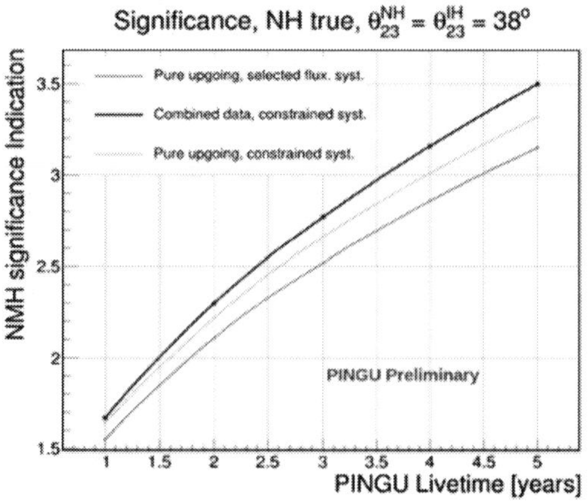

Figure 3. Contributions to the overall significance of the neutrino mass hierarchy as a function of time. The red curve is the significance in the pure upgoing sample when including the flux uncertainties. The yellow curve displays the significance in the pure upgoing sample when constraining the uncertainties based on the pure downgoing analysis. The blue curve is the total significance for the combined sample.

Table 1. NMH significance indication between samples for 5 years of livetime without uncertainties included.

Singnificance Indication $\sqrt{\Delta\chi^2_{min}}$		
Sample	Tracks	Cascades
Pure Up	7.67	9.11
Pure Down	0.55	0.75
Comb. Up	7.48	8.60
Comb. Down	2.24	5.04

parameters when analyzing the pure upgoing region. About half of the significance seem to come from upgoing events reconstructed as downgoing, while the other half seem to come from constraining the uncertainties. This is illustrated in Fig. 3. Note this is done at a mixing angle of $\theta_{23} = 38°$ for both hierarchies, so the overall significance is expected to be less than at the world average best-fit point. The red and blue curves in Fig. 3 have the same meaning as in Fig. 1, illustrating the pure upgoing data and the combined data respectively, while the yellow curve shows the pure upgoing sample with systematics constrained from the pure downgoing sample. We conclude the increase from the red to the yellow curve to be caused by better constraints on the uncertainties. The increase from the yellow to the blue curve stems from the oscillation signature carried above the horizon in mis-reconstructed events, which would otherwise have been lost. These two dominate over the negative effect of additional background in the up-going (signal) region.

4. Closing remarks

PINGU has started taking steps to quantify the impact of the systematic uncertainties of the atmospheric neutrino flux on the mass hierarchy measurement, and it turns out to be non-negligible. As a proof-of-concept, we have studied the effect of using the downgoing neutrino flux as a reference, since we expect little to no NMH sensitivity in that region of our parameter space. This allowed us to obtain a tighter constraint on the flux systematics, which increases the sensitivity of the NMH measurement. A part of the improvement indeed stems from the tighter constraint on the uncertainties, but an equally strong effect is due to upgoing events being reconstructed as downgoing, which are lost if the downgoing region is not included. However we have assumed an identical effective area and a similar background for the downgoing region as in the upgoing region. Further studies are needed to address these shortcomings, but our analysis seem to indicate improvements in sensitivity to the NMH are possible by using the downgoing atmospheric neutrino flux.

References

[1] IceCube Gen2 collaboration, "PINGU Letter of Intent", arXiv: 1401.2046. Update in preparation.
[2] Barr et. al., "Uncertainties in Atmospheric Neutrino Fluxes", Phys. Rev. **D74** (2006) 094009, arXiv: 0611266 [astro-ph]
[3] Honda, et al., "Atmospheric Neutrino Flux Calculation Using the NRLMSISE00 Atmospheric Model", Phys. Rev. **D75** (2007) 043006, arXiv: 1502.03916 [astro-ph.HE]
[4] R. A. Fisher, "On the mathematical foundations of theoretical statistics," Phil. Trans. Royal Soc. **A220** (1922) 309 {368}
[5] L. Verde, "Statistical Methods in Cosmology", Lect. Notes Phys. **800**, 147–177, 2010, arXiv: 0911.3105v1 [astro-ph.CO]
[6] E. K. Akhmedov, S. Razzaque and A. Y. Smirnov. "Mass hierarchy, 2–3 mixing and CP-phase with Huge Atmospheric Neutrino Detectors", JHEP **1302** (2013), p. 08

EPJ Web of Conferences **116**, 09001 (2016)
DOI: 10.1051/epjconf/201611609001

The MEUST deep sea infrastructure in the Toulon site

Patrick Lamare on behalf of the MEUST contributors

Aix-Marseille Université, CNRS/IN2P3, CPPM UMR 7346,13288 Marseille, France

Abstract. The MEUST infrastructure (Mediterranean Eurocentre for Underwater Sciences and Technologies) is a permanent deep sea cabled infrastructure currently being deployed off shore of Toulon, France. The design and the status of the infrastructure are presented.

1. Introduction

MEUST [1] is a second-generation follower of the pioneering ANTARES [2] submarine observatory which is being operated since a decade. It is designed as a shared platform between the neutrino physics and astrophysics communities in the context of the KM3NeT project [3], and the Earth and Environmental Science (ESS) communities as the West-Ligurian site of the EMSO European network of submarine observatories [4] but opened to all scientific domains. Compared to ANTARES the MEUST submarine network has a modular and extendable topology which will exceed the ANTARES capacities by an order of magnitude in most respects. The MEUST infrastructure is primarily dimensioned to host a large neutrino detector so-called ORCA for measurement of low energy neutrinos. A dense array of neutrino optical sensors distributed on mooring lines, the Detections Units (DU) installed on the seabed constitutes the ORCA detector designed by the KM3NeT collaboration. A DU consists of 18 Digital Optical Modules (DOM), equipped each with 31 photo-multipliers, and linked to each other as a vertical flexible line of about 250 m anchored on the seabed. The inter-DOM distance within a DU and the inter-DU distance on the seabed are ~9 m and ~20 m, respectively. The first set of instruments from ESS to be installed on MEUST is an Interface Instrumented Module (MII) which will be connected to the infrastructure. The MII hosts dedicated sensors and provide real time acoustic communication with an autonomous mooring line ("ALBATROSS") which instruments the whole water column on a height of 2000 m from the seabed.

2. MEUST site

The MEUST terrestrial base is located on the Brégaillon site of the Toulon Bay, where a new building is planned in 2018 to host the technical activities, including a control room. The MEUST submarine site has been selected after intensive characterization campaigns conducted in the past years at several possible locations, in order to find the best compromise between the site intrinsic qualities and external constraints such as existing cables and logistics costs. The MEUST final site (Fig. 1) is located ~40 km

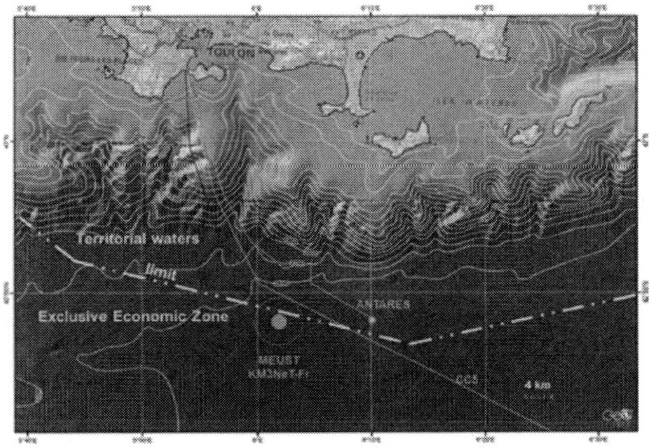

Figure 1. Geographic location of the MEUST site with the route of the MEUST cable to shore (red line).

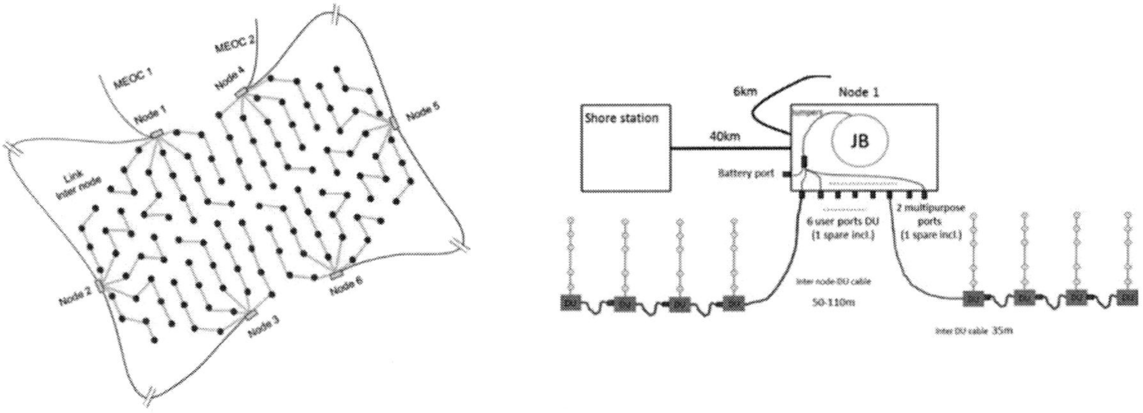

Figure 2. The MEUST generic submarine topology (left) and the layout of the 4-DU chain connected on the node (right).

offshore of Toulon at a depth of 2450 m and 15 km western of the ANTARES site. Acoustic and visual surveys of the seabed area have revealed a very flat and sandy area.

3. Submarine network topology

The full generic submarine network (Fig. 2 left) has a ring topology with up to 6 nodes connected to shore by two Main Electro Optical Cables (MEOC). The MEOCs are standard telecommunication cables with optical fibres for data transfer and one electrical conductor for power transfer. The network components (cables and nodes) are designed to allow a staged implementation using standard telecommunication marine deployment techniques and procedures.

The MEUST nodes are equipped each with 8 wet mateable connectors including 6 connectors usable for DUs only, and 2 multipurpose connectors suitable for both DUs and ESS instruments. The nominal instrumentation configuration connected to a node involves at least 20 DUs chained by 4 using 5 DU connectors and one set of ESS instruments using one multipurpose connector (Fig. 2 right). The 2 extras connectors are provisioned as spares in case of failures but may also allow to extend the available instrumentation.

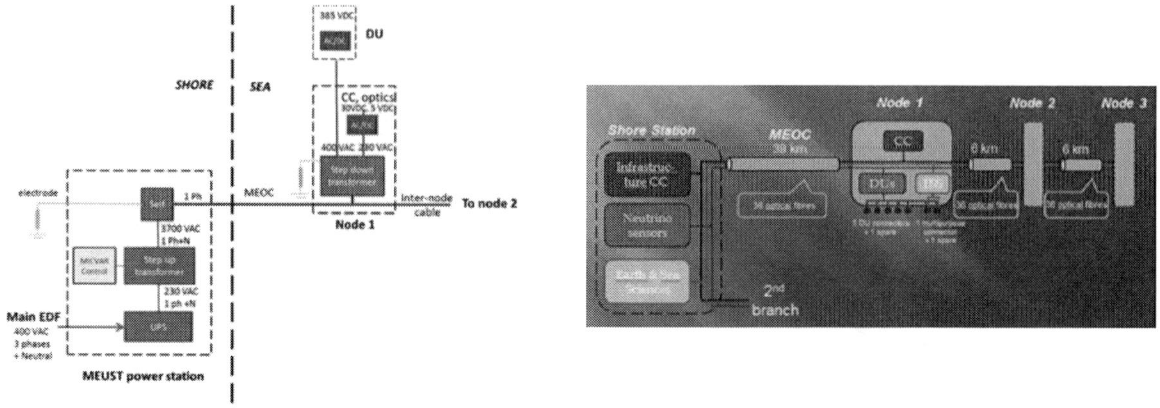

Figure 3. General schemes of the MEUST electrical power system for 1 node (left) and optical network (right).

4. Network functionalities

The MEUST infrastructure provides three main systems which are largely independent from each other: the electrical power distribution system, the optical network for data transfers and control of scientific instruments, and the Control Command for configuration and monitoring of the infrastructure network.

Proximity to the coast of the submarine site allows to perform the power transfer (Fig. 3 left) in High Voltage AC with sea return, as for ANTARES, using reliable electrical components with an overall acceptable offshore power loss. Power is transfer at 3700 VAC at maximum through the conductor of the MEOC and stepped down to 400 VAC by a transformer in each node. The DUs and ESS instruments include adequate AC/DC converters tailored to their DC current uses. The power system is primarily dimensioned by the DUs consumption and provides a usable power of ~8 kW on each node (~1 kW/connector).

The overall philosophy of the optical network is to provide an Ethernet network for a communication between the shore and the subsea instruments. The MEUST nodes include only passive optical components except optical amplifiers. The optical network (Fig. 3 right) comprises three independent sub systems distributed as following in the MEUST MEOC: the DU network using 30 optical fibres (10/node) to transfer and control the data of the neutrino detector, the ESS network using 2 optical fibres and the Control Command network using 2 optical fibres. Two extra optical fibres in the MEOC are reserved as spares. The DU optical fibres are distributed to the 8 node output connectors whereas the ESS optical fibres are distributed to the 2 multipurpose connectors only. The DU network uses Dense Wavelength Division Multiplexing (DWDM) with 50 GHz spacing (0.4 nm) in the C-band allowing up to 80 wavelengths in each optical fibre. Each DOM has its own wavelength with a point to point connection with the shore. A MEOC optical fibre can therefore transfer the data of 4 DUs (corresponding to 72 DOMs), which has guided the concept of the 4 DUs chained to one node connector. The ESS network uses bidirectional Coarse Wavelength Division Multiplexing (CWDM) with a wavelength grid of 20 nm, allowing the possibility of designing a sub network using a denser DWDM grid.

The Control Command manages the power, optical and monitoring systems of the infrastructure. It uses industrial Input-Output modules and switches, compatible with Linux for software developments, with a communication based on the Ethernet protocol. All the system is doubled from the sensors and actuators to provide a redundancy in case of failure of an individual component.

5. Infrastructure components

The components of the infrastructure have been prototyped and engineered from 2010 to 2015.

Figure 4. The node after assembly (left) and the MEOC cable under deployment (right).

A new power station, hosting an Uninterruptible Power Supply (UPS) with a Power Feed Equipment (PFE), is operational beside the ANTARES one on the Sablettes beach in La Seyne sur Mer where the MEOC cables land the shore.

The MEUST MEOC #1 (36 optical fibres) is a ~39 km long standard telecommunication cable from ALCATEL (reference OALC-7). The MEOC #2 is the ANTARES one, compliant with the MEUST specifications, which will be re-routed after the ANTARES decommissioning scheduled in 2017.

The first node (Fig. 4 left) was built and deployed. Its main active element is a Junction Box made of a titanium sphere containing the electrical step-down transformer, the optical components and the Control Command devices. The Junction box is inserted in a titanium frame hosting 2 panels of each 4 output wet mateable hybrid connectors (reference PRH/APC from the Teledyne ODI company). The connector panels include specific connection tools which have been developed in order to be able to use light commercial ROVs to connect the instruments.

6. Status and plan

The MEOC cable was deployed by Orange Marine on the seabed in December 2014 (Fig. 4 right). In April 215 with Orange Marine, the MEOC was recovered from the seabed, thanks to a dragging tail, and connected to the node on board using standard telecommunication jointing technique. The node was then successfully deployed with half of the inter-node cable (3.6 km) for future extension. The infrastructure was powered on and successfully operated one month before a failure which required a repair operation held in June 2015. The operation of the node confirmed the functionalities of the power and Control Command systems with the expected temperatures and humidity inside the JB. The repair operation didn't allow for a complete repair due to time constraints, so the node was brought back to shore for the repair of the MEOC penetrators. The repair is in progress and the re-deployment of the node is planned in early 2016. The first DU and the MII are ready for deployment, waiting for the node installation, and the ALBATROSS line is installed on the seabed.

In the coming years the submarine infrastructure is expected to be extended by additional nodes for connections of DUs to achieve the realization of a KM3NeT Building Block. Further ESS instruments are expected and in particular the transfer to MEUST of the ESS Secondary Junction Box of ANTARES.

The authors wish to thank R. Cocimano, A. Damico, E. Heine, G. Kieft, S. Mos and J-W. Schmelling for their contributions to specific aspects of the design of the MEUST infrastructure. The MEUST development was financially supported by the CNRS, Aix-Marseille Université, the town of Marseille and the European Regional Development Fund (ERDF).

References

[1] MEUST web page, available on `http://meust.cnrs.fr`

[2] J.A. Aguilar et al., *ANTARES: the first undersea neutrino telescope, Nuclear Instruments and Methods in Physics Research,* Section **A 656** (2011) pp 11–38

[3] KM3NeT web page, available on `http://www.km3net.org`

[4] EMSO web site, available on `http://www.emso-eu.org`

EPJ Web of Conferences **116**, 09002 (2016)
DOI: 10.1051/epjconf/201611609002

Design and mass production of the optical modules for KM3NeT-Italia project

Emanuele Leonora[a], Sebastiano Aiello, and Valentina Giordano on behalf of KM3NeT-Italia project

Istituto Nazionale di Fisica Nucleare, sezione di Catania, via S. Sofia 64, 95125 Catania, Italy

Abstract. The KM3NeT European project aims at constructing a km^3 underwater neutrino telescope in the depths of the Mediterranean Sea. The first phase that is under construction will comprise eight tower-like detection structures (KM3NeT-Italia), which will form the internal core of a km^3-scale detector. The detection element of KM3NeT-Italia, the optical module, is made of a 13-inch pressure-resistant glass-vessel that contains a single 10-inch photomultiplier and the relative electronics. The design of the whole optical module, the main results obtained from the massive photomultipliers measurements, and the foremost phases of the mass production procedure performed at the production site of Catania are also presented.

1. Introduction

The KM3NeT European project aims at constructing a km^3 underwater neutrino telescope in the depths of the Mediterranean Sea. Thanks to a dedicated funding by the Italian Ministry of Education, University and Research (MIUR) the construction of a first phase of the Italian KM3NeT node is under way. This detector will comprise eight tower-like detection units and 24 strings. The tower is composed by 14 storeys, each 8 m long, spaced out vertically of 20 m. The detection element of the telescope, called optical module (OM), is a pressure-resistant transparent glass vessel that contains a 10″ photomultiplier and the electronics devices for power supply, signal read-out and transmission.

Each storey of the tower hosts 6 optical modules, arranged into three couples downgoing and horizontally at each extremity and at the storey-centre looking 45° downwards. The optical module design, as well as each single component, was defined after an intense R&D work on photomultipliers, high voltage supply circuit, optical gel and mechanical supports. During the optical module production phase, 750 10-inch R7081 Hamamatsu photomultipliers have been characterized by means of dedicated test-benches. Following a well-established production procedure, 700 optical modules were assembled at the INFN production site located in Catania, and their electrical and mechanical functionalities were tested by using properly test-benches and also a hyperbaric chamber. The paper describes the design of the optical module and its main components in Sect. 2, the main results obtained from the massive

[a] e-mail: emanuele.leonora@ct.infn.it

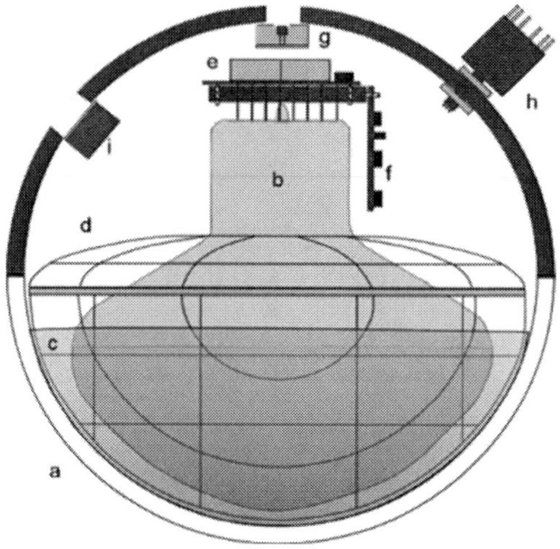

Figure 1. Schematic view of the main components of an optical module: (a) 13″ glass sphere, (b) PMT, (c) optical glue, (d) mu-metal cage, (e) HV base, (f) FEM (Front End Module), (g) LED beacon, (h) 8-pin connector, (i) Manometer.

photomultiplier measurements in Sect. 3, and also the main phases of the mass production procedure in the last Sect. 4.

2. The optical module design

The optical module is composed by a 13-inch pressure-resistant spherical glass-vessel produced by Benthos [1], which works as water and pressure protection for internal devices, ensuring also a good light transmission. Two hemi-spherical halves compose each vessel: a transparent one that supports the photo-detector, and a painted black one, which hosts the electronics. A single 10-inch photocathode photomultiplier, type R7081 produced by Hamamatsu [2], was chosen as photodetector, and it is optically and mechanically coupled to the transparent hemisphere by means of transparent glue Wacker SilGel 612 [3]. Inside the OM, a mu-metal cage, produced by ITEP [4], surrounds the PMT to shield it against the Earth's magnetic field. An active base produced by ISEG [5] supplies the PMT high voltage (HV). A custom designed [6] front-end electronic module (FEM), drives the HV base, digitizes the PMT signals, encodes and transmits the data to the external floor control module by means of an 8-pin SEA CON connector [7]. Inside the optical module there is a LED beacon system designed to emit calibrated light pulses for underwater calibrations and also a pressure meter to allow the check of the vacuum inside the sphere. Figure 1 shows schematically the OM components.

3. The photomultiplier characterization phase

The R7081 Hamamatsu photomultiplier has 10 stages and a 10-inch bialkali photocathode with a typical detection surface of about 500 cm^2 and a quantum efficiency of about 25% at the spectral peak of 400 nm. In the frame of the massive optical modules production, a batch of 750 photomultipliers was tested by means of a dedicated facility. The tests were performed on PMTs with the HV base already soldered, surrounded by the mu-metal cage, positioned into properly dark box where they were left for at least 8 hours before the measurements. The facilities developed to test the 750 PMTs consisted of 4 light-tight dark boxes, which allow automatized test on 8 PMTs per day. A software developed in LabView for acquisition and data analysis manages the DAQ system composed by two Digitizer

Table 1. Measurements from a batch of 750 R7081 Hamamatsu PMTs.

Parameters	Required values values	Mean values measured	Value Ranges measured
Voltage at Gain 5E7 [V]	<2000	1648	1460 ÷ 1900
Dark Count rate [Hz]	<8000	1128	274 ÷ 13200
P/V ratio	>2	3.2	1.76 ÷ 4.5
TTS FWHM [ns]	<3.5	2.4	1.98 ÷ 3.85
Pre-Pulse [%]	<1	0.18	0.04 ÷ 2
Delayed Pulse [%]	<8	5.65	4 ÷ 10.8
Type 1 after pulse [%]	<3	0.47	0.05 ÷ 4.71
Type 2 after pulse [%]	<10	7.3	3.1 ÷ 16.5

PCI-5153 produced by National Instruments [8], with a 500 MHz bandwidth, 2 GSample/seconds sampling and 8-bit resolution.

Table 1 summarizes all the measurements from the whole batch of PMTs, together with the project specifications. As it can be seen, the results confirm that such PMTs comply with the project's requirements. Among the whole batch of 750 PMTs, only 14 PMTs were rejected, because one or more parameters didn't comply with the range required, mainly the fraction of type 2 after pulses, the peak to valley ratio and the dark count rate.

3.1 Aging effects

A study of the aging effects of the PMTs to measure the variations of the parameters with the operating time was performed for almost 3 years (equivalent operating time of 45 years in water, with an optical background of 1 photo-electron (pe) at 200 kHz). The measurements consisted of cycles of two different phases: an accelerated aging phase, illuminating the PMT under test with a 3-pe light pulses at 1 MHz, interposed with a characterization phase, which consisted of measurements of gain, time and charge characteristics and spurious pulses. Results showed that all the PMT's parameters did not change substantially during long operating periods apart the gain, which had a maximal variation of about 30% [9].

3.2 The Earth's magnetic field effects on PMTs

Performances of large area photomultipliers are influenced by magnetic fields, because of the long trajectories of electrons from the photocathode to the anode. Variations on PMT parameters were measured with and without the mu-metal cage changing the PMT orientations with respect to the Earth's magnetic field. All the results confirmed that the performances of large area PMTs without any magnetic shield is affected by changes of its orientation with respect to the Earth's magnetic field, with variations of about 30% for the gain and over 40% for the detection efficiency and also for the peak to valley ratio. The use of a mu-metal cage as magnetic shielding considerably reduces these variations [9]. A magnetic shield is therefore mandatory in optical modules with large are photomultipliers.

4. The main phases of the production procedure

Following a well-established production procedure, almost 700 optical modules were assembled at the INFN integration site located in Catania. For the outgassing phase of the optical gel and for the closure of the two hemispheres of the optical modules, four vacuum boxes were constructed ($1 \times 1 \times 1\,m^3$), in which the absolute vacuum decreases to over 300 mbar in less than 2 min. To reduce the effects of humidity, each box was equipped with a system of Nitrogen flux. Schematically, the main phases of the assembly procedure can be described as follows:

- soldering of the high voltage supply base on the leads of the PMTs;
- testing of each PMT with the base already soldered, as described in Sect. 3. Only the PMTs that fully complied with the requirements were assembled into the optical modules;
- the HV base and the leads of each PMT were covered with a liquid coating (that dries in about 24 hours) to isolate all the electrical contacts;
- the transparent hemisphere where the PMT will be glued was cleaned carefully with optical paper and ethanol, and positioned into the vacuum box;
- the hemispherical part of the mu-metal cage was cleaned carefully with optical wipes and positioned into the glass hemisphere;
- one out-gassing cycle into the vacuum box with an under pressure of 400 mbar, followed by an air re-entry phase, made the elements ready for the optical gel pouring
- mixing of the two components of the optical gel, A and B, with a ratio of 100:60, with a controlled mixer with 120 turn/minutes, for 5 minutes, to prepare the required 1.5 litre of optical glue;
- pouring of the optical glue into the hemisphere, with the mu-metal cage already in position;
- two outgassing phases to accelerate the extraction of the air bubbles inside the gel;
- by means of proper tools, the PMT was positioned and immersed into the gel inside the hemisphere;
- three degassing phases, each one followed by a phase of Nitrogen re-entry to remove all the residual air bubbles and reduce humidity inside the glue;
- polimerization of the optical gel, performed at room temperature and atmospheric ambient for at least 12 hours;
- the Front End Electronic board was mounted by means of proper PVC structures, and connected to the PMT anode by a coaxial cable and to the HV base by a flexible ribbon cable;
- glueing of the LED beacon system by means of black silicone in a transparent window opened in the black hemisphere and its connection to the FEM board by a twisted cable;
- two sacks of Silica gel of 5 g were positioned into the OM, to reduce the humidity inside;
- to close the OM, the two hemispheres were positioned into the vacuum box. A wedge inserted trough the two hemispheres maintains open the OM for a 10-minute phase of Nitrogen flux, to reduce the internal humidity to less than 2%. When the hygrometer inside the box shows the expected value, an electric system manually driven removes the wedge, and the two hemispheres joined together;
- an under-pressure of 400 mbar and a following Nitrogen re-entry phase close the whole sphere;
- one turn of sealant Terostat-81 20×2 mm and three following turns of 50-mm tape Scotchrap 3 M all-weather corrosion protection applied at the junction of the two hemispheres sealed definitely the optical module. The under pressure inside is constantly checked by the manometer;
- a proper test-bench tested the functionality for each OM. A hyperbaric chamber also tested the watertight for random OMs, up to 350 bars.

5. Conclusions

A 13-inch diameter optical module with a large 10-inch PMT was designed as the detection element of the KM3NeT-Italia project. A batch of 750 R7081 Hamamatsu photomultipliers was characterized. Following a well-established production procedure, almost 700 optical modules were assembled at the INFN integration workshop located in Catania. The functionality was tested both from electrical and mechanical point of view by using proper test-benches and a hyperbaric chamber. They will be soon mounted in the tower-like detection units and deployed in the deep-sea site.

References

[1] Teledyne Benthos web site, http://www.benthos.com
[2] Hamamatsu web site, http://www.hamamatsu.com
[3] Wacker Chemie AG, http://www.wacker.com/cms/en/home/index.jsp
[4] Institute for Theoretical and Experimental Physics, Moscow, http://www.itep.ru
[5] SEG web site, http://www.iseg-hv.com
[6] C. Nicolau, proceeding of this conference
[7] SEA CON web-site: http://seaconword.com
[8] National Instrument web site: http://www.ni.com
[9] E. Leonora et al., 2015 JINST 10 T11003

EPJ Web of Conferences **116**, 09003 (2016)
DOI: 10.1051/epjconf/201611609003

Km3Net Italy – Seafloor network

Riccardo Papaleo for the KM3NeT collaboration

INFN – Laboratori Nazionali del Sud, via Santa Sofia 62, 95123 Catania, Italy

Abstract. The KM3NeT European project aims to construct a large volume underwater neutrino telescope in the depths of the Mediterranean Sea. INFN and KM3NeT collaboration, thanks to a dedicated funding of 21.000.000 € (PON 2007–2013), are committed to build and deploy the Phase 1 of the telescope, composed of a network of detection units: 8 towers, equipped with single photomultiplier optical modules, and 24 strings, equipped with multi-photomultipliers optical modules. All the towers and strings are connected to the main electro optical cable by means of a network of junction boxes and electro optical interlink cables. Each junction box is an active node able to provide all the necessary power to the detection units and to guarantee the data transmission between the detector and the on-shore control station. The KM3NeT Italia project foresees the realization and the installation of the first part of the deep sea network, composed of three junction boxes, one for the towers and two for the strings. In July 2015, two junction boxes have been deployed and connected to the new cable termination frame installed during the same sea campaign. The third and last one will be installed in November 2015. The status of the deep sea network is presented together with technical details of the project.

1. Introduction

The Km3NeT Italy is the first block of the KM3NeT neutrino telescope [1]. Figure 1 shows its underwater layout.

It is composed of 24 strings (red circle), 8 towers (green circle) and 3 Junction Boxes (JBs): 1 is dedicated to manage the towers (green square) and 2 will serve the strings (red square).

Each Detection Unit (DU), both string-like and tower-like structures, is connected to a JB which acts as an active underwater node for the exchange of data and power from and to the shore station.

The main electro optical cable (MEOC) is a standard telecommunication cable with 20 optical fibres (G655 type) for data transfer and one electrical conductor for power transfer. The cable is able to transport up to 100 kW to the detector. At its end a Medium Voltage Converter (MVC) transforms the power supply from 10 kV to 375 V.

The MVC is hosted on a Cable Termination Frame connected to the JBs by electro optical interlink cables equipped with 4 optical fibres and 2 electrical conductors.

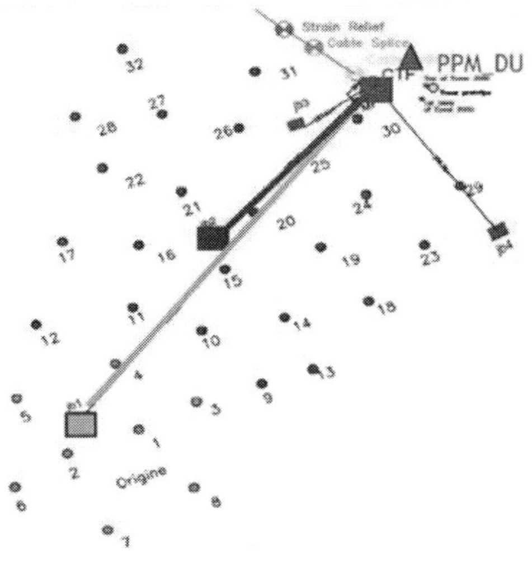

Figure 1. Underwater layout.

Figure 2. CTF during deployment operation (july 2015).

2. Cable Termination Frame (CTF)

The CTF (Fig. 2) is the termination system of the main electro optical cable. The design of the CTF includes a mechanical frame (Titanium), a Cable Termination Assembly system (for the splitting of the fibres and electrical connection), a patch panel with 5 hybrid ROV wet mateable connectors (each one able to manage 4 optical fibres) and a Medium Voltage Converter for the power supply of the underwater telescope.

The mechanical frame is realized in Titanium grade 5. The design of the CTF is optimized in order to guarantee the stability of the system during the deployment phase and after the touchdown on the seabed.

The first 3 connectors on the patch panel are used for the connection of the KM3NeT Italy JBs. The fourth connector is used for the connection of the EMSO JB (www.emso-eu.org).The fifth connector in the future could be used to connect another JB of the underwater detector.

Figure 3. The junction box.

3. Junction Box

The Junction Box (Fig. 3) is a key element of the detector and connects the MEOC to the detection units.

Each junction box is located near a group of up to 12 DU's to reduce as mch as possible the distance junction box – DU's.

Its main function are the distribution of the power to the units and the control of the data flux from/to shore.

The JB includes the following items:

- a mechanical frame to host all the PODs (Protective Ocean Devices) and the ROV wet mateable connectors. The frame has been designed in order to optimize the assembly, the maintenance and the sea operations of the JBs;
- 4 plastic panels to host the hybrid ROV wet mateable connectors: an input connector coming from the CTF and up to 12 connectors for the connection to the DUs;
- a vessel for fibre and conductor routing, named manifold, and three vessels with different functionality (Fig. 4):
    - a junction box power POD, to manage the power supply towards the DUs;
    - a junction box optical POD, to manage the optical data flux from and to the shore station;
    - a junction box electronic POD, to control and monitor the full system.

The mechanical frame is realized in Al Alloy 5083. The central part of the mechanical frame hosts the manifold and the PODs.

All the electronics devices (boards, optical amplifier, mux&demux filter, etc) have been grouped by function in order to optimize the reliability of the system and the connector and cabling system. The size and the weight of each POD is compatible with an easy handling and installation on board. This allows to replace a failed POD even on board, minimizing the costs for maintenance and the downtime of the apparatus. The material of the PODs is Al Alloy hard anodized in order to increase the protection against the corrosion.

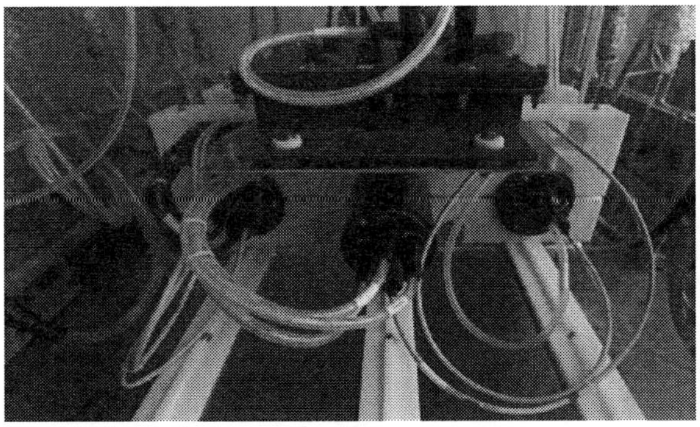

Figure 4. Detail of the manifold and the PODs installed on the mechanical frame.

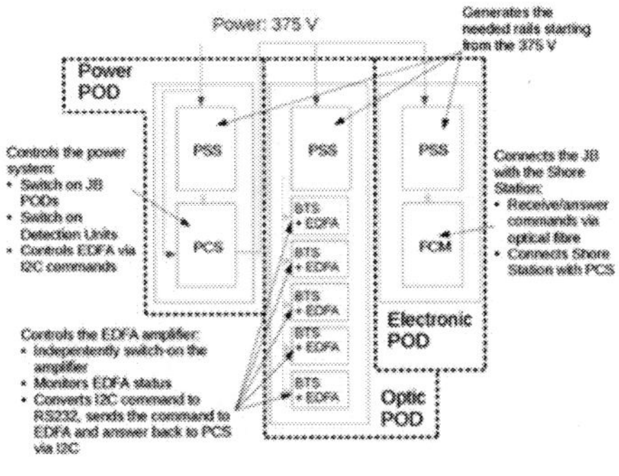

Figure 5. Scheme of the power system of the JB (string type).

From the electric point of view (Fig. 5), the JB simply acts as a switch of the main power line (375 V) provided by the DC-DC MVC in the CTF).

The power line which enters the JB, is converted by dedicated DC-DC converters and is split in 8 or 12 different lines; each of them is connected to a DU.

Those lines are controlled by switches located in the power POD of the JB. Dedicated commands from the Shore Station will power on the single DU, independently.

From the optical point of view, the JB will re-route all the optical fibres to the correspondent DU, to allow its direct communication with the shore station.

Optical amplifiers are necessary to guarantee the necessary optical budget for the communications. The amplifiers are located and managed in the optical POD of the JB.

4. Summary

The KM3NeT detector is under construction. On July 2015 a sea operation has been performed in order to install the first components of the underwater telescope: the new Cable Termination Frame and 2 Junction Boxes (1 for the tower and 1 for the string) has been installed and connected. The current plan of work foreseen the installation of the first DUs on December 2015. The KM3NeT seafloor network will be completed on spring 2016 with the installation of the last Junction Box.

Reference

[1] R. Coniglione, this conference

EPJ Web of Conferences **116**, 09004 (2016)
DOI: 10.1051/epjconf/201611609004

GRBNeT – A prototype for an autonomous underwater neutrino detector

K. Pikounis[1,2,a], C. Markou[1,b], E.G. Anassontzis[3], G. Androulakis[1], C. Bagatelas[1], K. Balasi[1], A. Belias[1], P. Damianos[1], E. Drakopoulou[1,2], E. Kappos[1], K. Manolopoulos[1], P. Rapidis[1], E. Tzamariudaki[1], and G. Voulgaris[3]

[1] Institute of Nuclear and Particle Physics, N.C.S.R. Demokritos, Patriarchou Grigoriou and Neapoleos 27, Agia Paraskevi, Greece
[2] National Technical University of Athens, Heroon Polytechniou 9, Zografou Campus, Greece
[3] Faculty of Physics, National and Kapodestrian University of Athens, Zografou Campus, Greece

Abstract. GRBNeT is a project aiming at the detection of ultra–high energy neutrinos, for example neutrinos originating from Gamma Ray Bursts. The goal is to design, construct and deploy a prototype unit of an autonomous (data/energy–wise) neutrino detector. Being autonomous is crucial since for the detection of ultra–high energy neutrinos a very large volume of water is required. Large scale facilities such as IceCube and KM3NeT are designed to be more sensitive to galactic and diffuse flux neutrinos rather than extragalactic ultra–high energy neutrinos. However, their sensitivity to such neutrinos could be increased by placing around and at larger distances detectors such as the one of the GRBNeT project. This extension would increase the instrumented volume of neutrino telescopes to several cubic kilometres. In addition to that, as no cable connection to the shore is required, GRBNeT detection units cost significantly less than regular detection units and can become a cost effective extension of large scale facilities. For the GRBNeT prototype unit ultra low power electronics have been developed. The response to high energy neutrinos from GRBs and to the atmospheric muon background has been simulated.

1. Introduction

The Gamma Ray Burst Neutrino Telescope (GRBNeT) is a project funded by the Greek National Strategic Reference Frame Work (THALES Initiative 2011–2015). It is a collaboration between the Institute of Nuclear and Particle Physics of N.C.S.R. Demokritos, the University of Athens and the Hellenic Center for Marine Research, consisting of more than 15 scientists and engineers. GRBNeT has been conceived and proposed in the spirit of earlier autonomous detector proposals [1–3].

Gamma Ray Bursts are short duration emissions of γ–rays associated with extremely violent explosions in distant galaxies, characterized by the largest luminous energy emissions known to date.

[a] e-mail: pikounis@inp.demokritos.gr
[b] e-mail: cmarkou@inp.demokritos.gr

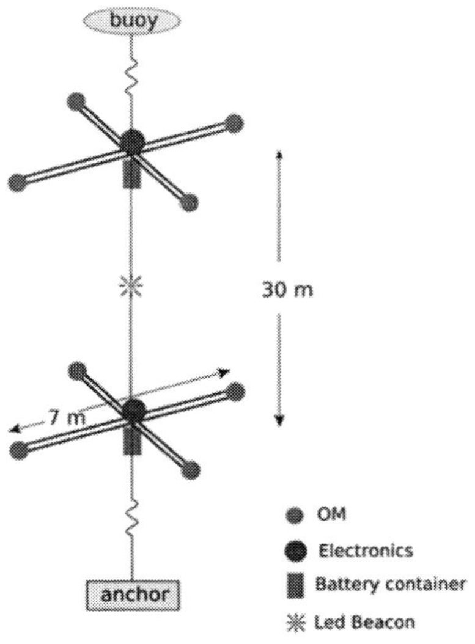

Figure 1. A schematic representation of the GRBNeT prototype detector.

It is predicted that during these explosions charged pions are produced through photon–hadron reactions, which in turn decay to ultra–high energy (PeV energies) neutrinos.

GRBNeT detection units can be placed independently from other infrastructure, creating a large neutrino detector. Using a multimessenger approach it is possible to pinpoint interesting astrophysical events in space and time that emit high energy neutrinos and then combine those in the offline analysis. GRBs are ideal candidates for this kind of analysis since their location in the sky and their exact time are precisely known by (γ–ray, x–ray, etc.) telescope observations. GRBNeT detection units can also be placed around large scale neutrino telescopes like KM3NeT/ARCA [4]. Since GRBNeT detection units do not depend on cables they can be deployed in various distances around existing underwater telescopes, the outcome would be the increase of the sensitivity in ultra–high energies of the said telescope in a very cost effective way.

2. Prototype design

The prototype unit consists of two cross–shaped titanium floors each with four Optical Modules (OMs) facing outwards. All four OMs are connected to the Control Board of their floor which is located in a separate sphere at the center of each cross. This sphere houses all the electronics necessary for the digitization, trigger and data acquisition. Each OM consists of a single 13″ Hamamatsu R8055 photomultiplier (PM), surrounded by a mu–metal mesh and placed in a 17″ glass sphere. Studies have been made to optimise the detection of ultra-high energy neutrinos ($E_\nu > 1$ PeV) with respect to the OMs' orientation. It was found that OMs facing the horizontal perform better, as compared to any other direction. In addition, having a detection unit in the Mediterranean Sea with the PMs facing horizontally gives a full sky coverage during the day.

The two floors are seperated by a 30 m long rope and therefore their relative azimuthal position is varying. As there is no cable connection between the two floors, each of them is operated independently from the other; a led beacon is placed in equal distance between the two floors producing artificial events every seven hours for testing purposes, triggered by an autonomous timer. The detector is anchored on the sea bottom by an appropriate weight and held vertical by a top buoy. A schematic representation of the detector is shown in Fig. 1.

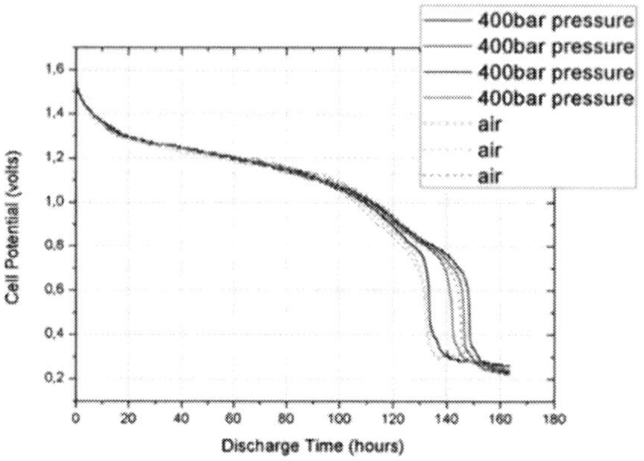

Figure 2. The voltage (V) of seven batteries versus the discharge time of each cell. Dotted/solid lines correspond to batteries discharging in air/under pressure of 400 bar respectively.

3. Power autonomy

The main idea behind the GRBNeT project is the autonomous character of the detector. By eliminating the dependence on cables, the detector is cost effective and easier to deploy. For this purpose, very low power electronics were designed and implemented. Power is provided by D type Varta alkaline batteries, arranged in clusters of 13 batteries in series. For each floor, 43 clusters in parallel provide enough power for at least six months. The batteries are contained in two plastic containers (one per floor) exposed to full depth pressure of 3000 m. The containers are filled with paraffin oil, which is an insulator and practically incompressible for that ambient pressure.

The behaviour of the batteries discharging under pressure has been studied in detail. Several tests were performed at pressures around 400 bar and no significant deviation has been observed. The discharge times of batteries on a 10Ω resistance at 1 and 400 bar are shown in Fig. 2, where the voltage drop of seven batteries is plotted against the discharge time of each cell.

4. Electronics – DAQ

The Data Acquisition and Slow Control are autonomous and all the data produced are written in SD–Cards. The DAQ has been designed with low power FPGA and the Slow Controls are implemented in custom made electronics [5]. They are both housed inside a glass sphere located at the center of each cross–shaped floor. The Control Board is the system that handles and distributes power (from the batteries) to the PMs and the other electronics (FPGA and Atomic clock), sets the high voltages on the PMs, reads the analog pulses produced by the PMs, digitizes them according to 4 thresholds and feeds the digital pulses to the FPGA for the coincidence logic to be applied. The four thresholds correspond to pulse charges of 1, 20, 30 and 50 photo–electrons (PEs), therefore from each analog pulse a maximum of four digital pulses are created. It also records the voltage of the PMs, the readings of a compass and a thermometer. All readings are recorded in SD–cards with a time stamp produced by an internal clock. The internal clock of each Control Board was synchronized with a GPS clock before deployment.

The FPGA implements the coincidence logic and records the single photoelectron rate. It stores the data produced in two separate SD–Cards, one for the single photoelectron rate and one for the events satisfying the trigger. For the single photoelectron rate, the number of pulses per second arriving from each PMT above the first threshold is being written. The coincidence thresholds were tuned in order to reduce the background and preserve battery energy, so a high threshold of 20 PEs was used. The trigger requires at least two pulses above threshold in different OMs within a time window of 200 ns.

Figure 3. Left: A floor lifted from the deck of R/V AEGAEO. Right: A floor just before submersion.

When such a coincidence is detected, the time window opens for an additional 100 ns. In these 300 ns the rising times of the pulses that pass the second threshold, the ToTs of the pulses that pass each one of the three higher thresholds and the PM id are recorded. All the ToTs of the pulses passing each one of the three higher thresholds are recorded to better approximate the actual shape of the analog pulse. Each floor has its own FPGA recording coincidences independently from the other.

For the timing of the FPGAs two different techniques are applied. In the upper floor a low power atomic clock is attached to the FPGA. The atomic clock gives the actual time for the time stamps and also provides stable pulses per second used for the triggering by the FPGA. The time of the atomic clock was synchronized in the same way as the internal clocks of the Slow Controls before deployment. The lower floor does not have an atomic clock so the clock attached to the FPGA will be solely responsible for the time stamping and the timing needed for the coincidences. Thus, in the offline analysis, it will be possible to assess the performance of the embedded clock, and measure any potential long term drift, using the artificial events produced by the led beacons (placed in equal distance between the two floors) as time reference.

5. Simulations

Simulations studies have been made on the expected performance of the prototype to charge current muon neutrino and anti–neutrino events, and to atmospheric muon background events corresponding to a livetime of approximately 3 months, using the latest tools developed for KM3NeT [4]. The noise from the decay of the radioactive ^{40}K which is contained in sea water has also been added in both samples. For this prototype, given it's small effective volume, special focus is given on atmospheric muon events. This is done because the atmospheric muons triggered through the livetime of the prototype, can provide the statistics needed to validate its response. The trigger concept mentioned in Sect. 4 was also tested for neutrino events following an E^{-2} spectrum. It was found that more than 79% of the events that satisfy the trigger are high energy neutrino events ($E_\nu > 100$ TeV), which indicates that such a threshold setting for the prototype triggers high energy events.

6. Deployment

The deployment of the prototype took place on the 28th of October 2015 off the coast of Pylos in Peloponnese with the oceanographic ship "R/V AEGAEO" of the Hellenic Center for Marine Research. Figure 3 shows two photographs taken during deploymnet. Pylos site has ideal water properties for neutrino detection experiments, i.e. low velocity currents, low bioluminescence and low rate of sedimentation [6]. The prototype is deployed at a depth of 2960 m and the coordinates are lat: 36° 50.012′ and long: 21° 30.003′, and it will be retrieved shortly after the six month operational period. In future implementation the operational time is expected to be a factor of 2 or 3 larger.

References

[1] H.J. Crawford et al, *NuBE proposal*, 2006, `https://www.yumpu.com/en/document/view/38665069/`

[2] P. Rapidis, Nucl. Instr. and Meth. A, **602**, 54–57 (2009)

[3] Mou Roy et al, *The Prediction and Detection of UHE Neutrino Bursts*, `arXiv:astro-ph/9903231v1`

[4] KM3NeT/ARCA, LOI (under preparation)

[5] K. Manolopoulos et al, *Digital and Analog Electronics for an autonomous, deep–sea, Gamma Ray Burst Neutrino protype detector* (on these proceedings)

[6] E.G. Anassontzis et.al., Astroparticle Physics **34**, *Issue 4*, 187–197 (11/2010)

EPJ Web of Conferences **116**, 10001 (2016)
DOI: 10.1051/epjconf/201611610001

AMON: Transition to real-time operations

D.F. Cowen[1,2], A. Keivani[2], and G. Tešić[2]

[1] Dept. of Astronomy and Astrophysics, the Pennsylvania State University, Univ. Park, PA 16802, USA
[2] Dept. of Physics, the Pennsylvania State University, Univ. Park, PA 16802, USA

Abstract. The Astrophysical Multimessenger Observatory Network (AMON) will link the world's leading high-energy neutrino, cosmic-ray, gamma-ray and gravitational wave observatories by performing real-time coincidence searches for multimessenger sources from observatories' subthreshold data streams. The resulting coincidences will be distributed to interested parties in the form of electronic alerts for real-time follow-up observation. We will present the science case, design elements, current and projected partner observatories, status of the AMON project, and an initial AMON-enabled analysis. The prototype of the AMON server has been online since August 2014 and processing archival data. Currently, we are deploying new high-uptime servers and will be ready to start issuing alerts as early as winter 2015/16.

1. Introduction

Current and future high-energy observatories, representing a joint effort by thousands of scientists, aim to solve longstanding questions about origins of cosmic neutrinos and cosmic rays, as well as the possible existence of gravitational waves. However, the current non-electromagnetic observatories have limited sensitivities for detection of cosmic sources. The number of standalone statistically significant detections by these facilities is expected to be low (e.g., about ten per year for IceCube neutrinos [1–3]). Data streams from these facilities contain a large number of signals that are indistinguishable from background events. These "subthreshold" events from any single observatory cannot be used alone to claim a statistically significant source discovery. However, if subthreshold events from distinct observatories are found in temporal and spatial coincidence, then jointly these events may rise to discovery-level significance. Since many multimessenger sources are expected to be transient, it is also important to detect coincidences in real-time and promptly distribute electronic alerts for follow-up EM observations.

The Astrophysical Multimessenger Observatory Network (AMON) is designed to enable these types of coincidence searches both in real-time and with archival data. It will integrate heterogeneous data from many of the world's leading high-energy (neutrino, cosmic-ray and gamma-ray) and gravitational wave observatories into a single virtual system with combined sensitivity greater than that of any individual facility. More details about the scientific motivation, the design elements and the anticipated discovery potential of AMON are summarized in Ref. [4]. In this proceeding, we present an overview

Table 1. Triggering and follow-up observatories that have signed AMON Memorandum of Understanding (MoU) or Letters of Collaboration (LoC).

Observatory	Messenger type	Role	Status
ANTARES	neutrinos	triggering	MoU
FACT	gamma rays	triggering/follow-up	MoU
Fermi LAT	gamma rays	triggering	LoC
Fermi GBM	gamma rays	triggering	LoC
HAWC	gamma rays	triggering	MoU
IceCube	neutrinos	triggering	MoU
LIGO	gravitational waves	triggering	LoC
LMT	radio waves	follow-up	LoC
MASTER	optical photons	follow-up	MoU
Pierre Auger	neutrons, neutrinos and gamma rays	triggering	MoU
PTF	optical photons	follow-up	LoC
Swift BAT	X rays	triggering	MoU
Swift XRT	X rays	follow-up	MoU
Swift UVOT	UV and optical photons	follow-up	MoU
VERITAS	gamma rays	triggering/follow-up	MoU

of AMON, its current status including its readiness for the real-time operations, and initial results from an AMON-enabled analysis.

2. AMON overview and status

The AMON network consists of triggering and follow-up observatories which have signed the AMON Memorandum of Understanding (MoU) [5], summarized in Table 1. The triggering observatories are sensitive to one or more messengers and typically have large fields of view (FoV) and high duty cycles. Ground based triggering observatories send their candidate events to AMON via direct links in real-time. Events from satellite experiments (e.g., Swift BAT and Fermi LAT) are received by utilizing the Gamma-ray Coordinates Network (GCN) [6]. AMON also uses GCN to distribute its electronic alerts (in the VOEvent format [7]) to the follow-up observatories in real-time. The follow-up observatories are pointing telescopes situated on Earth (e.g., FACT, MASTER, VERITAS, etc.), as well as orbital telescopes (Swift XRT & UVOT) that respond in near real-time to AMON alerts and perform follow-up observations for electromagnetic counterparts of the multimessenger signal.

Triggering facilities linked by AMON collect heterogeneous data used for detecting multimessenger coincidences. The collected data are not transmitted to AMON in their raw format, but in the higher level "trigger event" form that is specific for each observatory type, in the form of XML files in VOEvent format. Events from each observatory stream contain the following minimal information needed for coincidence searches: a trigger time, an event position with a positional error or probability density function and a false positive rate density in the vicinity of the event ($sr^{-1}s^{-1}$). The information from these trigger events is combined statistically by algorithms that search in real-time for temporal and spatial coincidences. For each coincident signal found, an AMON alert is created. Information in the AMON alert includes the best fitted common source position, burst onset time, positional errors, false alarm rate, number of events and type of messengers contributing to a given alert.

The AMON database will store events received in real-time from the AMON member observatories as well as their archival data. In addition, the database stores AMON alerts, observatory configuration models, selection criteria used in alert search algorithms, multiple analysis stream configurations including astrophysical sources, and information from follow-up observations of the AMON alerts. The database underwent successful beta testing by using multiple observatory simulated, and more recently real data (blinded), both using archival data, as well as using data in real-time. Currently, the database holds public data from IceCube-40 and IceCube-59 string configurations, Fermi LAT and Swift BAT data. It also stores private archival data from Pierre Auger and ANTARES 2008, and public data from LIGO runs S5 and S6 is in theprocess of being stored.

2.1 Real-time operations

In May, 2015, AMON established a connection with GCN [6, 8]. The AMON alert streams at GCN are established as private streams that are initially available only to collaborators. In the future, any of these streams can be made public with the approval of the corresponding observatories whose events participate in that particular alert stream. AMON has already started issuing test alerts (from scrambled data) via GCN to some of its network members. There are two event streams from IceCube that are transmitted to AMON in real-time: the ν_μ singlet stream and the high-energy starting event (HESE) stream. The ν_μ stream contains subthreshold data at the rate of about 3 mHz. The HESE stream will provide events at a much lower rate, but with a much high signal-to-noise ratio. As early as winter 2016, AMON will start issuing real-time alerts via GCN to its collaborators and follow-up facilities.

3. Neutrino + gamma search

Here we present the results of archival coincidence analyses using public neutrino data from the 40-string configuration of IceCube (IC40) and contemporaneous public gamma-ray data from Fermi LAT ($\approx$41 weeks). The IC40 public data only contain neutrinos from the northern hemisphere ($\approx$14,000 neutrinos). The Fermi LAT data that are being used is selected by imposing additional criteria using Fermi LAT purest (i.e. lowest instrumental background) analysis class (Pass 7-V6 Ultraclean; 4.1 million photon events for the period overlapping IC40). We calculate the angular separation between each photon event and each neutrino and identify all pairs with smaller than 10° separation. We also apply a temporal cut (ΔT) of $\pm$50 s between neutrino and photon arrival times.

To perform the archival coincidence analysis, an unbinned log-likelihood function (λ) is used. This metric depends on event arrival direction and its uncertainty as well as a background rejection term and is defined as $\lambda = 2\ln\left(P_{\rm LAT}(\vec{x}|\hat{x}_\gamma)P_{\rm IC}(\vec{x}|\hat{x}_\nu)\right) - 2\ln\left(B(\hat{x}_\gamma)\right)$, where λ is the likelihood parameter, $\hat{x}_\gamma$ and $\hat{x}_\nu$ are the arrival direction of the photon and neutrino events, respectively, $\vec{x}$ is the best fit position, $B(\hat{x}_\gamma)$ is the Fermi LAT background rejection term, $P_{\rm LAT}(\vec{x}|\hat{x}_\gamma)$ and $P_{\rm IC}(\vec{x}|\hat{x}_\nu)$ are the energy dependent point spread functions (PSF) of the LAT and IceCube observatories, respectively. In this metric, large values of λ indicate a higher probability correlated pair.

We generate a series of 10,000 scrambled datasets to estimate the null distribution. We scramble time and right ascension of each neutrino event while keeping declination and intrinsic neutrino properties unchanged. Applying the spatial and temporal cuts on all of the $\nu-\gamma$ pairs, on average 2207$\pm$40 pairs in each of the 10,000 null datasets are left. Then a log-likelihood (λ) value is calculated for each $\nu-\gamma$ pair from the equation above. Additionally, a series of 10,000 signal tests are created by injecting a limited number of forced coincidences into the null data. Spatial and temporal cuts are applied and λ is calculated for each pair.

Unblinding the IC40 data and performing the coincidence analysis with Fermi LAT data result in 2138 coincidences applying the cuts of $\Delta\theta < 10°$ and $|\Delta t| < 50$ s. The AD test statistic on this dataset provides a p-value of about 4%. This p-value is consistent with simulations including the injection of 70–100 signal events. The λ distribution plot for unblinded data is presented in Fig. 1. The null hypothesis and a theoretical signal model with 70 injected signals ($N_{\rm sig}$=70) are also plotted for comparison. The bottom plot in Fig. 1 shows the residuals of data and $N_{\rm sig}$=70 versus the null hypothesis to better represent the differences between the distributions.

Three different further tests were performed to explore whether real cosmic $\nu-\gamma$ pairs are present in the data at $N_{\rm sig}$=70 level. We perform the tests only on events with high-λ values which are more likely to be signal. We checked the mean number of photons in coincidence with each single neutrino,

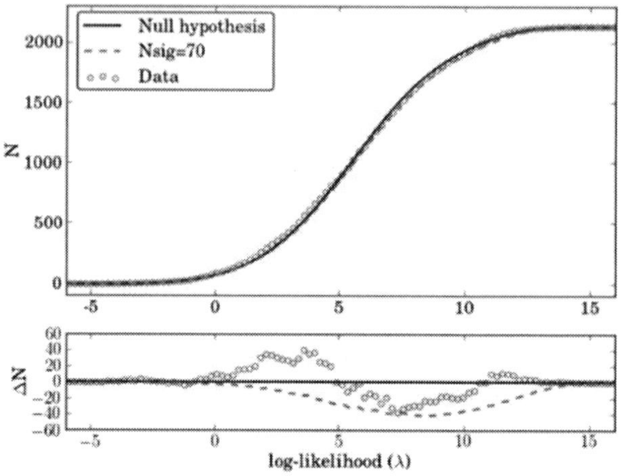

Figure 1. Top: cumulative distribution of log-likelihood values of the IC40 and Fermi LAT coincidences after unblinding the data. A theoretical signal model with 70 injected signals (N_{sig}=70) and the null hypothesis are plotted for comparison. All distributions are normalized and multiplied by 2138, which is the total number of coincidences after unblinding the data. Bottom: The residuals of data and N_{sig}=70 versus the null hypothesis.

the time difference between the photon event and the neutrino, and the clustering of detected pairs. In all cases, we found consistency with background and no evidence for a contribution from cosmic $\nu - \gamma$ pairs.

4. Conclusion

Multimessenger astroparticle physics is a field that has grown rapidly since the current generation of non-electromagnetic observatories started continuously monitoring large portions of the sky. The AMON cyberinfrastructure enables real-time and archival searches for possible multimessenger sources by conducting joint subthreshold analysis of multimessinger data. AMON has made significant progress in both real-time and archival analysis. In addition to the ν+γ analysis described above, AMON-enabled searches for primordial black hole evaporation [9], γ+(gravitational wave), and ν+(blazar TeV γ) are underway. AMON will start issuing electronic alerts via GCN in the winter of 2016. Jointly, the AMON partner facilities may probe the high-energy universe via all four forces of nature for the first time in human history.

This research is supported by grants from the U.S. National Science Foundation (003403953) and the Institute for Gravitation and the Cosmos of the Pennsylvania State University.

References

[1] M.G. Aartsen et al. (IceCube), Science **342**, 1242856 (2013), `1311.5238`
[2] M.G. Aartsen et al. (IceCube), Phys. Rev. **D88**, 112008 (2013), `1310.5477`
[3] M.G. Aartsen et al. (IceCube), Phys. Rev. Lett. **113**, 101101 (2014), `1405.5303`
[4] M.W.E. Smith et al., Astropart. Phys. **45**, 56 (2013), `1211.5602`
[5] *The AMON Memorandum of Understanding*, `http://amon.gravity.psu.edu/mou_aug2012.shtml`

[6] *Gamma-ray Coordinates Network (GCN)*, http://gcn.gsfc.nasa.gov/
[7] *International Virtual Observatory Alliance, VOEvent*, http://www.ivoa.net/documents/VOEvent
[8] S. Barthelmy, *Private communication*
[9] G. Tešić, PoS **ICRC2015**, 328 (2015)

EPJ Web of Conferences **116**, 10002 (2016)
DOI: 10.1051/epjconf/201611610002

Capabilities of IceCube's gamma-ray, optical and X-ray follow-up programs

Thomas Kintscher[a] for the IceCube Collaboration

Deutsches Elektronen-Synchrotron (DESY), Platanenallee 6, 15738 Zeuthen, Germany

Abstract. The IceCube neutrino observatory is a $1\,\mathrm{km}^3$ detector for Cherenkov light in the ice at the South Pole. Although the presence of a diffuse astrophysical neutrino flux has been confirmed, its origin has yet to be resolved. Given the current constraints on continuous point source searches, transient and variable objects emerge as promising, detectable source candidates. IceCube boosts the sensitivity to these types of sources by alerting third-party observatories of neutrino events clustered in direction and time. This paper will showcase several neutrino-triggered multi-messenger programs in IceCube along with their results and prospects.

1. Introduction

Understanding the origin of cosmic rays has been a long standing goal in astrophysics. Multi-messenger observations, here using neutrinos and cosmic rays, can provide a viable tool in locating and understanding cosmic acceleration sites. Hadronic interactions producing cosmic rays can also produce high-energetic neutrinos [1], which reach earth unabsorbed and undeflected and whose arrival directions point back to their origin likely coinciding with the sources of cosmic rays.

The IceCube experiment has shown evidence for the existence of a diffuse astrophysical neutrino flux [2], the sources of which have yet to be identified. In addition to the search for steady sources [3], transient and variable objects present further possible source candidates. Simultaneous observations of neutrinos and photons during their various states of activity can improve the sensitivity to these sources and yield a more complete picture of their acceleration mechanisms.

In gamma rays the availability of observations is limited to certain points of time in the past due to the field of view of optical, Cherenkov and X-ray telescopes, covering typically just several square degrees. In contrast, neutrino telescopes are able to observe the full sky around the clock with a duty cycle close to 100%. Using them as a trigger for optical, gamma-ray and X-ray telescopes can therefore yield more observations in e.g. gamma rays and correlated neutrinos.

So called Neutrino triggered Target of Opportunity (NToO) programs need to cover several order of magnitude w.r.t. the time-scale on which sources' activities vary: They range from milliseconds (e.g. fast radio bursts) over days to years in the case of active galactic nuclei (AGN) [4]. Thus, the

[a] e-mail: `thomas.kintscher@desy.de`

analysis chains are highly automated and alerts of neutrino flares can be generated without human intervention.

Currently, IceCube operates two of these NToO programs: the gamma-ray follow-up (GFU) and the optical and X-ray follow-up (OFU, XFU). After the description of the detector in Sect. 2, the operation of both programs is explained in Sect. 3. Section 4 closes with a summary.

2. The IceCube detector

The IceCube neutrino observatory is a cubic-kilometer detector for Cherenkov light, installed in the ice at the geographic South Pole [5]. The construction of the detector, i.e. deployment of its 5160 optical modules, was completed in 2010. Of the possible kinds of interactions, the selection of muon tracks from charged-current muon neutrino interactions in the detector or nearby bedrock is of particular interest to this analysis. The track-like signature of the muon can be used to reconstruct the direction of the muon and corresponding neutrino with a median precision of one degree.

The muon filter at the South Pole selects these events with a rate of 40 Hz. In order to suppress the background from atmospheric muons, the Online Level 2 filter uses likelihood reconstructions, which take into account photon propagation and the ice properties. As computing power at the South Pole is limited, computationally demanding reconstructions are only applied to events passing further cuts. After cuts on the likelihood fit quality, the event rate is reduced to 5 Hz. Afterwards, multivariate classifiers are employed to improve the neutrino purity to 90% at an event rate of 5 mHz.

3. Follow-up programs

As mentioned before, two follow-up programs are currently in operation. In the past, their respective event selection and clustering analysis was run at the South Pole and only the most significant clusters triggering an alert were sent to the North through a low-latency satellite link. Such an alert then triggered an automatically generated email to the follow-up observatories.

In order to improve the flexibility of the system and facilitate maintenance the design is about to be changed: Information (e.g. direction, energy, fit quality and uncertainty estimates) about every well-reconstructed, track-like event passing a common event selection will be sent immediately through the Iridium satellite network. After a median delay of 22 seconds after the neutrino interaction, events are received in the North and multiple analyses with different physics goals run on the events from the stream.

3.1 Gamma-ray follow-up

Catching flares from known, variable sources such as AGN is the aim of the so-called gamma-ray follow-up program. Upon observing an event in IceCube, a time-dependent point source analysis is run on a list of pre-agreed upon source candidates. This list is comprised of AGN from the Fermi catalog [6] in the northern hemisphere, which have shown variable behavior in gamma rays in the past and are visible (in terms of direction and redshift) to the respective follow-up telescope. An unbinned maximum likelihood analysis considering the quality of the angular reconstruction as well as the energy spectrum of the events fits two quantities: The spectral index of the events and the most likely number of signal events among the events in the time window.

The significance of the result is determined using a likelihood ratio test of the likelihood given the best fitted spectral index and number of signal events versus the likelihood assuming only background events. By testing multiple, increasing time windows starting from the latest (triggering) event and reaching up to three weeks in the past, the most likely temporal extent of a cluster of neutrinos can be determined.

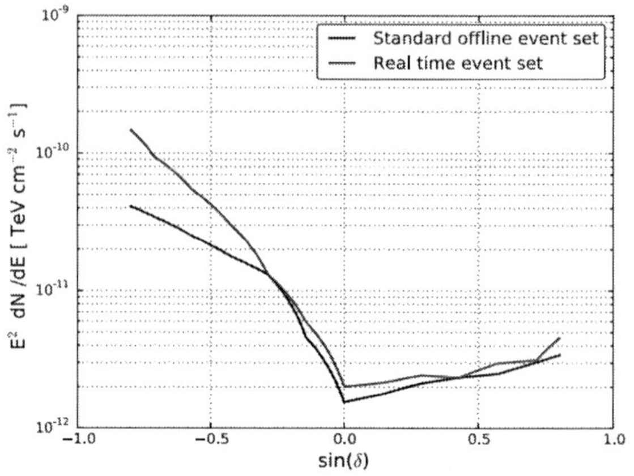

Figure 1. Sensitivity comparison for online and offline event selections as a function of the declination δ. The offline selection is performed after transferring the data to the North, while the online selection runs within the limited resources at the South Pole.

In case the significance passes a threshold and permitting visibility (which may be impaired due to the moon or the zenith angle of the source w.r.t. the telescope up to 24 hours after the alert) the alert is forwarded to the MAGIC and/or VERITAS telescope.

Since the inauguration of the program in March 2012, the significance threshold was passed 14 times. Three alerts have been sent to MAGIC and five to VERITAS. One alert has been followed up by each telescope. In both cases, no significant gamma-ray emission was observed.

In Fig. 1 the sensitivity of the likelihood analysis is compared to that of the offline point source analysis. For the majority of the sky, the sensitivity of both analyses is comparable, with the offline event selection being slightly more refined to the availability of more time-consuming reconstructions.

3.2 Optical and X-ray follow-up

The optical and X-ray follow-up programs share the same event selection and alert generation technique. Similar to the GFU, the selection of track-like muon events is using a multivariate classifier, but this program is limited to the northern hemisphere and optimized equally for E^{-1}, E^{-2} and E^{-3} spectra (whereas the GFU is optimized for an E^{-2} spectrum). As the majority of the events is still of atmospheric origin (muons and neutrinos), a further cut is placed allowing only cluster of neutrinos to pass that are separated by less than 3.5 degrees and arrive within 100 seconds of each other. The very short timescale of the clustering is targeted at observing neutrinos from gamma-ray bursts (GRBs) and supernovae.

Clusters of two events matching these criteria occur one or twice per week, but a cluster of three or more events passing the cut has never been observed. The alert rate is reduced further by additional cuts involving the angular separation, the reconstruction quality, the follow-up telescope's field of view and the temporal distance of the events, yielding a rate of nine alerts per year to the Palomar Transient Factory (PTF [7]) and seven alerts per year to the SWIFT X-Ray Telescope [8].

On March 30th, 2012, the most significant alert from the OFU program was sent to PTF. Two neutrinos were recorded within 1.7 seconds with a spatial separation of roughly one degree. In the follow-up observation performed by PTF ten days later, the previously unknown core-collapse supernova PTF12csy was discovered 0.14 degrees away from the mean neutrino direction. Figure 2 shows the location of the neutrinos and the supernova, as well as the result of the follow-up observations. Archival searches of the Pan-STARRS1 survey concluded the age of PTF12csy to be at least 169 days at the time of the alert. Considering the probability to find a random core-collapse supernova within

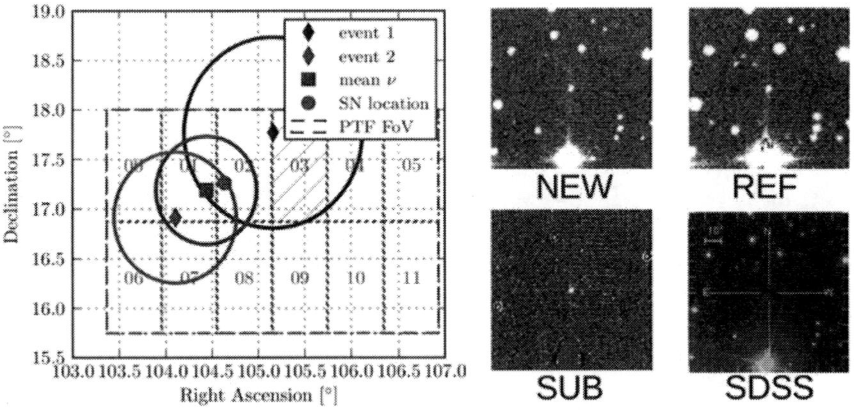

Figure 2. Left: Location of the two neutrinos that triggered the alert to PTF. The field of view of the PTF CCDs is given by the gray boxes (CCD 03 was broken). Right: Result of the PTF observation after the IceCube trigger and earlier image from the SDSS.

the neutrino error circle and the probability of the alert itself, a combined significance of $2.2\,\sigma$ cannot exclude a chance detection and the neutrinos are considered to be unrelated to the supernova [9].

4. Summary and outlook

The IceCube experiment is able to reconstruct and transmit events, upon which alerts can be generated almost immediately. In the past, the gamma-ray and optical/X-ray follow-up programs have demonstrated reliably that these alerts can successfully trigger multi-messenger observations. In the future, a more refined event selection and updated reconstruction algorithms at the South Pole will improve the sensitivity further. Real-time selections of other event classes than through-going muons (such as starting events) will provide new follow-up opportunities.

References

[1] T.K. Gaisser, T. Stanev, Astropart. Phys. **39–40**, 120 (2012), 1202.0310
[2] M.G. Aartsen et al. (IceCube), Astrophys. J. **809**(1), 98 (2015), 1507.03991
[3] M.G. Aartsen et al. (IceCube), Astrophys. J. **796**(2), 109 (2014), 1406.6757
[4] A.M. Atoyan, C.D. Dermer, New Astron. Rev. **48**, 381 (2004), astro-ph/0402646
[5] A. Achterberg et al. (IceCube), Astropart. Phys. **26**, 155 (2006), astro-ph/0604450
[6] M. Ackermann et al., Astrophys. J. **743**, 171 (2011), 1108.1420
[7] N.M. Law et al., PASP **121**, 1395 (2009), 0906.5350
[8] N. Gehrels et al. (Swift Science), Astrophys. J. **611**, 1005 (2004), [Erratum: Astrophys. J. 621, 558 (2005)], astro-ph/0405233
[9] M.G. Aartsen et al. (IceCube, Pan-STARRS1 Science Consortium, Swift, PTF), Astrophys. J. **811**(1), 52 (2015), 1506.03115

EPJ Web of Conferences **116**, 10003 (2016)
DOI: 10.1051/epjconf/201611610003

Follow-up of high energy neutrinos detected by the ANTARES telescope

Aurore Mathieu[a], on behalf of the ANTARES Collaboration

Aix Marseille Università, CNRS/IN2P3, CPPM UMR 7346, 13288 Marseille, France

Abstract. The ANTARES telescope is well-suited to detect high energy neutrinos produced in astrophysical transient sources as it can observe a full hemisphere of the sky with a high duty cycle. Potential neutrino sources are gamma-ray bursts, core-collapse supernovae and flaring active galactic nuclei. To enhance the sensitivity of ANTARES to such sources, a detection method based on follow-up observations from the neutrino direction has been developed. This program, denoted as TAToO, includes a network of robotic optical telescopes (TAROT, Zadko and MASTER) and the *Swift*-XRT telescope, which are triggered when an "interesting" neutrino is detected by ANTARES. A follow-up of special events, such as neutrino doublets in time/space coincidence or a single neutrino having a very high energy or in the specific direction of a local galaxy, significantly improves the perspective for the detection of transient sources. The analysis of early and long term follow-up observations to search for fast and slowly varying transient sources, respectively, has been performed and the results covering optical and X-ray data are presented in this contribution.

1. Introduction

High energy neutrinos are unique messengers to study the high energy Universe, as they are neutral, stable and weakly interacting. They are thought to be produced in violent astrophysical phenomena when accelerated hadronic cosmic rays interact with matter or radiation fields surrounding their astrophysical sources. Promising candidate sources are gamma-ray bursts (GRBs), active galactic nuclei or core-collapse supernovae (CCSNe). However, even with the recent detection of high energy cosmic neutrino events by the IceCube experiment [1, 2], no high energy neutrino source has been identified up to now.

The ANTARES telescope [3], completed in 2008 in the Mediterranean Sea, aims at detecting these high energy neutrino sources with a 3-dimensional array of 885 photomultiplier tubes distributed on twelve lines. The detection of Cherenkov light induced by the propagation of up-going muons in water is used as a signature of muon neutrino interaction close or within the instrumented volume.

The identification of steady neutrino sources is challenging due to the very small number of expected events and a large background contamination from atmospheric muons and neutrinos. A way to overcome this difficulty is to search for neutrinos from flaring astrophysical sources of photons. Searches for transient phenomena offer very promising opportunities for high energy neutrino telescopes, as

[a] e-mail: amathieu@cppm.in2p3.fr

the combination of information in a limited space/time window around the transient event allows the background to be significantly reduced. Furthermore, as a neutrino telescope has an instant coverage of at least a hemisphere with a high duty cycle, a neutrino detection can be used to trigger electromagnetic observations at various wavelengths, which enhance the detection sensitivity.

2. TAToO

TAToO (Telescopes-ANTARES Target Of Opportunity) [4] is the multi-wavelength follow-up program operating within the ANTARES experiment since 2009. This alert system allows a network of robotic optical telescopes (TAROT [5], Zadko [6], MASTER [7], and ROTSE [8] until the end of 2014) and, since mid 2013, the X-ray telescope on board the *Swift* satellite [9] to be triggered soon after the detection of a special neutrino event. Three online criteria, tuned to match the rate of ~25 alerts per year accepted by the optical telescopes, are currently implemented to trigger the system:

- High energy: detection of a single high energy neutrino event ($< E >\sim 7$ TeV).
- Directional: detection of a single neutrino event having direction consistent ($< 0.4°$) with the position of galaxies within 20 Mpc of Earth.
- Multiplet: detection of at least two neutrinos spatially coincident ($< 3°$) within a predefined time window ($<$ 15 minutes).

Since the beginning of TAToO, more than 160 neutrino triggers have been sent to the telescope network with a minimum delay of ~5 seconds after the neutrino detection, and with the incoming neutrino direction reconstructed to a precision better than 0.5° for high energy events. Among them, 10 alerts have been forwarded to the *Swift*-XRT, following the very high energy trigger criterion that selects a subsample of high energy events to comply with the rate of 6 alerts per year accepted by *Swift*. Follow-up observations of ANTARES neutrino alerts have been performed according to two strategies: a prompt strategy dedicated to the search for fast transient sources, such as gamma-ray bursts (GRBs), with images taken in a maximum delay of 24 hours after the neutrino detection, and a long term follow-up strategy with images taken up to two months after the trigger and well-suited to the search for slowly varying transient sources, such as core-collapse supernovae (CCSNe).

3. Early follow-up analysis

Early follow-up observations have been performed by optical and X-ray telescopes for 42 and 10 neutrino triggers, respectively. Optical observations have been performed by the TAROT and ROTSE telescopes with a minimum delay of 17 seconds after the neutrino detection. For each observation, between 6 and 30 images have been taken with 60 or 180 seconds exposure. All these images have then been analysed with a dedicated pipeline based on a subtraction method [10].

For each trigger followed by the *Swift*-XRT, four observations lasting 2000 seconds each have been made around the neutrino position after a mean delay of 6.3 hours. X-ray data have been automatically analysed at the UK Swift Science Data Centre with dedicated pipelines.

In both optical and X-ray wavelengths, no transient source has been identified as the counterpart to a neutrino trigger. Therefore, upper limits on the magnitude and on the flux density from a potential optical and X-ray counterpart have been derived [10]. These limits correspond to the sensitivity reached by the instruments for each observation.

3.1 Limit on a GRB association

Among the population of fast transient sources, GRBs are one of the most promising sources for high energy neutrino emission. Because in this analysis no optical or X-ray counterpart has been observed in

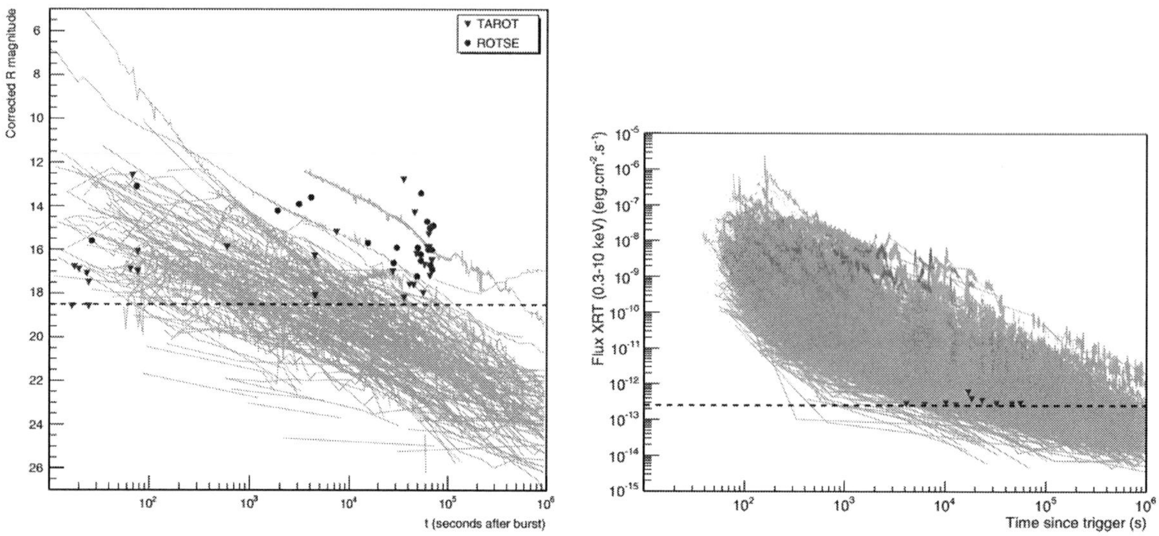

Figure 1. Left: upper limits (red and blue) on optical GRB magnitudes based on a comparison with 301 GRB afterglow light curves (grey) observed from 1997 to 2014 by optical telescopes. Right: upper limits (red) on X-ray afterglow fluxes compared to 689 X-ray light curves (grey) of GRBs detected by the *Swift*-XRT from 2007 to 2015. On each plot, the horizontal dashed line corresponds to the maximum sensitivity reached by the optical and X-ray telescopes for this analysis.

coincidence with the neutrino alerts, the probability to reject a GRB origin for these neutrinos can be estimated. To do so, a comparison with optical and X-ray upper limits obtained for each neutrino alerts and GRB afterglow light curves has been made (Fig. 1).

In the optical band, a GRB association is rejected, with a confidence level above 80%, if the optical observation is started within one minute after the neutrino trigger, while for follow-up observations started later than a few minutes after the trigger, the derived upper limits do not constrain a GRB origin for the detected neutrinos, due to the limited sensitivity of the TAROT and ROTSE telescopes.

Concerning the X-ray follow-up, the probability to reject a GRB origin is as high as 71% only for observations carried out within 1.1 hours after the trigger. For later observations, a GRB origin is very unlikely for the neutrino events, as few GRBs are known to have weaker X-ray afterglows than the upper limits.

4. Long term follow-up analysis

Follow-up images have been taken up to two months after the neutrino detection by optical telescopes according to the long term strategy. The optical counterpart search has been performed with the dedicated image subtraction pipeline for 71 alerts followed from October 2009 to January 2015. Up to 12 nights have been processed for each alert with limiting magnitudes ranging from 14.0 to 19.4. No transient source has been identified as the counterpart associated with one of the 71 detected neutrinos.

4.1 Limit calculation in the case of CCSNe

The non-observation of an optical transient in coincidence with a neutrino event allows constraints on the Ando & Beacom model parameters [11] to be placed. This model predicts the production of high energy neutrinos in mildly relativistic jets of CCSNe, with a flux depending on the jet energy E_{jet} and the Lorentz boost factor of the jet Γ.

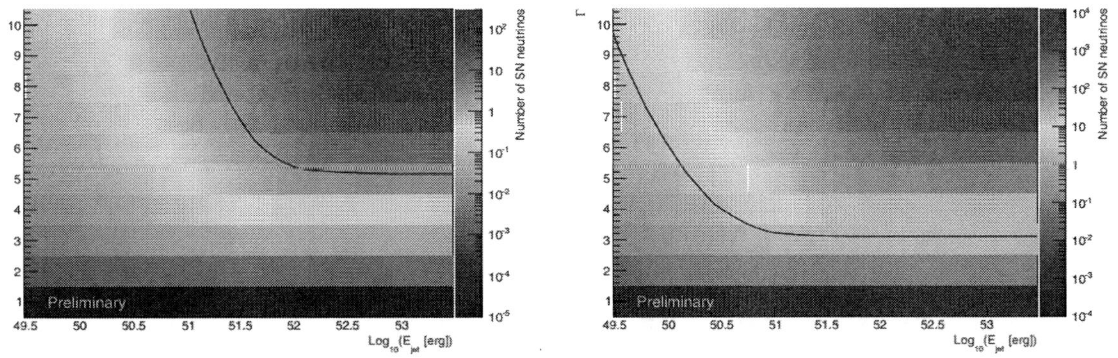

(a) d = 3 Mpc.
(b) d = 460 kpc.

Figure 2. Expected number of neutrinos detectable with ANTARES from a CCSN at different distances for different Ando & Beacom model parameter combinations. The red line indicates the limit above which combinations of $E_{\mathrm{jet}} - \Gamma$ are excluded at 90% C.L. as no CCSN neutrinos have been detected.

Figure 2 shows the number of expected neutrinos in ANTARES for a CCSN with a jet pointing towards Earth at a distance of 3 Mpc and 460 kpc. The number of neutrinos is calculated for Γ from 1 to 10, and for jet energies in the range 3×10^{49} erg $< E_{\mathrm{jet}} < 3 \times 10^{53}$ erg. As no optical CCSN has been observed for the 71 alerts, combinations of model parameters producing a detectable signal in ANTARES can be excluded. A red line on each plot indicates the limit above which parameter combinations are excluded at the 90% C.L.. Figure 2 (b) shows that a CCSN with standard values for mildly relativistic jets ($E_{\mathrm{jet}} = 3 \times 10^{51}$ erg and $\Gamma = 3$) is excluded up to 460 kpc.

5. Conclusion

TAToO is a promising multi-messenger approach to detect transient sources, as it is able to send alerts within a few seconds after the neutrino detection and with an angular resolution better than 0.5°. The image analysis has not yet allowed transient sources associated with the selected neutrino events to be discovered. However, stringent limits on the GRB origin of individual neutrinos have been placed, as well as constraints on the CCSN parameters from the Ando & Beacom model.

References

[1] M.G. Aartsen et al., Science, **342**, 1, 2013
[2] M.G. Aartsen et al., Phys. Rev. Lett., **113**(10), 101101, 2014
[3] M. Ageron et al., Nucl. Instr. Meth. A, **656**, 11–38, 2011
[4] M. Ageron et al., Astropart. Phys., **35**, 530–536, 2012
[5] A. Klotz, M. Boer, J.L. Atteia, B. Gendre, AJ, **137**, 4100–4108, 2009
[6] D.M. Coward et al., Astron. Soc. of Australia, **27**, 331–339, 2010
[7] V. Lipunov et al., Advances in Astronomy, **30**, 2010
[8] C.W. Akerlof et al., PASP, **115**, 132–140, 2003
[9] D.N. Burrows et al., Space Sci. Rev., **120**, 165–195, 2005
[10] S. Adrián-Martínez et al., submitted to JCAP, `arXiv:1508.01180`, 2015
[11] S. Ando and J.F. Beacom, Phys. Rev. Lett., **95**(6), 061103, 2005

EPJ Web of Conferences **116**, 10004 (2016)
DOI: 10.1051/epjconf/201611610004

Correlation between the UHECRs measured by the Pierre Auger Observatory and Telescope Array and neutrino candidate events from IceCube

A. Christov[1], G. Golup[2], T. Montaruli[1], M. Rameez[1,a] for the IceCube Collaboration[b], J. Aublin[3], L. Caccianiga[3], P.L. Ghia[3], E. Roulet[2], M. Unger[4] for the Pierre Auger Collaboration[c], H. Sagawa[5], and P. Tinyakov[6] for the Telescope Array Collaboration[d]

[1] Département de physique nucléaire et corpusculaire, Université de Genève, 24 Quai Ernest Ansermet, 1211 Genève, Switzerland
[2] Centro Atómico Bariloche, Av. Bustillo 9500, S. C. de Bariloche 8400, Argentina
[3] Laboratoire de Physique Nucléaire et de Hautes Energies (LPNHE), Universités Paris 6 et Paris 7, CNRS-IN2P3, 4 place Jussieu, 75252 Paris, France
[4] Karlsruhe Institute of Technology – Campus North – Institut fër Kernphysik, Karlsruhe, Germany and New York University, New York, USA
[5] Institute for Cosmic Ray Research, University of Tokyo, Kashiwa, Chiba, Japan
[6] Service de Physique Théorique, Université Libre de Bruxelles, Boulevard du Triomphe (Campus de la Plaine), Ixelles 1050, Belgium

Abstract. We present the results of three searches for correlations between ultra-high energy cosmic ray events measured by Telescope Array and the Pierre Auger Observatory and high-energy neutrino candidate events from IceCube. Two cross-correlation analyses of ultra-high energy cosmic rays are done: one with 39 "cascades" from the IceCube "high-energy starting events" sample and the other one with 16 high-energy "tracks". The angular separation between the arrival directions of neutrinos and UHECRs is scanned. The same events are also used in a separate search stacking the neutrino arrival directions and using a maximum likelihood approach. We assume that UHECR magnetic deflections are inversely proportional to the energy with values 3°, 6° and 9° at 100 EeV to account for the uncertainties in the magnetic field strength and UHECR charge. A similar analysis is performed on stacked UHECR arrival directions and the IceCube 4-year sample of through-going muon-track events that was optimized for neutrino point source searches.

1. Introduction

The sources of ultra-high energy cosmic rays (UHECRs) have so far not been identified. The CRs do not point back to their sources as they are charged particles and are deflected en-route to Earth. Since both CR composition at ultra-high energies as well as the intervening magnetic field strength are poorly

[a] e-mail: `Mohamed.Rameez@unige.ch`
[b] `http://icecube.wisc.edu/collaboration/authors/current`
[c] `http://www.auger.org/archive/authors_2015_12.html`
[d] `http://www.telescopearray.org/index.php/research/collaborators`

known, this deflection cannot be computed precisely. If the CR composition is light i.e. mainly protons, the magnetic deflection may be only a few degrees at energies above a few tens of EeV. Secondary particles including neutrinos (νs) are produced in the sources by the interactions between the CRs and ambient photon and matter fields. Neutrinos have no charge and interact only through the weak force, so their arrival directions do point back to where they originated from, even though they are hard to detect. This work describes a joint analysis by the IceCube, Pierre Auger and Telescope Array Collaborations to search for angular correlations between the arrival directions of high-energy νs and UHECRs.

2. The observatories and data sets

2.1 The IceCube Neutrino Telescope

IceCube is a cubic-kilometer neutrino detector embedded in the ice at the geographic South Pole [1] between depths of 1450 m and 2450 m. Different IceCube data sets are considered in this work. The first is a set of cascades (ν_e and ν_τ charged-current, and all flavour neutral-current interactions) that have been detected in a search for high-energy events where the interaction occurs within the detector [2]. These 39 cascades, which are part of the HESE ("High-Energy Starting Events") set (deposited energy range: ~30–2000 TeV) are derived from data taken between May 2010 and May 2014 and are hereby called "high-energy cascades". A second set of events referred to as "high-energy tracks" (energy above ~70 TeV) consists of the 7 tracks (ν_μ charge current interactions) in the HESE sample [2] that have energies and directions which make them more likely to be of extraterrestrial origin than the other track events in that sample and 9 muon tracks found in a search of a diffuse up-going ν_μ flux [3].

The third data set used is called the "4-year point-source sample" [4] and consists of tracks detected between May 2008 and May 2012 with sub-degree median angular resolution. The set includes about 400,000 events, mostly up-going atmospheric νs from the Northern hemisphere and high-energy atmospheric muons from the Southern hemisphere.

2.2 The Pierre Auger Observatory

The Pierre Auger Observatory, located in Malargüe, Argentina (35.2° S, 69.5° W, 1400 m a.s.l.) [5, 6], consists of surface detectors and air fluorescence telescopes designed to perform complementary measurements of air showers created by UHECRs. The data set used for the present analysis includes 231 events with $E > 52$ EeV and zenith angles smaller than 80° recorded by the surface detector array from January 2004 to March 2014 [7]. The exposure determined by geometrical considerations for the period analyzed amounts to 66,452 km^2 sr yr. The angular resolution is better than 0.9°. The absolute energy scale has a systematic uncertainty of ~14% and the energy resolution is ~12%.

2.3 Telescope Array

The Telescope Array (TA) is located in Utah, USA (39.3° N, 112.9°W, 1400 m a.s.l.) [8] and similarly detects extensive air showers generated by UHECRs. The UHECR sample considered in the present analysis consists of 87 events with $E > 57$ EeV and zenith angles smaller than 55° collected between May 2008 and May 2014 by the surface detector array. A subset of events has been published in [9]. The total exposure is around 9,500 km^2 sr yr. The angular resolution is better than 1.5°. As with the PAO the energy scale of the surface detector array is cross-calibrated with fluorescence telescopes. The energy resolution is better than 20% with a systematic uncertainty on the absolute energy scale of 21%.

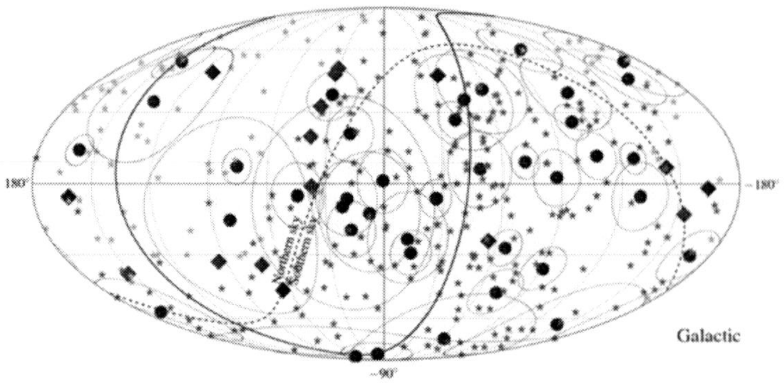

Figure 1. Map in Galactic coordinates showing the arrival directions of the IceCube cascades (black dots) and tracks (diamonds), as well as those of the UHECRs detected by the Pierre Auger Observatory (magenta stars) and Telescope Array (orange stars). The circles around the showers indicate angular errors. The black diamonds are the HESE tracks while the blue diamonds stand for the tracks from the through-going muon sample. The blue curve indicates the Super-Galactic plane.

3. The searches

This section describes three different analyses. A cross-correlation and a stacking likelihood analysis are done on the sample of high-energy cascades and high-energy tracks and the UHECRs detected by Auger and TA. Cascade and track-like events are considered separately since, due to their different angular resolutions, the angular distance at which a signal (if any) can be detected would be different. A third analysis is performed on stacked UHECRs and the IceCube 4-year point-source sample. The magnetic deflections of CRs have to be accounted for in the likelihood tests. For simplicity, we model individual deflections as a random variable 2-dimensional Gaussian distribution with the energy dependent standard deviation $\sigma_{MD}(E_{CR}) = \mathrm{D} \times 100\ \mathrm{EeV}/E_{CR}$, and we consider the representative values D = 3°, 6° and 9° (the latter is just used for the likelihood test with the high-energy cascades and high-energy tracks). These values are reasonable test values as shown by a backtracking simulation of the detected UHECRs in the galactic magnetic field models of Pshirkov et al. [12] and Jansson and Farrar [13], presented in [14].

3.1 UHECR correlation searches with high-energy cascades and tracks

The arrival directions of the high-energy tracks and cascades in IceCube, and of the UHECRs measured by Auger and TA are shown in Fig. 1 in Galactic coordinates. Two different analyses are performed with these data sets: a cross-correlation and a stacking likelihood analysis.

Cross-correlation analyses: The cross-correlation method consists of computing the number of UHECR-ν pairs as a function of their angular separation α, $n_p(\alpha)$, and comparing it to the expectation from an isotropic distribution of arrival directions of CRs. The angular scan performed in this case is between 1° and 30° with a step of 1° and, due to this scan, the method does not rely on any assumption about the exact value of the strength of the magnetic deflections, unlike the likelihood method.

Likelihood stacking analyses: Stacking a set of sources is a conventional way of enhancing the statistical weight of multiple weak signals to enhance the discovery potential. Since νs are not deflected on their way to Earth, ν arrival directions are stacked. An unbinned likelihood method is used, with the

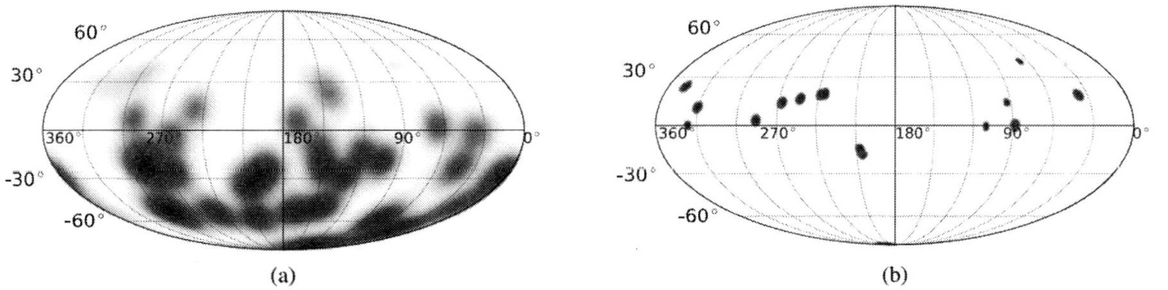

Figure 2. Neutrino likelihood maps convoluted with the declination dependent exposure of Auger in Equatorial coordinates. The plot on the left is for the high-energy cascades, while the one on the right is for the high-energy tracks. Similar figures for the exposures of TA can be seen in [14].

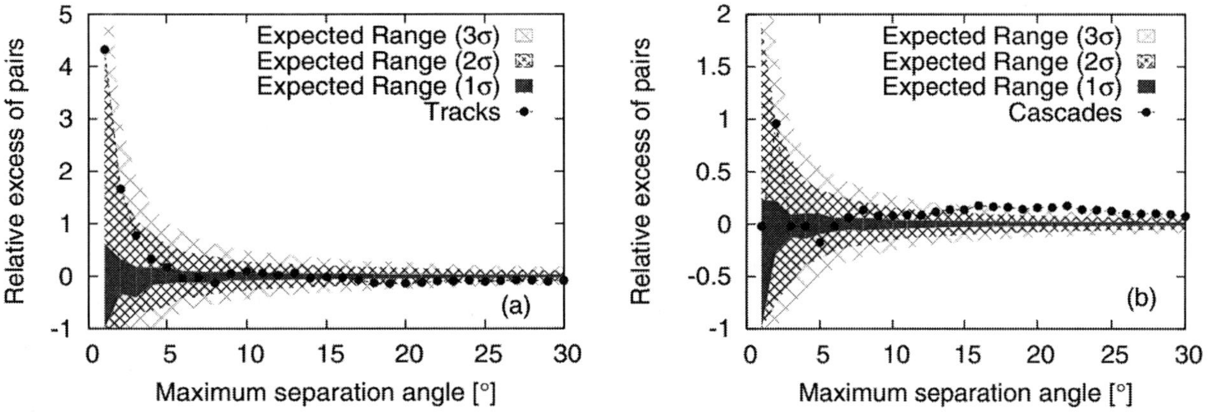

Figure 3. Relative excess of pairs, $[n_p(\alpha)/\langle n_p^{iso}(\alpha)\rangle] - 1$, as a function of the maximum angular separation between the neutrino and UHECR pairs, for the analysis done with the track-like events (a) and with the cascade events (b). The 1σ, 2σ and 3σ fluctuations expected from an isotropic distribution of arrival directions of CRs are shown in red, blue and grey, respectively.

log of the likelihood function defined as:

$$\log \mathcal{L}(n_s) = \sum_{i=1}^{N_{\text{Auger}}} \log\left(\frac{n_s}{N_{\text{CR}}}\mathcal{S}^i_{\text{Auger}} + \frac{N_{\text{CR}} - n_s}{N_{\text{CR}}}\mathcal{B}^i_{\text{Auger}}\right) + \sum_{i=1}^{N_{\text{TA}}} \log\left(\frac{n_s}{N_{\text{CR}}}\mathcal{S}^i_{\text{TA}} + \frac{N_{\text{CR}} - n_s}{N_{\text{CR}}}\mathcal{B}^i_{\text{TA}}\right), \quad (1)$$

where n_s, the number of signal events, is the only free parameter and $N_{\text{CR}} = N_{\text{Auger}} + N_{\text{TA}}$ is the total number of UHECRs. S^i_{Auger} and S^i_{TA} are the signal probability density functions (PDFs) for Auger and TA respectively and are Gaussians centered at the arrival directions of the UHECRs with widths accounting for their magnetic deflection and angular uncertainty [14]. B^i_{Auger} and B^i_{TA} are the corresponding background PDFs representing the probabilities of observing a CR from a given direction assuming an isotropic flux. Consequently they are taken to be the Auger and TA normalized exposures. The log-likelihood of Eq. (1) is maximized with respect to n_s and the ratio of the maximum likelihood w.r.t the case where $n_s = 0$ is taken as the test statistic [14].

Figure 2 shows the normalized neutrino likelihood maps convoluted with the exposure of Auger to demonstrate the spread of the different observed neutrino arrival directions visible from the site.

Results: In Fig. 3 we show the results obtained upon applying the cross-correlation method to the data. For the case of the sample of high-energy tracks, the maximum departure from the isotropic expectation of CRs (fixing the positions of the νs) is at an angular distance of 1°, where 0.38 pairs are expected on average and 2 pairs are detected. The post-trial p-value is 28%. For the analysis done using the high-energy cascade events, the smallest pre-trial p-value occurs at an angular distance of 22°, for which 575 pairs are observed while 490.3 are expected on average. The post-trial p-value is 5×10^{-4} with respect to expectations of an isotropic flux of CRs. As an a posteriori study, we also evaluated the significance under the hypothesis of an isotropic distribution of neutrinos, fixing the UHECR arrival directions (note that this alternative hypothesis preserves the degree of anisotropy in the arrival directions of CRs that is suggested by the TA "hot spot" [15] or the excess around Cen A reported by Auger [7]). The post-trial p-value is 8.5×10^{-3}. For the likelihood stacking analysis, the most significant deviation from the isotropic flux is found for the magnetic deflection parameter $D = 6°$ for the cascade sample. The observed pre-trial p-value is 2.7×10^{-4}. Using a conservative method to estimate the trial factor [14], the post-trial p-value is found to be 8.0×10^{-4}. For this search the post-trial p-value with respect to the isotropic neutrino flux hypothesis is 1.3×10^{-3} ($\sim 3\sigma$). We see that for both the cross-correlation and the likelihood stacking analyses, the p-values obtained under the null hypothesis of isotropic neutrinos turn out to be larger than the ones obtained under the null hypothesis of isotropic CRs, the differences reflecting the extent to which the original p-values, from the isotropic cosmic-ray hypothesis, are due to an accidental alignment of the neutrinos with the known clustering of the cosmic rays.

3.2 Stacking search for neutrino point-sources in the 4 year point-source sample

The ν data set used for this analysis is the IceCube point-source data set. A stacking analysis is done but in this case (as opposed to the previous one) the stacked sources are the measured positions of UHECRs. An unbinned likelihood method is performed where the log likelihood is defined as:

$$\log \mathcal{L}(n_s, \gamma) = \sum_{i=1}^{N_\nu} \log \left(\frac{n_s}{N_\nu} \mathcal{S}_i^{\text{tot}} + \left(1 - \frac{n_s}{N_\nu}\right) \mathcal{B}_i \right), \tag{2}$$

where n_s is the number of signal events in the sample and γ is the spectral index of the neutrino source candidates, assumed to collectively follow an unbroken power-law spectrum $\propto E^{-\gamma}$. N_ν is the total number of astrophysical neutrino candidate events in the sample. $\mathcal{S}_i^{\text{tot}}$ is the signal PDF for the stacked sources and $\mathcal{B}_i$ is the background PDF. The log-likelihood, $\log(\mathcal{L})$ is maximized w.r.t. n_s and γ and the TS is defined as in Sect. 3.1.

Only UHECRs above a certain energy $E_{th} = 85$ EeV have been considered. The value of E_{th} was decided using the procedure described in [14].

Applying the method to the actual data, all observations are found to be compatible with the background only hypothesis. The smallest post-trial p-value is 25% for the hypothesis of $D = 3°$, with a fitted excess of $\sim$123 events and $\gamma = -3.24$. The analysis with $D = 6°$ yields a p-value larger than 50%.

4. Conclusions

Three analyses have been performed to investigate correlations between UHECRs detected by the Pierre Auger Observatory and Telescope Array with various samples of IceCube ν candidates. The results obtained are all below 3.3σ. There is a potentially interesting result in the analyses performed with the set of high-energy cascades when compared to assumed isotropic arrival directions of CRs. If we compare the result to an isotropic flux of neutrinos (fixing the positions of the CRs) to consider the effect of anisotropies in the arrival directions of CRs (such as the TA hot spot), the significance is $\sim 2.4\sigma$. These results were obtained with relatively few events and we will update these analyses in future with increasing statistics to follow their evolution.

References

[1] A. Achterberg et al. [IceCube Coll.] Astropart. Phys. **26**, 155 (2006)
[2] M.G. Aartsen et al. [The IceCube Collaboration], Proc. of the 34th Int. Cosmic Ray Conf. 1081 (2015)
[3] M.G. Aartsen et al. [The IceCube Collaboration], Phys. Rev. Lett. **115**(8), 081102 (2015)
[4] M.G. Aartsen et al. [The IceCube Collaboration], Astrophys. J. **796**, 109 (2014)
[5] J. Abraham et al. [The Pierre Auger Collaboration] Nucl. Instrum. Meth. **A523**, 50 (2004)
[6] A. Aab et al. [The Pierre Auger Collaboration] Nucl. Instrum. Meth. A **798**, 172 (2015)
[7] A. Aab et al. [The Pierre Auger Collaboration], Astrophys. J. **804**, 1 (2015)
[8] T. Abu-Zayyad et al. [Telescope Array Collaboration], Nucl. Instrum. Meth. A **689**, 87 (2012)
[9] T. Abu-Zayyad et al. [The Telescope Array Collaboration], Astrophys. J. **777**, 88 (2013)
[10] J. Abraham et al. [The Pierre Auger Collaboration], Phys. Rev. Lett. **101**, 061101 (2008)
[11] T. Abu-Zayyad et al. [The Telescope Array Collaboration], Astrophys. J. **768**, L1 (2013)
[12] Pshirkov M.S., Tinyakov P.G., Kronberg P.P., Newton-McGee K.J., Astrophys. J., 738, 192, 10.1088/0004-637X/738/2/192
[13] R. Jansson, G.R. Farrar, Astrophys. J. **757**, 14 (2012)
[14] M.G. Aartsen et al. [The IceCube, Pierre Auger and Telescope Array Collaborations], Proc. of the 34th Int. Cosmic Ray Conf. 1082 (2015)
[15] R.U. Abbasi et al. [The Telescope Array Collaboration], Astrophys. J. **790**, L21 (2014)

EPJ Web of Conferences **116**, 11001 (2016)
DOI: 10.1051/epjconf/201611611001

Multi-messenger aspects of cosmic neutrinos*

Markus Ahlers

WIPAC & Department of Physics, University of Wisconsin–Madison, Madison, WI 53706, USA

Abstract. The recent observation of TeV-PeV neutrinos by IceCube has opened a new window to the high-energy Universe. I will discuss this signal in the context of multi-messenger astronomy. For extragalactic source scenarios the corresponding gamma-rays are not directly observable due to interactions with the cosmic radiation backgrounds. Nevertheless, the isotropic sub-TeV gamma ray background observed by *Fermi*-LAT contains indirect information from secondary emission produced in electromagnetic cascades. On the other hand, observation of PeV gamma rays would provide a smoking-gun signal for Galactic emission. Interestingly, the overall energy density of the observed neutrino flux is close to a theoretical limit for neutrino production in ultra-high energy cosmic ray sources and might indicate a common origin of these phenomena. I will highlight various multi-messenger relations and their implications for neutrino source scenarios.

1. Introduction

The recent discovery of cosmic TeV-PeV neutrinos by the IceCube observatory [2–5] has added an important new pillar to high-energy multi-messenger astronomy. Neutrino emission corresponds to a unique probe of the high-energy Universe. Unlike cosmic rays (CRs), neutrinos are not deflected in cosmic magnetic fields over their long propagation distances and unlike gamma-rays they are not absorbed by pair production via $\gamma\gamma$ interactions with radiation in the source or cosmic radiation backgrounds. In the rather wide energy range from about 10 TeV to 10 EeV neutrinos are thus the only pointing probes that can cover cosmic distances.

The origin of the high-energy neutrino flux is presently unknown. Possible Galactic contributions to super-TeV neutrinos are the diffuse neutrino emission of galactic CRs, the joint emission of galactic PeV sources, and extended galactic structures like the *Fermi Bubbles* or the Galactic halo. Extragalactic source candidates include galaxies with intense star formation, cores of active galactic nuclei (AGN), low-luminosity AGN, blazars, low-power GRBs, cannonball GRBs, intergalactic shocks, and active galaxies embedded in structured regions. We refer to the recent review [6] for a list of references.

High-energy astrophysical neutrinos are expected to be produced by hadronic interactions of high energy CR nucleons with gas ("pp") and radiation ("$p\gamma$"). The individual mesons (mostly pions) produced in these interactions typically carry 20% of the initial CR nucleon energy and produce a high-energy flux of neutrinos via their decay $\pi^+ \to \mu^+\nu_\mu$ followed by $\mu^+ \to e^+\nu_e\bar{\nu}_\mu$, and the charge-conjugate processes. In this decay chain each neutrino receives on average 1/4th of the initial pion

*This article is an excerpt from an ICRC 2015 proceedings contribution [1].

energy. The initial ratio of the three neutrino flavors, $\nu_e : \nu_\mu : \nu_\tau \simeq 1:2:0$, is then expected to be visible via their oscillation-averaged composition, approximately $\nu_e : \nu_\mu : \nu_\tau \simeq 1:1:1$. Neutral pions produced in the same hadronic CR interactions are expected to produce a high-energy gamma-rays via the decay $\pi^0 \to 2\gamma$.

The production of high-energy neutrinos is thus intrinsically tied to the presence of CRs and the associated emission of gamma-rays from neutral pion production. In the following sections we will discuss multi-messenger relations between cosmic rays, gamma rays and neutrinos and highlight the implications for various neutrino emission models and future tests.

2. Neutrinos and Cosmic Rays

Hadronic interactions of CRs with gas (pp) or radiation ($p\gamma$) produce a flux of neutrinos with an energy of about 5% of the initial CR nucleon. Hence, the CRs responsible for the observed neutrino flux have corresponding energies reaching 20 PeV (proton) to 1 EeV (iron) depending on composition. This composition-dependent energy range is above the CR "knee" around 3 PeV and the "ankle" around 3 EeV. This region is expected to be the transition region of the spectrum between low-energy Galactic and high-energy extragalactic CRs. It is hence not entirely clear from an energetics point of view if Galactic or extragalactic sources are responsible for the neutrino flux.

Interestingly, the overall energy density of the observed neutrino flux is close to a theoretical limit for neutrino production in the sources of ultra-high energy (UHE) CRs [7]. The neutrino and CR nucleon (N) emission rates Q (in units of $\mathrm{GeV}^{-1}\mathrm{s}^{-1}$) are related via

$$\frac{1}{3}\sum_\alpha E_\nu^2 Q_{\nu_\alpha}(E_\nu) \simeq \frac{1}{4}\frac{f_\pi K_\pi}{1+K_\pi} E_N^2 Q_N(E_N), \tag{1}$$

where $f_\pi \leq 1$ is the pion production efficiency in hadronic interaction and neutrino and CR nucleon energy are related as $E_\nu \simeq 0.05 E_N$. The factor K_π corresponds to the ratio of charged to neutral pions, which can be approximated as $K_\pi \simeq 2$ for pp and $K_\pi \simeq 1$ for $p\gamma$ interactions.

After production in hadronic interactions the flux of neutrinos is not further affected by absorption mechanisms and only subject to redshift losses. In the following we assume a flat universe dominated by vacuum energy with $\Omega_\Lambda \simeq 0.7$ and cold dark matter with $\Omega_\mathrm{m} \simeq 0.3$ [8]. The Hubble parameter at earlier times is then given by its value today of $H_0 \simeq 70$ km s^{-1} Mpc^{-1} and the relation $H^2(z) = H_0^2(\Omega_\Lambda + \Omega_\mathrm{m}(1+z)^3)$. The observable diffuse neutrino flux from a homogeneous distribution of extragalactic sources with co-moving density $\mathcal{H}(z)$ can then be readily estimated as (see, e.g. Ref. [9]

$$\phi_\nu(E_\nu) = \frac{1}{4\pi}\int_0^\infty \frac{\mathrm{d}z}{H(z)}\mathcal{H}(z)Q_\nu((1+z)E). \tag{2}$$

The local emission rate density of UHE CRs depend on spectrum and composition. For an E^{-2} flux of protons it can be estimated as $E_p^2 Q_p(E_p)\mathcal{H}(0) \simeq (1-2)\times 10^{44}\,\mathrm{erg\,Mpc^{-3}\,yr^{-1}}$ [9]. Hence, using Eq. (2) the diffuse neutrino flux can be estimated as

$$E_\nu^2\phi_\nu(E_\nu) \simeq \frac{\xi_z f_\pi K_\pi}{1+K_\pi}(2-4)\times 10^{-8}\,\mathrm{GeV\,cm^{-2}\,s^{-1}\,sr}, \tag{3}$$

where we define the evolution factor (assuming constant f_π)

$$\xi_z = \int_0^\infty \mathrm{d}z \frac{(1+z)^{-\Gamma}}{\sqrt{\Omega_\Lambda + (1+z)^3\Omega_\mathrm{m}}}\frac{\mathcal{H}(z)}{\mathcal{H}(0)}. \tag{4}$$

For a spectral index $\Gamma = 2$ and no source evolution ($\mathcal{H}(z) = \mathcal{H}(0)$) this evolution factor is $\xi_z \simeq 0.5$, whereas evolution following the star-formation rate [10, 11] gives $\xi_z \simeq 2.4$.

Since the pion production efficiency is limited, $f_\pi \leq 1$, the estimate (3) provides a theoretical upper limit on neutrino production, the *Waxman-Bahcall* (WB) bound [7] (see also Ref. [12]). It is intriguing that the observed flux [2–5] is very close to this upper limit. This might be just by chance, or it could indicate a deeper multi-messenger relation. It is important to keep in mind that the observed neutrino flux does not directly relate to UHE CRs with energies above EeV and for proton-dominated scenarios that we assumed in the previous estimate an extrapolation over two orders of magnitude is required.

Neutrino fluxes close to the WB bound would require very efficient CR production with optical thickness $\tau_{p\gamma/pp} \gg 1$, such that $f_\pi \simeq 1$, i.e. CR reservoirs [13] such as starburst galaxies [14, 15] or clusters of galaxies [16–18]. Interestingly, the energy density of Galactic CRs require a similar energy density. Assuming that 1% of the kinetic energy of 10^{51} erg of a supernova (SN) explosion is converted to CRs and assuming normal galaxies with densities $\mathcal{H}_0 \simeq 10^{-3}\mathrm{Mpc}^{-3}$ and a SN rate of 10^{-2} yr^{-1} we arrive at $E_p^2 Q_p(E_p) \simeq 10^{44}\,\mathrm{erg\,Mpc^{-3}\,yr^{-1}}$. This coincidence together with the saturation of the WB bound has let to speculations that Galactic and extragalactic CRs might be produced in the same transient sources [13].

If the sources of UHE CRs are responsible for the neutrino emission it might be possible to cross-correlate these events [19]. The strong interaction of UHE CRs with the cosmic microwave background (CMB) limit the survival length to 200 Mpc above the Greisen-Zatsepin-Kuzmin [20, 21] (GZK) cutoff at around 50 EeV. UHE CR sources that appear at larger distance will still be visible by their neutrino emission, but not in UHE CRs. We can make a simple estimate what fraction of neutrinos would be expected to correlate with the local population of UHE CRs which can be sampled by the observed UHE CR arrival direction: The Hubble horizon at about 4.4 Gpc is about 20 times larger than the GZK horizon. That means that we only expect that about one neutrino out of 20 should correlate with local sources. In comparison, the most recent event sample of the high-energy starting event (HESE) analysis [5] comprises 53 events with a signal-to-background ratio close to 1. Hence, we expect only 1-2 of these events to correlate to local UHE CR sources and thus with the arrival direction of observed UHE CR events. This estimate is consistent with the absence of strong cross-correlations in a dedicated analysis of this data [19].

3. Neutrinos and gamma-rays

Hadronic interactions of CRs will not only produce charged pions that on decay release a flux of high-energy neutrinos, but also neutral pions that decay via $\pi^0 \to 2\gamma$. The production rates are simply related by

$$\frac{1}{3}\sum_\alpha E_\nu^2 Q_{\nu_\alpha}(E_\nu) \simeq \frac{K_\pi}{4} E_\gamma^2 Q_\gamma(E_\gamma), \tag{5}$$

where the gamma-ray and neutrino energies are related as $E_\gamma \simeq 2E_\nu$. This relation does not depend on the pion production efficiency and thus the gamma-to-neutrino multi-messenger relation is less affected by the mechanisms that lead to the initial production of pions, except for the relative charged-to-neutral pion rate K_π already introduced earlier.

However, the production rate described by Eq. (5) is not necessarily the emission rate of the sources. In particular, sources that efficiently produce neutrinos via $p\gamma$ interactions require a strong photon target spectrum. The gamma-rays produced via neutral pion decay can in this case undergo e^+e^- pair production before they escape. These leptons can scatter with the same background photons via inverse-Compton scattering or emit synchrotron emission in magnetic fields. The electromagnetic emission initiated by hadronic interactions might thus look much different from the initial production spectrum (5), although the total energy is conserved.

The situation is less complicated for optically thin sources, in particular those sources that are dominated by pp interactions. In this case the gamma-rays described by Eq. (5) are expected to correspond to the source emission spectrum. The pion production efficiency via pp interactions does only weakly depend on the CR energy and the neutrino and gamma-ray spectra are expected to follow the (time-integrated) CR spectrum, cf. Eq. (1), typically a broken power-law up to a maximal energy limited by the CR source type.

However, once released from distant sources it is inevitable that high-energy gamma rays interact with cosmic radiation backgrounds, in particular the CMB. The pair production interaction length of PeV gamma rays in the CMB is of the order of 10 kpc, which makes it impossible to observe this emission over extragalactic distances. However, pair production and subsequent inverse-Compton scattering of the high energy leptons will lead to electromagnetic cascades that shift the initial radiation into the sub-TeV gamma-ray band and supplement the direct sub-TeV emission of the source. The observed gamma-ray background in this energy region provides hence a general upper limit on the diffuse hadronic emission [22], which also applies to the production of cosmogenic neutrinos produced via the GZK interaction [23–26].

These general considerations allow to constrain various extragalactic source scenarios of the TeV-PeV neutrino flux. As discussed earlier, hadronic γ-ray and neutrino emission produced in pp sources, e.g., in starburst galaxies [14] is expected to follow the parent CR spectrum and can thus be extrapolated as a power-law to lower energies. It was shown in Ref. [27] that an unbroken power-law emission with a spectral index much softer than $\Gamma \simeq 2.2$ would be inconsistent with the isotropic diffuse gamma-ray background (IGRB) inferred by *Fermi* [28]. This places some tension on pp scenarios since the best-fit power-law index of the TeV-PeV neutrino data is significantly softer, $\Gamma \simeq 2.50 \pm 0.09$ [4].

Moreover, the total extra-galactic gamma-ray background (EGB) above 50 GeV is expected to be largely dominated by the emission of AGN (see, e.g., Ref. [29]). In a recent analysis of pixel-by-pixel photon count statistics [30] it was estimated that blazars comprise about $86^{+16}_{-14}\%$ of the total EGB intensity. This places even stronger bounds on pp scenarios that are expected not to coincide with AGN [31]. Recently, it was argued that limits derived from the cross-correlation of *Fermi* gamma-rays with galaxy catalogues exclude the pp scenarios with soft emission and weak redshift evolution [32]. Other electromagnetic emission bands can also become important in constraining pp emission scenarios, for instance radio emission of galaxy clusters [18].

The situation for $p\gamma$ scenarios is somewhat different. Firstly, the production spectrum of gamma-rays and neutrinos are obtained via a folding of the initial CR spectrum and the photon target spectrum. For typical target spectra present in astrophysical sources these emission spectra are not well described as power laws. Hence, from the spectral fit of the IceCube data in the TeV-PeV range we can not make a simple extrapolation towards lower energies. However, even the cascade emission associated with gamma rays emitted in the TeV-PeV is already at a level of 10% in the 0.1–1 TeV gamma-ray range. Thus, a careful decomposition of the IGRB into contributions of known populations can also constrain this emission scenario.

However, the presence of strong photon target fields in $p\gamma$ sources is also expected to enhance the gamma-ray opacity with respect to pair production. In fact, a high pion production efficiency of more than 0.01, which can be motivated by the high neutrino intensity at around 10 TeV [33], implies that $p\gamma$ sources are opaque to 1–100 GeV gamma rays. Depending on the low energy tail of the photon target spectrum this opacity can also extend into the TeV-PeV gamma-ray region and hence these sources would not be constrained by either direct observation by *Fermi* or indirect emission in the IGRB. Neutrinos could be the unique messenger of high-energy processes for these sources.

The arguments in the previous paragraphs focused on the emission of gamma-rays from extragalactic sources, which are strongly attenuated by the cosmic radiation backgrounds. However, with the limited angular resolution and statistics of the neutrino observation it is possible that extended emission from Galactic sources can contribute to the data (see, e.g., the study [34]). These emission includes the diffuse gamma-ray emission of the Galactic Plane [35], quasi-diffuse emission from the sum of Galactic sources, or extended emission from *Fermi Bubbles* [36, 37] or Galactic halo [38]. Exotic contributions like decaying heavy dark matter could also produce an extended emission in gamma-rays and neutrinos (see the recent review [6] for a list of references).

The observation of PeV gamma-rays with an attenuation length of about 10 kpc via pair production with CMB photons would be a *smoking gun* for this Galactic production [39, 40]. Various CR observatories already provide strong TeV-PeV gamma-ray limits, in particular KASCADE-Grande [41]. Studies of the isotropic diffuse high-energy gamma-ray emission in the near future with the HAWC observatory [42] and the far future with the air shower arrays LHAASO [43] or HiSCORE [44] can greatly improve the present limits. However, all these experiments are located in the Northern Hemisphere and have only a limited view of the Galactic Center region or the *Fermi Bubbles*.

4. Conclusions

The recent observation of TeV-PeV neutrinos has added an important new pillar to multi-messenger astronomy. Presently, the observed neutrino events have limited angular resolution and statistics and the underlying source population remains elusive, see e.g. Ref. [45]. However, multi-messenger relations between cosmic rays and gamma rays can already provide hints and constraints for possible source scenarios.

Presently, one particularly puzzling observation is the particularly high intensity of the IceCube observation in the (10–100) TeV range, which is indicated by a recent combined data analysis [4]. The underlying source population is required to have a high CR luminosity and pion-production efficiency while at the same time it has to avoid strong limits implied by the observed level of the extragalactic gamma-ray background seen by *Fermi*.

The continued observation of the neutrino flux with IceCube will help to identify possible substructures in the spectrum that provide indirect evidence of the source emission type. An improved study of neutrino flavor composition will provide an important test of astrophysical neutrino emission and atmospheric background models. Subdominant contributions from Galactic emission could become evident by the manifestation of anisotropies or the detection of hadronic PeV gamma-ray emission.

I would like to thank the organizers of *VLVνT 2015* for a very enjoyable conference. This work is supported by the National Science Foundation under grants OPP-0236449 and PHY-0236449.

References

[1] M. Ahlers, in *Proceedings, 34th International Cosmic Ray Conference (ICRC 2015)* (2015)
[2] M.G. Aartsen et al. (IceCube), Science **342**, 1242856 (2013), 1311.5238
[3] M. Aartsen et al. (IceCube), Phys. Rev. Lett. **113**, 101101 (2014), 1405.5303
[4] M.G. Aartsen et al. (IceCube), Astrophys. J. **809**, 98 (2015), 1507.03991
[5] M.G. Aartsen et al. (IceCube), in *Proceedings, 34th International Cosmic Ray Conference (ICRC 2015)* (2015), 1510.05223
[6] M. Ahlers, F. Halzen, Reports on Progress in Physics **78**, 126901 (2015)
[7] E. Waxman, J.N. Bahcall, Phys. Rev. **D59**, 023002 (1999), hep-ph/9807282
[8] K. Olive et al. (Particle Data Group), Chin. Phys. **C38**, 090001 (2014)
[9] M. Ahlers, F. Halzen, Phys. Rev. **D86**, 083010 (2012), 1208.4181
[10] A.M. Hopkins, J.F. Beacom, Astrophys. J. **651**, 142 (2006), astro-ph/0601463
[11] H. Yuksel, M.D. Kistler, J.F. Beacom, A.M. Hopkins, Astrophys. J. **683**, L5 (2008), 0804.4008
[12] K. Mannheim, R.J. Protheroe, J.P. Rachen, Phys. Rev. **D63**, 023003 (2001), astro-ph/9812398
[13] B. Katz, E. Waxman, T. Thompson, A. Loeb (2013), 1311.0287
[14] A. Loeb, E. Waxman, JCAP **0605**, 003 (2006), astro-ph/0601695

[15] I. Tamborra, S. Ando, K. Murase, JCAP **1409**, 043 (2014), 1404.1189
[16] V. Berezinsky, P. Blasi, V. Ptuskin, Astrophys. J. **487**, 529 (1997), astro-ph/9609048
[17] K. Murase, S. Inoue, S. Nagataki, Astrophys. J. **689**, L105 (2008), 0805.0104
[18] F. Zandanel, I. Tamborra, S. Gabici, S. Ando, Astron. Astrophys. **578**, A32 (2015), 1410.8697
[19] M.G. Aartsen et al. (IceCube, Pierre Auger, Telescope Array), in *Proceedings, 34th International Cosmic Ray Conference (ICRC 2015)* (2015), 1511.02109
[20] K. Greisen, Phys.Rev.Lett. **16**, 748 (1966)
[21] G. Zatsepin, V. Kuzmin, JETP Lett. **4**, 78 (1966)
[22] V. Berezinsky, A.Y. Smirnov, Phys. Lett. **B48**, 269 (1974)
[23] V. Berezinsky, G. Zatsepin, Phys. Lett. **B28**, 423 (1969)
[24] V. Berezinsky, A. Gazizov, M. Kachelriess, S. Ostapchenko, Phys. Lett. **B695**, 13 (2011), 1003.1496
[25] M. Ahlers, L. Anchordoqui, M. Gonzalez-Garcia, F. Halzen, S. Sarkar, Astropart. Phys. **34**, 106 (2010), 1005.2620
[26] G. Decerprit, D. Allard, Astron. Astrophys. **535**, A66 (2011), 1107.3722
[27] K. Murase, M. Ahlers, B.C. Lacki, Phys. Rev. **D88**, 121301 (2013), 1306.3417
[28] M. Ackermann et al. (Fermi), Astrophys. J. **799**, 86 (2015), 1410.3696
[29] M. Ajello et al., Astrophys. J. **800**, L27 (2015), 1501.05301
[30] M. Ackermann et al. (Fermi-LAT) (2015), 1511.00693
[31] K. Bechtol, M. Ahlers, M. Di Mauro, M. Ajello, J. Vandenbroucke (2015), 1511.00688
[32] S. Ando, I. Tamborra, F. Zandanel, Phys. Rev. Lett. **115**, 221101 (2015), 1509.02444
[33] K. Murase, D. Guetta, M. Ahlers (2015), 1509.00805
[34] M. Ahlers, Y. Bai, V. Barger, R. Lu (2015), 1505.03156
[35] M. Ackermann et al. (Fermi), Astrophys. J. **750**, 3 (2012), 1202.4039
[36] M. Su, T.R. Slatyer, D.P. Finkbeiner, Astrophys. J. **724**, 1044 (2010), 1005.5480
[37] M. Ackermann et al. (Fermi-LAT), Astrophys. J. **793**, 64 (2014), 1407.7905
[38] A.M. Taylor, S. Gabici, F. Aharonian, Phys. Rev. **D89**, 103003 (2014), 1403.3206
[39] N. Gupta, Astropart. Phys. **48**, 75 (2013), 1305.4123
[40] M. Ahlers, K. Murase, Phys. Rev. **D90**, 023010 (2014), 1309.4077
[41] G. Schatz et al., in *Proceedings, 28th International Cosmic Ray Conference (ICRC 2015)* (2003), pp. 2293–2296
[42] T. DeYoung (HAWC), Nucl. Instrum. Meth. **A692**, 72 (2012)
[43] M. Zha (LHAASO, ARGO-YBJ), Int. J. Mod. Phys. Conf. Ser. **10**, 147 (2012)
[44] M. Tluczykont, D. Horns, D. Hampf, R. Nachtigall, U. Einhaus, M. Kunnas, T. Kneiske, G.P. Rowell, Nucl. Instrum. Meth. **A692**, 246 (2012)
[45] M. Ahlers, F. Halzen, Phys. Rev. **D90**, 043005 (2014), 1406.2160

EPJ Web of Conferences **116**, 11002 (2016)
DOI: 10.1051/epjconf/201611611002

A discussion of IceCube neutrino events, circa 2015

Arman Esmaili[1], Andrea Palladino[2], and Francesco Vissani[1,2]

[1] Laboratori Nazionali del Gran Sasso, INFN, Assergi, L'Aquila
[2] GSSI, Gran Sasso Science Institute (INFN), L'Aquila

Abstract. IceCube has changed the rules of the game and continues to progress. Their observations are compatible with cosmic neutrinos undergoing 3 flavor oscillations. The topologies of the events have been used to probe ordinary and exotic physics. Still, we need independent confirmations of this assumption; the astrophysical connections heavily rely on speculations (excepting special cases, such as GRB); the amount of prompt events is not known precisely; double bang and/or Glashow resonance events are still to be seen; the energy and the angular distributions are not well-known, even if the simplest picture (isotropic flux, power law distributed in energy) is still compatible with the data. In this talk, we select specific topics concerning expectations, inferences and prospects.

1. Remarks on prompt neutrinos

Cosmic rays of sufficiently high energies produce charm mesons interacting in the atmosphere. Their decays, differently from those of pion and kaon mesons, are very fast and produce the so called prompt neutrinos. These decays do not suffer any effect of the atmosphere, thus the signal is not yearly modulated. Prompt neutrinos have a peculiar flavor structure, $\nu_e \approx \nu_\mu$ and $\nu_\tau \approx 0$, within 10%. Thus, the signal from charm is rich in ν_e that are just $\sim 1/20$ of the conventional component. The normalization of prompt neutrinos is not precisely predicted, but at few 10 of TeV, they could dominate atmosperic neutrino spectrum making it harder, till $\alpha \sim 2.7$. The analyses of IceCube experiment deplete prompt events coming from above exploiting neutrino-muon correlation, as suggested in [1]. These analyses probed the normalization predicted by [2], but they did not see trace of prompt events. It is interesting that the cosmic neutrino signal has a slope similar (slightly below) $\alpha \sim 2.7$ and it is certainly important that, as claimed by IceCube, that prompt events, even if present, do not explain the bulk of observations. However, this leaves us the task of finding prompt events, or to understand "why IceCube lacks of charm". The answer could be quite simple but it is important to know it.

2. Expectations on oscillations of cosmic neutrinos

Assume cosmic neutrinos have been seen. Ordinary 3 neutrinos oscillations modify their fluxes and thanks to IceCube's capability to classify different topologies of events, these type of effects can be probed. We clarify the expectations and compare with the data, discussing residual uncertainties.

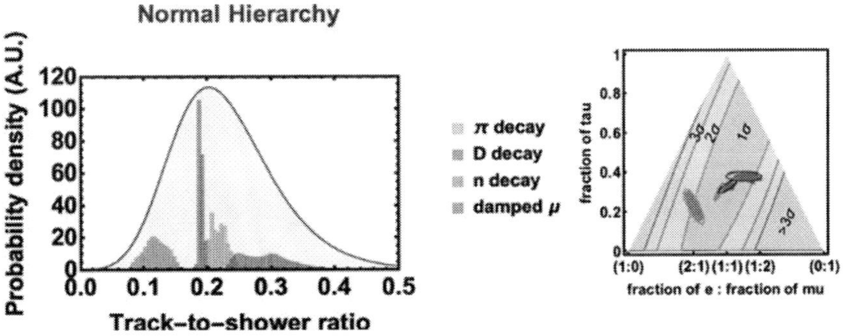

Figure 1. Right panel, the N_T/N_S (track-to-shower) ratio; the large smooth curve derives from IceCube data analysis, the 4 small histograms are the expectations [4]. Left panel, the flavor triangle; the wide oblique areas are the outcome of data analysis, whereas the 4 dotted areas enclosed by ellipses are the expectations [5]. In both plots it is assumed that the slope of the power spectrum is $\alpha = 2.3$ and we consider 4 scenarios for neutrino production mechanism: 1) The (standard) pion decay; 2) the decay of heavy mesons (charm); 3) the neutron decay; 4) the case when the pion decays, but muons are instead trapped in the medium, the *damped muon* case.

The expectations can be summarized in a few points: *1.* Cosmic neutrinos are supposedly produced in environments with small particle densities. *2.* During their propagation, they are subject to vacuum oscillations, also known as oscillations in Gribov-Pontecorvo (averaged) regime. *3.* Three flavor oscillations have an important impact on cosmic neutrinos [3] and have been precisely measured in terrestrial laboratories. *4.* Residual uncertainties can be treated with MonteCarlo techniques [4] and are most neatly described by adopting the three natural parameters of vacuum oscillations [5].

To date, we have probed the flavor of the incoming neutrinos by measuring two types of event "topologies", namely *track* events, largely corresponding to charged current muon neutrinos (or antineutrinos) interactions on nucleons and *shower* events, corresponding to charged ν_e, ν_τ instead, or to neutral currents. The assumption of a power law distribution of the cosmic neutrino fluxes, $F \propto E_\nu^{-\alpha}$ is plausible for a non-thermal population and allows us, along with the effective areas of IceCube, to compute the expected number of tracks and showers, N_T and N_S. Evidently they depend linearly upon the flux normalizations and have a mild dependence upon the assumed slope α. The observed number of events can be treated in Poisson statistics and thus compared with the expectations.

The first comparison between data and observations uses the track-to-shower ratio, N_T/N_S [6]. This is a very neat way to display IceCube's observations, however, the predictions depend a bit upon the slope [4]. The second manner uses the fact that in an equilateral triangle, the sum of the 3 distances of a inner point from the sides is constant (=Viviani theorem) applied to the 3 fraction of fluxes, e.g. $F_{\nu_e}/(F_{\nu_e} + F_{\nu_\mu} + F_{\nu_\tau})$ for the electronic fraction [7]. This is called *flavor triangle* and it is used e.g. in [8], see [5] for the study of the theoretical errors. In this way, the predictions become independent from the slope of the power law but the observations are not; moreover, the fractions of the fluxes are not directly observable. Expectations and observations are found to be in agreement, as evident from Fig. 1. This is true in particular for the conventional mechanism of neutrinos production, i.e. pion decay. The speculative cases when neutrinos at the source derive from neutron decay are somewhat disfavored, while the flavor ratio expected from prompt events in atmosphere is not, instead.

These results are significant extensions of the directly available knowledge. In fact, vacuum oscillations become perceivable when the product of the difference of neutrino masses squared Δm^2 times the distance between production and detection L times the inverse of the neutrino energy E_ν obeys the condition: $\Delta m^2 c^4 L/(hcE_\nu) \gg 1$ i.e., $0.81 \Delta m^2 c^4/\mathrm{eV}^2\ L/\mathrm{km}/(E_\nu/\mathrm{GeV}) \gg 1$, obtained from the 1st oscillation maximum. This is less than few hundreds for atmospheric neutrinos crossing the whole Earth and with $E_\nu = 100\,\mathrm{MeV}$. For a hypothetical source of the Milky Way, this is at least 10^6, even if we use the smallest Δm^2 with $L = 1$ kpc and $E_\nu = 1$ PeV. Accepting the most plausible option,

Table 1. Expected number of events per year in IceCube as a function of the spectral index [10].

	$\alpha = 2$	$\alpha = 2.3$	$\alpha = 2.6$
Double pulse	0.2	0.1	< 0.1
Glashow pp	0.5	0.2	< 0.1
Glashow $p\gamma$	0.2	0.1	< 0.1

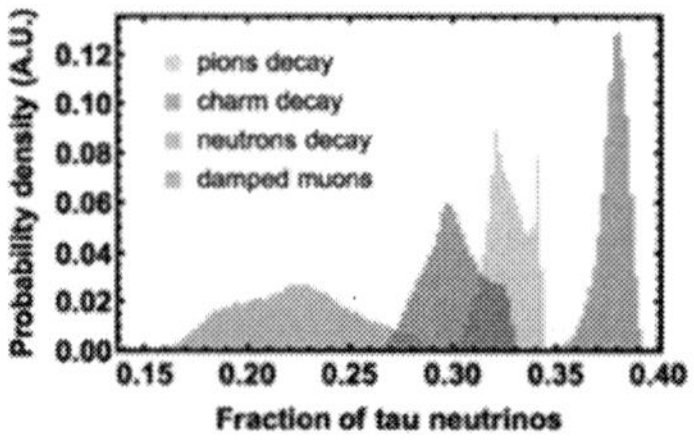

Figure 2. The amount of tau neutrinos depends a little bit on the production mechanism and it is not perfectly known due to residual uncertainties in oscillations parameters, but the main point to emphasize is simply that: **If tau neutrinos are observed, the cosmic origin is proved.**

that these neutrinos are extragalactic, the distances increase at least by 10^3 times and consequently also the product! This means that we have attempted to rely on what we have learned in the terrestrial laboratories by endorsing a brave extrapolation... and, till now, it works.

3. Inferences on cosmic neutrinos

We elaborate on hypothesis cosmic neutrino. A large amount of tau neutrinos should be present, and if the spectrum does not deviate strongly from a power law, the Glashow resonance signal will be observable, and its amount will inform us on the type of cosmic source; see Table 1 for a summary. Moreover, we can begin to explore (or exclude) exotic possibilities such as neutrino decay.

Tau neutrinos:– To date, no tau event has been observed in the existing dataset but the expectations have been clarified [9, 10]. Due to oscillations tau neutrinos are unavoidable [3], see Fig. 2. A tau neutrino below few 100 TeV produces a signal similar to the one produced by an electron neutrino or a neutral current event. At high energies, the primary vertex (where hadrons deposit energy) and the secondary vertex (where tau decays) are separated; this is the same principle used by the OPERA detector on a microscopic scale. If the energy is larger than few PeV, the two vertices are visible in two separated sets of phototubes, and this is called *double-bang* [3]. The single phototubes of IceCube, able to resolve the time structure of the energy deposited, can identify a *double-pulse* [9] already above half a PeV. Double pulse events are expected to be contributed from a region where we have already observed neutrinos: If neutrinos are power law distributed, we are confident at 90% CL to observe a tau when the exposure of IceCube will be tripled, i.e., in about 10 years.

Glashow resonance:– The production of real W^- bosons by $\bar{\nu}_e$ impinging on electrons, i.e., the Glashow resonance, requires antineutrinos of 6.32 PeV, namely with energies 3 times larger than the highest one to date observed. Interestingly, given the observed signal at lower energies, the intensity of Glashow resonance does depend upon the production mechanism; indeed, if we have $p\gamma$ collisions, we produce mostly π^+ and thus $\bar{\nu}_e$ arise only when we consider neutrino oscillations [11, 12]. The signal is about two times less in this case than for pp collisions, see e.g. [5]. Again, no signal is observed yet. Plausible explanations are, i) the spectral index is larger than 2; ii) the slope of the cosmic neutrino component is close to 2, but we have another component with larger slope; iii) there is a cutoff in the spectrum before 6 PeV; iv) or we need to collect more exposure. The calculations of [10] authorize a moderate optimism toward the possibilities to observe a signal, even if it emphasizes the difficulties to distinguish the $p\gamma$ and pp cases; compare with the conclusions of [12, 13].

Neutrino decay:– The idea that neutrinos could be unstable is old. A radical possibility is that the heavier neutrinos decay to the lighter one and an almost invisible, massless particle. After IceCube data, this hypothesis has been investigated, relying again on the obseved track-to-shower ratio. Since the lightest neutrino mass has a different composition in the normal hierarchy and the inverse hierarchy, the predictions are different. More precisely, in the first case the probability to receive a muon neutrinos $|U^2_{\mu 1}|$ is smaller than in the usual case, whereas in the second case, this probability is $|U^2_{\mu 3}|$ that is, instead, larger than in the usual case. By using the four years data [14], it was shown that in any case in both mass hierarchies, we can claim that this hypothesis on neutrino decay is disfavored at least at the 2σ CL. Note incidentally that an important additional information derives from the neutrinos from SN1987A, whose mere detection witnesses against inverse mass hierarchy, since the lightest antineutrino $\bar{\nu}_3$ contains a very tiny amount of the species that was detected, namely, $\bar{\nu}_e$ [15].

4. Prospects for astronomical identification

Neutrino-gamma connection at the source:– The most reasonable hypothesis for the origin of cosmic neutrinos is that they come from regions of the Universe where cosmic rays energies collide with gas or with some photonic target, producing copiously mesons. In first approximation, the collisions with gas will produce π^+, π^-, π^0 in similar proportions whereas the collisions with photons $p\gamma \to \Delta^+$ will produce two π^0 per π^+, owing to isospin selection rules. The charged pions, upon decay, yield in turn neutrinos but we expect to have also gamma rays from π^0 decay. We conclude that,

γ-rays are four times more abundant for $p\gamma$ sources than for pp sources

Moreover, a quantitative estimation of the connection of γ-ray and neutrino fluxes [16], assuming pp sources, shows that the two are quite similar (equal within 20% or so depending on the slope).

Destiny of γ-rays and the window of Fermi-LAT:– Neutrinos and γ-rays differ, since the latter might interact after being produced. E.g., collisions with the photons that permeate the Universe lead to e^+e^- production, for energies above $E_\gamma \approx m_e^2/\epsilon_\gamma = 250$ GeV if $\epsilon_\gamma = 1$ eV. Thus, if the source is transparent and the leptonic emission is not overwhelming, there is *a chance* to observe γ-rays directly produced with neutrinos. It is in the energy window below 100 GeV or so, studied by Fermi-LAT. The very high energy γ-rays that correspond to IceCube's neutrinos are necessarily modified, if not else, by their propagation over cosmological distances. They will be degraded till they reach energies below 100 GeV or so, when Universe is (approximatively) transparent. These γ-rays, once again, contribute to the window of energy monitored by Fermi-LAT.

BL Lacertae objects as neutrino sources:– We have argued the importance of γ-ray observations below 100 GeV. Fermi-LAT proved that the BL Lac are the brightest extragalactic sources above 10 GeV. These are AGNs with jets pointing toward the Earth and featureless optical spectra (i.e., lacking strong emission/absorption lines). Many more of these BL Lac are not visible as point sources, being too faint and/or too far from us. They contribute to the unresolved γ-ray radiation, again observed by Fermi-LAT see Table 2. The sum of resolved and unresolved BL Lac *almost* saturates the observed diffuse emission, with a margin of uncertainty of some ten percents, that should be attributed to other sources. Figure 3 compares the total BL Lac γ-rays spectrum with the one of neutrinos.

The angular distribution:– In first approximation, IceCube's events are distributed isotropically but a new significant component with a different distribution, possibly correlated with some galactic structure,

Table 2. The 1st column is obtained summing the intensity of 211 identified BL Lac [17] (5 BL Lac have more than 1% of the total γ-ray flux each; the brightest has 2%); the 2nd extrapolates to the low galactic latitude region $|b| < 15°$, avoided in the analysis, by assuming isotropy; the last two are theoretical estimations available from [17]. We checked that models in [17] imply total numbers in reasonable agreement with those obtained summing resolved and unresolved BL Lac. The theoretical models named LDDE and PLE are defined in [17].

BL Lac class	resolved	resolved (all)	unresolved (LDDE)	unresolved (PLE)
Flux 10^{-7}**ph/(cm^2 s sr)**	6.3	8.5	$8^{+2.0}_{-1.3}$	$10^{+2.1}_{-1.7}$

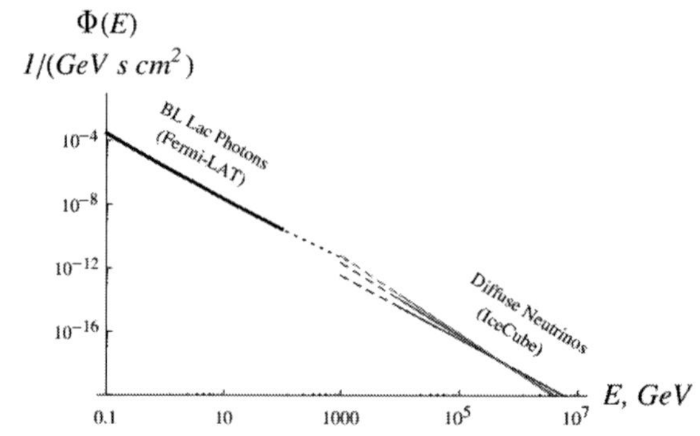

Figure 3. Total emission in γ-rays from BL Lac, as summarized by the best fit LDDE model [17], compared with the neutrino emission of IceCube for the 3 representative fits of [18]. Dotted and dashed lines are simply extrapolations. There is no strong incompatibility with the hypothesis that γ and neutrino emissions come from BL Lac, but the shape of the spectra is quite different especially if the neutrino spectrum is steep. The importance to understand the neutrino emission at the lowest energies is evident.

is not excluded. No evident point source or class of sources corresponds to these events. The direction of the showers is identified with a precision of $\delta\theta = 10$–$20°$, whereas the tracks are known with an angular resolution 10 times better. This is important, since the fraction of solid angle is proportional to the square of $\delta\theta$,[1] despite the fact that the tracks are less and subject to background contamination especially at low energies.

5. Unconventional sources

Motivated by the IceCube events, several new physics scenarios have been proposed as possible sources of these events, whether the entirety or at least a part of the events. Among them the scenario of interpreting the events by decaying PeV-scale dark matter gained attention. In this scenario the neutrinos are originated from the Galactic and extra-galactic components of the flux from the dark matter decay to standard model particles, say leptons and quarks, with the lifetime $\sim 10^{28}$ s [19]. Figure 4 shows the energy distribution of events in dark matter scenario. The angular distribution of the events also marginally prefer the expected distribution in dark matter scenarios rather than the pure isotropic extra-galactic astrophysical scenario [20]. However, larger statistics are required to make any conclusion. The dark matter scenario can be tested in the near future by high energy gamma-ray experiments, both by the IGRB data [21], or extensive air-shower detectors [22]. Particularly, in this scenario an accompanying flux of high energy gamma-rays with peculiar anisotropy pattern is expected, at a level detectable by near future detectors, such as LHAASO [23].

[1] For this reason, high energy neutrino telescopes in water have a strong scientific advantage, being able to obtain $\delta\theta \approx 0.2°$.

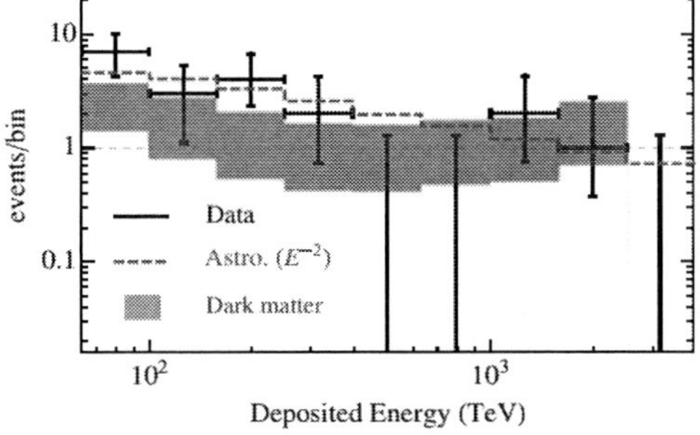

Figure 4. The shaded region shows the energy distribution of events in a toy model for decaying dark matter scenario. The width of the shaded regions corresponds to variation of dark matter lifetime within 1σ range around the best-fit point obtained from the fit to IceCube data. The dark matter mass assumed equal to 4 PeV. The green dashed line shows the expected events from astrophysical E_ν^{-2} flux. The black points show the IceCube data.

6. Discussion

The case made by IceCube for a new population of very high energy neutrinos, on top of conventional atmospheric neutrinos is very convincing, even if the role of charm (prompt) contribution is still to be elucidated. The new population can be attributed to cosmic neutrinos, that are searched since more than half a century. It is consistent with ordinary vacuum oscillations. A definitive proofs of this hypothesis could come from tau and/or Glashow resonance. The conventional expectation of a hard $\sim E^{-2}$ distributions would imply that they are observable; tau events are less but largely contributed by neutrinos of the energies that have been already seen. However, a fit of the IceCube data with a power law distribution prefers steeper slopes. The main needs are: to confirm the case for cosmic neutrinos independently; to search deviations from an isotropic distribution; to locate (some of) its source(s); to understand well the spectrum; to remind that this could be just the beginning.

References

[1] S. Schonert, T.K. Gaisser, E. Resconi and O. Schulz, Phys. Rev. D **79** (2009) 043009
[2] R. Enberg, M.H. Reno and I. Sarcevic, Phys. Rev. D **78** (2008) 043005
[3] J.G. Learned and S. Pakvasa, Astropart. Phys. **3** (1995) 267
[4] A. Palladino, G. Pagliaroli, F. L. Villante and F. Vissani, Phys. Rev. Lett. **114** (2015) 17, 171101
[5] A. Palladino and F. Vissani, Eur. Phys. J. C **75** (2015) 433
[6] A. Esmaili and Y. Farzan, Nucl. Phys. B **821** (2009) 197
[7] G. Barenboim and C. Quigg, Phys. Rev. D **67** (2003) 073024
[8] M.G. Aartsen et al. [IceCube Collaboration], Phys. Rev. Lett. **114** (2015) 17, 171102
[9] M.G. Aartsen et al. [IceCube Collaboration], `arXiv:1509.06212`
[10] A. Palladino, G. Pagliaroli, F.L. Villante and F. Vissani, `arXiv:1510.05921`
[11] V.S. Berezinsky and A.Z. Gazizov, JETP Lett. **25** (1977) 254
[12] L. A. Anchordoqui, H. Goldberg, F. Halzen and T.J. Weiler, Phys. Lett. B **621** (2005) 18
[13] V. Barger, J. Learned and S. Pakvasa, Phys. Rev. D **87** (2013) 3, 037302
[14] G. Pagliaroli, A. Palladino, F. Vissani and F.L. Villante, `arXiv:1506.02624`, accepted in PRD
[15] P. Baerwald, M. Bustamante and W. Winter, JCAP **1210** (2012) 020
[16] F.L. Villante and F. Vissani, Phys. Rev. D **78** (2008) 103007
[17] M. Ajello et al. [Fermi-LAT Collaboration], Astrophys. J. **780** (2014) 73
[18] T.K. Gaisser [IceCube Collaboration], `arXiv:1507.07871`

[19] A. Esmaili and P.D. Serpico, JCAP **1311** (2013) 054
[20] A. Esmaili, S.K. Kang and P.D. Serpico, JCAP **1412** (2014) 054
[21] K. Murase, R. Laha, S. Ando and M. Ahlers, Phys. Rev. Lett. **115** (2015) 071301
[22] A. Esmaili and P.D. Serpico, JCAP **1510** (2015) 014
[23] S. Vernetto and P. Lipari, `http://pos.sissa.it/archive/conferences/236/740/ICRC2015_740.pdf`

EPJ Web of Conferences **116**, 11003 (2016)
DOI: 10.1051/epjconf/201611611003

KM3NeT-ARCA project status and plan

R. Coniglione[a] on behalf of the KM3NeT Collaboration

INFN, Laboratori Nazionali del Sud, via S. Sofia 62, Catania 95123, Italy

Abstract. The KM3NeT Collaboration aims at building a research infrastructure in the depths of the Mediterranean Sea hosting a cubic kilometre neutrino telescope. The KM3NeT/ARCA detector is the ideal instrument to look for high-energy neutrino sources thanks to the latitude of the detector and to the optical characteristics of the sea water. The detector latitude allows for a wide coverage of the observable sky including the region of the Galactic centre and the optical sea water properties allow for the measure of the neutrino direction with excellent angular resolution also for cascade events. The technologically innovative components of the detector and the status of construction will be presented as well as the capability it offers to discover neutrinos.

1. Introduction

The main aims of the KM3NeT Collaboration are the observation of cosmic high energy neutrinos and the study of their origins and the determination of the neutrino mass hierarchy. To reach these goals the Collaboration started the construction of a network of two telescopes in the depth of the Mediterranean Sea: ARCA, with an instrumented volume of about 1 km^3 that will be installed near the Sicilian (Italy) coast, and ORCA, a smaller and more densely instrumented detector for the study of the neutrino mass hierarchy that will be installed near the Toulon coast (France). In this paper the status and performances of the KM3NeT/ARCA detector will be described while the KM3NeT/ORCA detector is discussed in [1]. The detection principle of the under sea high-energy neutrino detectors is based on the measurement of the Cherenkov light induced in water by the secondary charged particles produced in neutrino interactions inside or in the vicinity of the detector.

The study of high energy neutrinos produced in galactic and extragalactic sources can shed light on the production mechanisms of high energy particles and can give important constraints on the acceleration models being neutrinos a strong signature of the presence of hadronic processes. The IceCube observation of the high-energy astrophysical neutrino flux [2, 3] has marked a turning point in astro-particle physics. The presence of neutrino of cosmic origin states that high energy particles are produced in the Cosmos with a neutrino flux of intensity detectable by the present generation of high energy neutrino telescopes. The sources producing high energy neutrinos have not been yet determined

[a] e-mail: coniglione@lns.infn.it

and the spectral shape and flavour composition have not been precisely defined [4]. Also the claimed isotropic distribution of the measured neutrino flux [3] requires further experimental confirmation.

KM3NeT/ARCA is a complementary instrument to IceCube: located in the Northern hemisphere and exploiting sea water as medium for the Cherenkov light propagation it has a different field of view and presents different systematics uncertainties. The good angular resolution, both for cascade and track events, and the large visibility of the Galactic Plane makes ARCA the ideal instrument to investigate Galactic sources and in general all sources located in the Southern sky. This unique characteristics has pushed the collaboration to optimize the detector design for the detection of neutrinos from Galactic sources which have an expected energy from a few TeV to a few 10 TeV. The confirmation of the diffuse neutrino flux measured by IceCube is expected as one of the first physics goal of the Collaboration.

2. The KM3NeT/ARCA detector design

The arrival times of the light collected by optical sensors and their geometrical positions are used to reconstruct the tracks of the particles produced in neutrino interactions and from that also the neutrino direction and its energy. Large detectors with a size of a cubic-kilometer arranged in a 3-dimensional array of photo sensors are required.

One of the concept that has been validated in the past years by the KM3NeT Collaboration is that the overall sensitivity is not reduced if the neutrino telescope is split into separate detector building blocks [5]. The consequent detector modularity complies with the technical specifications for the construction and maintenance of the infrastructure. KM3NeT/ARCA will be constituted by two building blocks, each one of 115 Detection Units (DU) having an instrumented volume of about 0.5 km^3. The DUs are flexible structures anchored at the sea floor, kept vertical by buoys and connected to shore by an electro-optical cable network. Each DU hosts 18 Digital Optical Modules (DOM). Each DOM consists of a pressure resistant glass sphere that contains 31 3 inches photomultipliers and the readout electronics. The horizontal distance between strings is about 90 m and the vertical distance between adjacent optical modules is 36 m.

A brief description of two technologically innovative elements of the detector, the DU and the DOM, is given hereafter. A description of the other parts of the detector as the trigger and data acquisition system, the time calibration, the positioning system, the electronics, the data transmission system, the live monitoring and the GRID computing system is available in the other KM3NeT contributions to this workshop.

The DOM [6, 7] is a pressure-resistant glass sphere of 17-inch diameter hosting 31 3-inch photomultiplier tubes (PMTs) with low power high-voltage bases and read-out electronics. Each tube is surrounded by a reflector that effectively increases the collection efficiency per PMT by about 27%. The lower hemisphere of each DOM contains 19 of the PMTs, downward-looking, whereas the other 12 PMTs in the upper hemisphere are up-looking. The optical contact between the PMTs and the glass sphere is assured by optical gel. The front-end electronics [8] amplify the PMT signals and transform them into digital time-over-threshold information that is fed into the readout via optical fibres. All PMT signals above an adjustable threshold (typically the equivalent of 0.3 photo-electrons) are sent to shore, where event candidates are selected by online filters running on a computer farm. The DOM contains three calibration sensors: a LED nano-beacon for time calibration, a compass and tiltmeter for orientation calibration and an acoustic piezo sensor, glued to the inner surface, for position calibration. A DOM mounted on a DU is shown in Fig. 1 left.

This innovative photon sensor, which is the result of an intense R&D activity, has several advantages with respect to the conventional optical modules with one 10 inch PMT. Its photocathode area is 3 to 4 times the area of a single PMT module and has an almost uniform angular coverage. The photocathode segmentation allows for a more efficient identification of more than one photon arriving at the DOM. This improves the background rejection and provides directional information. These features have

Figure 1. Left – The DOM fixed on the two parallel Dyneema© ropes. Right – The deployment of the LOM vehicle.

been confirmed by the analysis of data collected with a DOM prototype [9] hosted on the ANTARES instrumented line [10] and with the DU prototype [11] deployed at the Italian site and operated for 9 months. A procedure for time calibration which exploits ^{40}K decays and the LED beacons inside the DOM has been developed for the DU prototype and tested, reaching a precision at the level of nanosecond.

The DU [12] is a slender structure consisting of two parallel Dyneema© ropes to which the DOMs are attached via a titanium collar (see Fig.1 left). Attached to the ropes is the vertical electro-optical cable, a pressure balanced, oil-filled, plastic tube that contains two copper wires and 18 optical fibres for the power and data transmission. At each DOM two power conductors and a single fibre are branched out via the breakout box. The breakout is the black box mounted on the top of the glass sphere in Fig. 1 left.

For the deployment, a detection unit is wound around a spherical frame with diameter of about 2.2 m (Launcher of Optical Modules, LOM [13]), which is posed on the seabed and after an acoustic release command starts unwinding the DU with a rotating upwards movement. At the end of operation the LOM reaches the sea surface, where it is collected for reuse. A picture of the LOM ready for the deployment is shown in Fig. 1, right.

3. The status of the KM3NeT/ARCA detector

The KM3NeT/ARCA detector will consist of 2 building blocks, each with 115 DUs. The design of the detector allows for a staged construction. Funding for the first phase of the KM3NeT (Phase-1) is already available. It foresees the deployment of:

- 24 string-like DUs and 8 DUs based on a flexible tower design [14] at the Italian site.
- 7 string-like DUs at the French site with a denser distribution of the optical sensors, representing the first nucleus of the ORCA detector [1].

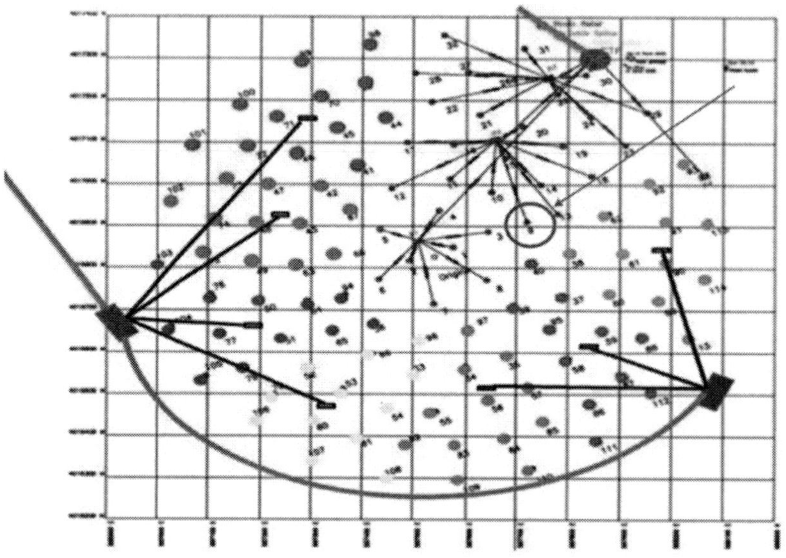

Figure 2. Footprint of one ARCA building block. The position of the first DU is indicated with the red arrow.

The volume of the ARCA Phase-1 detector is about 0.1 km^3 and, when completed, it will be the largest neutrino telescope in the Northern hemisphere.

DOMs and DUs integration sites have been already established. Two DUs have been already integrated and fully tested. One of them has been successfully deployed on December 3rd 2015 at the Italian site. The deployment, at 3500 m under the sea level smoothly validated the entire procedure and the functionality of the sea floor infrastructure. Data taking and analysis started a few hours after switching on the apparatus. In Fig. 2 the layout of the ARCA detector is shown and the position of the first DU indicated with a red arrow.

4. Physics performances

Two classes of events can be identified in a high energy neutrino telescope: track-like events and cascade-like events. The track-like events are generated mainly by muons that are produced in the matter inside or surrounding the detector through ν_μ CC interactions. The cascade-like events are produced in the matter near or inside the detector volume mainly through ν_e CC interactions and NC interactions of neutrinos of all flavours.These two categories of events produce very different time-space hit patterns in the detector. Dedicated reconstruction algorithms exploiting the characteristic hit patterns have been developed in order to get the direction and the energy of the neutrinos. The good optical properties of the sea-water and the directional response of the DOM make possible to achieve angular resolutions of about 0.2° for track events for ν_μ above 10 TeV and 2° for cascade events with ν_e energy above 50 TeV. A very promising track reconstruction algorithm, that include detailed knowledge of the detector, is in its final developing stage and the first tests show that the angular resolution can be improved to a level of 0.1°.

The primary strategy for the search of neutrino sources is the identification of up-going muons produced by neutrinos crossing the Earth. Thanks to the location in the Northern hemisphere, the field of view of the KM3NeT detector includes a large part of the Galactic plane. In this region of the sky the most promising neutrino sources are expected [15]. In addition to its favourable position, the good angular resolution makes the ARCA detector the ideal tool to search for neutrino sources. Its sensitivity

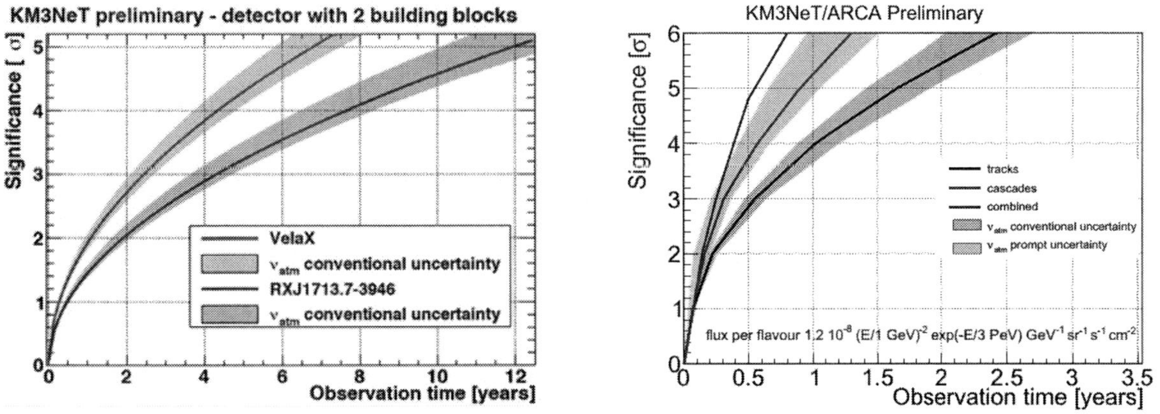

Figure 3. Significance as a function of the observation time for the detection of: Left – the Galactic sources RX J1713 and Vela-X. The bands represent the uncertainties due to the conventional component of the atmospheric neutrino flux (see Ref. [16]); Right – a diffuse flux of neutrinos corresponding to the signal reported by IceCube for the cascade channel (red line) and muon channel (black line). The black and red bands represent the uncertainties due to the conventional and prompt component of the neutrino atmospheric flux (see Ref. [18]). The blue line represents the results of the combined analysis.

to Galactic point-like source has been estimated by means of Monte Carlo simulations [16] looking for up-going muon neutrino tracks. Some Galactic sources that are the most intense high energy gamma ray sources and possible neutrino emitters, the Super Nova Remnants RXJ1713 and Vela Junior and the Pulsar Wind Nebula VelaX have been studied in particular. The RXJ1713 and the Vela X neutrino spectrum have been estimated considering the hypothesis of transparent sources and a fully hadronic production mechanism [16, 17]. The measured extension of the source in γ-rays has been taken into account in the simulation. Results from simulations show that a significance of about 3 sigma is reached in 3 years (Fig. 3 left).

The confirmation of the measurement of the diffuse flux by IceCube, with different methodology, improved resolution and complementary field of view, is an important goal for KM3NeT. Moreover, the better angular resolution to cascade events can constraint the region of the sky from where these neutrinos origin. The sensitivity of the ARCA detector to the neutrino flux measured by IceCube has been evaluated using both track-like events reconstructed up to 10° above the horizon and cascade-like events in the full angular range [18]. The results of this analysis are summarized in Fig. 3 right. A significance of 5 sigma can be reached in less than one year.

5. Summary

In this contribution the KM3NeT/ARCA project has been briefly presented, the main technical components of the detector and performance have been described. The construction of the detector, which in its final configuration consists of 2 building blocks of 115 DUs with a total instrumented volume of about 1 km^3, has begun. KM3NeT-Phase1 is fully funded and will consist of 24 strings and 8 towers, at the Italian site near Capo Passero (Sicily), and 7 strings at the French site, close to the Toulon coast.

In December 2015 the first DU of the ARCA detector has been deployed and the first data files have been analyzed. All the DOMs are functional.

The detector sensitivities to the neutrino fluxes from Galactic sources and to the diffuse flux measured by IceCube have been discussed for the full ARCA detector.

References

[1] D. Samtleben (KM3NeT collaboration), in this workshop (2015)
[2] M.G. Aartsen et al. (IceCube collaboration), Science **342**, 1242856 (2013)
[3] M.G. Aartsen et al. (IceCube collaboration), Astrophys. J. **809**, 98 (2015)
[4] M.G. Aartsen et al. (IceCube collaboration), PRL **114**, 171102 (2015)
[5] M. de Jong (KM3NeT collaboration), Proc. 33rd Int. Cosmic Ray Conference **1, 0891** (2013)
[6] R. Bruijn, D. van Eijk (KM3NeT collaboration), Proc. 34th Int. Cosmic Ray Conference **001, 1157** (2015)
[7] D. Vivolo (KM3NeT collaboration), in this workshop (2015)
[8] D. Real (KM3NeT collaboration), in this workshop (2015)
[9] S. Adrián-Martínez et al. (KM3NeT collaboration), Eur. Phys. J. **C 74**, 3056 (2014)
[10] M. Ageron et al., NIM **A 656**(1), 11 (2011), ISSN 0168-9002, `http://www.sciencedirect.com/science/article/pii/S0168900211013994`
[11] S. Adrián-Martínez et al. (KM3NeT collaboration) (2015), `arXiv:1510.01561[astro-ph.IM]`
[12] P. Kooijman (KM3NeT collaboration), Proc. 34th Int. Cosmic Ray Conference **001, 1173** (2015)
[13] E. de Wolf et al. (KM3NeT collaboration), NIM **A 725**, 241 (2013)
[14] S. Aiello et al., Astroparticle Physics **66**, 1 (2015), ISSN 0927-6505, `http://www.sciencedirect.com/science/article/pii/S0927650514001960`
[15] A. Kappes, J. Hinton, C. Stegmann, F. Aharonian, Astrophys. J. **656**, 870 (2007), erratum in Astrophys. J. 661 (2007) 1348, `arXiv:astro-ph/0607286`
[16] A. Trovato et al. (KM3NeT collaboration), in this workshop (2015)
[17] S. Kelner, F. Aharonian, V. Bugayov, Phys. Rev. **D 74**, 034018 (2006), erratum in Phys. Rev. **D 78** 039901 (2009), `arXiv:0802.0562 [physics.ins-det]`
[18] L. Fusco et al. (KM3NeT collaboration), in this workshop (2015)

EPJ Web of Conferences **116**, 11004 (2016)
DOI: 10.1051/epjconf/201611611004

Results from IceCube

Tyce DeYoung[a] for the IceCube Collaboration

Dept. of Physics and Astronomy, Michigan State University, 567 Wilson Rd., East Lansing, MI 48824, USA

Abstract. Data from the IceCube Neutrino Observatory have revealed the existence of a flux of high energy neutrinos of extraterrestrial origin, which is observed in a number of analyses spanning different energy ranges, fields of view, and neutrino flavors. The current data are consistent with an isotropic, equal-flavor flux described by a simple power law spectrum, but deviations from this simple model cannot yet be constrained with high precision. The existing observations in this area are reviewed, along with recent results on dark matter searches and observations of cosmic rays.

1. Introduction

The IceCube Neutrino Observatory is a gigaton-scale neutrino detector located at the U.S. Amundsen-Scott South Pole Station. Completed in 2011, it instruments a cubic kilometer of the Antarctic ice cap with 5,160 Digital Optical Modules (DOMs) consisting of 10" photomultiplier tubes and associated readout electronics, deployed at depths of 1,450 m to 2,450 m below the surface of the ice on 86 electrical cables, or strings. Most of these strings are deployed on a triangular grid with a characteristic spacing of 125 m, while eight of the strings form a denser array in the center of IceCube with a lower energy threshold, known as DeepCore. An additional 324 DOMs are installed in an array of 81 two-tank stations on the surface above the deep detector, known as IceTop. Detector operations have been extremely stable, with an overall duty factor of 99.6% since 2014. Subtracting maintenance periods and transient issues with individual DOMs, the 'clean' uptime of the detector is over 98%. No DOMs have failed during the past 2.5 years, and we project that more than 98% of the DOMs originally deployed will remain operational in 2020, ten years after completion.

The primary goal of IceCube is the detection of high energy neutrinos from astrophysical particle accelerators such as the sources of the cosmic rays. Secondary goals include cosmic ray studies, searches for dark matter and other exotica, and measurements of neutrino oscillations. Current and planned neutrino oscillation measurements are summarized elsewhere in these proceedings; highlights from the other areas are discussed here.

---

[a] e-mail: deyoung@pa.msu.edu

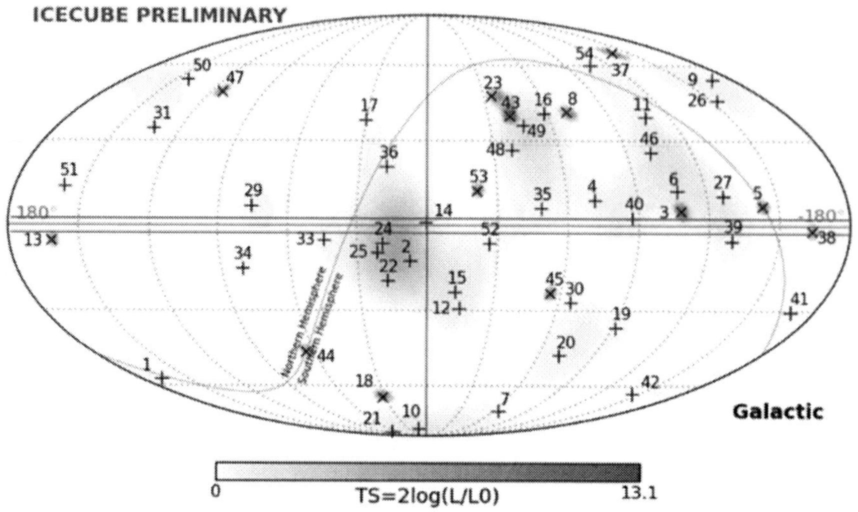

Figure 1. Arrival directions of the 52 high energy contained neutrino candidate events, in galactic coordinates. Crosses indicate cascade-like events while X's show tracks. The color scale indicates the significance of the event excess at each point, which depends on the reconstruction uncertainty of each event. Events 28 and 32 are identified as background and are omitted.

2. Observations of astrophysical neutrinos

The first astrophysical neutrinos observed with IceCube [1] were discovered as anomalous high-energy cascade-like events in a search for ultrahigh-energy neutrinos produced through the Berezinsky-Zatsepin effect [2]. Track-like events are those with a muon track long enough to be separated from the hadronic shower produced at a neutrino interaction vertex; at extremely high energies, τ leptons could also be classified as track-like. Cascade-like events include ν_e charged-current (CC) events, all neutral current (NC) events, and ν_τ CC events at energies low enough that the τ lepton decays before emerging from the vertex shower (unless the τ decays to μ).

A follow-up analysis designed to detect high energy neutrinos interacting with nucleons inside the IceCube volume, sensitive to neutrinos of all flavors and incident from all directions, yielded 28 neutrino candidates with energies above 30 TeV [3], observed in the data taken from May 2010 to May 2012. With the inclusion of two more years of data, a total of 54 event were observed in this analysis (two of which can be clearly identified as background), with expected backgrounds of 12.6 ± 5.1 atmospheric muons and $9.0^{+8.0}_{-2.2}$ atmospheric neutrinos. A purely atmospheric origin of these events is now excluded at a significance of 6.5σ. A sky map of those events is shown in Fig. 1.

Several other analyses [4, 5] focused on events interacting in or near IceCube have extended these observations to lower energies or increased the sensitivity to cascade-like events. In parallel, a separate analysis searching for high energy up-going muon tracks produced primarily by neutrinos interacting outside of IceCube also produced 3.7σ evidence for astrophysical neutrinos in two years of data [6]. The excess of through-going muons at high energies is shown in Fig. 2.

A wide range of explanations for the astrophysical neutrino flux have been proposed. The observation of high energy neutrinos well separated from the Galactic plane implies that at least some of the neutrinos originate in extragalactic sources. Gamma-ray bursts (GRBs) have long been considered candidate cosmic ray accelerators [7]. Present observations [8] constrain but do not rule out models which posit GRBs as the sources of the cosmic rays. However, the neutrino flux predicted by such models is around two orders of magnitude lower than the observed flux, as illustrated in Fig. 3 – GRBs may produce cosmic rays, but other sources dominate the astrophysical neutrino flux. Active galactic nuclei are

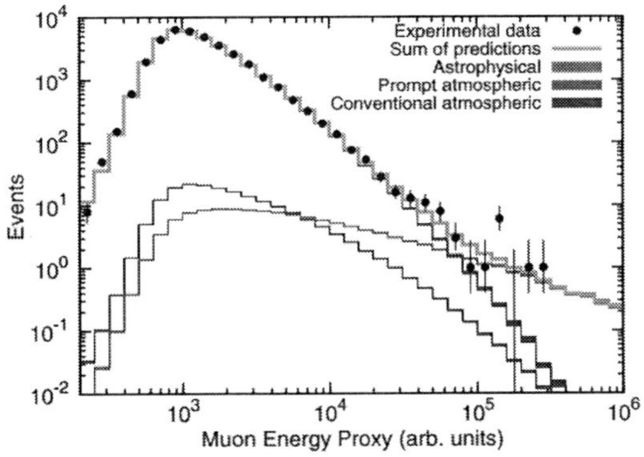

Figure 2. Spectrum of the up-going muon events, in arbitrary units. An excess above the atmospheric expectation is visible above 10^5. The green line indicates the best-fit power-law astrophysical neutrino spectrum, with a spectral index of -2.2.

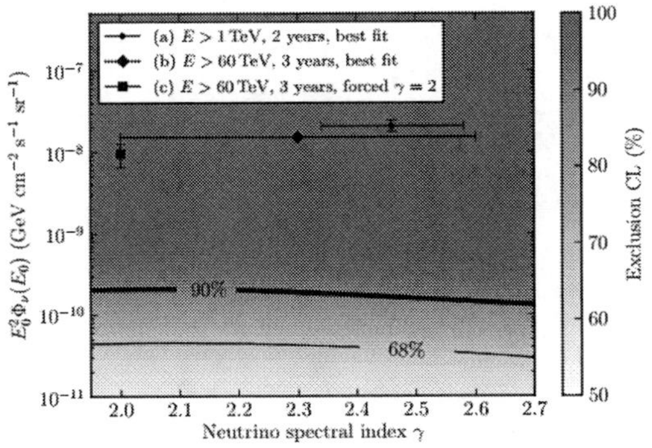

Figure 3. Upper limits on the ν flux from known GRBs normalized at $E_0 = 100$ TeV, assuming an unbroken power law spectrum, compared to measurements of the diffuse astrophysical neutrino flux [3, 4]. The observed flux does not appear to originate from GRBs.

another popular class of candidate cosmic ray sources. A stacking analysis of blazars detected by Fermi-LAT suggests that less than 30% of the observed neutrino flux comes from these sources. This limit on blazars is considerably more model-dependent, however, due to the inherent variability of these sources.

Observations from ANTARES limit the contribution from a single Galactic source, such as the Galactic center [9], but do not rule out the possibility of a population of Galactic sources or a spatially extended source associated with the Galaxy contributing to the observed flux. A weak correlation of neutrino arrival directions with the Galactic plane is observed, with a p-value of 2.5%, but only when the definition of the plane is extended to $\pm 7°$, substantially wider than would be expected in standard astrophysical models. At present, there is no compelling evidence for a Galactic contribution to the neutrino flux, but the possibility of a sub-dominant Galactic component cannot be ruled out.

A global fit to the data observed in the various IceCube analyses has been performed [10]. The observed rates of the different neutrino topologies are consistent with a $\nu_e : \nu_\mu : \nu_\tau = 1{:}1{:}1$ flavor ratio, although there is a substantial degeneracy between ν_e and ν_τ rates. A pure ν_e flux at the sources (e.g. only neutrons escaping) is disfavored at greater than 95% C.L., but either the generic 1:2:0 meson-decay flux or a purely ν_μ source flux (e.g. due to synchrotron cooling of muons in the sources) is allowed at 1σ. Although the fluxes observed by the various analyses are generally compatible, there is some weak tension between the spectra inferred from different data sets, with analyses sensitive primarily to higher energy muon neutrinos preferring harder spectra and those sensitive to lower energies and cascades

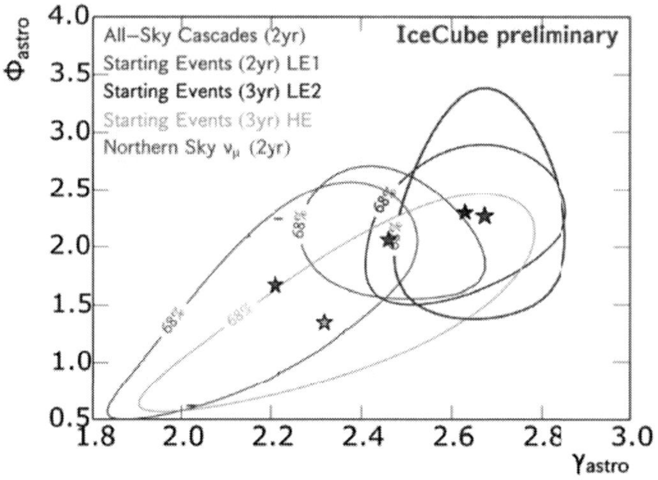

Figure 4. Fitted normalization (in units of 10^{-8} GeV cm^{-2} s^{-1} sr^{-1}) and spectral index from several measurements of the astrophysical neutrino flux targeting different neutrino flavors and energy ranges.

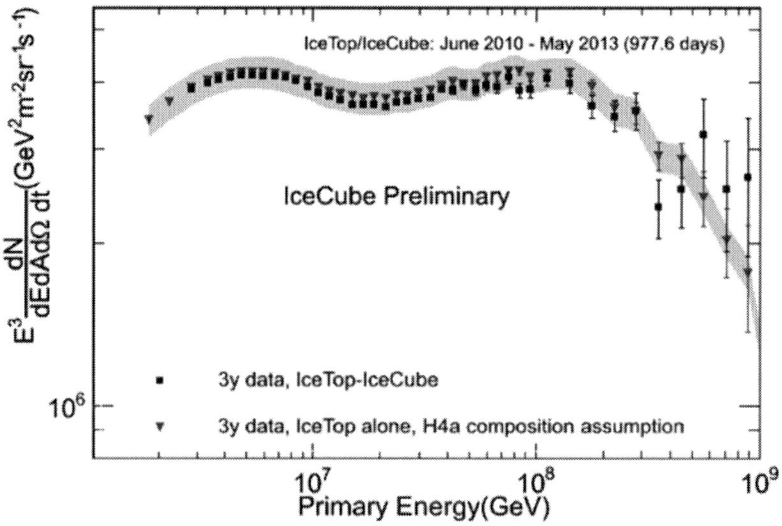

Figure 5. Comparison between the energy spectrum obtained by IceTop alone, with its 7% composition uncertainty indicated by the shaded band, and the energy spectrum obtained by the IceTop-IceCube analysis.

preferring softer ones, as indicated in Fig. 4. This tension is not statistically significant, but if confirmed with additional data, it could be due to spectral features, anisotropies in the flux (e.g. due to a Galactic component), or differences between neutrino flavors due e.g. to source dynamics. It is quite possible that the flux discovered by IceCube is produced by multiple types of sources, but considerably more data will be required to explore such a complicated scenario.

3. Measurements of cosmic rays

The IceTop extensive air shower (EAS) array installed at the surface above the deep IceCube detector measures the cosmic ray spectrum above the knee. The spectra measured by IceTop in two parallel analyses are shown in Fig. 5. The first analysis combined IceTop and IceCube data, enabling a unique

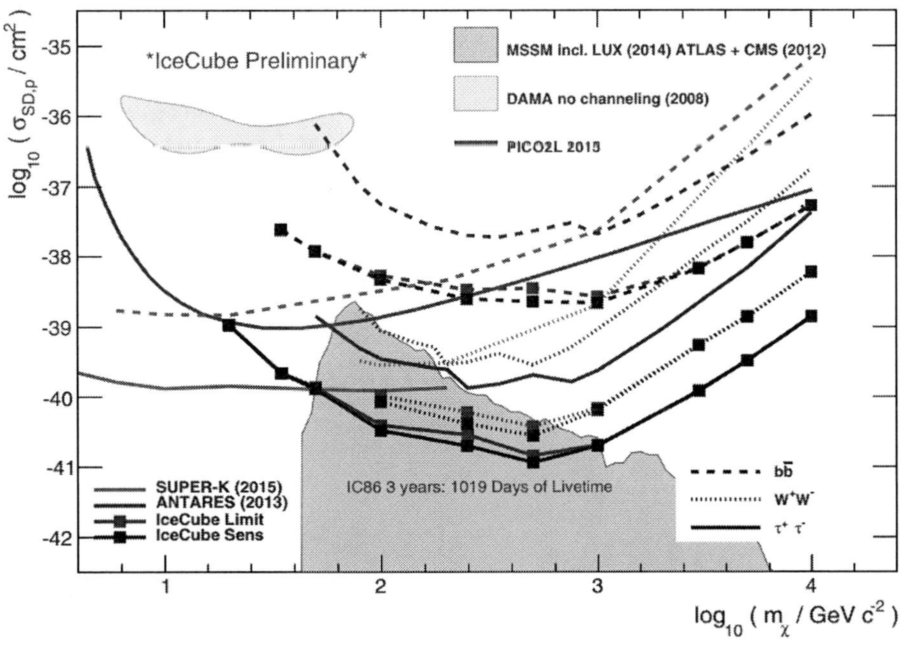

Figure 6. Sensitivities and upper limits on the spin-dependent WIMP-nucleon scattering cross section from a three-year IceCube search for solar dark matter annihilation. Black lines indicate IceCube sensitivities and red lines the actual upper limits, while blue and purple indicate limits from other indirect searches. In all cases, dotted, dashed, and solid lines indicate the assumed annihilation channel. The shaded regions indicate the allowed regions corresponding to the DAMA [20] measurement and the MSSM including direct detection and collider constraints on both SI and SD scattering.

measurement of the chemical compositions by comparison of the muonic and electromagnetic energy deposited in each shower. The measured composition becomes gradually heavier up to approximately 100 PeV, where there is a change in the trend. The second analysis used IceTop data alone, assuming a composition following the H4a model of Gaisser [11], which shares these features. The inferred spectra are very consistent, and clearly show a hardening of the spectrum around 20 PeV followed by a spectral break slightly above 100 PeV.

IceTop and IceCube data have also been used to measure the anisotropies in the cosmic ray flux first reported by EAS arrays [12, 13] and Super-Kamiokande [14] in the northern hemisphere. The origin of these anisotropies remains uncertain. IceCube and IceTop provide a unique view of these anisotropies in the southern sky, as well as the statistical power to investigate their energy dependence. At middle latitudes, the IceCube/IceTop observations [15, 16] agree well with those from the north. The anisotropies extend at least to several PeV, and the pattern shifts as the energy increases from the 10 TeV scale to the PeV scale; regions of the sky which show a relative excess at lower energies have a deficit at higher energies and vice versa.

4. Searches for dark matter

In addition to observations of astrophysical neutrinos and cosmic rays, IceCube has a broad program of indirect searches for dark matter. The general strategy is to search for neutrinos produced as byproducts of the annihilation or decay of dark matter that has accumulated in nearby gravitational wells. These include the wells of the Galaxy, local dwarf galaxies, the Earth, and the Sun.

Solar dark matter searches are of particular interest because of the Sun's size, proximity, and light nuclear composition. Dark matter must scatter off regular matter to become trapped, and in general the scattering cross section may have both spin-independent (SI) and spin-dependent (SD) terms. For the latter, underground direct dark matter detectors lose the benefit of coherent scattering off heavy nuclei, and IceCube places some of the most stringent limits, particularly for heavier dark matter.

Two parallel searches for neutrinos from solar WIMP annihilation have been conducted with the IceCube-DeepCore detector. The results of the analyses are similar; limits from one, using three years of data from the full detector, are shown in Fig. 6. The limits depend strongly on the assumed annihilation channel. Three scenarios are shown in Fig. 6, where the WIMPs are assumed to annihilate exclusively to $b\bar{b}$, W^+W^-, and $\tau^+\tau^-$, respectively. The latter two channels produce harder neutrino spectra which are more favorable for IceCube detection, and IceCube limits on these models provide interesting constraints on supersymmetric models not probed by current direct detection and collider searches. For WIMP masses above roughly 50 GeV, IceCube limits are more stringent than those from ANTARES [17], Super-Kamiokande [18], and (for several channels) PICO2L [19].

5. Outlook

The IceCube detector continues to take high-quality data, with excellent stability and longevity. The existence of a high-energy astrophysical neutrino flux has been confirmed via several analyses targeting different energy ranges, fields of view, and neutrino flavors. These observations are consistent and there is no compelling evidence for deviations from the simplest scenario of an isotropic, equal-flavor power-law flux. However, constraints on more complex scenarios are relatively weak and additional data may reveal additional details. IceCube also provides interesting measurements of the spectrum and composition of the cosmic rays and limits on dark matter scenarios.

References

[1] M. G. Aartsen et al. [IceCube Collaboration], Phys. Rev. Lett. **111**, 021103 (2013)
[2] V. S. Berezinsky and G. T. Zatsepin, Phys. Lett. B **28**, 423 (1969)
[3] M. G. Aartsen et al. [IceCube Collaboration], Science **342**, 1242856 (2013)
[4] M. G. Aartsen et al. [IceCube Collaboration], Phys. Rev. D **91**, no. 2, 022001 (2015)
[5] M. G. Aartsen et al. [IceCube Collaboration], Phys. Rev. Lett. **114**, no. 17, 171102 (2015)
[6] M. G. Aartsen et al. [IceCube Collaboration], Phys. Rev. Lett. **115**, no. 8, 081102 (2015)
[7] For a recent review, see K. Asano and K. Murase, Adv. Astron. **2015**, 568516 (2015)
[8] M. G. Aartsen et al. [IceCube Collaboration], Astrophys. J. **805**, no. 1, L5 (2015)
[9] S. Adrián-Martínez et al. [ANTARES Collaboration], Astrophys. J. **786**, L5 (2014)
[10] M. G. Aartsen et al. [IceCube Collaboration], Astrophys. J. **809**, no. 1, 98 (2015)
[11] T. K. Gaisser, Astropart. Phys. **35**, 801 (2012)
[12] M. Amenomori [Tibet AS-γ Collaboration], Science **314**, 439 (2006)
[13] A. A. Abdo et al. [Milagro Collaboration], Phys. Rev. Lett. **101**, 221101 (2008)
[14] G. Guillian et al. [Super-Kamiokande Collaboration], Phys. Rev. D **75**, 062003 (2007)
[15] R. Abbasi et al. [IceCube Collaboration], Astrophys. J. **746**, 33 (2012)
[16] M. G. Aartsen et al. [IceCube Collaboration], Astrophys. J. **765**, 55 (2013)
[17] S. Adrián-Martínez et al. [ANTARES Collab.], J. Cosmol. Astropart. Phys. **1311**, 032 (2013)
[18] K. Choi et al. [Super-Kamiokande Collaboration], Phys. Rev. Lett. **114**, no. 14, 141301 (2015)
[19] C. Amole et al. [PICO Collaboration], Phys. Rev. Lett. **114**, no. 23, 231302 (2015)
[20] R. Bernabei et al. [DAMA Collaboration], Eur. Phys. J. C **56**, 333 (2008)

EPJ Web of Conferences **116**, 11005 (2016)
DOI: 10.1051/epjconf/201611611005

Baikal-GVD: Results, status and plans

A.D. Avrorin[1], A.V. Avrorin[1], V.M. Aynutdinov[1], R. Bannash[7], I.A. Belolaptikov[2], D.Yu. Bogorodsky[3], V.B. Brudanin[2], N.M. Budnev[3], I.A. Danilchenko[1], G.V. Domogatsky[1], A.A. Doroshenko[1], A.N. Dyachok[3], Zh.-A.M. Dzhilkibaev[1], S.V. Fialkovsky[5], A.R. Gafarov[3], O.N. Gaponenko[1], K.V. Golubkov[1], T.I. Gress[3], Z. Honz[2], K.G. Kebkal[7], O.G. Kebkal[7], K.V. Konischev[2], A.V. Korobchenko[2], A.P. Koshechkin[1], F.K. Koshel[1], A.V. Kozhin[4], V.F. Kulepov[5], D.A. Kuleshov[1], V.I. Ljashuk[1], M.B. Milenin[5], R.A. Mirgazov[3], E.R. Osipova[4], A.I. Panfilov[1], L.V. Pan'kov[3], E.N. Pliskovsky[2], M.I. Rozanov[6], E.V. Rjabov[3], B.A. Shaybonov[2], A.A. Sheifler[1], M.D. Shelepov[1], A.V. Skurihin[4], A.A. Smagina[2], O.V. Suvorova[1], V.A. Tabolenko[3], B.A. Tarashansky[3], S.A. Yakovlev[7], A.V. Zagorodnikov[3], and V.L. Zurbanov[3]

[1] Institute for Nuclear Research, 60th October Anniversary pr. 7A, Moscow 117312, Russia
[2] Joint Institute for Nuclear Research, Dubna 141980, Russia
[3] Irkutsk State University, Irkutsk 664003, Russia
[4] Skobeltsyn Institute of Nuclear Physics MSU, Moscow 119991, Russia
[5] Nizhni Novgorod State Technical University, Nizhni Novgorod 603950, Russia
[6] St. Petersburg State Marine University, St. Petersburg 190008, Russia
[7] EvoLogics GmbH, Berlin, Germany

Abstract. The future next-generation neutrino telescope Baikal-GVD will be a km3-scale array aimed at the detection of astrophysical neutrino fluxes. It will have modular structure and consist of functionally independent sub-arrays – clusters of strings of optical modules. The prototyping phase of the project has been concluded in 2015 with the deployment of the first cluster of Baikal-GVD in Lake Baikal. We discuss the current status and perspectives of the Baikal-GVD project.

1. Introduction

The next generation neutrino telescope Baikal-GVD is a kilometer-scale deep underwater Cherenkov detector currently under construction in Lake Baikal [1]. Baikal-GVD is designed to detect astrophysical neutrino fluxes at energies from a few TeV up to the highest energies of 10^6 TeV and particularly mapping the high-energy neutrino sky in the Southern Hemisphere including the region of the Galactic center.

The site chosen for the experiment is in the southern basin of Lake Baikal. Here, the combination of hydrological, hydro-physical, and landscape factors is optimal for deployment and operation of the neutrino telescope. The water depth is about 1360 m at distances beginning from about three kilometers

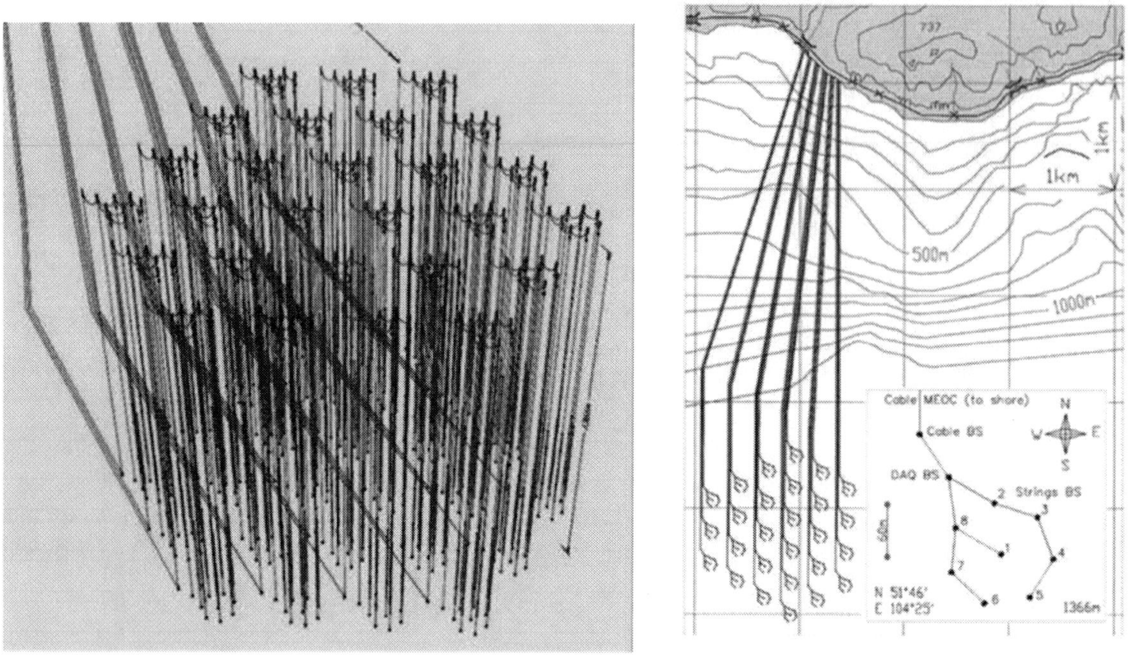

Figure 1. Left: Artist's view of the GVD-telescope. Right: Layout of GVD. Inner box: layout of a single GVD-cluster.

from the shore. The water transparency is characterized by an absorption length of about 20–25 m and a scattering length of 30–50 m [2]. The water luminescence is moderate at the detector site.

Baikal-GVD is a 3-dimensional lattice of optical modules (OMs) – photomultiplier tubes housed in transparent pressure spheres. The OMs are arranged at vertical load-carrying cables to form strings. The telescope has a modular structure and consists of clusters of strings – functionally independent sub-arrays, which are connected to shore by individual electro-optical cables (see Fig. 1).

The first prototype of the GVD electronics was tested in Lake Baikal in April 2008. It was a reduced-size section with six optical modules. This detection unit provided the possibility to study basic elements of the future detector: new optical modules and the Flash Analog-to-Digital Converter (FADC) based measuring system [3]. During the next 2 years different versions of prototype string were tested in Lake Baikal as a part of the old NT200+ detector. The operation of these prototype strings in 2009 and 2010 allowed a first assessment of the DAQ performance [4, 5].

The prototyping/early construction phase of the Baikal-GVD project, which aims at construction and operation of the first demonstration cluster of GVD, has been started in 2011 by deployment and operation of autonomous engineering arrays [6]. The next important step in the realization of the GVD project was made in 2013 by deployment of an enlarged engineering array – the first stage of the first GVD-cluster, which comprised 72 OMs arranged on three 345 m long strings and operated from April 2013 to February 2014 [1, 7]. In 2014 the second stage of the demonstration cluster, which comprised 112 OMs arranged on five strings, as well as an instrumentation string with array calibration and environment monitoring equipment, was deployed. It was operated during 2014 in data taking and testing modes. One of the main goals of the operation in testing mode was to study the quality of in-situ calibration procedures. The performance and quality of these procedures have been verified by reconstructing position and intensity of a distant calibration laser source. The position of the laser source was reconstructed using the arrival times of photons detected by the PMTs, taking into account the timing calibration of the PMTs by LEDs. Results of the reconstruction were compared with the laser

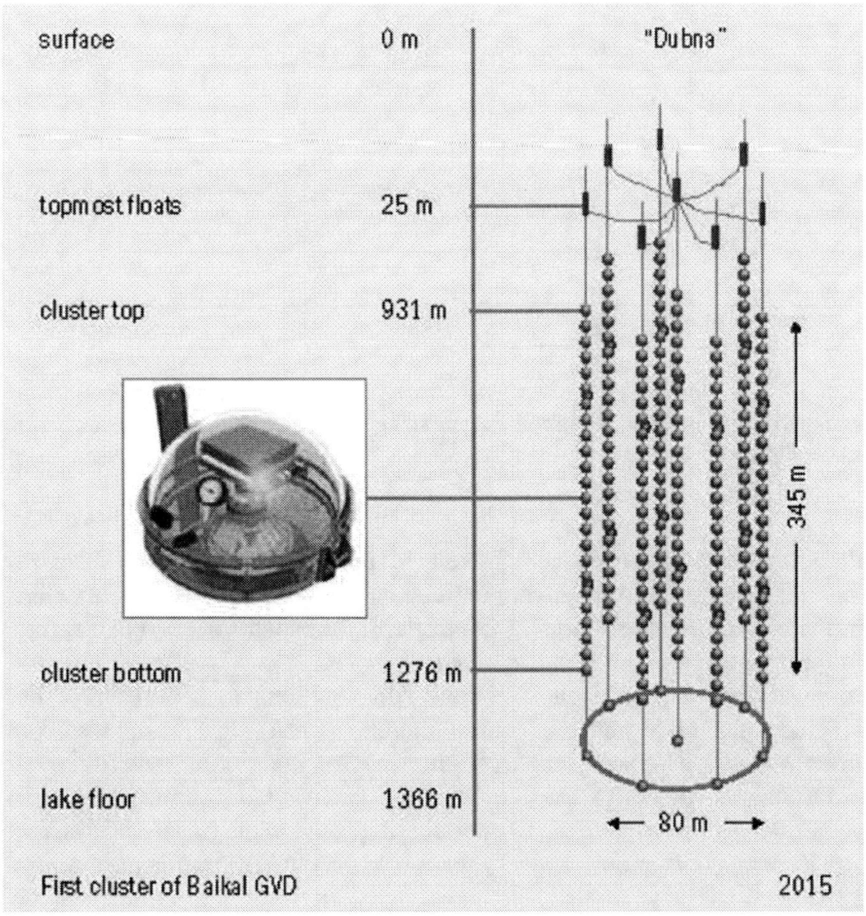

Figure 2. Schematic view of the "Dubna" cluster, the first cluster of the Baikal-GVD neutrino telescope.

coordinates obtained by the acoustic positioning system [8]. The obtained results prove the expected quality of the array calibration procedures.

2. The first GVD-cluster

In April 2015 the first demonstration cluster of Baikal-GVD named "Dubna" was deployed and commissioned in Lake Baikal (see Fig. 2). It encloses 1.7 Megatons of the clear waters of Lake Baikal. It comprises a total of 192 optical modules arranged at eight 345 m long strings, as well as an acoustic positioning system and an instrumentation string with equipment for array calibration and monitoring of environmental parameters. An artist's view of the cluster is shown in Fig. 2. Each string comprises 24 OMs spaced by 15 m at depths of 900 m to 1250 m below the surface. In 2015 seven side strings have been located at a reduced radius of 40 m around a central one (compared to 60 m for the baseline configuration). The reason is to increase the sensitivity to low-energy atmospheric muons and neutrinos which are used for array calibration. In the 2016, strings will be moved to the baseline distances.

Each optical module consists of a pressure-resistant glass sphere of 43.2 cm diameter which holds the OM electronics and the PMT which is surrounded by a high permittivity alloy cage for shielding it against the Earth magnetic field. A large photomultiplier tube Hamamatsu R7081-100 with a 10-inch hemispherical photocathode and quantum efficiency up to 35% has been selected as light sensor.

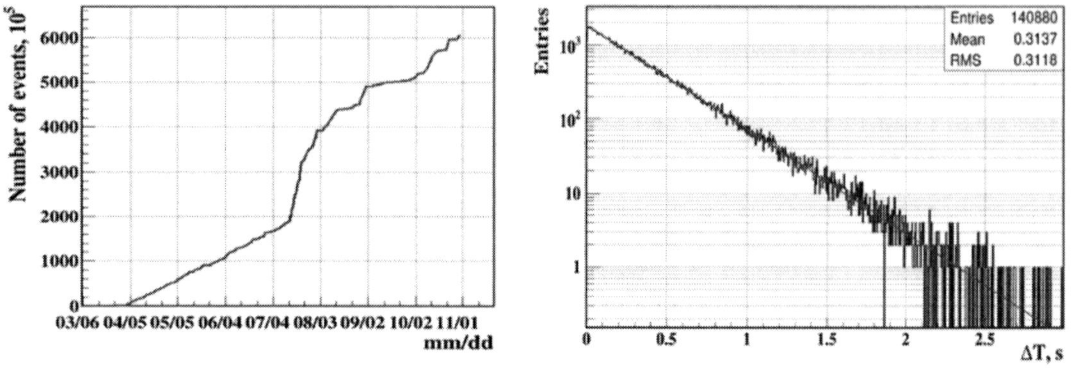

Figure 3. Left: Integrated number of recorded events since April 2015. Right: The time difference between subsequent events is shown for one run.

Besides the PMT, an OM comprises a high voltage power supply unit (HV), a fast two-channel preamplifier, and a controller. For time and amplitude calibration of the measuring channel, two LEDs are installed in the optical module. The OM controller is intended for HV control, monitoring of the PMT noise and for time and amplitude calibration [9].

The OMs along a string are combined in sections – the functional detection unit of the telescope [10]. Each section includes 12 OMs and the central module (CeM). PMTs signals from all OMs are transmitted through 90 m long coaxial cables to the CeM of the section, where they are digitized by custom-made ADC boards with 200 MHz sampling rate. The waveform information from all measuring channels of the section is transferred to the Master board located in the CeM. The Master board provides readout of the ADC data, connection via local Ethernet to the cluster DAQ-center, control of the section operation and the section trigger logic. A request analyzer forms the section trigger request (local trigger) on the basis of channel requests L (low channel threshold, ~0.3 p.e.) and H (high threshold, ~3 p.e.) from 12 ADC channels. This unit contains a programmable coincidence matrix (12H×12L), which provides a simple way to generate the section trigger request. The basic trigger modes are: (A) coincidences of > N L-requests within a selectable time window (L > N-trigger), or (B) coincidences of L and H requests from any neighbouring OMs within a section (L&H-trigger). A request of the section trigger is transferred from the Master board through a string communication module (CoM) to the cluster DAQ-center, where a global trigger for all sections is generated. The cluster DAQ-center is located at the 30 m depth on central string. Data from the strings are transferred through DSL-modem Ethernet channel to the cluster center. The data transmission between the cluster DAQ-center and shore station is provided through optical fiber lines of about 6 km length.

The cluster has been operating from April 2015 in several testing and data taking modes. A total of 6.1×10^8 events has been recorded from April till November 2015. Figure 3 (left) gives an indication of the data taking efficiency during 2015. The change of the slope of the curve in Fig. 3 (left) is caused by the lowering of the used trigger thresholds in July. Shown in Fig. 3 (right) is the time di?erence between subsequent events for one run. The exponential behaviour of this distribution is consistent with the expectation for randomly distributed experimental events and illustrates the quality of the data. As an illustration, one event observed in June 2015 is shown in Fig. 4. The photon distribution of the event is consistent with the pattern of Cherenkov photons from a down-going high-energy muon.

The first cluster of Baikal-GVD has the potential to record astrophysical neutrinos with a flux value measured recently by IceCube [11, 12]. The search for high-energy neutrinos is based on the selection of cascade events generated by neutrino interactions in the sensitive volume of array. After applying an iterative procedure of vertex reconstruction followed by the rejection of hits contradicting

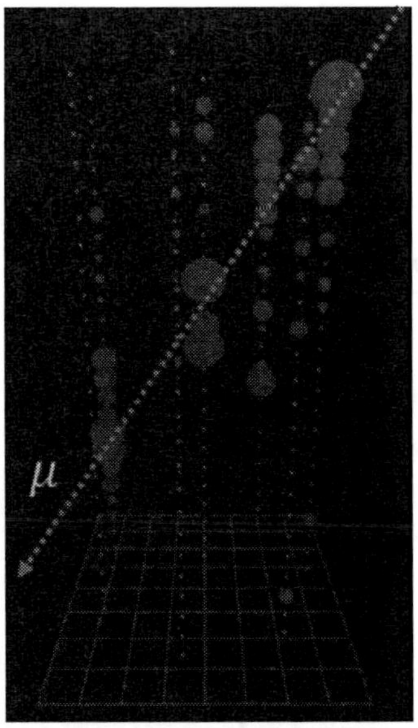

Figure 4. Event view of the high-energy down-going muon event observed in June 2015. Each sphere represents a OM. Colors represent the arrival times of the photons where red indicates early and blue late times. The size of the spheres is a measure for the recorded number of photoelectrons. The dotted line denotes the reconstructed track.

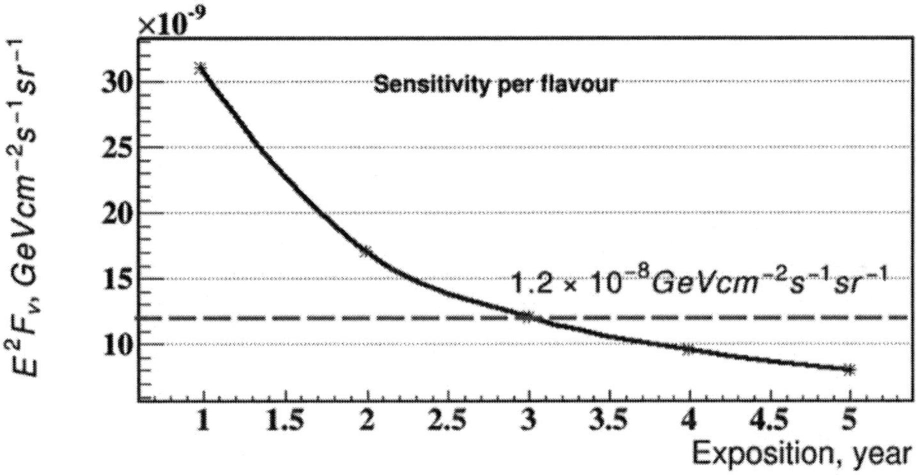

Figure 5. Cluster sensitivity for one flavor neutrino flux with an E^{-2} spectrum as function of the observation years. The long-dashed line indicates the one-flavor neutrino flux value obtained by IceCube.

the cascade hypothesis on each iteration stage, events with a final multiplicity of hit OMs $N_{hit} > 20$ are selected as high-energy neutrino events. The accuracy of shower direction reconstruction is about 4 degree (median value), which is substantially better than the 10–15 degree accuracy for IceCube [11]. We expect about one event per year with $E_{sh}>100$ TeV from an all-flavor astrophysical flux with the normalization $E^2F=3.6\cdot10^{-8}$ GeV cm^{-2} s^{-1} sr^{-1} in the cluster, compared to about 10 events in IceCube. The preliminary estimation of one cluster sensitivity for one flavor neutrino flux with an E^{-2} spectrum and flavor ration 1:1:1 for all flavor flux as function of the observation years is shown in Fig. 5. Three year exposition will result in a sensitivity corresponding to the level measured by IceCube.

The baseline configuration of the Baikal-GVD cluster will comprise 288 OMs arranged at 8 strings. Each string will consist of 3 sections (12 OMs per section) arranged at depths from 700 m to 1250 m. In 2016 the demonstration cluster will be upgraded to the baseline configuration with 8 new sections with a total of 96 OMs. By 2020, Baikal-GVD is set to consist of 8 baseline clusters covering a total volume of about 0.4 km^3(GVD phase-1).

3. Conclusion

The ambition of the Baikal collaboration is the construction of a km^3-scale neutrino telescope – the Gigaton Volume Detector in Lake Baikal. During the R&D phase of the GVD project in 2008–2010, its basic elements – new optical modules, FADC readout units, underwater communications and trigger systems – have been developed, produced and tested in situ by long-term operating prototype strings. The prototyping phase of the GVD project has been performed since April 2011 and aimed at comprehensive in-situ tests of all elements and systems of the future detector. This phase has been concluded with deployment and commissioning of the first demonstration cluster of GVD in April 2015. For the years 2016-2020, the deployment of an array with about 2000 OMs is envisaged.

This work was supported by the Russian Found for Basic Research (grants 13-02-12221, 14-02-00175, 14-02-00972).

References

[1] A. Avrorin et al., The prototyping/early construction phase of the BAIKAL-GVD project, NIM **A742**, 82
[2] A. Avrorin et al., Asp-15 – A stationary device for the measurement of the optical water properties at the NT200 neutrino telescope site, Nucl. Instr. Meth. **A693**, 186–194
[3] V. Aynutdinov et al., The prototype string for the km3-scale Baikal neutrino telescope, NIM **A602**, 227
[4] V. Aynutdinov et al., The BAIKAL neutrino experiment, NIM **A626**, 13
[5] V. Aynutdinov et al., The gigaton volume detector in Lake Baikal, NIM **A639**, 30
[6] A. Avrorin et al., Current status of the Baikal-GVD project, NIM **A725**, 23
[7] A. Avrorin et al., Status and recent results of the Baikal-GVD project, Phys. of Part. and Nucl. **46**, 211
[8] A. Avrorin et al., A hydroacoustic positioning system for the experimental cluster of the cubic-kilometer-scale neutrino telescope at Lake Baikal, Instr. Exp.Tech. **56**, 449
[9] A. Avrorin et al., The optical module of the Baikal-GVD neutrino telescope, these Proceedings
[10] A. Avrorin et al., Data acquisitions system of the NT1000 Baikal neutrino telescope, Instr. and Exp. Tech. **57**, 262
[11] M. G. Aartsen et al., Evidence for high-energy extraterrestrial neutrinos at the IceCube detector, Science **342**, 1242856
[12] M. G. Aartsen et al., Observation of high-energy astrophysical neutrinos in three years of IceCube data, Phys. Rev. Lett. **113**, 101101

EPJ Web of Conferences **116**, 11006 (2016)
DOI: 10.1051/epjconf/201611611006

Results from the ANTARES neutrino telescope

M. Spurio, on behalf of the ANTARES Collaboration[1,2,a]

[1] Dipartimento di Fisica e Astronomia Università di Bologna, Italy
[2] INFN- Sez. di Bologna, Viale Berti Pichat 6/2, 40127 Bologna, Italy

Abstract. ANTARES is the largest neutrino telescope in the Northern hemisphere, running in its final configuration since 2008. After the discovery of a cosmic neutrino diffuse flux by the IceCube detector, the search for its origin has become a key mission in high-energy astrophysics. The ANTARES sensitivity is large enough to constrain the origin of the IceCube excess from regions extended up to 0.2 sr in the Southern sky. The Southern sky has been studied searching for point-like objects, for extended regions of emission (as the Galactic plane) and for signal from transient objects selected through multimessenger observations. Upper limits are presented assuming different spectral indexes for the energy spectrum of neutrino sources. In addition, ANTARES provides results on studies of the sky in combination with different multimessenger experiments, on atmospheric neutrinos, on the searches for rare particles in the cosmic radiation (such as magnetic monopoles and nuclearites), and on Earth and Sea science. Particularly relevant are the searches for Dark Matter: the limits obtained for the spin-dependent WIMP-nucleon cross section overcome that of existing direct-detection experiments. The recent results, widely discussed in dedicated presentations during the 7th edition of the Very Large Volume Neutrino Telescope Workshop (VLVνT-2015), are highlighted in this paper.

1. Introduction

The ANTARES underwater neutrino telescope is located 40 km off the coast of Toulon, France and anchored to the seabed at a depth of 2.5 km. It consists of an array of 885 photomultiplier tubes (PMTs) of 10 inches covering an instrumented volume of approximately 0.01 km^3, and it is designed primarily to search for charged-current interactions of neutrinos with $E_\nu > 100$ GeV. Results from different analyses are reported here and in [1]. These include several measurements which are used to constrain both point-like (Sect. 2) and extended (Sect. 3) origins of the astrophysical flux observed by IceCube [2]; a new reconstruction of showers which allows a point-source search with cascade events (Sect. 4); limits on indirect searches for dark matter (Sect. 5).

[a]e-mail: spurio@bo.infn.it

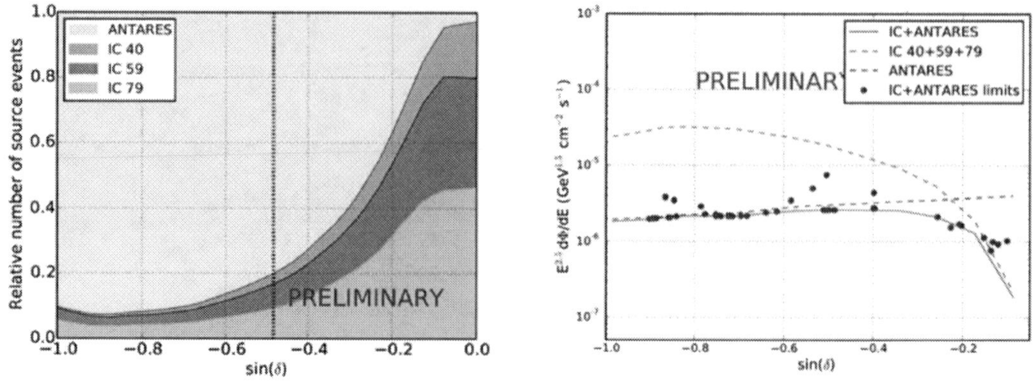

Figure 1. Left: Fractional contributions of each data set to the total number of signal events passing cuts in the joint ANTARES-IceCube analysis, for sources with an $E^{-2.5}$ spectrum, as a function of sinus of the declination δ. Right: Sensitivities (lines) and limits (dots) to an $E^{-2.5}$ flux, using ANTARES (blue), IceCube (red), and combined (green) data, as a function of $\sin\delta$.

2. Searches for astrophysical neutrino point sources

Astrophysical point-like sources of neutrinos can be individuated looking for an excess of muons from the interactions of ν_μ in the proximity of the detector. The high rate of downgoing muons from the interactions of cosmic rays (CRs) in the Earth's atmosphere restricts such searches to events coming from below, or only a few degrees above the horizon. The primary background to such searches is due to atmospheric neutrinos and those few atmospheric muons mis-reconstructed as up-going. The long scattering length of light in seawater provides an excellent directional resolution on the ν_μ of $\sim 0.4°$ for an E^{-2} source [3]. This allows a strong suppression of both backgrounds in the signal direction, and a correspondingly good sensitivity to neutrino sources located in the Southern Hemisphere.

A joint ANTARES and IceCube search for a neutrino excess from selected sources in the Southern hemisphere is detailed in [4]. The relative number of source events expected to be present in each data set is shown in Fig. 1 (left) for an $E^{-2.5}$ spectrum, the current best-fit to the cosmic neutrino flux from IceCube [2]. The ANTARES contribution is dominant for declination $< -15°$. In fact, ANTARES is more sensitive to tracks produced by (relatively) low-energy ν_μ, while IceCube requires high-energy events to distinguish them from the huge background due to atmospheric muons. The overall sensitivities of the two detectors is also a function of the background rates, and of the angular and energy resolutions, which are not shown. The results of the combined search are shown in Fig. 1 (right), for an $E^{-2.5}$ source spectrum. No significant cluster is found and the combined analysis improves limits set by each experiment.

The cosmic ν signal observed by IceCube is due to high-energy events starting in the fiducial volume. Most of these events (shortened as HESE) are showers for which the angular determination is poor (10–20 degrees). No significant fluctuation over a diffuse flux has been observed so far. However, the clustering of some events is intriguing, suggesting the possibility of a Galactic component in the cosmic signal [5]. In [6], it was argued that the HESE cluster near the Galactic plane [7] could be due to a single source, not identified because of the poor angular resolution. ANTARES searched for a possible excess in a wide region near the Galactic center, without any positive result. These limits the flux of such a source as a function of spectral index, shown by the solid lines in Fig. 2.

AGN have long been proposed as a source of high-energy CRs and, hence, neutrinos. Blazars (AGN with jets pointed towards the Earth), exhibit bright flares which dominate the extragalactic γ-ray sky observed by Fermi-LAT and IACTs. Using multi-wavelength observations, several bright blazars have

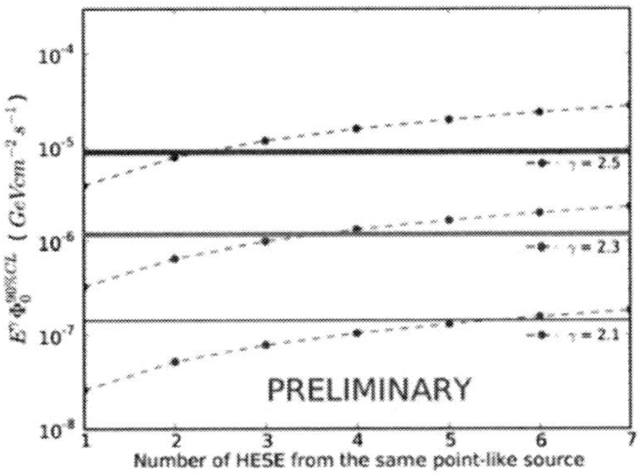

Figure 2. ANTARES limits (solid lines) at 90% C.L. on the contribution of point-like sources to the IceCube HESE sample for different spectral indices, shown for a source at $\delta = -29°$. These are compared to (dashed lines) the flux required to produce a given expected number of HESE, according to the computation in [5]. The range where the latter is greater than the former is excluded. ANTARES can thus rule out any single point-source of neutrinos in the region of the Galactic Centre with spectral index of -2.5 as having a flux corresponding to more than 2 HESE.

been reported by the TANAMI collaboration [8] to lie within the 50% error bounds of the reconstructed arrival directions of the PeV-scale events IC 14 and IC 20 observed by IceCube. ANTARES observes signal-like events from the two brightest blazars in the field of IC 20 [9], although this is also consistent with background fluctuations. The highest-energy "Big Bird" event (IC 35) was detected during an extremely bright flare from the blazar PKS B1424-418, which lies within the 50% error region of the IC 35 arrival direction. No ANTARES event was found from this source during the flaring period, constraining severely the model in [8].

Another analysis [10] targets a sample of 41 blazar flares observed by Fermi-LAT and 7 by the IACTs. The lowest pre-trial p-value of 3.3% was found for the blazar 3C 279, which comes from the coincidence of one event with a 2008 flare. However, the post-trial p-value is not significant. Similar methods were also used to search for neutrino emission during the flares from galactic x-ray binaries. A total of 34 x-ray- and γ-ray-selected binaries were studied, with no significant detections, allowing some of the more optimistic models for hadronic acceleration in these sources to be rejected at 90% C.L. (paper in preparation). Independent searches for spatial (through two-point correlation method) and temporal collective effects in the ANTARES data have been presented by R. Gracia and A. Coleiro in [1] and by A. Kouchner at this conference.

Under particular conditions, a dedicated program (denoted as TAToO) sends an alert message to trigger robotic optical telescopes, and/or the Swift-XRT. The very short alert-generation time (a few seconds) and half-sky simultaneous coverage of ANTARES makes it ideal for detecting transient signals. Results from 42 optical and 7 x-ray alerts occurred from 2011 and the beginning of 2015 are reported in [11]. No associated transient event was detected, and this non-observation was used to place limits on the astrophysical origin of the detected neutrinos. It should be mentioned that on September 1st, 2015, the XRT onboard Swift followed an ANTARES high-energy neutrino, detecting an uncatalogued x-ray source at 8 arcmin from the neutrino direction [12]. Different multi-wavelength observations (from IR, optic, radio, x-ray and TeV devices) have followed the alerts. The x-ray follow-up of this source during five days has shown an x-ray flare of few days large. Optical data have indicated a bright star in the same location of the XRT source without significant variability. Additional multi-wavelength data have permitted to identify it as a young stellar object. There is therefore a very low probability that the x-ray source is associated with the observed neutrino.

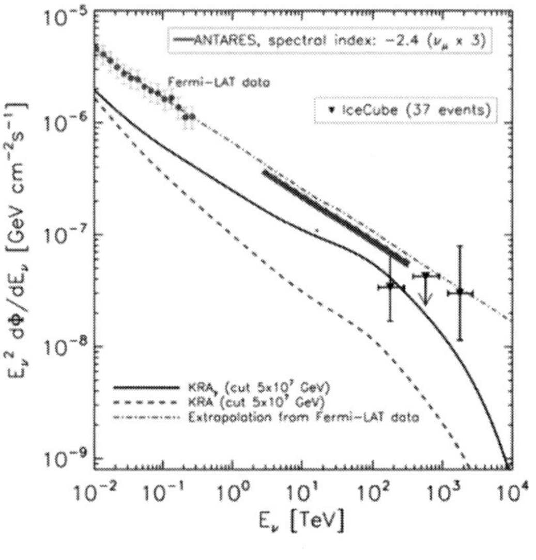

Figure 3. ANTARES upper limit for the flux of neutrinos from the Galactic Center, $|l| < 40°$ and $|b| < 3°$ (see L. Fusco in [1]). The result is compared to the expected neutrino flux (dot-dashed line) extrapolated from the Fermi-LAT diffuse flux up to IceCube energies [2] and to the neutrino expectations as computed in [14] in the same Galactic Plane region.

3. Searches for extended sources in the Southern sky

Several extended galactic regions (Fermi Bubbles, Galactic Plane, and the IceCube cluster) have been proposed as hadronic acceleration sites. Some models involve CRs that produce pions and, hence, neutrinos from collisions with the interstellar medium during propagation in our galaxy. Some evidence for these processes comes from observations by Fermi-LAT. In particular, the Fermi Bubbles are giant regions of γ-ray emission extending out of the galactic center, and are proposed as hadronic acceleration sites with neutrinos expected from p-p collisions. ANTARES searches for an excess in the neutrino flux from these regions use "on-zones" defined by specific templates. The number of events from on-zones is compared to that from "off-zones" of exactly the same size and shape, but offset in right ascension. Thus, the off-source regions give an unbiased estimate of the background in the source region in a way that is independent of simulations. The background from the atmospheric neutrinos is suppressed using the energy estimate of the neutrino candidates.

Upper limits from the neutrino emission from the Fermi Bubbles were already set in [13]. The recent update of this analysis (see S. Hallmann in [1]) shows a slight excess in the source region, corresponding to a 1.9 σ significance.

Direct evidence of CR interactions with interstellar matter comes again from Fermi-LAT, which observed a diffuse galactic γ-ray emission. The ANTARES location in the Northern latitude is ideally suited to study the expected neutrino flux from the inner galactic plane. A search has been performed searching in the regions of galactic longitude $|l| < 40°$ and latitude $|b| < 3°$ with respect to nine off-zones. No excess in the on-zone region was found and the resulting limits are shown in Fig. 3. In particular, the hypothesis of a 1–1 relation between the γ-ray and neutrino flux from the Galactic Ridge is ruled out at 90% C.L., showing that ANTARES is already testing the well-established multimessenger $\gamma - \nu -$CR paradigm in our Galaxy. The present limit cannot rule out however models from more-detailed simulations of galactic CR propagation.

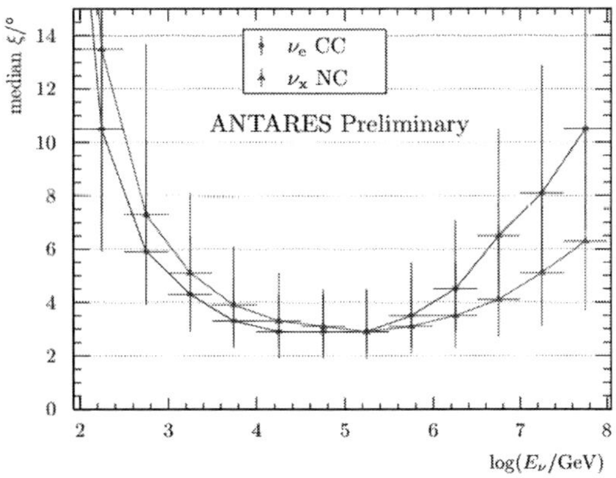

Figure 4. Angular resolutions for Monte Carlo NC (blue) and CC (red) events reconstructed with the new shower algorithm. Below 10 TeV, the resolutions worsen due to a decreasing number of photons being detected, while above 300 TeV, the events begin to saturate the detector. Over the 10–300 TeV range, arrival directions are reconstructed by the algorithm with a median angular error of 3°, and a resolution on deposited energy of 5%, although the latter is limited by the total ANTARES systematic energy uncertainty of approximately 10%. From T. Michael in [1].

4. Reconstruction of cascade events

Cascade events are mainly due to neutral-current (NC) interactions, and ν_e and ν_τ charged-current (CC) interactions. With respect to ν_μ CC interactions, these events have worse angular resolution but the energy deposited in the detector is more-strongly correlated with the energy of the primary neutrino. In addition, cascades suffer from a smaller background from atmospheric events. These latter advantages allowed the diffuse cosmic neutrino flux detected by IceCube to be first observed in the cascade channel [2].

Cascade identification and reconstruction has been in development in ANTARES for several years. Recently, an algorithm with an unprecedented angular resolution, of typically 3°–4° accuracy has been released, see Fig. 4. This for the first time enables a point-source search using the cascade channel. A combined point-source search using both muon-track and cascade events has been performed using 1622 days of effective livetime from 2007 to 2013. The sample consisted of 6261 muon-track events, and 156 cascade events, with an estimated contamination of 10% mis-reconstructed atmospheric muons in each. The sensitivity on point-like sources are given in Fig. 5. No significant excess was observed.

5. Dark matter and Exotics

Neutrino telescopes can set limits on different WIMP dark-matter scenarios by limiting the neutrino flux expected from WIMP interactions in the Sun, Earth, Galactic Centre, and dwarf galaxies. Since the expected dark-matter density tends to be strongly peaked near the centers of these objects, and ANTARES has an excellent angular resolution, competitive limits can be set.

Limits on the spin-dependent (WIMP-proton) interaction cross section σ_{SD}^p from ANTARES observations of the Sun using the $\tau\bar{\tau}$ channel are given in Fig. 6 (left). Neutrino telescopes produce the strongest limits in the $M_{\rm WIMP} > 200$ GeV range, surpassing even the direct-detection experiments. Limits on the WIMP-WIMP velocity-averaged self-annihilation cross section $\sigma_A v$ from the Galactic Centre from [15] are reported in Fig. 6 (right).

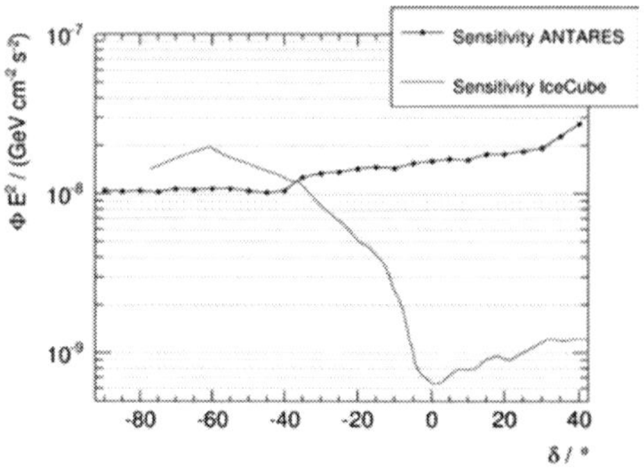

Figure 5. Sensitivity of the ANTARES targeted search to flavor-uniform neutrino point sources with E^{-2} spectra in terms of flux per flavor, compared to the IceCube results from Ref. [2]. While the atmospheric background produces predominantly muon-track events, an E^{-2} point source with a flavor-uniform flux would be expected to produce a cascade-to-track ratio of 3:10, significantly increasing the sensitivity of the search. This ratio is lower to one because of the long range of muons, which allows to detect ν_μ interacting far away the instrumented volume, and the different response of the detector for tracks and cascades. The achieved sensitivity is approximately 10^{-8} GeV^{-1} cm^{-2} s^{-1} for $\delta < -40°$. From T. Michael in [1].

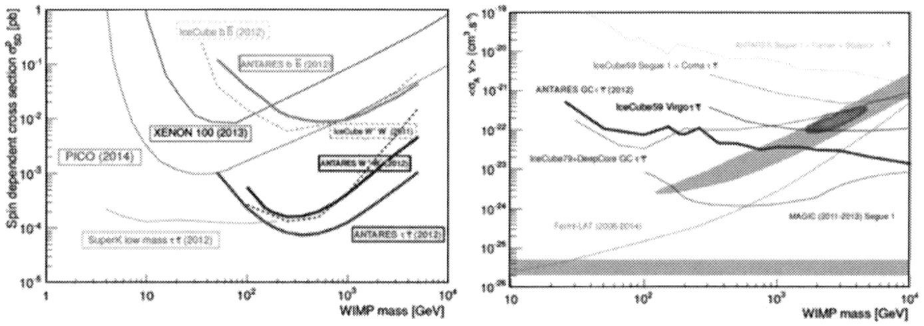

Figure 6. ANTARES limits $\sigma_{\rm SD}^{p}$ from the Sun (left) and on $\langle\sigma_A v\rangle$ from the Galactic Centre (right) as a function of the WIMP mass [15]. From C. Tönnis in [1].

6. Conclusion

The scientific framework of neutrino astrophysics has significantly changed after the IceCube detection of PeV neutrinos. The ANTARES neutrino telescope has proved to be a highly successful instrument for performing a wide range of physics analyses. The present focus is to help to understand the origin (galactic, extragalactic, mixed?) of the cosmic neutrino signal. In particular, ANTARES has an excellent angular resolution on both muon-track and cascade events, facilitated by the optical properties of deep-sea water. This has allowed constraining a wide range of source scenarios, from galactic plane emission to blazars, point-like or extended objects in the Southern sky.

A new era in neutrino astronomy will begin in 2017, with the decommissioning of ANTARES, and the completion of KM3NeT Phase 1, with a unique design of multi-PMT optical modules. Phase-1 will have an instrumented volume a factor of ~ 3 larger than that of ANTARES [16].

References

[1] The ANTARES Collaboration: Contributions to ICRC 2015, The Hague. `arXiv:1510.04508`
[2] M.G. Aartsen et al., ApJ **809**, 98 (2015)

[3] S. Adrián-Martínez, et al., ApJ **786** (2014) L5
[4] S. Adrián-Martínez, et al. (ANTARES and IceCube Collaborations), arXiv:1511.02149
[5] M. Spurio, Phys. Rev. D **90**, 103004 (2014)
[6] M.C. Gonzalez-Garcia, F. Halzen, V. Niro. Astropart.Phys. **57-58** (2014) 39-48
[7] M.G. Aartsen, et al., Phys. Rev. Lett. **113**, 101101 (2014)
[8] F. Krauß et al., A&A **566**, L7 (2014)
[9] S. Adrián-Martínez et al., A&A **576**, L8 (2015)
[10] S. Adrián-Martínez et al., arXiv:1506.07354 (2015) (Accepted by JCAP)
[11] S. Adrián-Martínez et al., arXiv:1508.01180 (2015) (Accepted by JCAP)
[12] GCN circular # 18231 and an Atel telegram # 7987. Paper in preparation
[13] S. Adrián-Martínez et al., European Physical Journal C **74**, 2701 (2014)
[14] D. Gaggero et al., arXiv:1504.00227 (2015)
[15] S. Adrián-Martínez et al., JCAP10 (2015) 068
[16] R. Coniglione, on the behalf of KM3NeT. These proceedings

EPJ Web of Conferences **116**, 11007 (2016)
DOI: 10.1051/epjconf/201611611007

Global 3ν oscillation analysis: Status of unknown parameters and future systematic challenges for ORCA and PINGU

Francesco Capozzi[1,2], Eligio Lisi[2,a], and Antonio Marrone[1,2]

[1] Dipartimento Interateneo di Fisica, Università di Bari, via Amendola 173, 70126 Bari, Italy
[2] Istituto Nazionale di Fisica Nucleare, Sezione di Bari, via Orabona 4, 70126 Bari, Italy

Abstract. Within the standard 3ν oscillation framework, we illustrate the status of currently unknown oscillation parameters: the θ_{23} octant, the mass hierarchy (normal or inverted), and the possible CP-violating phase δ, as derived by a (preliminary) global analysis of oscillation data available in 2015. We then discuss some challenges that will be faced by future, high-statistics analyses of spectral data, starting with one-dimensional energy spectra in reactor experiments, and concluding with two-dimensional energy-angle spectra in large-volume atmospheric experiments. It is shown that systematic uncertainties in the spectral shapes can noticeably affect the prospective sensitivities to unknown oscillation parameters, in particular to the mass hierarchy.

1. Introduction

The 2015 Nobel Prize in Physics, awarded jointly to Takaaki Kajita and Arthur B. McDonald "for the discovery of neutrino oscillations, which shows that neutrinos have mass" [1], represents both the crowning point of a long quest for neutrino flavor transformations, and an incitement to further proceed in understanding the physics of ν masses and mixings. Indeed, world neutrino data have established a simple 3ν framework, where the three known flavor states ν_α ($\alpha = e, \mu, \tau$) are mixed with three mass states ν_i ($i = 1, 2, 3$), separated by two squared mass gaps δm^2 and Δm^2, via three mixing angles θ_{12}, θ_{13}, θ_{23} and a possible CP-violating phase δ [2]. Although global ν data analyses such as [3] provide precise constraints on five parameters (the squared mass gaps and the three mixing angles), we still do not know the octant of θ_{23} ($< \pi/4$ or $> \pi/4$), the sign of Δm^2 which tells the mass spectrum hierarchy (normal NH for $+\Delta m^2$ or inverted IH for $-\Delta m^2$), and the value of the Dirac phase δ that could induce CP violating effects in the leptonic sector (if $\delta \neq 0, \pi$). Further unknowns, not considered herein, include the absolute ν mass scale and the Dirac or Majorana nature of the neutrino fields [2], as well as possible new neutrino states or interactions beyond the standard 3ν framework.

In this contribution we illustrate the status of the unknown parameters $\theta_{23} - \pi/4$, sign(Δm^2), and δ, in the light of a recent global neutrino data analysis. We then argue that future analyses of high-statistics (and large volume) neutrino experiments will face new and unprecedented challenges, related

[a]e-mail: eligio.lisi@ba.infn.it

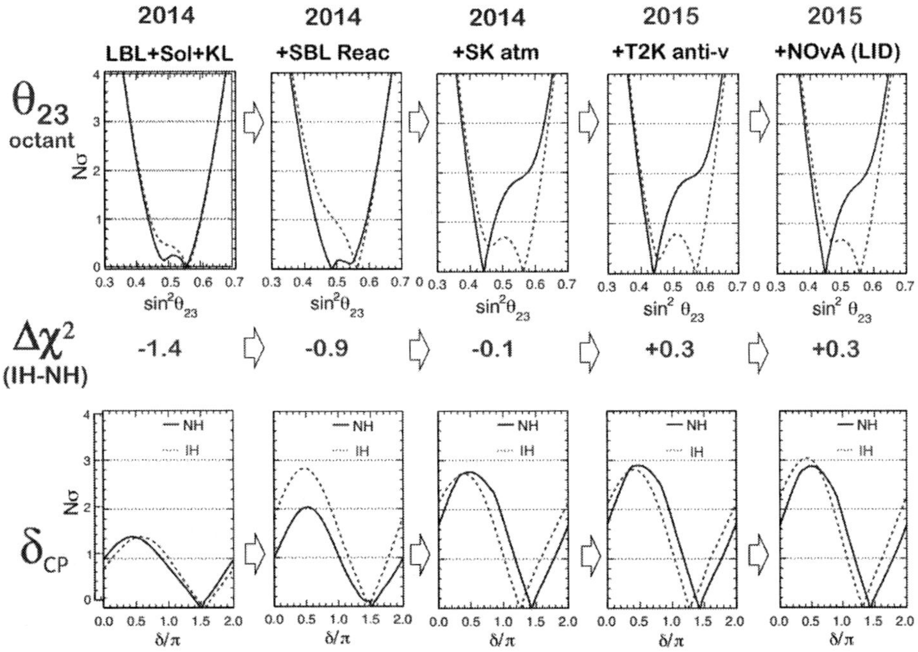

Figure 1. Global neutrino data analysis, with preliminary updates with respect to [3]. Upper panels: Contraints on $\sin^2\theta_{23}$ in terms of $N_\sigma = \sqrt{\Delta\chi^2}$ for increasingly rich data sets. Lower panels: Corresponding constraints on the CP-violating phase δ. Middle row: Corresponding values of the $\Delta\chi^2$ for hierarchy difference (IH vs NH).

to spectral shape systematics which may hinder, e.g., mass hierarchy signatures. We shall first discuss these issues in the relatively simple context of medium-baseline reactor projects, and then in connection with prospective atmospheric neutrino experiments in ice (or water).

2. Global 3ν oscillation data analysis: Status of unknowns

Our latest published global ν data analysis, presented in terms of χ^2 best-fit values and allowed regions, was reported in [3]. In the context of unknown oscillation parameters, we illustrate here preliminary updated results including short-baseline (SBL) reactor ν data from the Neutrino 2014 Conference [4], as well as recent long-baseline (LBL) accelerator data from the T2K ($\overline{\nu}$ run) and NOvA experiments in the $\nu_\mu \to \nu_e$ appearance channel, as appeared in 2015 [5]. A more complete analysis, including recent atmospheric data from Super-Kamiokande and IceCube [6], will be reported elsewhere [7].

Figure 1 shows the values of $\Delta\chi^2$ or, equivalently, $N_\sigma = \sqrt{\Delta\chi^2}$ for three unknown parameters: the θ_{23} octant, the mass hierarchy, and the CP-violating phase δ. From left to right, the first three panels refer to the combination of LBL, solar and KamLAND data, plus SBL reactor data, plus SK atmospheric data, all updated to 2014 with respect to [3]. The last two panels contain, in addition, 2015 data from T2K and then from NOvA in the default (so-called LID) analysis [5]. It can be seen that, at 90% C.L. ($\sim 1.6\sigma$), and even in the most complete data set (right panels), it is not possible to assign a specific octant to θ_{23}, nor to select a preferred hierarchy: both octants and hierarchies are allowed. On the other hand, an intriguing hint emerges in favor of $\delta \neq 0, \pi$ and, in particular, in favor of $\sin\delta < 0$, with a best fit close to $\delta \simeq 3\pi/2$. This value corresponds to a case of "maximal CP violation" ($\sin\delta \simeq -1$), while the opposite maximal case ($\sin\delta \simeq +1$) is disfavored at $\simeq 3\sigma$.

Time will tell if this interesting hint for leptonic CP violation will be confirmed (or not) by future LBL data [5]. In the meantime, the mass hierarchy and θ_{23} octants might remain undetermined. The

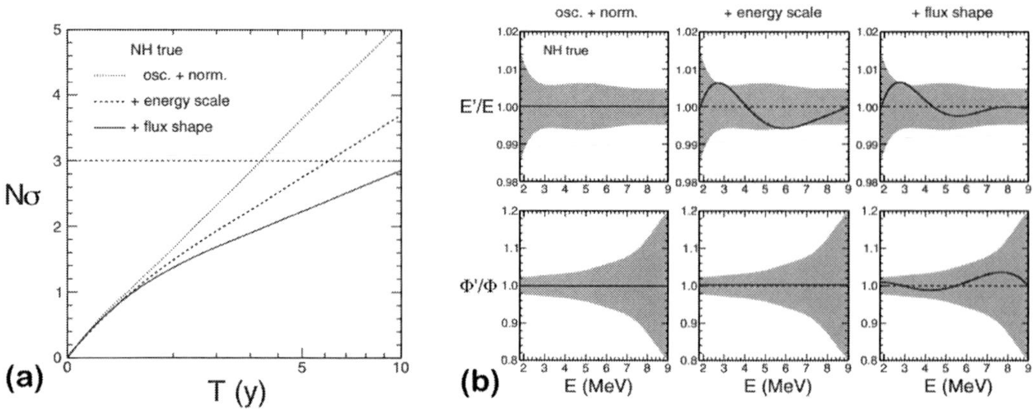

Figure 2. JUNO prospective data analysis for true NH (from [8]). Panel (a): Rejection power for the wrong IH, in terms of N_σ as a function of time T (with tics scaling as $\sqrt{T}$). From top to bottom, the curves include: oscillation parameters and flux normalization errors (osc.+norm), plus energy scale uncertainties E'/E (+energy scale), plus flux shape uncertainties Φ'/Φ (+ flux shape). Panel (b): Corresponding best-fits of the E'/E and Φ'/Φ energy profiles for prospective 5-yr data-taking, overlaid to the current $\pm 1\sigma$ error bands (in blue). The case of true IH (not shown, see [8]) leads to similar results.

JUNO (reactor) and PINGU (atmospheric) neutrino projects have the potential to address the mass hierarchy issue (and PINGU also the θ_{23} octant) with future high-statistics measurements which, however, will present new challenges for global ν data analyses, as shown in the next sections.

3. JUNO and systematics of one-dimensional neutrino spectra

Unoscillated reactor neutrino spectra $\Phi(E)$, which depend only on one variable (the ν energy E), were thought to be accurately known at a few % level until recently. However, a "bump" structure at ~ 5 MeV in prompt energy was discovered in the three independent experiments RENO, Daya Bay and Double Chooz [4], leading to $O(10\%)$ local variations with respect to previous expectations, likely due to poorly known nuclear reaction spectral components. Conservatively, one should thus consider the possibility that the reactor spectra may be affected by continuous "functional" uncertainties on both the energy scale (spectrum x-axis, $E \to E'(E)$), and the flux shape (spectrum y-axis, $\Phi(E) \to \Phi'(E)$), especially in future experiments which will measure the oscillated spectra with high statistics and very good resolution, such as in the JUNO (China) and RENO-50 (South Korea) projects [4], which aim at determining the neutrino mass hierarchy via precision oscillometry.

In [8] we systematically performed the exercise of letting both ratios E'/E and Φ'/Φ vary smoothly, within their currently allowed error bands, in a JUNO-like experimental setting. Representative results for the mass hierarchy discrimination are shown in Fig. 2. In the left panel (a), the dotted black curve shows the increase of the mass hierarchy sensitivity in time, assuming solely the systematic uncertainties related to the oscillation parameters and to the reactor flux and geoneutrino background normalizations. Such curve basically scales as $\sqrt{T}$, showing that such systematics do not spoil significantly the intrinsic statistical sensitivity. However, including first the energy-scale errors and then the flux-shape uncertainties (as smooth variations of E'/E and Φ'/Φ from unity, respectively) the sensitivity is significantly degraded, and can eventually be $< 3\sigma$ even after 10 years of data taking. Panel (b) of the same figure shows the best-fit functional forms of E'/E and Φ'/Φ, which are well within the colored 1σ error bands. By halving such error bands, e.g., via proper calibrations, one could reach a sensitivity $> 3\sigma$ in a reasonable 5-yr time (not shown, see [8] for details).

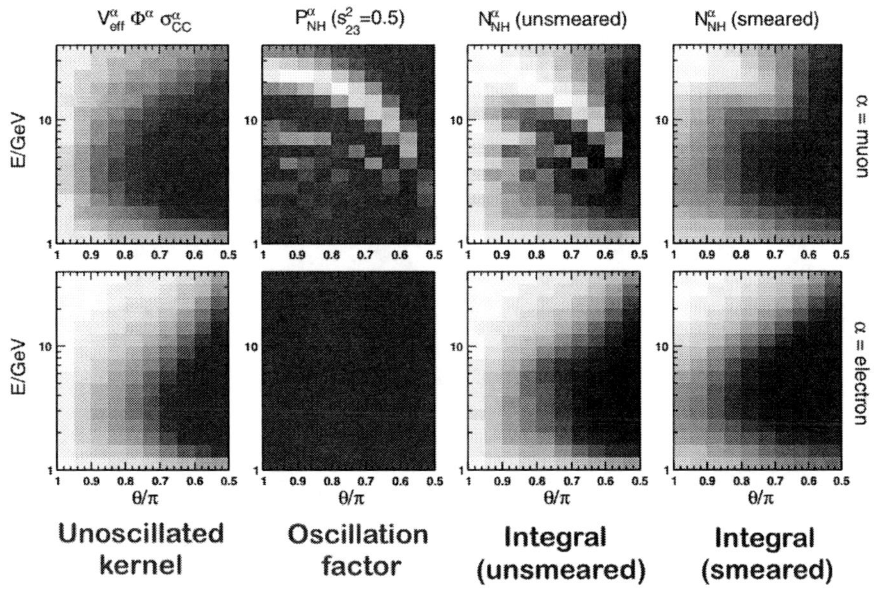

Figure 3. Breakdown of the main ingredients of muon (upper panels) and electron (lower panels) spectra in PINGU (taken from [9]), in an arbitrary color scale ranging from white (null event rate) to dark colors for increasing event rates (with pixels representing data bins). From left to right: Unoscillated kernel (flux times target volume times cross-section); flavor oscillation factors for $\sin^2\theta_{23} = \pi/4$; resulting integral spectrum for perfect energy-angle resolutions (unsmeared) and finally for realistic resolutions (smeared). The final smeared spectra for NH (this figure) and for IH (not shown) would not be distinguishable by eye.

In conclusion, even for a relatively simple one-dimensional spectrum (such as $\Phi(E)$ at reactors), shape systematics may play a relevant role. The next section will deal with the more challenging case of two-dimensional event spectra in energy and angle.

4. PINGU (ORCA) and systematics of two-dimensional neutrino spectra

In this Workshop, extensive discussions have been focussed on future large-volume, high-statistics atmospheric neutrino experiments such as PINGU [6] and ORCA [10]. Here we report some results from our recent analysis of prospective data and sensitivities to the mass hierarchy and to θ_{23} in a PINGU-like experimental setting [9]. (Qualitatively, our results also apply to an ORCA-like setting.)

PINGU will measure the incoming energy and direction of muons and electrons (but with no lepton charge discrimination), which represent proxies for the unknown incoming energy E and zenith angle direction θ of the parent neutrinos. The observable two-dimensional (energy-angle) spectra of μ-like and e-like events are obtained as convolutions of several ingredients related to the source (atmospheric neutrino fluxes), to flavor change in propagation (oscillations), to the interaction process (cross section and target volume), and to detection characteristics (energy and angular resolutions).

Figure 3 shows how these ingredients determine the final spectral shapes. The convolution of an unoscillated kernel with oscillation factors, and then with energy-angle resolution functions, leads to observable spectra which have very broad structures and, in particular, do not display at a glance the oscillation patterns which encode the mass hierarchy signatures (see [9] for further details). However, the high statistics reachable by PINGU in a few years may be sufficient to discriminate the hierarchy via spectral analyses, provided that the sensitivity is not spoiled by systematic effects, which are unavoidably associated with any of the above ingredients. In this workshop, particular attention has been paid to uncertainties related to input atmospheric neutrino fluxes and neutrino cross sections [11],

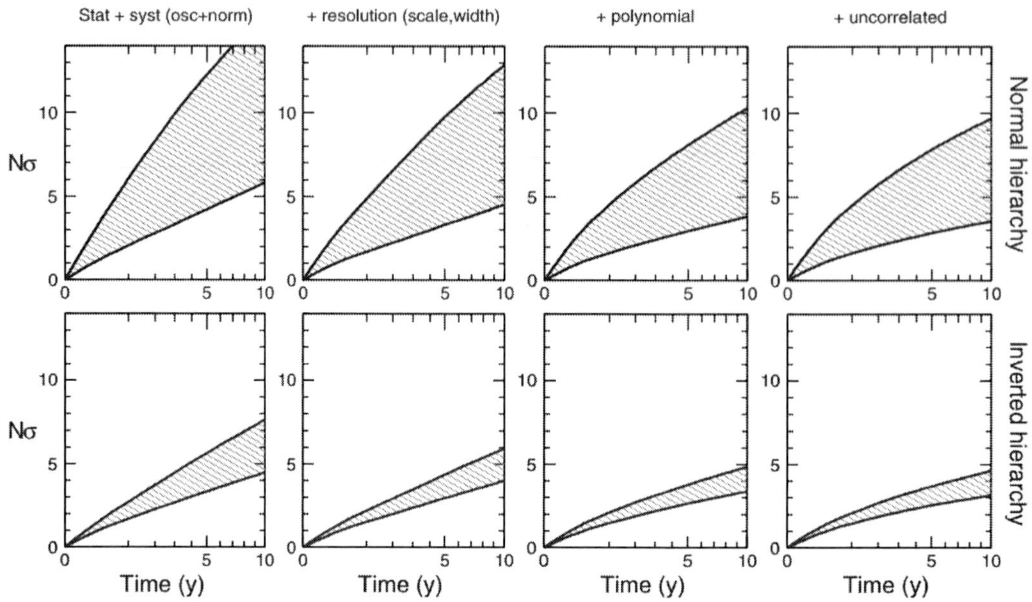

Figure 4. PINGU sensitivity to the hierarchy (N_σ), for either true NH (top panels) or true IH (bottom panels), as a function of the live time T in years. The abscissa is scaled as $\sqrt{T}$, so that the sensitivity bands would grow linearly for purely statistical errors. From left to right, the fit includes the following systematic errors: oscillation and normalization uncertainties, energy scale and resolution width errors, polynomial shape systematics (with up to quartic terms) at the 1.5% level, and uncorrelated systematics at the 1.5% level. See [9] for details.

but there are also other systematics related to the oscillating factors, to the energy-angle resolution functions, and to the effective target volume as a function of energy [9].

In [9] we have performed a thorough analysis of the effects of several systematic errors, both correlated and uncorrelated within bins, and with both known and uncertain energy-angle shape, in the context of PINGU. Figure 4 shows the main results of such work, in terms of sensitivity bands as a function of time for the cases of true NH (top) and true IH (bottom). The width of the bands is determined by the relatively large error on the θ_{23} mixing angle, which we take in the range $\sin^2\theta_{23} \in [0.4, 0.6]$. From left to right, the fit includes the following systematic errors: oscillation and normalization uncertainties, energy scale and resolution width errors, polynomial shape systematics (with up to quartic terms), and uncorrelated systematics. The last two error sources are kept at the default level of 1.5% (see [9] for a detailed definition of such uncertainties). Note that polynomial shape systematics are meant to parametrize smooth deviations from the expected shapes, due to a variety of energy-angle spectral uncertainties (atmospheric fluxes, differential cross sections, detector efficiencies, etc.) Including only normalization and systematic errors, the sensitivity N_σ grows almost linearly with T, i.e., the experiment is not limited by these systematics, even after 10 years of data taking. However, the progressive inclusion of correlated shape systematics, both "known" (resolution scale and widths) and "unknown" (ad hoc polynomial deviations), and eventually of uncorrelated shape systematics, provide an overall decrease in the 5-year (10-year) PINGU sensitivity by up to $\sim 35\%$ ($\sim 40\%$). In particular, in the worst case of true IH, a 3σ discrimination of the hierarchy might take ~ 10 yr.

However, it is possible to improve upon this situation by a better control of spectral shape systematics, which appears as a feasible goal by optimizing the PINGU configuration, and by studying in greater detail the effects of atmospheric flux and cross-section uncertainties [11]. As shown in [9], a reduction of the systematic errors in Fig. 4 by a factor of about two would definitely make it possible to discriminate the hierarchy at $> 3\sigma$ in less than 10 years in the worst case. Further benefits might come from a reduction of the θ_{23} uncertainty (see [9]).

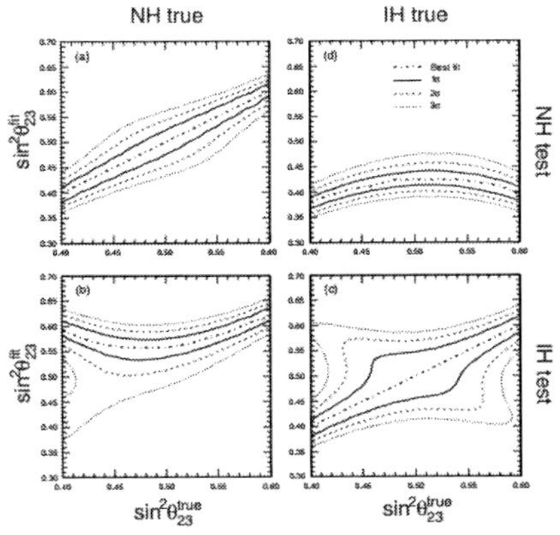

Figure 5. Fitted value $\sin^2\theta_{23}^{\rm fit}$ (at 1, 2 and 3σ) versus the true value $\sin^2\theta_{23}^{\rm true}$, for the four possible cases where the test hierarchy (i.e., the one assumed in the fit) is either the true or the wrong one: (a) NH = true, NH = test; (b) NH = true, IH = test; (c) IH = true, IH = test; (d) IH = true, NH = test.

Although PINGU itself will contribute to improve the accuracy on θ_{23}, the interpretation of its prospective θ_{23} bounds depends, in turn, upon the mass hierarchy determination: until the NH/IH issue it is not settled and remains ambiguous, the θ_{23} bounds in PINGU may be significantly biased. Figure 5 shows the subtle interplay between hierarchy discrimination and θ_{23}, in the four cases where the test hierarchy is equal (or not) to the true hierarchy. The cases where there is an hierarchy mismatch clearly lead to biased results for the θ_{23} bounds; see [9] for a more extensive discussion.

In conclusion, in the context of the standard 3ν framework and of a global data analysis, we have illustrated the status of the unknown oscillation parameters: the θ_{23} octant, the mass hierarchy, and the possible CP phase δ. We have then discussed the challenges related to spectral shape uncertainties in some future experiments, which aim at determining the mass hierarchy via one-dimensional event spectra (e.g., JUNO and RENO-50 with reactor neutrinos) or two-dimensional event spectra (e.g., PINGU and ORCA with atmospheric neutrinos). Improvements in our understanding and control of spectral shape systematics are needed to safely reach $> 3\sigma$ sensitivities for the NH/IH discrimination.

E.L. thanks the VLVNT 2015 organizers for kind hospitality and support. This work is supported by INFN and MIUR through the "Theoretical Astroparticle Physics" project.

References

[1] `www.nobelprize.org/nobel_prizes/physics/laureates/2015/`
[2] K.A. Olive et al. (Particle Data Group), Chin. Phys. C **38**, 090001 (2014)
[3] F. Capozzi et al., Phys. Rev. D **89**, 093018 (2014) [arXiv:1312.2878 [hep-ph]]
[4] Neutrino 2014, XXVI International Conf. on ν Physics and Astrophysics, `neutrino2014.bu.edu/`
[5] A. Weber, talk at this Workshop
[6] J.-P. Yanez, talk at this Workshop
[7] F. Capozzi, E. Lisi, A. Marrone, D. Montanino, and A. Palazzo (2016), in preparation.
[8] F. Capozzi et al., Phys. Rev. D **92**, 093011 (2015) [arXiv:1508.01392 [hep-ph]]
[9] F. Capozzi et al., Phys. Rev. D **91**, 073011 (2015) [arXiv:1503.01999 [hep-ph]]
[10] D. Samtleben, talk at this Workshop
[11] J. Nieves, J. Sandroos, A. Gazizov, talks at this Workshop

EPJ Web of Conferences **116**, 11008 (2016)
DOI: 10.1051/epjconf/201611611008

KM3NeT/ORCA status and plans

Dorothea F.E. Samtleben[1,2,a] on behalf of the KM3NeT Collaboration

[1] Leiden University, The Netherlands
[2] Nikhef, National Institute for Subatomic Physics, The Netherlands

Abstract. Neutrinos created in interactions of cosmic rays with the atmosphere can serve as a powerful tool to unveil the neutrino mass hierarchy (NMH). At low energies, around a few GeV, matter effects from the transition through the Earth are expected to imprint a distinct but also subtle signature on the oscillation pattern, specific to the ordering of the neutrino masses. KM3NeT/ORCA (Oscillations Research with Cosmics in the Abyss), a densely instrumented building block of the upcoming KM3NeT neutrino telescope, will be designated to measuring this signature in the Mediterranean Sea. Using detailed simulations the sensitivity towards this signature has been evaluated. The multi-PMT detectors allow in the water for an accurate reconstruction of GeV neutrino event signatures and distinction of neutrino flavours. For the determination of the mass hierarchy a median significance of 2-6σ has been estimated for three years of data taking, depending on the actual hierarchy and the oscillation parameters. At the same time the values of several oscillation parameters like θ_{23} will be determined to unprecedented precision.

1. Introduction

Neutrino physics provides for an exciting opportunity to advance our current understanding of fundamental particles and interactions, as several open questions are yet to be answered for these elusive particles. Though there exists, by now, significant evidence for their non-zero masses from oscillation measurements, the ordering of the masses has not yet been determined and also the Majorana/Dirac nature of the masses as well as the possible CP-phase in the mixing matrix are unknown. Two modes of possible mass ordering are distinguished whereby one is commmonly called the *normal hierarchy* (NH, $m_{\nu 3} > m_{\nu 1}$) and the other the *inverted hierarchy* (IH, $m_{\nu 1} > m_{\nu 3}$). The sensitivity of upcoming experiments will within the next decade improve to a level that the neutrino mass hierarchy (NMH) can be determined.

Abundant neutrinos are created in interactions of cosmic rays with the atmosphere and measured with current neutrino telescopes. An expansion of the sensitivity of the neutrino telescopes towards low energies of a few GeV will give access to an energy range in which the NMH influences the oscillation pattern sufficiently to be identified [1]. Matter effects during the passage of the neutrinos through the Earth affect the transition probabilities for the different neutrino flavours depending on the NMH.

---

[a] e-mail: dosamt@nikhef.nl

However, the impact on the oscillations is for NH and IH opposite for neutrinos and antineutrinos, which cannot be distinguished in a neutrino telescope. Still since the neutrino and antineutrino fluxes from the atmosphere are not exactly equal and also the neutrino cross section is about a factor of two larger than the antineutrino cross section [2] a net signature specific to the NMH remains in the oscillation pattern and can be measured.

2. Detector

The feasibility of neutrino studies with an underwater neutrino telescope has already successfully been demonstrated by the ANTARES neutrino telescope which is operational in the Mediterranean Sea since 2007 [3]. Building on this experience the KM3NeT research infrastructure [4] has been planned and is currently under construction. It will host two neutrino telescope installations: ARCA (Astroparticle Research with Cosmics in the Abyss) at an Italian site located about 100 km from Capo Passero (Sicily, Italy) at a depth of 3500 m, focussed on the measurement of high energetic astrophysical neutrinos and ORCA (Oscillations Research with Cosmics in the Abyss) at a French site located about 40 km from Toulon (France) at a depth of 2475 m, focussed on low energetic atmospheric neutrinos. The same technology will be used at both sites but different detector densities will be implemented optimized for the different addressed energy ranges.

A pressure resistant glass sphere of 17 inch diameter (Digital Optical Module – DOM), which hosts 31 3″ photomultipliers (PMTs) distributed to cover all directions, is the active sensor of the telescope [5]. For the PMT signals a nanosecond timing precision is achieved and all time-over-thresholds of these signals are sent to shore for further processing and filtering. Calibration tools are implemented to determine in real-time the position of the detectors to the precision of a few centimeters and the orientation within a few degrees.

The small PMTs allow for single photon counting and signal correlations enable powerful background rejection of photons from bioluminiscence or from the radioactive decays of ^{40}K in the seawater. The homogeneously distributed ^{40}K signal is not only considered a background but serves also as a constant calibration source to determine the time offsets on a DOM and enables continuous monitoring of the PMT performance [6].

The DOMs are distributed on strings which are fixed on the floor and kept taut by a buoy. One string contains 18 DOMs whose distance determines the accessible energy range. For ORCA the DOM distance on a string is currently being optimized in the range of 6 m to 18 m. The results shown in this contribution have been determined for a distance of 6 m. The distance between the strings is for ORCA fixed to 20 m according to deployment constraints.

In December 2014 the first main electro-optical cable was successfully deployed at the French Site followed by the deployment of the first junction box for the connection of strings in the summer 2015. The first 6 strings will be built and deployed by the end of 2016. The full ORCA array of 115 strings will then be constructed and become operational within three years depending on the availability of funds.

3. Simulations and event reconstruction

The possible measurement of the oscillation pattern has been studied using detailed simulations [7]. The neutrino and antineutrino interactions were generated with a software based on the GENIE [8] neutrino event generator. The weighting of the events follows the Bartol atmospheric neutrino flux model [9]. The background of down-going muons was generated using the MUPAGE [10] program. All particles in the vicinity of the detector are propagated with a GEANT based software package. This generates Cherenkov light from primary and secondary particles whereby the effects of absorption and scattering in water as well as the characteristics from the PMTs are taken into account in the propagation and signal creation. Also photons from ^{40}K decays are added in the simulations.

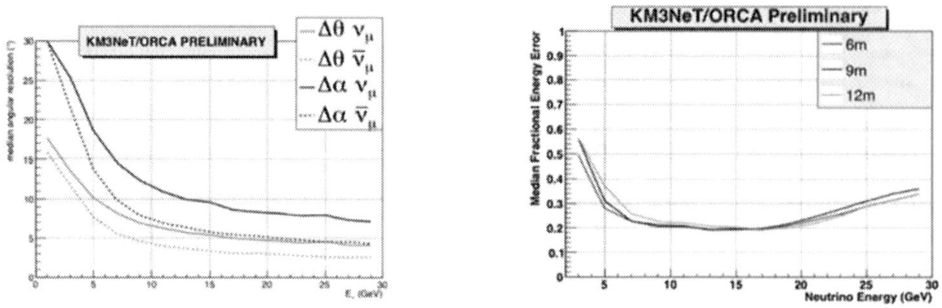

Figure 1. Left: The median angular resolution is shown for ν_μ events as a function of the neutrino energy for the neutrino direction α and the zenith angle θ. Right: The median energy resolution is shown as a function of the neutrino energy for different vertical DOM distances.

The main task of the reconstruction is to provide accurate information on the direction and energy of the neutrino as well as on the flavour. A neutrino interaction in the detector results in a track-like or a cascade-like signature. A charged current (CC) interaction of a ν_μ leads to the creation of a muon which leaves signals along its path in the detector and thus shows a track-like signature. The length of a 5 GeV muon track amounts to $\sim$25 m. The ν_e and ν_τ CC interactions create an electron and a tau respectively, which both lead to the creation of a cascade and thus a cascade-like signature. The distinction of the track-like and the cascade-like signatures allows thus some flavour identification for the CC interactions. The neutral current (NC) interactions of neutrinos of all flavours result in a cascade at the interaction vertex and thus also a cascade-like signature. In this case no flavours can be distinguished and the NC interactions represent a background to the oscillation measurement.

The event signatures also depend on the inelasticity of the interaction as not all energy of the neutrino will be transferred to the final lepton which is taken into account in the reconstruction.

Both event signatures (track/cascade) are addressed with designated multi-step algorithms. Hits are selected based on timing correlations and for the track-like signature a grid of directions is evaluated with likelihood fits whereby the candidate with the best likelihood is chosen. The energy of the neutrino is estimated in the case of the track signature from the length of the reconstructed track as well as the associated hits. For the cascade, first the interaction vertex is determined, then the direction, energy and inelasticity using a likelihood fit.

The directional and energy resolution for the track reconstruction can be seen in Fig. 1. In the relevant energy range of 5–12 GeV the directional resolution is better than 10 degrees and the energy resolution better than 25%. The performance varies only slightly with the DOM distance and is similar for cascades.

The inelasticity y for ν_e CC interactions is determined in the likelihood fit from the amount of photons found at the Cherenkov angle with respect to the reconstructed direction as compared to the amount of photons at other angles. A comparison of the angle distributions of photons is shown in Fig. 2 separately for events from different inelasticity ranges. The Cherenkov photons from the electron lead to a prominent peak at the Cherenkov angle for events with low inelasticities.

For the ν_μ CC interactions the inelasticity is reconstructed from time residual distributions of the hits. The distributions are created with respect to the expectation of Cherenkov light from a track hypothesis and with respect to the expectation from shower light from the vertex. The dependence of these distributions on the inelasticity of the event is then exploited for the inelasticity determination. For both cascade and track interactions some sensitivity on the inelasticity could be gained but is also still being optimized. It is not yet used in the NMH determination but will be exploited in the future since it can increase the sensitivity as has been shown for ν_μ interactions [11].

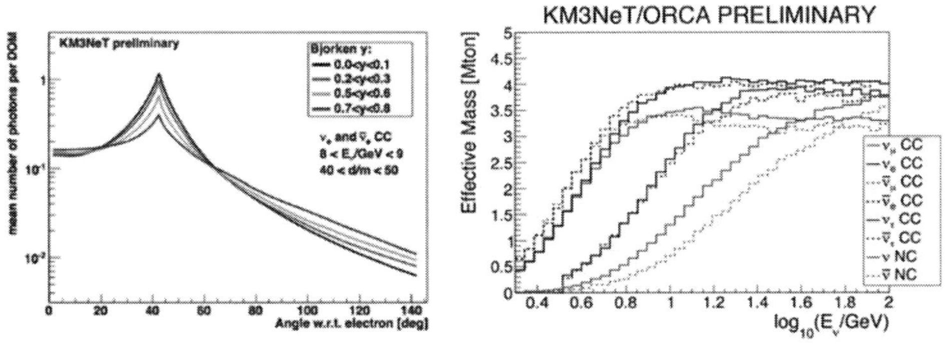

Figure 2. Left: The distribution of photon angles with respect to the electron direction is shown for ν_e events. Right: The effective mass is shown for the different (anti-)neutrino interactions.

An important parameter for evaluating the reconstruction performances and the sensitivity for the measurement of the neutrino mass hierarchy is the effective volume, defined as the volume of a perfectly efficient detector to observe neutrinos interacting within that volume. Figure 2 shows on the right side the effective mass from the reconstruction of the different neutrino interactions, obtained by multiplying the effective volume by the seawater density, equal to 1.025 g/cm^3. For an energy around 5 GeV the effective mass reaches the instrumented detector mass for ν_μ and ν_e CC interactions.

The identification of a track-like or cascade-like event is implemented using a random decision forest (RDF) classification [12]. Many characteristics of the event signatures like the likelihood values of the reconstruction or time residuals are exploited in this classification. For the important energy range between 5–12 GeV the identification succeeds in 80% of the antineutrino events, whereby this fraction rises with energy. The performance for neutrinos is slightly worse due to the different inelasticities of neutrino events compared to antineutrinos as on average less of the energy of the neutrinos will go to the muon. Cascades from ν_e interactions are misidentified as tracks in less than 10% of the cases and this fraction decreases strongly with energy.

The background from the abundant down-going muons is suppressed at first with basic cuts on the reconstructed zenith angle and vertex position. Only up-going events are selected and the vertex position is constrained to reside approximately in the instrumented volume. This decreases the amount of atmospheric muons by three orders of magnitude. Further suppression, by two orders of magnitude, is achieved by using a boosted decision tree (BDT) for the identification of these events. Only a few percent of misreconstructed muon events remain in the final sample of reconstructed events.

4. Sensitivity calculation

Using the atmospheric neutrino fluxes, the neutrino interaction cross sections and a model for the matter distribution in the Earth [13] the expected oscillation patterns can be calculated for the different mass hierarchies. From these the distinction of the hierarchies can be evaluated. More than ten thousand neutrino events are expected to be detected per year. The differences between the NH and IH oscillation patterns are of the order of a few percent in small energy-zenith regions, so that unprecedented control of systematics is mandatory.

The reconstruction performance together with the particle identification and background contamination is used to generate the oscillation pattern as a function of zenith and energy as pseudo-experiments for different oscillation parameters and hierarchies. For each oscillation pattern the likelihood for NH or IH is determined and a likelihood ratio calculated. The oscillation parameters are included in the fit. Parameters which had been shown to have little impact are kept fixed $(\theta_{12}, \theta_{13}, \Delta m^2)$

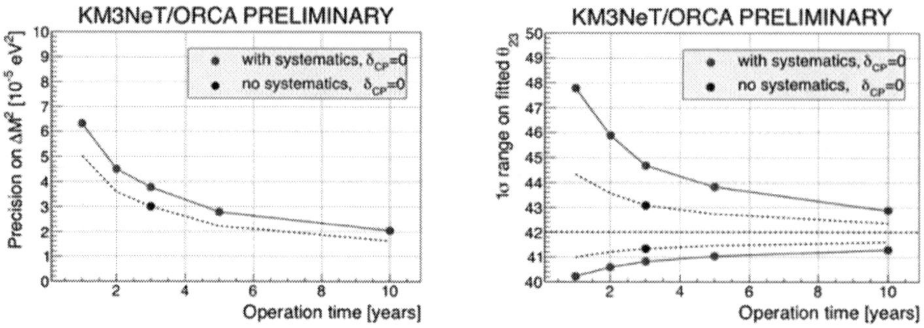

Figure 3. Left: The precision in ΔM^2 as determined in the measurement is shown as a function of the observation time. Right: Similar presentation for the precision on θ_{23}.

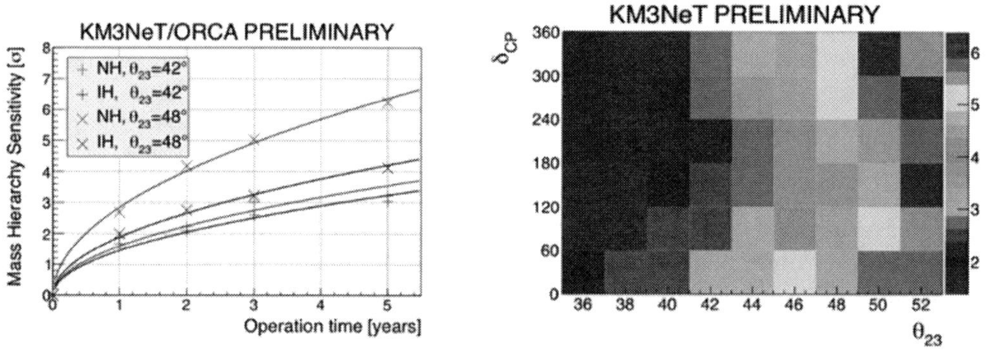

Figure 4. Left: The median significance of the NMH distinction is shown as a function of time for the different hypotheses and two different θ_{23} values. Right: The median significance of the measurement is shown for the NH as color scale as a function of the true δ_{CP} and θ_{23}.

while the others are fitted unconstrained ($\Delta M^2 = \frac{1}{2}(\Delta m^2_{32} + \Delta m^2_{31}), \theta_{23}, \delta_{CP}$). Figure 3 shows the development of the precision reached in the oscillation parameters ΔM^2 and θ_{23} as a function of the observation time. After three years the uncertainty on ΔM^2 will already only be half of the present uncertainty and also the uncertainty on θ_{23} will shrink similarly.

Aside from the oscillation parameters also the neutrino fluxes, cross sections and the hadronization impact the pattern. Many of these external uncertainties are parametrized and incorporated as nuisance parameters in the global fit, e.g. the overall normalization, the $\nu/\bar{\nu}$ ratio, the e/μ ratio, the NC scaling and the energy slope. The largest average fluctuation on these parameters with respect to a nominal simulation is found to be 11% for the NC scaling while for all other parameters fluctuations smaller than 4% are found. None of the considered systematics compromises the determination of the NMH.

The sensitivity towards the NMH is evaluated from the resulting likelihood ratio distributions in the pseudo experiments for NH and IH as a median significance, which is the significance for excluding the wrong hierarchy when the measured likelihood ratio is at the median of the true hierarchy distribution. The development of the sensitivity with observation time is shown in Fig. 4 for two different values of θ_{23} and the different hierarchies. On the right side of the figure the dependence of the significance on θ_{23} and δ_{CP} is shown for the NH, which ranges from 2–6σ in the currently most likely range for θ_{23}.

5. Conclusions

KM3NeT/ORCA will measure the flavour oscillations of atmospheric neutrinos to unprecented precision. The neutrino mass hierarchy impacts the oscillation pattern and can thus be determined with these measurements. The reconstruction of the neutrino event signatures and distinction of different flavour interactions were evaluated in detailed simulations and also a powerful background rejection could be demonstrated. The median significance to determine the NMH was evaluated to range from 2 to 6σ in three years depending on the oscillation parameters and hierarchy (NH or IH).

The sea infrastructure for ORCA is already being set up at the foreseen French site in the Mediterranean Sea and the first six ORCA strings will be deployed by the end of 2016. After this proof of the technology the full foreseen detector of 115 strings can then be completed within three years.

References

[1] E.Kh. Akhmedov, S. Razzaque and A. Yu. Smirnov, JHEP **1302**, 082 (2013)
[2] J. Formaggio and G. Zeller, Rev. Mod. Phys. **84**,1307–1341 (2012)
[3] M. Ageron et al., NIM A **656**, 11–38 (2011)
[4] M. de Jong, PoS (NEUTEL2015) 055
[5] R. Bruijn and D. van Eijk, PoS (ICRC2015) 1157
[6] S. Adrian-Martinez et al., `arXiv:1510.01561`, accepted for publication by EPJC
[7] A. Margiotta for the ANTARES Collaboration, Nucl. Instr. Meth. A **725**, 98–101 (2013)
[8] C. Andreopoulos et al., NIM A **614**, 87–104 (2010)
[9] G.D. Barr et al., Phys. Rev. D **70**, 023006 (2004)
[10] G. Carminati et al., Comp. Phys. Comm. **179**, 915–923 (2008), M. Bazzotti et al., Comp. Phys. Comm. **181**, 835–836 (2010)
[11] M. Ribordy and A.Yu. Smirnov, Phys. Rev. D **87**, 113007 (2013)
[12] L. Breiman, Machine Learning **45**, 5–32 (2001)
[13] A.M. Dziewonski and D.L. Anderson, Phys. Earth Planet. Inter., **25**, 297–356 (1981)

EPJ Web of Conferences **116**, 11009 (2016)
DOI: 10.1051/epjconf/201611611009

From DeepCore to PINGU

Measuring atmospheric neutrino oscillations at the South Pole

J.P. Yáñez[a] for the IceCube-Gen2 Collaboration

DESY, 15735 Zeuthen, Germany

Abstract. Very large volume neutrino telescopes (VLVNTs) observe atmospheric neutrinos over a wide energy range (GeV to TeV), after they travel distances as large as the Earth's diameter. DeepCore, the low energy extension of IceCube, has started making meaningful measurements of the neutrino oscillation parameters θ_{23} and $|\Delta m^2_{32}|$ by analyzing the atmospheric flux at energies above 10 GeV. PINGU, a proposed project to lower DeepCore's energy threshold, aims to use the same flux to further increase the precision with which these parameters are known, and eventually determine the sign of Δm^2_{32}. The latest results from DeepCore, and the planned transition to PINGU, are discussed here.

1. Introduction

Evidence that neutrinos change flavor (oscillate) as they travel, and are thus massive, has accumulated in the last decades (see [1] and references therein). Although the phenomenon has been confirmed from a variety of neutrino sources, atmospheric (anti-)neutrinos, ν_μ and ν_e, remain a very powerful tool to study it.

The amplitude of oscillation probabilities depends on the mixing of flavor and mass eigenstates, encoded in mixing angles θ_{ij}. The oscillation phase is proportional to $\Delta m^2_{ij} L/E$, where L is the distance traveled by the neutrino, E is its energy and Δm^2_{ij} is the squared mass difference of the mass eigenstates. Atmospheric neutrinos are produced with energies from MeV to the TeV-scale and can travel distances as large as the Earth's diameter before detection. This means that experiments measuring these neutrinos are sensitive to the mass-splitting Δm^2_{32} and the mixing angles θ_{23} and θ_{13} over a wide L/E range [2].

Current VLVNTs can measure atmospheric neutrinos starting at approximately 10 GeV [3, 4], and future proposed telescopes aim to lower the threshold to a few GeV [5, 6]. Neutrinos crossing the Earth at these energies undergo neutrino-electron coherent forward scattering, which affects how they oscillate [7]. Between $E_\nu \approx 1 - 12$ GeV, this leads to large deviations from vacuum oscillations for neutrinos or antineutrinos, depending on the ordering of the neutrino masses [8–10]. Above this energy range saturation occurs and, while oscillation probabilities still differ from those in vacuum, they are the same for neutrinos and antineutrinos regardless of the neutrino mass ordering.

[a]e-mail: `juan.pablo.yanez@desy.de`

IceCube's DeepCore [4] detects neutrinos in the saturation regime. The experiment is thus suited for measuring the atmospheric parameters θ_{23} and $|\Delta m^2_{32}|$ [11, 12]. The Precision IceCube Next Generation Upgrade (PINGU) [5], a proposed successor to DeepCore, will lower the energy threshold to access the resonant regime, identify the sign of Δm^2_{32} and, with that, will place the final piece of the neutrino mass ordering puzzle.

2. Detector performance of DeepCore and PINGU

IceCube is an ice Cherenkov neutrino detector that consists of 5160 digital optical modules (DOMs) deployed over a volume of 1 km^3, at depths between 1450 and 2500 m at the geographic South Pole [13]. The DOMs are arranged in 86 strings, with each string holding 60 of them. Most strings are situated over a quasi-hexagonal grid, with an inter-string separation of 125 m, and a vertical spacing between DOMs of 17 m. Towards the middle of the detector, eight strings are located in between the regular grid, reducing the inter-string spacing to 40–70 m. The DOMs on these strings have 35% higher quantum efficiency than the standard DOMs and are only 7 m apart. In this volume, known as DeepCore, it is possible to detect neutrinos with energies as low as 10 GeV [4].

PINGU will consist of 40 additional strings deployed inside the DeepCore volume with an inter-string spacing of approximately 20 m [5, 14]. Each string will host 96 high quantum efficiency DOMs 3 m apart. The additional optical modules result in a factor of 10 increase in photocathode area density with respect to DeepCore. The closer spacing between the DOMs in PINGU lowers the energy threshold for detection to a few GeV, and improves the reconstruction of the energy and arrival zenith angle of the neutrinos interacting in its volume.

2.1 Signal and background

There are eight distinct neutrino-nucleon interactions that DeepCore can, and PINGU will, observe: charged current (CC) ν_e, ν_μ and ν_τ, and neutral current (NC) interactions of all flavors, both of neutrinos and antineutrinos. Only two of them, CC ν_μ and $\bar{\nu}_\mu$, are expected to produce muons with a range similar to the DOM spacing[1]. All of these interactions below $E_\nu \leq 100$ GeV constitute the signal for oscillation measurements in DeepCore and PINGU. Muon neutrino disappearance is the strongest effect expected, and since most muon neutrinos oscillate into tau neutrinos, ν_τ appearance is also testable. Transitions involving electron neutrinos are relevant in the resonance region. The identification and reconstruction of these events are explained in the following sections.

Atmospheric muons, produced in the same cosmic ray showers as the neutrinos, are a dominant source of background events. In DeepCore data analyses, atmospheric muons are rejected by a variety of veto algorithms that try to correlate the signal observed in DeepCore with the arrival time of photons observed outside the DeepCore volume. The same algorithms are also used to extract a pure atmospheric muon sample from data. These data are used as a background template in the final fit of the simulation. A similar strategy is being implemented for PINGU. After the veto information has been exhausted, the remaining atmospheric muon background is rejected by requiring the events to have a minimum of 5 direct photons (see below) and by applying cuts on the reconstruction quality.

Background events are also introduced by PMT thermal noise and radioactive decays in the glass of the DOM, which become relevant when looking for faint events. Events produced by this noise are also removed by quality cuts on the reconstruction. A more important effect of this background is that it can bias the energy reconstruction and affect the efficiency of veto algorithms. In order to model it correctly, noise simulation is tuned to the data [15].

[1] While 18% of CC ν_τ interactions will also produce muons from tau decay, the expected number of events of this type at energies relevant for atmospheric neutrino oscillations is negligible.

2.2 Reconstruction of neutrino events

All neutrino events are reconstructed using the general hypothesis that at the point of interaction a hadronic shower is produced together with a collinear muon. A muon hypothesis is fit to the event with track lengths ranging from zero (i.e. ultimately a cascade-like event) to the detector's size. Only a handful of the photons emitted in low energy neutrino interactions are observed by DeepCore, and these photons may scatter multiple times before being detected. The latest DeepCore results [12] relied on finding unscattered photons to aid the event selection and event reconstruction. Unscattered, or direct, photons are identified by restricting the patterns formed by their arrival time to a string. While requiring 5 or more direct photons results in a 30% signal efficiency, this criterion keeps the subset of the detected neutrinos whose direction can be easily reconstructed. The muon track direction is fit using these photons, following [16]. Since the light produced by a cascade is also preferentially emitted in the Cherenkov angle events without a muon are also fit, although the resolution obtained is significantly worse. The neutrino interaction position, the energy of the hadronic cascade at the interaction point, and the range of the muon produced are fit in a second step using all photons recorded by the detector while the direction along which these parameters are varied is kept fixed. At $E_\nu = 15\,\mathrm{GeV}$, the resulting median zenith angle resolution is $10°$ for ν_μ and $25°$ for ν_e, while the median energy resolution is 25% for both.

A new method for reconstructing neutrino interactions has been successfuly developed in the last few years. It uses a maximum likelihood estimator where the expected time of arrival of photons to a DOM given a particle hypothesis is finely binned. The effects of the optical properties of the South Pole ice are taken into account. In contrast to the method described above, this approach has the advantages of being able to reconstruct nearly all neutrino events (no direct photons are required) and of producing a significantly better directional reconstruction for events without a muon present. Other resolutions obtained are comparable. The maximum likelihood reconstruction is applied to study the performance of PINGU, where the energy and zenith angle resolutions are expected to improve by approximately 30%.

2.3 Particle identification

Muons with long tracks are used to identify CC ν_μ interactions. In the latest results from DeepCore, neutrino-induced muons are identified by comparing the quality of the muon track directional reconstruction to a cascade fit that assumes the light observed has been randomized by scattering and travels spherically outwards from its emission point. The classification stabilizes at $E_\nu \geq 30\,\mathrm{GeV}$, and above this energy it correctly identifies 60% of CC ν_μ, while misidentifying 30% of cascades. A similar strategy is being developed for the maximum likelihood reconstruction.

The particle identification in PINGU is more complex, as ν_e and ν_μ are expected to contribute to the sensitivity to discriminate the neutrino mass ordering [5]. A multivariate method has been put in place to classify the events according to their topology taking advantage of the reduced inter-string spacing. The classifier shows a stable behavior at $E_\nu \geq 15\,\mathrm{GeV}$, correctly identifying about 80% of CC ν_μ and misidentifying less than 20% of cascade-like events.

3. Analysis methods and systematic uncertainties

Current and future analysis of DeepCore data are aimed at measuring the atmospheric neutrino flux and extracting the oscillation parameters θ_{23} and $|\Delta m^2_{32}|$. The simulation is fit to the data under the assumption of normal and inverse mass ordering, and both results are reported. The CP-violating phase δ is held to zero. Only marginal sensitivity to the neutrino mass ordering is expected from DeepCore data. A main goal of PINGU analyses, on the other hand, is to discriminate between the two possible

Table 1. List of the sources of error included in neutrino oscillation analyses in the latest DeepCore result (DC), ongoing studies of DeepCore data (DC+) and PINGU (P). DIS stands for deep inelastic scattering. BY refers to the six parameters used in the Bodek-Yang model implemented in GENIE.

Source of error		Nominal value	Uncertainty	DC	DC+	P	
Neutrino interactions	Total cross section scaling	GENIE model [18]		Free	x	x	x
	Linear energy dependence		$E^{\pm 0.03}$	x	x	x	
	DIS low Q^2 tuning		BY $\pm[25, 40]\%$		x	x	
	NC scaling		$\pm 10\%$		x	x	
	Quasi-elastic axial mass		$-15\% + 25\%$	x	x	x	
	Resonance axial mass		$\pm 20\%$	x	x	x	
Atmospheric ν and μ flux	Overall scaling	Honda 2015 [19]	Free	x	x	x	
	Spectral index		$E^{\pm 0.04}$	x	x	x	
	Flux ratio $\nu/\bar{\nu}$		$\pm 20\%$		x	x	
	ν_e/ν_μ relative scaling		$\pm 3\%$	x	x	x	
	Atm. μ contamination	From data	Free	x	x		
Detection process	DOM overall efficiency	Muons and flashers	$\pm 10\%$	x	x		
	DOM angular acceptance	Flashers, laser	Up to 50%	x	x		
	Bulk ice model	Flashers	Models in [20, 21]	x	x		
	Hadronic energy scaling	Geant4 [22]	$\pm 5\%$		x	x	

mass orderings. In this measurement, the atmsopheric oscillation parameters would also be measured while the solar parameters and θ_{13} would be taken from global analyses of all available oscillation data [17].

A likelihood ratio method is used both for DeepCore and PINGU to find the best fit between data and simulation, and to estimate the error of the measurement. In DeepCore the method relies on the production of several high statistics simulation sets that span the space of the systematic uncertainties. Sources of these systematic uncertainties are implemented as nuisance parameters and the data is left to select the set of parameters in the simulation that describe it best. The uncertainties on the oscillation parameters are estimated from scanning the likelihood landscape around the best fit point and using a χ^2 approximation. The same method is used for studies which aim at improving the precision of current DeepCore results.

PINGU uses two methods for estimating its sensitivity to the neutrino mass ordering. The likelihood ratio approach is implemented as outlined in [23]. In this method, pseudo-experiments are drawn from two simulation templates, each produced under a different mass ordering. The oscillation parameters requried for the two templates are selected so that they result in the case of "maximum confusion" between hierarchies. Each pseudo-experiment is fit under the assumption of both hierarchies and the likelihood ratio is calculated. The separation between the resulting distributions serves as an estimate of the confidence with which the two hypotheses can be discriminated. The second method used in PINGU is a simplified $\Delta\chi^2$ approach. This is a fast parametric analysis of the Fisher information matrix [24] and relies on the partial derivatives of the event counts in each bin with respect to the parameters under study. The results obtained from the simplified $\Delta\chi^2$ and the likelihood ratio approaches have been found to be in agreement, and both are used to report the projected sensitivity of PINGU.

In all methods discussed, sources of uncertainty are implemented as additional parameters in the fit to the data. Table 1 contains a list of the uncertainties considered for the latest DeepCore result, those being studied for future DeepCore measurements, and those included in the PINGU studies. An additional fine-grained study of the impact of the uncertainties of the neutrino flux predicions for PINGU analyses, not listed in Table 1, can be found in [25]. We note that, directly impacting the measurement of the oscillation parameters, simulation tests indicate that the detector energy scaling is strongly correlated with $|\Delta m^2_{32}|$. The value of $\sin^2\theta_{23}$ shows a dependence on the assumed angular acceptance of the DOM and also on the parameters that modify the angular flux of atmospheric neutrinos. The sensitivity to the neutrino mass ordering depends strongly on the true value of $\sin^2\theta_{23}$.

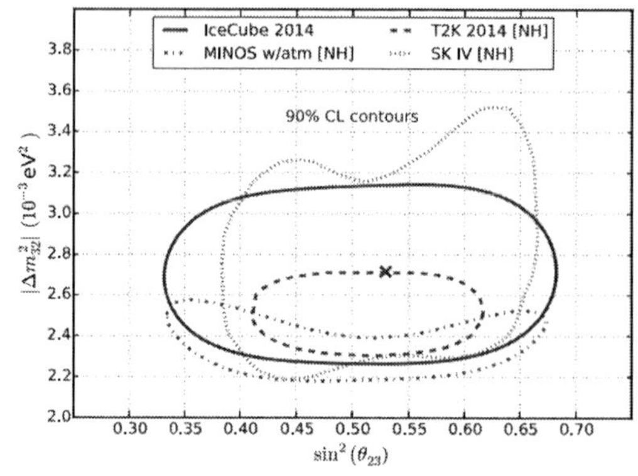

Figure 1. Comparison of best fit and confidence intervals of atmospheric oscillation parameters obtained with DeepCore [12] with those of SuperKamoikande [26], T2K [27] and MINOS [28].

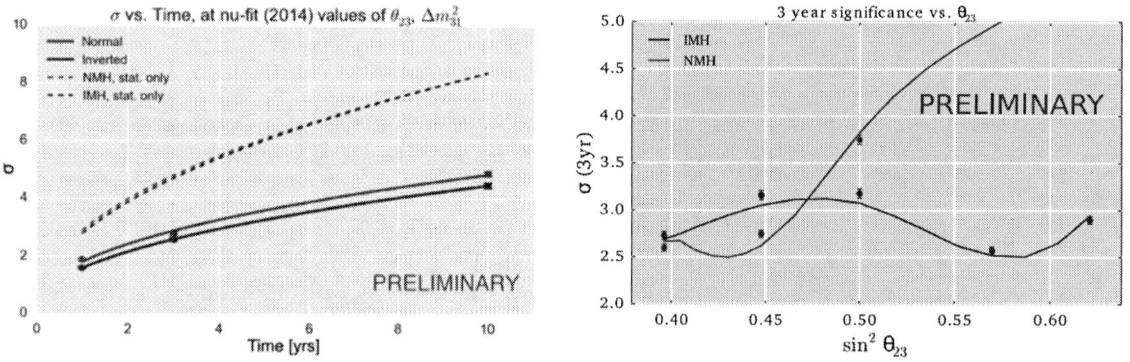

Figure 2. PINGU projections. Left: Expected significance for discriminating the mass ordering as a function of time. Right: Significance as a function of the true value of $\sin^2 \theta_{23}$ at the three year benchmark.

4. Results and projections for DeepCore and PINGU

The latest published results of DeepCore are a muon disappearance analysis of three years of data considering only track-like events with $\cos\theta_{\rm reco} < 0$ and $E_{\rm reco} = [6, 56]$ GeV. Some 5174 events are found in that range. The best fit point is found at $\sin^2\theta_{23} = 0.53^{+0.09}_{-0.12}$ and $\Delta m^2_{32} = 2.72^{+0.19}_{-0.20} \times 10^{-3}\,{\rm eV}^2$. Figure 1 shows the best fit point and the 90% confidence regions as a function of the atmospheric oscillation parameters.

The sensitivity of an eventual re-analysis of the three years data set has been investigated in a simulation study. This study includes a less restrictive event selection (including cascade-like events), events from the complete zenith range, and the maximum likelihood event reconstruction described in Sect. 2.2. The senstivity calculation using these modifications indicates the possibility to reduce the uncertainty on $|\Delta m^2_{32}|$ to half of the currently reported value, and that the uncertainty on $\sin^2\theta_{23}$ may be decreased by 20%.

The predicted sensitivity of PINGU to the neutrino mass ordering as a function of time is shown in Fig. 2 (left), where the accumulated impact of the systematic uncertainties considered is also shown. These studies indicate that PINGU will be able to identify the neutrino mass ordering with a significance of 3σ after 3–3.5 years of operation. The sensitivity dependence on the true value of $\sin^2\theta_{23}$ at the 3−years benchmark is shown in Fig. 2 (right). A favorable combination of true physics parameters results in a faster identification of the mass ordering.

5. Outlook

Atmospheric neutrinos are a valuable tool to study neutrino oscillations. IceCube's DeepCore has demonstrated that very large volume neutrino telescopes can use atmospheric neutrinos to make meaningful contributions to the field. Current results have started to approach the precision of dedicated experiments. Studies indicate that the full potential has not been reached and the addition of the PINGU array can grealty improve on these measurements by accumulating large samples of very well reconstructed GeV-scale neutrinos and ultimately identify the neutrino mass ordering.

References

[1] K. Olive et al. (Particle Data Group), Chin. Phys. **C38**, 090001 (2014)
[2] J.P. Yañez, A. Kouchner, Adv. High Energy Phys. **2015**, 271968 (2015), 1509.08404
[3] S. Adrian-Martinez et al. (ANTARES Collaboration), Phys. Lett. **B714**, 224 (2012), 1206.0645
[4] R. Abbasi et al. (IceCube Collaboration), Astropart. Phys. **35**, 615 (2012), 1109.6096
[5] M. Aartsen et al. (IceCube-PINGU Collaboration) (2014), 1401.2046
[6] J. Brunner (KM3NeT Collaboration), in *ICRC* (2015)
[7] L. Wolfenstein, Phys. Rev. **D17**, 2369 (1978)
[8] S.P. Mikheev, A.Yu. Smirnov, Sov. J. Nucl. Phys. **42**, 913 (1985), [Yad. Fiz.42,1441(1985)]
[9] V. Ermilova et al., Short Notices of the Lebedev Institute **5**, 26 (1986)
[10] E.K. Akhmedov, Sov. J. Nucl. Phys. **47**, 301 (1988)
[11] M. Aartsen et al. (IceCube Collaboration), Phys. Rev. Lett. **111**, 081801 (2013), 1305.3909
[12] M. Aartsen et al. (IceCube Collaboration), Phys. Rev. **D91**, 072004 (2015), 1410.7227
[13] A. Achterberg et al. (IceCube Collaboration), Astropart. Phys. **26**, 155 (2006)
[14] K. Hanson (IceCube-Gen2 Collaboration), in *VLVnT Workshop, Rome* (2015)
[15] M. Larson, Master's thesis, University of Alabama (2013)
[16] J.A. Aguilar et al. (ANTARES Collaboration), Astropart. Phys. **34**, 652 (2011), 1105.4116
[17] M. Gonzalez-Garcia, M. Maltoni, T. Schwetz, JHEP **1411**, 052 (2014), 1409.5439
[18] C. Andreopoulos et al., Nucl. Instrum. Methods **A614**, 87 (2010), 0905.2517
[19] M. Honda et al., Phys. Rev. **D92**, 023004 (2015), 1502.03916
[20] M.G. Aartsen et al. (IceCube Collaboration), Nucl. Instrum. Methods **A711**, 73 (2013)
[21] D. Chirkin (IceCube Collaboration), in *ICRC* (2013), 1309.7010
[22] S. Agostinelli et al., Nucl. Instrum. Methods A **506**, 250 (2003)
[23] D. Franco et al., JHEP **1304**, 008 (2013), 1301.4332
[24] R.A. Fisher, Phil. Trans. Roy. Soc. Lond. **A222**, 309 (1922)
[25] J. Sandroos (IceCube-Gen2 Collaboration), in *VLVnT Workshop, Rome* (2015)
[26] M. Nakahata (Super-Kamiokande Collaboration), in *XVI International Workshop on Neutrino Telescopes, Venice* (2015)
[27] K. Abe et al. (T2K Collaboration), Phys. Rev. **D91**, 072010 (2015), 1502.01550
[28] P. Adamson et al. (MINOS Collaboration), Phys. Rev. Lett. **110**, 251801 (2013), 1304.6335

EPJ Web of Conferences **116**, 11010 (2016)
DOI: 10.1051/epjconf/201611611010

Phenomenology of atmospheric neutrinos

Anatoli Fedynitch[a]

Karlsruher Institut für Technologie, Institut für Kernphysik, Postfach 3640, 76021 Karlsruhe, Germany

Abstract. The detection of astrophysical neutrinos, certainly a break-through result, introduced new experimental challenges and fundamental questions about acceleration mechanisms of cosmic rays. On one hand IceCube succeeded in finding an unambiguous proof for the existence of a diffuse astrophysical neutrino flux, on the other hand the precise determination of its spectral index and normalization requires a better knowledge about the atmospheric background at hundreds of TeV and PeV energies. Atmospheric neutrinos in this energy range originate mostly from decays of heavy-flavor mesons, which production in the phase space relevant for prompt leptons is uncertain. Current accelerator-based experiments are limited by detector acceptance and not so much by the collision energy. This paper recaps phenomenological aspects of atmospheric leptons and calculation methods, linking recent progress in flux predictions with particle physics at colliders, in particular the Large Hadron Collider.

1. Introduction

Hadronic interactions of high energy cosmic rays (CR) with the atmosphere produce extensive cascades of secondary particles. The electromagnetic component of electrons and γ-rays dissipates its energy in form of bremsstrahlung and pair production. In a particle cascade, hadrons can either interact, producing a sub-cascade similar to the original one but at lower energy, or decay into leptons or other hadrons. Since leptons interact very weakly with the atmosphere, they reflect the history of the air-shower development and easily reach down to the surface.

A Monte Carlo program seems to be the most reasonable calculation method, because by using random numbers it can generate an infinite number of cascade topologies and also employ detailed physical models of sub-leading processes, such as multiple scattering. The high degree of realism is, on the other hand, a drawback of the Monte Carlo calculation, since the history of every particle has to be traced, and there are billions of particles in a high energy particle shower. Due to the requirements on computing time and storage, one has to deal with a limited number of repetitions constraining the achievable precision. Rare processes or subtle signatures might therefore be smeared out by the statistical error. The flux of long-lived mesons and the resulting conventional lepton flux component attenuate quickly at very high energy due to re-interactions with air nuclei. Biasing techniques in

---

[a] e-mail: anatoli.fedynitch@cern.ch

simulation, as implemented in FLUKA [1, 2] or in upcoming CORSIKA [3] versions, overcome this problem to some extent.

Additional challenges stem from the inclusion of very rare processes such as the decay of heavy flavor, charm or bottom, hadrons inside air-showers. Heavy mesons, mostly $D^{\pm}$ and D^0, decay immediately and produce prompt muons and neutrinos. This component does not suffer from re-interactions with air up to energies of several tens of PeV. But a drop in calculation efficiency comes from difference in inclusive production cross-sections. The ratio of production cross-sections in a p-air collision at 100 PeV and high $x = \frac{E_{\text{secondary}}}{E_{\text{CR}}} \sim 0.5$ is $\sim \sigma_{\pi^{\pm}}/\sigma_{D^{\pm}} \sim 10^{-4} - 10^{-5}$, i.e. 10^4 simulated showers yield on average a few prompt leptons. At this point one can start introducing additional approximations or optimizations to further reduce computational time in exchange for precision, loosing at some point the justification to run complex Monte Carlo codes at all.

So, why should we be interested in a very efficient calculation? First of all, because the models used as ingredients in an atmospheric flux calculation carry sizable uncertainties coming from models of primary cosmic ray spectrum and composition, from the seasonal and geographic variability of the atmosphere and from the modeling of hadronic interactions. By having the possibility to easily re-run calculations with varying model assumptions we can propagate the effect of each model uncertainty on the input side to the flux prediction.

In the next section a numerical method is introduced, which competes with full Monte Carlo calculations in terms of precision, but requires only a tiny fraction of computational time.

2. Matrix cascade equations

The flux of particles in air is described by the coupled cascade integro-differential equations

$$\begin{aligned}\frac{\mathrm{d}\Phi_h(E,X)}{\mathrm{d}X} = &-\frac{\Phi_h(E,X)}{\lambda_{int,h}(E)}\\ &-\frac{\Phi_h(E,X)}{\lambda_{dec,h}(E,X)}\\ &+\sum_l \int_E^{\infty} dE_l \, \frac{\mathrm{d}N^{int}_{l(E_l)\to h(E)}}{\mathrm{d}E} \frac{\Phi_l(E_l,X)}{\lambda_{int,l}(E_l)}\\ &+\sum_l \int_E^{\infty} dE_l \, \frac{\mathrm{d}N^{dec}_{l(E_l)\to h(E)}}{\mathrm{d}E} \frac{\Phi_l(E_l,X)}{\lambda_{dec,l}(E_l,X)}. \end{aligned} \tag{1}$$

Particle interactions and decays are the two competing processes and it is convenient to write the equation as a function of the slant depth X [g/cm^2]. Continuous energy losses arising from the passage of charged particles through air can be neglected at energies above several tens of GeV. Index h determines the type of particle, like π, K, η, $D^{\pm}$, μ, ν, *etc*. The sums over all cascade members l define the couplings between the evolution equations for each particle type. The inclusive particle production spectra or redistribution functions $\frac{dN}{dE}$ describe the spectrum of secondaries created in particle interactions or decays. The process rates are defined in terms of interaction and decay lengths λ in units of X. More details about the equation and solutions can be found in [4, 5].

In a typical semi-analytical approach one tries to decouple the system by neglecting rare processes, for example, the production of nucleons in secondary pion interactions or the production of K^- in interactions of K^+. Often these approximations are valid under certain restrictions, such as under the assumption of steep power-law primary spectra or Feynman-scaling of particle production spectra.

One can criticize that additional uncertainties due to the calculation method are introduced (and need quantification), on top of those from the physical models themselves.

An approach aimed to avoid these calculation uncertainties is based on numerical integration of the coupled system by discretization on an energy grid. Eq. (1) becomes

$$\frac{\mathrm{d}\Phi^h_{E_i}}{\mathrm{d}X} = -\frac{\Phi^h_{E_i}}{\lambda^h_{int,E_i}} + \sum_{E_k \geq E_i}^{E_N} \sum_l \frac{c_{l(E_k)\to h(E_i)}}{\lambda^l_{int,E_k}} \Phi^l_{E_k}$$
$$-\frac{\Phi^h_{E_i}}{\lambda^h_{dec,E_i}(X)} + \sum_{E_k \geq E_i}^{E_N} \sum_l \frac{d_{l(E_k)\to h(E_i)}}{\lambda^l_{dec,E_k}(X)} \Phi^l_{E_k}. \tag{2}$$

where the interaction coefficients

$$c_{l(E_l)\to h(E_h)} = \Delta E_l \frac{\mathrm{d}N_{l\to h}(E_l)}{\mathrm{d}E_h}(E_h) \tag{3}$$

absorb the finite interval size and the inclusive production cross-sections. The energy grid extends from 50 to 10^{10} GeV using 8-9 bins per decade. Analogously to Eq. (3) the inclusive decay coefficients

$$\frac{\mathrm{d}N^{dec}_{l\to h}(E_l)}{\mathrm{d}E} = \sum_i BR_{i,l\to h} \frac{\mathrm{d}N_{i,h}}{\mathrm{d}E} \quad \text{and} \quad d_{l(E_l)\to h(E_h)} = \Delta E_l \frac{\mathrm{d}N^{dec}_{l\to h}(E_l)}{\mathrm{d}E_h}(E_h) \tag{4}$$

are obtained by summing over all spectra of exclusive branching channels i that contain the secondary particle h.

By combining the solutions for each particle type h and each energy bin E_i in a vector

$$\vec{\Phi} = \left(\vec{\Phi}^p \quad \vec{\Phi}^n \quad \vec{\Phi}^{\pi^+} \quad \ldots \vec{\Phi}^{\bar{\nu}_\mu} \ldots\right)^T, \quad \text{where} \quad \vec{\Phi}^p = \left(\Phi^p_{E_0} \quad \Phi^p_{E_1} \quad \cdots \quad \Phi^p_{E_N}\right)^T. \tag{5}$$

the coupled system of equations can be expressed in matrix notation

$$\frac{\mathrm{d}}{\mathrm{d}X}\vec{\Phi} = \left[(-\mathbf{1} + \boldsymbol{C})\boldsymbol{\Lambda}_{int} + \frac{1}{\rho(X)}(-\mathbf{1} + \boldsymbol{D})\boldsymbol{\Lambda}_{dec}\right]\vec{\Phi}. \tag{6}$$

The interaction and decay lengths are arranged in diagonal rate ($\sim 1/\lambda$) matrices $\boldsymbol{\Lambda}$, where the air density ρ dependence has been factorized out. The interaction or decay coefficients from Eqs. (3) and (4) are arranged in sparse matrices $\boldsymbol{C}$ and $\boldsymbol{D}$.

By numerically integrating Eq. (6), the coupled cascade equations are evolved collectively for all contributing particle types. Numerical issues arise when short-lived hadrons are included, such as D mesons or Λ_C baryons, since these particles decay on a much shorter timescale compared to conventional hadrons. By introducing a special type of steady-state approximation, the so-called resonance approximation, these stiffness problems can be resolved. The full system of the order 6000×6000 can be solved on an ordinary PC in a few seconds. More details about this approach and general results can be found in [6, 7]. The code "MCEq" is available online[1].

2.1 Connection with LHC

The Large Hadron Collider (LHC) at CERN with its center of mass energies between 900 GeV and 13 TeV covers an energy regime which is very relevant for inclusive lepton fluxes. Figure 1 shows that the LHC reaches cosmic ray interaction energies at which 1 PeV to 10 PeV atmospheric neutrinos are produced. However, the data do not constrain significantly atmospheric flux calculations.

---

[1] `https://github.com/afedynitch/MCEq`

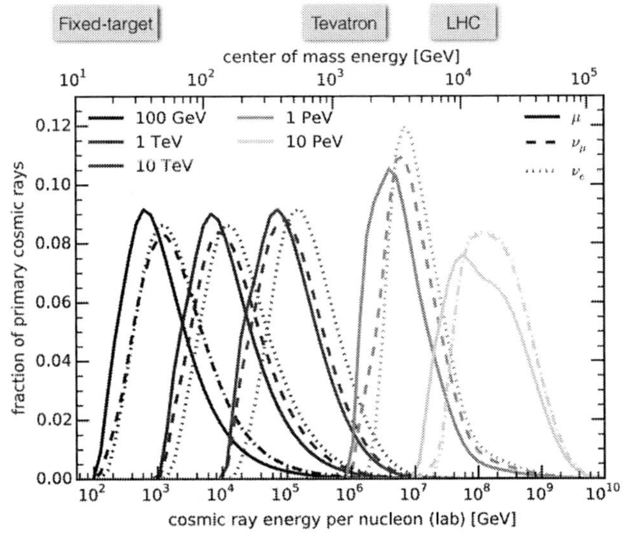

Figure 1. Energy spectrum for priamry cosmic rays that lead to the production of mono-energetic secondary leptons. Line styles represent different leptons and colors the lepton energy.

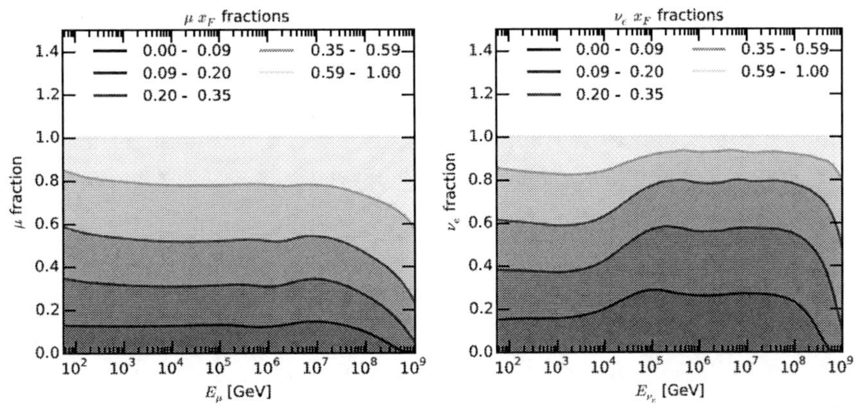

Figure 2. Contribution of (Feynman-)x ranges in the calculation of the inclusive atmospheric lepton flux for muons (left) and electron neutrinos (right).

One of the reasons is that the LHC is running in pp, pPb or $PbPb$ mode. Lead's high mass produces an extreme energy density emphasizing collective effects of Quark-Gluon-Plasma rather than the role of identified high-x mesons which are of higher importance for cosmic ray applications. These data is therefore of limited use for cosmic ray interaction model developers who aim to constrain nuclear effects up to the iron mass. This situation can be significantly improved by running the LHC with carbon or nitrogen nuclei.

A more serious constraint is the phase space coverage of current LHC experiments. Except a few dedicated roman pots or zero-degree calorimeters, none of the general purpose detectors covers the x-range relevant for inclusive fluxes. Typical Feynman-x_{F} of secondary particles rarely exceed 0.05 before leaving the acceptance region. Using the matrix code it is possible to study the contribution of ranges in $x \approx x_{\mathrm{F}}$ (at high energies) to the total inclusive flux. Bands in Fig. 2 illustrate the contribution of different kinematical regions to the inclusive lepton flux by using the ability of the numerical integrator to apply cuts on ranges in x. Less than 10% of conventional muons and muon neutrinos originate from

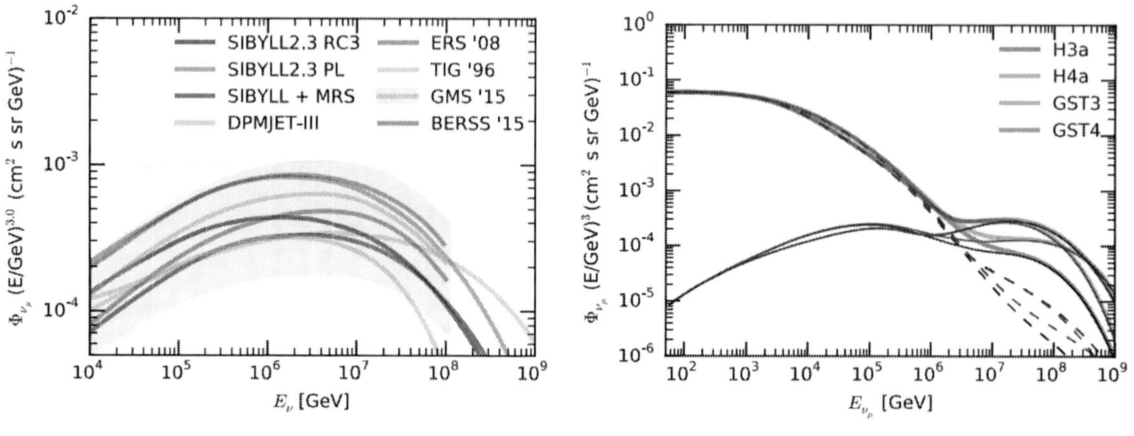

Figure 3. Prompt muon or electron neutrino flux calculated for a simple broken power-law proton spectrum vs. other calculations described in the text (left). Muon neutrino flux calculated using different composition scenarios of ultra-high energy CR (right). Dashed lines represent the conventional flux.

secondaries created at an x value < 0.09, giving an idea to what extent an ordinary LHC detector with particle identification capabilities would experimentally constrain the flux prediction. Muon neutrinos exhibit a pronounced shift towards lower x above 100 TeV, where prompt neutrinos dominate. They are sensitive to the production properties of charmed mesons which are predicted to be distributed more centrally in SIBYLL-2.3 [8, 9]. These results are model dependent and by exchanging the interaction model some shifts in the importance of x-ranges can be observed. The central result is that the interaction models are weakly constrained by LHC observations in the phase space relevant for atmospheric leptons.

2.2 Comments on prompt neutrino calculations

The prompt component is one of the dominant backgrounds in the determination of the spectral index of the astrophysical neutrinos with IceCube [10]. For an unambiguous identification of an up-going astrophysical flux a precise determination of the contribution from prompt neutrinos is even of crucial importance. So far, IceCube has not claimed any a signal from prompt leptons. However, we should be certain that this flux exists, since heavy flavor mesons are created in all particle collisions of sufficient energy.

We have estimated this flux using a preliminary version of the SIBYLL 2.3 Monte Carlo generator, which has been cross-checked against LHC data, and a new, preliminary, version of DPMJET-III [11]. In the left panel of Fig. 3 is shown an updated version of our plot from [6]. It compares more popular calculations ERS [12], TIG [13] and MRS [14] with the more recent BERSS [15] and GMS [16] calculations. The band around the GMS line represents uncertainties from the inputs of the next-to-leading order (NLO) perturbative calculation, i.e. renormalization and factorization scale, parton distribution function, charm mass and parton shower matching. A similar calculation [17], aiming for extraction of state-of-the-art knowledge about forward charm production in accordance with LHCb results, arrives at comparable errors. On one hand, the high uncertainty is expected given the limited acceptance of LHC experiments with respect to the phase space relevant for atmospheric leptons. On the other hand, none of the listed calculations, although carried out with different methods and models, contradicts the accelerator-based estimation. This is encouraging, since a measurement of the flux normalization by neutrino telescopes, like IceCube or ANTARES, will give insight into small-x physics inaccessible to the current collider generation.

However, the flux variation due to the cosmic-ray composition is of similar magnitude, as it can be seen by comparing with the right panel of Fig. 3. H3a [18] and GST3 [19] are mixed composition scenarios, while H4a and GST4 are model versions with only protons at energies above the ankle of cosmic ray energy spectrum. With more data and higher precision from the upgraded versions of the Pierre Auger Observatory and Telescope Array it should be possible to further constrain this particular uncertainty.

I am grateful to Ralph Engel, Felix Riehn, Thomas K. Gaisser and Todor Stanev for inspiring discussions.

References

[1] A. Ferrari, P.R. Sala, A. Fassò, J. Ranft, *FLUKA: A multi-particle transport code (program version 2005)* (CERN, Geneva, 2005)
[2] T.T. Böhlen, F. Cerutti, M.P.W. Chin, A. Fassò, A. Ferrari, P.G. Ortega, A. Mairani, P.R. Sala, G. Smirnov, V. Vlachoudis, Nuclear Data Sheets **120**, 211 (2014)
[3] J., Kyle, *CORSIKA Modifications for Faster Background Generation* (Presented at VLVνT, 2015)
[4] T.K. Gaisser, *Cosmic rays and particle physics* (Cambridge Univ Pr, 1990)
[5] P. Lipari, Astropart. Phys. **1**, 195 (1993)
[6] A. Fedynitch, R. Engel, T.K. Gaisser, F. Riehn, T. Stanev, PoS(ICRC2015)1129 (2015)
[7] A. Fedynitch, R. Engel, T.K. Gaisser, F. Riehn, T. Stanev, EPJ Web Conf. **99**, 08001 (2015), 1503.00544
[8] F. Riehn, R. Engel, A. Fedynitch, T.K. Gaisser, T. Stanev, PoS(ICRC2015)558 (2015)
[9] F. Riehn, R. Engel, A. Fedynitch, T.K. Gaisser, T. Stanev, EPJ Web Conf. **99**, 12001 (2015), 1502.06353
[10] M.G. Aartsen et al. (IceCube), Science **342**, 1242856 (2013), 1311.5238
[11] S. Roesler, R. Engel, J. Ranft (2000), hep-ph/0012252
[12] R. Enberg, M.H. Reno, I. Sarcevic, Physical Review D **78**(4), 43005 (2008)
[13] M. Thunman, G. Ingelman, P. Gondolo, Astroparticle Physics **5**(3-4), 309 (1996)
[14] A.D. Martin, M.G. Ryskin, A.M. Stasto, Acta Physica Polonica B **34**(6), 3273 (2003), hep-ph/0302140
[15] A. Bhattacharya, A. Stasto, I. Sarcevic, R. Enberg, M.H. Reno, JHEP **1506**(6), 110 (2015), 1502.01076
[16] M.V. Garzelli, S. Moch, G. Sigl, JHEP **10**, 115 (2015), 1507.01570
[17] R. Gauld, J. Rojo, L. Rottoli, J. Talbert, JHEP **11**, 009 (2015), 1506.08025
[18] T.K. Gaisser, Astroparticle Physics **35**(12), 801 (2012)
[19] T.K. Gaisser, T. Stanev, S. Tilav, Frontiers of Physics **8**(6), 748 (2013)

EPJ Web of Conferences **116**, 11011 (2016)
DOI: 10.1051/epjconf/201611611011

Neutrino-nucleus interactions in the few-GeV region

J. Nieves[a]

Instituto de Física Corpuscular (IFIC), Centro Mixto CSIC-Universidad de Valencia, Institutos de Investigación de Paterna, 46071 Valencia, Spain

Abstract. Knowledge of neutrino interaction cross sections is an important and necessary ingredient in any neutrino measurement and it is crucial to achieve the precision goals of the neutrino oscillation program, making possible new discoveries like the CP violation in the leptonic sector. We briefly review the major features of the neutrino cross sections in the few-GeV region, paying special attention to nuclear effects.

1. Introduction

Recent interest in neutrino interactions in the few GeV energy region comes from neutrino oscillation experiments and their need to reduce systematic errors. This is because neutrinos are detected through their interactions with the nuclei that form part of the detectors. Therefore, a better understanding of weak interaction mechanisms in dense matter is actively pursued [1–3]. However, the relevance of neutrino interactions with matter extends over a large variety of topics in astrophysics, physics beyond the Standard Model, hadronic physics and nuclear physics. Thus, for instance, the interactions of electroweak probes with nuclei are interesting as a tool to discover properties of hadrons and their modification in the nuclear medium, since they penetrate deep inside the nucleus assessing the densest regions. This allows to study various aspects of the many-body problem such as nuclear correlations, meson exchange currents and relativistic effects. At intermediate energies, it is also possible to investigate the excitation of baryonic resonances and their in-medium modifications. Moreover, the medium effects in these processes can be linked to chiral symmetry restoration in nuclear matter, and shed light to some parts of the QCD phase diagram.

At low energies the neutrino interacts with composite entities such as nucleons or nuclei. Given enough energy, the neutrino can actually begin to resolve the internal structure of the target: the neutrino can scatter off an individual quark inside the nucleon: DIS (deep inelastic scattering) and it manifests in the creation of a hadronic shower.

We will focus on the neutrino charged-current (CC) or neutral current (NC) inclusive reaction:

$$\nu_l + A_Z \to l^- + X, \quad \nu_l + A_Z \to \nu_l + X. \tag{1}$$

---

[a] e-mail: jmnieves@ific.uv.es

The double differential cross section, with respect to the outgoing lepton kinematical variables is given in the Laboratory (LAB) frame by

$$\frac{d^2\sigma_{\nu_l}}{d\Omega(\hat{k}')dE'_l} = \frac{|\vec{k}'|}{|\vec{k}\,|}\frac{G_F^2}{4\pi^2\eta}L^{(\nu)}_{\mu\sigma}W^{\mu\sigma} \tag{2}$$

with $\vec{k}$ and $\vec{k}'$ the LAB lepton momenta, E'_l the energy of the outgoing lepton, $G_F = 1.1664 \times 10^{-11}\,\mathrm{MeV}^{-2}$, and L and W are the leptonic and hadronic tensors, respectively. Besides, $\eta = 1(4)$ for CC (NC) processes. The leptonic tensor reads ($\epsilon_{0123} = +1$)

$$L^{(\nu)}_{\mu\sigma} = L^s_{\mu\sigma} + \mathrm{i}L^a_{\mu\sigma} = k'_\mu k_\sigma + k'_\sigma k_\mu - g_{\mu\sigma}k\cdot k' + \mathrm{i}\epsilon_{\mu\sigma\alpha\beta}k'^\alpha k^\beta. \tag{3}$$

The hadronic tensor includes a collection of non-leptonic vertices and corresponds to the charged or neutral electroweak transitions of the target to all possible final states. It is completely determined by six independent, Lorentz scalar and real, structure functions. After contracting with the leptonic tensor, one obtains that for massless leptons[1] only three of them, $F^\nu_{1,2,3}(x,q^2)$, are relevant, namely

$$\frac{d^2\sigma_{\nu_l}}{dxdy} = \frac{G_F^2 ME_\nu}{\eta\pi}\left\{\left(1 - y - \frac{Mxy}{2E_\nu}\right)F^\nu_2 + xy^2F^\nu_1 + y(1-y/2)xF^\nu_3\right\} \tag{4}$$

with E_ν the incoming neutrino energy, M the nucleon mass, $q = k - k'$, $x = -q^2/2Mq^0$, $y = q^0/E_\nu$. The cross section for the CC antineutrino induced nuclear reaction is easily obtained by i) changing the sign of the parity-violating term, proportional to F_3, in the differential cross section (this is because $L^{(\bar\nu)}_{\mu\sigma} = L^{(\nu)}_{\sigma\mu}$.), Eq. (4), and ii) using the appropriate weak current in the definition/computation of the hadron tensor. In the case of antineutrino NC driven processes, it is only needed to flip the sign of the term proportional to F_3 in the differential cross section, since the hadron NC is not affected.

The hadronic tensor is determined by the W or Z gauge boson selfenergy in the nuclear medium. To evaluate this latter object requires a theoretical scheme, where the relevant degrees of freedom and nuclear effects could be taken into account.

2. High energy cross sections $E_\nu \gg 1-5$ GeV, $Q^2 = -q^2 \gg 1$ GeV^2

This kinematic region is mostly out of the resonance production region and it is fully dominated by DIS processes. DIS offers an opportunity to probe the partonic structure of the nucleon both in its free state and when the nucleon is bound in a nucleus. Actually DIS has been extensively used to validate the standard model and probe nucleon structure. The structure of the nucleon is described in terms of parton distribution functions (PDFs) giving the longitudinal, transverse, and spin distributions of quarks within the nucleon. Neutrino scattering can play an important role in extraction of these fundamental PDFs since only neutrinos via the CC weak interaction can resolve the flavor of the nucleon's constituents: ν interacts with $d, s, \bar{u}$ and $\bar{c}$ while the $\bar\nu$ interacts with $u, c, \bar{d}$ and $\bar{s}$. High-statistics measurement of the nucleon's partonic structure, using neutrinos, could complement studies with electromagnetic probes.

In DIS processes, the neutrino scatters off a quark in the nucleon via CC or NC interactions producing a lepton in the final state, together with a hadronic shower. The Bjorken scaling variable, x introduced in Eq. (4) plays an important role in deep inelastic neutrino scattering, where the quark can carry a portion of the incoming energy momentum of the struck target. The variable y is related to the energy of the hadronic system.

[1] One should bear in mind that in the case of CC ν_τ interactions, the $\tau-$lepton mass produces large differences [2].

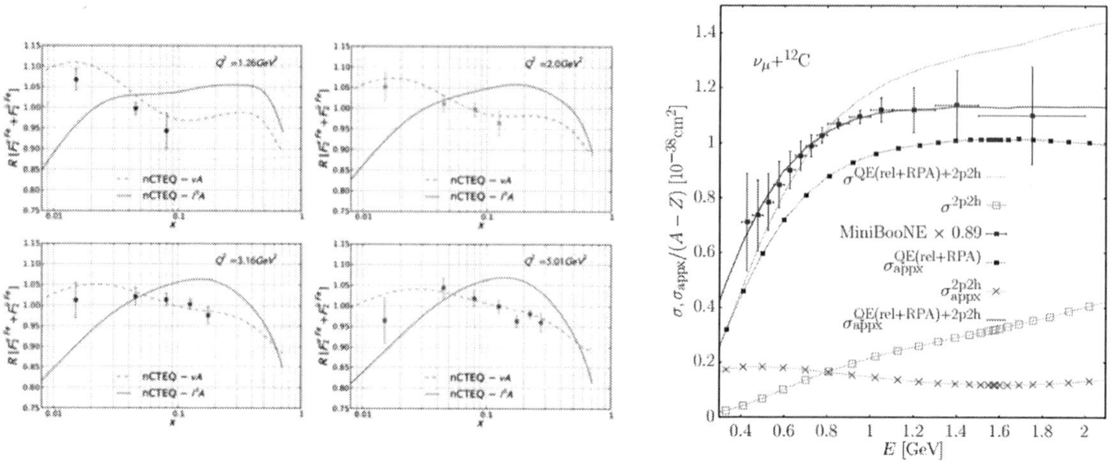

Figure 1. Left: Nuclear correction factor R for the average F_2 structure function in CC ν, $\bar{\nu}$ Fe scattering at $Q^2 =$ 1.26, 2.0, 3.16, and 5.01 GeV^2 compared to the measured NuTeV points [6]. The dashed lines show the result of the nCTEQ [5] analysis of νA (CHORUS [7], CCFR/NuTeV [8]) differential cross sections plotted in terms of the average $F_2^{\rm Fe}$ divided by the results obtained with the reference fit (free proton) PDFs. For comparison, the solid curve stands for the nCTEQ fit [5] to the charged-lepton data. Right: ν_μ+^{12}C cross sections. Theoretical (σ) and approximate (σ_{appx}) CCQE-like integrated cross sections as a function of the ν energy (see Ref. [9] for further details). The MiniBooNE data [10] and errors (shape) have been re-scaled by a factor 0.9, as a result of the analysis carried out in Ref. [11]. All theoretical results have been obtained with the model of Refs. [12, 13] and $M_A = 1.05$ GeV.

Assuming the quark parton model, the structure functions $F_i(x, Q^2)$ in Eq. (4), are expressed in terms of the quark composition of the target. In the case of scattering off nucleons,

$$F_2(x, Q^2) = 2 \sum_{i=u,d,...} \left[xq(x, Q^2) + x\bar{q}(x, Q^2)\right], \quad xF_3(x, Q^2) = 2 \sum_{i=u,d,...} \left[xq(x, Q^2) - x\bar{q}(x, Q^2)\right]. \tag{5}$$

The struck quark carries a fraction x of the nucleon's momentum, such that xq ($x\bar{q}$) is the probability of finding the quark (antiquark) with a given momentum fraction (PDFs). Besides,

$$2xF_1(x, Q^2) = \frac{\left(1 + 4M^2x^2/Q^2\right) F_2(x, Q^2)}{1 + R_L(x, Q^2)} \tag{6}$$

where $R_L(x, Q^2) = \sigma_L/\sigma_T$ is the ratio of cross sections for scattering off longitudinally and transversely polarized exchange bosons. In the pursuit of precision measurements of neutrino oscillation parameters, large data samples and dedicated effort to minimize systematic errors have allowed charged lepton and neutrino experiments to independently isolate all six of the CC weak structure functions $F_1^{\nu N}(x, Q^2)$, $F_1^{\bar{\nu} N}(x, Q^2)$, $F_2^{\nu N}(x, Q^2)$, $F_2^{\bar{\nu} N}(x, Q^2)$, $F_3^{\nu N}(x, Q^2)$, $F_3^{\bar{\nu} N}(x, Q^2)$ over a wide range of x and Q^2 values [4]. Taking differences and sums of these structure functions, specific PDFs in a given x, Q^2 bin can in turn be better isolated. Extracting this full set of structure functions relies on the $y-$variation of the structure function coefficients in the expression for the cross-section.

To achieve realistic descriptions of DIS processes, additional effects, like non-vanishing lepton masses, QCD corrections, heavy quark or nuclear and radiative corrections need also to be considered. These corrections are typically well understood and do not produce large uncertainties in the cross sections [2]. We illustrate this for the case of nuclear corrections in the left panels of Fig. 1. There, we see nuclear corrections turn out to be small, of the order of 5%, and are successfully described by the nCTEQ analysis [5]. In general, there exist some tension between charged-lepton ($l^{\pm}A$) and the neutrino

(νA) data, as these plots suggest, and the size of the nuclear corrections extracted from neutrino data are smaller than those obtained from charged lepton scattering.

3. Intermediate energy cross sections

Below the DIS region, the neutrino scatters off nucleons rather than its constituent partons. We distinguish between inelastic and elastic or quasielastic processes.

3.1 Inelastic processes

Given enough energy, neutrinos can excite the struck nucleon to an excited state. In this case, the neutrino interaction produces a baryon resonance N^*. The baryon resonance quickly decays, most often to a nucleon and single pion final state. Resonances can also decay to multipion final states, and in general neutrino interactions in this energy range can also produce final states involving kaons and hyperons. DIS is also a source of multipion and kaon final states depending on the neutrino energy.

Because of the inherent complexity of reconstructing multiple pions final states, there are not many existing experimental measurements of this process. All of the existing measurements have been performed strictly using deuterium-filled bubble chambers.

The main reaction channels involving strange quarks are K $(\Delta S = 1)$ and $\bar{K}$ $(\Delta S = -1)$ production

$$\nu_l\, p \to l^-\, K^+\, p, \qquad \bar{\nu}_l\, p \to l^+\, K^-\, p, \tag{7}$$

$$\nu_l\, n \to l^-\, K^+\, n, \qquad \bar{\nu}_l\, n \to l^+\, K^-\, n, \tag{8}$$

$$\nu_l\, n \to l^-\, K^0\, p, \qquad \bar{\nu}_l\, p \to l^+\, \bar{K}^0\, n, \tag{9}$$

and CC and NC associated strangeness production $(\Delta S = 0)$

$$\nu_l\, n \to l^-\, K^+\, (\Lambda, \Sigma^0), \qquad (\nu_l, \bar{\nu}_l)\, p \to (\nu_l, \bar{\nu}_l)\, K^+\, (\Lambda, \Sigma^0), \tag{10}$$

$$\nu_l\, n \to l^-\, K^0\, \Sigma^+, \qquad (\nu_l, \bar{\nu}_l)\, n \to (\nu_l, \bar{\nu}_l)\, K^+\, \Sigma^-, \tag{11}$$

$$\nu_l\, p \to l^-\, K^+\, \Sigma^+, \qquad (\nu_l, \bar{\nu}_l) n \to (\nu_l, \bar{\nu}_l)\, K^0\, (\Lambda, \Sigma^0), \tag{12}$$

$$\bar{\nu}_l\, p \to l^+\, K^+\, \Sigma^-, \qquad (\nu_l, \bar{\nu}_l)\, p \to (\nu_l, \bar{\nu}_l)\, K^0\, \Sigma^+, \tag{13}$$

$$\bar{\nu}_l\, p \to l^+\, K^0\, (\Lambda, \Sigma^0), \tag{14}$$

$$\bar{\nu}_l\, n \to l^+\, K^0\, \Sigma^-. \tag{15}$$

Associated strangeness production is the dominant process but it has a high threshold because both a kaon and a hyperon are produced. This leaves single kaon production as the main source of kaons at low energies, in spite of being Cabibbo suppressed. These reactions typically have small cross sections because the kaon channels are not enhanced by any dominant resonance.

Weak pion production off nucleons cross sections grow to become one of the main reaction mechanisms for neutrinos of few-GeV energies. The first requirement to put neutrino induced pion production on nuclear targets on a firm ground is to have a realistic model at the nucleon level. Most of the models describe the pion production process by means of the weak excitation of the $\Delta(1232)$ resonance and its subsequent decay into $N\pi$, and in some occasions incorporate background terms. The major part of the models includes also the weak excitation of higher resonances as intermediate states. Vector form factors are fixed from helicity amplitudes extracted in the analysis of pion electroproduction data, while the axial couplings are obtained from PCAC [3]. Last years have witnessed an intense experimental and theoretical activity in this topic (see for instance Refs. [1–3] and references therein). Here, we will only make two remarks. The first one is that the Rein–Sehgal model [14], used by almost all Monte Carlo generators, provides a really poor description of the pion electroproduction data on protons [15, 16]. Indeed, the model underestimates significantly the electron data, and it also reveals itself unsatisfactory in the axial sector at $q^2 = 0$, where the divergence of axial current can be related to the πN amplitude by PCAC. Secondly, we would like to stress that, in sharp contrast to the previous reactions, weak pion production in dense matter is strongly affected by nuclear corrections, which might not be under control. Among these corrections, those that account for the interactions with the nucleons of the particles produced inside the nucleus on their way out are important. These corrections are commonly referred as final state interaction (FSI) effects. As example of the theoretical difficulties faced we refer the reader to the MiniBooNE flux-folded differential $d\sigma/dp_\pi$ cross section data in mineral oil reported in Ref. [17], which cannot be described by the well theoretically founded models of Refs. [18] and [19]. The latter approach is based in the model of Ref. [20] for weak pion production reaction off nucleons, which includes background non-resonant chiral terms and deuteron corrections [21], and it uses a Monte Carlo cascade model to account for FSI of the outgoing pion. The approach of Ref. [18] considers the weak excitation of several resonance contributions as intermediate states, and it implements a transport model to include FSI effects. Both models provide similar predictions for the MiniBooNE flux-folded differential $d\sigma/dp_\pi$ cross section, where FSI effects are clearly visible in the distribution. Because of the FSI some pions are absorbed, but other ones are scattered and lose to nucleons part of their energy. FSI is essential to fill the low momentum part of the distribution. On the other hand, the combined effect of pion quasi-elastic scattering and pion absorption through Δ excitation, depletes the $p_\pi \sim 250 - 450$ MeV region producing a distortion of the differential cross section shape that significantly worsens the description of the data. The artificial exclusion of the FSI effects leads to a better description of the $d\sigma/dp_\pi$ distribution. We refer the reader to Refs. [3, 18, 19] for a more detailed discussion on this open problem.

3.2 Quasielastic and elastic processes

Below 2 GeV, neutrino-nucleus interactions are predominantly quasielastic (QE), where the neutrino scatters off a nucleon bound in the nucleus. In the case of CC interactions at the nucleon level the possible reactions are $\nu_l n \to p l^-$ and $\bar{\nu}_l p \to n l^+$, while for NC, neutrinos elastically scatter from nucleons, $\nu N \to \nu N$. In these reactions nuclear effects are quite important [1, 3] and they need to be accounted for when determining the neutrino oscillation parameters. Indeed, uncertainties on these nuclear corrections are an important limitation in the oscillation analysis and a source of systematics.

As an example of the importance of the nuclear corrections, we will briefly describe here the so called MiniBooNE CCQE puzzle. This collaboration reported a CCQE cross section measurement [10] using a high-statistics sample of ν_μ interactions on ^{12}C. The size of the cross section was found to be unexpectedly large, and within the relativistic global Fermi gas model employed in the analysis of Ref. [10], a large nucleon axial mass of $M_A = 1.35 \pm 0.17$ GeV was needed to describe the data. This value was in conflict with that extracted from CCQE measurements on deuterium and, to a lesser extent, hydrogen targets $M_A = 1.016 \pm 0.026$ GeV [22], which is in excellent agreement with the pion electro-production result, $M_A = 1.014 \pm 0.016$ GeV, obtained from the nucleon axial radius [3, 23]. Furthermore, NOMAD also reported a small value of $M_A = 1.05 \pm \pm 0.06$ GeV [24]. The MiniBooNE data-set (CCQE-like) accounted for events with no pions in the final state (events in which only one muon is detected), but Monte Carlo correcting for those cases where CC1π production was followed by pion absorption. It was customary to take for granted that most of those events could be attributed to the QE scattering of the weak probe on a nucleon. This choice discards pions coming off the nucleus, since they will give rise to additional leptons after their decay. A natural solution to this puzzle came from the incorporation of RPA nuclear corrections to the QE cross section [11, 12] and multi-nucleon effects [11, 13, 25, 26] (see right panel of Fig. 1). The latter effects account for events where the gauge boson is absorbed by two interacting nucleons (excitation of a 2p2h nuclear component). Up to re-scattering processes which could produce secondary pions, 2p2h events will give rise to only one muon in the final state, and thus they are included in the MiniBooNE data-set. Thus, they could be experimentally misidentified as QE events. Moreover, these events affect the reconstruction of the incoming ν energy [9] and the flux-unfolded cross section (Fig. 1). The same scheme [27] describes also the MiniBooNE CCQE-like antineutrino data sample of Ref. [28].

4. Conclusions

Nuclear effects lead to sizable corrections on neutrino nucleus cross sections at low $Q^2 < 1\text{GeV}^2$, which should be incorporated in neutrino event generators.

This research has been supported by the Spanish Ministerio de Economía y Competitividad under the contracts FIS2011-28853-C02-02, FIS2014-51948-C2-1-P, FIS2014-57026-REDT and SEV-2014-0398, by Generalitat Valenciana under contract PROMETEOII/2014/0068.

References

[1] J.G. Morfin, J. Nieves, J.T. Sobczyk, Adv.High Energy Phys. **2012**, 934597 (2012)
[2] J. Formaggio, G. Zeller, Rev.Mod.Phys. **84**, 1307 (2012)
[3] L. Alvarez-Ruso, Y. Hayato, J. Nieves, New J.Phys. **16**, 075015 (2014)
[4] K.A. Olive et al. (Particle Data Group), Chin. Phys. **C38**, 090001 (2014)
[5] K. Kovarik, I. Schienbein, F.I. Olness, J.Y. Yu, C. Keppel, J.G. Morfin, J.F. Owens, T. Stavreva, Phys. Rev. Lett. **106**, 122301 (2011)
[6] M. Tzanov et al. (NuTeV), Phys. Rev. **D74**, 012008 (2006)
[7] G. Onengut et al. (CHORUS), Phys. Lett. **B632**, 65 (2006)
[8] I. Schienbein, J.Y. Yu, C. Keppel, J.G. Morfin, F. Olness, J.F. Owens, Phys. Rev. **D77**, 054013 (2008)
[9] J. Nieves, F. Sanchez, I. Ruiz Simo, M. Vicente Vacas, Phys.Rev. **D85**, 113008 (2012)
[10] A. Aguilar-Arevalo et al. (MiniBooNE Collaboration), Phys.Rev. **D81**, 092005 (2010)
[11] J. Nieves, I. Ruiz Simo, M. Vicente Vacas, Phys.Lett. **B707**, 72 (2012)
[12] J. Nieves, J.E. Amaro, M. Valverde, Phys.Rev. **C70**, 055503 (2004)
[13] J. Nieves, I. Ruiz Simo, M. Vicente Vacas, Phys.Rev. **C83**, 045501 (2011)
[14] D. Rein, L.M. Sehgal, Annals Phys. **133**, 79 (1981)
[15] K.M. Graczyk, J.T. Sobczyk, Phys.Rev. **D77**, 053001 (2008)
[16] T. Leitner, O. Buss, U. Mosel, L. Alvarez-Ruso, PoS **NUFACT08**, 009 (2008), 0809.3986
[17] A. Aguilar-Arevalo et al. (MiniBooNE Collaboration), Phys.Rev. **D83**, 052009 (2011)
[18] O. Lalakulich, U. Mosel, Phys.Rev. **C87**, 014602 (2013)
[19] E. Hernandez, J. Nieves, M.J. Vicente Vacas, Phys.Rev. **D87**, 113009 (2013)
[20] E. Hernandez, J. Nieves, M. Valverde, Phys.Rev. **D76**, 033005 (2007)
[21] E. Hernandez, J. Nieves, M. Valverde, M. Vicente Vacas, Phys.Rev. **D81**, 085046 (2010)
[22] A. Bodek, S. Avvakumov, R. Bradford, H.S. Budd, Eur.Phys.J. **C53**, 349 (2008)
[23] V. Bernard, N. Kaiser, U.G. Meissner, Phys.Rev.Lett. **69**, 1877 (1992)
[24] V. Lyubushkin et al. (NOMAD Collaboration), Eur.Phys.J. **C63**, 355 (2009)
[25] M. Martini, M. Ericson, G. Chanfray, J. Marteau, Phys.Rev. **C80**, 065501 (2009)
[26] M. Martini, M. Ericson, G. Chanfray, J. Marteau, Phys.Rev. **C81**, 045502 (2010)
[27] J. Nieves, I. Ruiz Simo, M. Vicente Vacas, Phys.Lett. **B721**, 90 (2013)
[28] A. Aguilar-Arevalo et al. (MiniBooNE Collaboration), Phys.Rev. **D88**, 032001 (2013)

EPJ Web of Conferences **116**, 11012 (2016)
DOI: 10.1051/epjconf/201611611012

(Very)-high-energy gamma-ray astrophysics: The future

Alessandro De Angelis[a]

INFN Padova, via Marzolo 8, 35141 Padova, Italy; LIP/IST, Av. Elias Garcia 14, 1000 Lisboa, Portugal

Abstract. Several projects planned or proposed can significantly expand our knowledge of the high-energy Universe in gamma rays. Construction of the Cherenkov telescope array CTA is started, and other detectors are planned which will use the reconstruction of extensive air showers. This report explores the near future, and possible evolutions in a longer term.

1. Introduction

Among cosmic rays, photons are particularly important. Being neutral they can travel long distances without being deflected by galactic and extragalactic magnetic fields, hence they allow us to directly study their emission sources. The gamma emission is also related to the emission of charged cosmic rays, and to the emission of neutrinos. Finally, gamma-rays can be the signature of new physics at fundamental scales, new particles in particular. These facts have pushed the scientific community during the last years to study high-energy gamma rays; many new results have been obtained since the beginning of the millennium, in particular thanks to the imaging atmospheric Cherenkov telescopes (IACTs) H.E.S.S., MAGIC and VERITAS. See [1] for a summary of the main achievements.

Still many problems are however open, in particular:

- The study of the origin of cosmic rays. Thanks to the first generation of IACTs and to the AGILE and Fermi satellites, it is well established nowadays that supernova remnants (SNRs) can accelerate cosmic rays up to some 100 TeV. The mechanism of acceleration of particles up to the PeV is still to be established. Above some PeV, it is common belief that the acceleration sites should be extragalactic, but the mechanism, possibly related to the behavior of matter near supermassive black holes (SMBH), is still to be understood.
- The study of the propagation of extragalactic gamma rays, which is a probe of cosmology and of the interaction of gamma rays with the cosmic background.

  Once produced, photons must travel towards the observer. Electron-positron (e^-e^+) pair production in the interaction of VHE photons off extragalactic background photons is the main source of opacity of the Universe to γ-rays whenever the corresponding photon mean free path

[a] e-mail: alessandro.deangelis@pd.infn.it

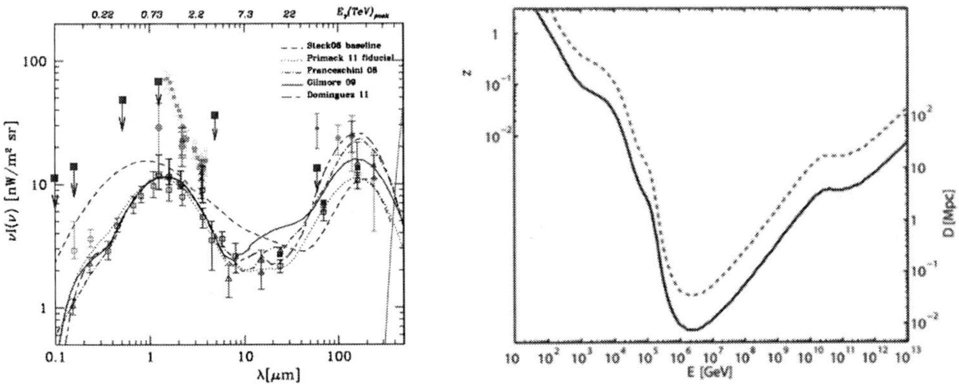

Figure 1. Left: Spectral energy distribution of the EBL as a function of the wavelength. Open symbols correspond to lower limits from galaxy counts while filled symbols correspond to direct estimates. The curves show a sample of different recent EBL models, as labeled. On the upper axis the TeV energy corresponding to the peak of the $\gamma\gamma$ cross section is plotted. From L. Costamante, IJMPD 22 (2013) 1330025. Right: Curves corresponding to the so-called gamma-ray horizon $\tau(E, z) = 1$ (lower) and to a survival probability of $e^{-\tau(E,z)} = 1\%$ (upper). Adapted from A. de Angelis, G. Galanti, M. Roncadelli, MNRAS 432 (2013) 3245.

is of the order of the source distance, or smaller. The cross-section is maximized ($\sigma_{max} \simeq 1.7 \times 10^{-25}\,\mathrm{cm}^2$) for background photons of energy:

$$\epsilon(E) \simeq \left(\frac{900\,\mathrm{GeV}}{E}\right)\,\mathrm{eV}\,. \tag{1}$$

The maximum photon density corresponds to the CMB; its total density is about 400 photons per cubic centimeter.

A region of particular interest is the so-called extragalactic background light (EBL), i.e., the light in the visible and near infrared regions. It consists of the sum of starlight emitted by galaxies throughout their whole cosmic history, plus possible additional contributions, like, e.g., light from hypothetical first stars that formed before galaxies were assembled. Therefore, in principle the EBL contains important information about both the evolution of baryonic components of galaxies and the structure of the Universe in the pre-galactic era. Note that the light is redshifted by a factor $(1 + z)$ due to the expansion of the Universe, and thus, the visible light from old sources is detected as infrared. The density of EBL photons in the region near the visible can be derived by direct deep-field observations, and by constraints on the propagation of VHE photons. A plot of the present knowledge on the density of photons in the EBL region is shown in Fig. 1, left. The corresponding opacity of the Universe is illustrated in Fig. 1, right.

- The examination of the ultimate nature of matter and of physics beyond the Standard Model, in particular through searches for dark matter, for new particles in general in particular by the determination of the energy density of the vacuum, for the effects of quantum gravity. High-energy gamma astrophysics is sensitive to energy scales important for particle physics, in particular the 100 GeV scale expected for cold dark matter; the TeV scale, where supersymmetric particles might appear; and finally, it might be possible to access indirectly the unification scale and the Planck scale: an energy $\sim 10^{19}$ GeV, corresponding to a mass $\sqrt{\hbar c/G}$.

We shall define in the following as high-energy (HE) photons the photons above 30 MeV; as very-high-energy (VHE) the photons above 30 GeV.

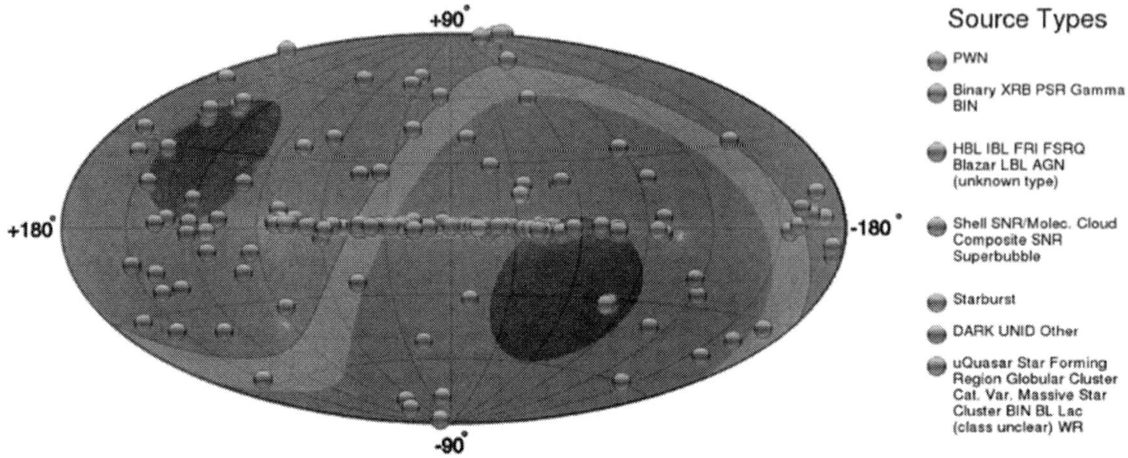

Figure 2. Sources of VHE emission plotted in galactic coordinates. The background represents the high-energy gamma-rays detected by Fermi-LAT. The clear blue and pink regions correspond respectively to the visible regions within 30° of the zenith from a detector at the tropic in the Northern hemisphere (approximately like MAGIC or VERITAS) and at the Southern hemisphere (approximately like H.E.S.S.). From `http://tevcat.uchicago.edu/`, January 2016.

A look to the sources of cosmic gamma-rays in the HE region shows a diffuse background, plus a set of localized emitters. Roughly 3000 HE emitters have been identified up to now, mostly by the Fermi-LAT, and some 200 of them are VHE emitters as well (Fig. 2).

About half of the VHE gamma-ray emitters are objects in our Galaxy; most of them can be associated to supernova remnants (SNR). The remaining half are extragalactic, and the space resolution of present detectors (slightly better than 0.1°) is not good enough to associate them with particular points in the host galaxies; we believe, however, that they are produced in the vicinity of supermassive black holes in the galaxies. The abundance of galactic objects can be explained by the fact that, being closer, they suffer a smaller attenuation.

The sky exhibits in particular transient events from steady emitters ("flares") and burst of gamma-rays from previously dark regions ("gamma-ray bursts").

2. How to detect photons at different energies: Present and near future

Due to the conversion probability in the atmosphere (whose thickness is about 28 radiation length at sea level) only satellite-based detectors can detect primary X/γ-rays. Satellites are anyway small, being of about 1 m^2 their maximum dimensions because of the cost of space technology. If the energy of the primary particle is large enough, some of the products of the shower generated by the interaction with the atmosphere, acting as a calorimeter, can reach ground-based detectors. It is appropriate to build such detectors at high altitudes, where atmospheric dimming is lower. The area sampled by ground-based detectors can be much larger than the typical areas of satellites. Since the fluxes of high-energy photons are low and decrease rapidly with increasing energy, TeV and PeV gammas can be detected only from the atmospheric showers they produce, i.e., by means of ground-based detectors.

2.1 Satellites

Satellite gamma telescopes can detect the primary photons at energies lower than ground-based telescopes.

Main figures of merit for a satellite-borne detector are its effective area (i.e., the product of the area times the detection efficiency), the energy resolution, the space or angular resolution (called as well point-spread function, or PSF).

Detectors on satellites have a small effective area, of order of 1 m^2 maximum, which limits the sensitivity. They have a large duty cycle, and they suffer a low rate of background events, since they can be coupled to anticoincidence systems rejecting the charged cosmic rays. They have a large cost, dominated by the costs of launch and by the strong requirements of instruments which must be sent in space, with little or no possibility of intervention to fix possible bugs.

The Fermi satellite, in orbit since 2008, has an effective area of about 1 m^2; its function is double, since it can also operate as a trigger for ground-based telescopes.

2.2 Ground-based gamma detectors

Ground-based VHE telescopes such as HAWC, H.E.S.S., MAGIC and VERITAS, detect the atmospheric showers produced by primary photons and cosmic rays of energy higher than those observed by satellites.

The two kinds of detectors (on satellite and at ground) are complementary. At energies below 1 GeV or so, the showers generated by photons do not have the time to develop properly, and thus the only way to detect such photons below this energy is related to the use of satellites. Ground-based detectors have a huge effective area, so their sensitivity is high; they detect a huge amount of background events, but they have low cost.

There are two main classes of ground based VHE gamma detectors: the EAS arrays and the Cherenkov telescopes.

2.2.1 EAS detectors

The EAS detectors, such as HAWC which is presently in operation, are large arrays of detectors sensitive to charged secondary particles generated in the atmospheric showers. They have a high duty cycle and a large field-of-view, but a relatively low sensitivity. The energy threshold of such detectors is rather large, in the order of a few hundred GeV – a shower initiated by a 1 TeV photon typically has its maximum 8 km a.s.l.

The principle of operation is the same as the one for the UHE cosmic rays detectors like Auger, i.e., direct sampling of the charged particles in the shower. This can be achieved:

- either using a sparse array of scintillator-based detectors, Resistive-Plate Chambers (RPC), or water Cherenkov pools;
- or by effective covering of the ground, to ensure efficient collection and hence lower the energy threshold.

Possibly a combination of the techniques above can be used, with a central region more dense, to allow discrimination between hadron-generated and photon-generated showers, and a peripheral region.

The energy threshold of EAS detectors is at best in the 0.5 TeV–1 TeV range, so they are built to detect UHE photons as well as the most energetic VHE gammas. At such energies fluxes are small and large surfaces of order of 10^4 m^2 are required.

Concerning the discrimination from the charged cosmic ray background, muon detectors devoted to hadron rejection may be present. Otherwise, it is based on the reconstructed shower shape. The direction of the detected primary particles is computed from the arrival times with an angular precision of about 1°. The calibration can be performed by studying the shadow in the reconstructed directions caused by the Moon. Energy resolution is poor.

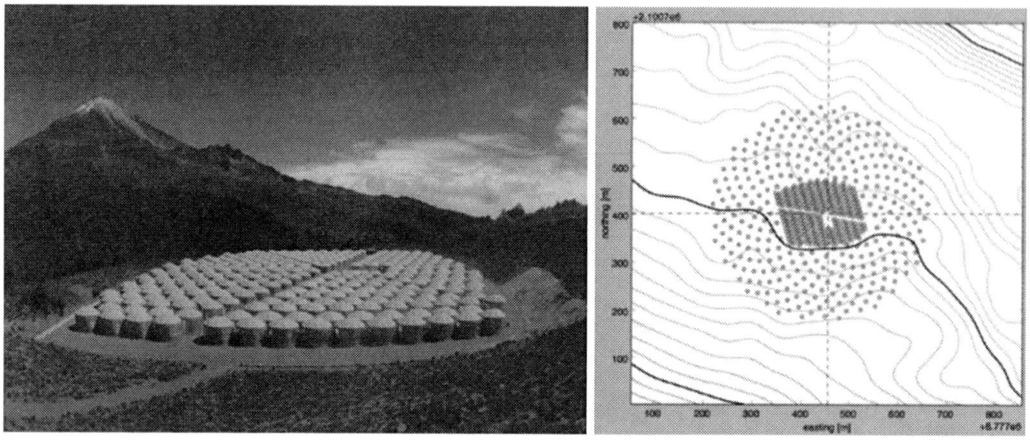

Figure 3. Left: the HAWC detector. Right: its possible extension.

Somehow, the past generation EAS detectors were not sensitive enough and just detected a handful of sources. This lesson lead to a new EAS observatory with much larger sensitivity: the High Altitude Water Cherenkov detector HAWC, inaugurated in 2015.

HAWC (Fig. 3) is a very high-energy gamma-ray observatory located in Mexico at about 4100 m asl and latitude 19° N. It consists of 300 steel tanks of 7.3 m diameter and 4.5 m deep, covering an instrumented area of about 22 000 m^2. Each tank is filled with purified water and contains four PMT of 3.8 inch diameter and one high-quantum efficiency 10-inch PMT, which observe the Cherenkov light emitted in water by superluminal particles in atmospheric air showers. Photons traveling through the water typically undergo Compton scattering or produce an electron-positron pair, also resulting in Cherenkov light emission. This is an advantage of the water Cherenkov technique, as photons constitute a large fraction of the electromagnetic component of an air shower at ground.

HAWC improves the sensitivity for a Crab-like point spectrum by a factor of 15 in comparison to MILAGRO. The sensitivity should be also such to possibly detect gamma-ray burst emissions at high energy.

2.2.2 Cherenkov telescopes

Most of the experimental results on VHE photons are presently due to Imaging Atmospheric Cherenkov Telescopes (IACTs).

IACTs such as the first successful cosmic detectors, called WHIPPLE, and then the second generation instruments HEGRA and CANGAROO, and presently the third generation H.E.S.S., MAGIC and VERITAS, detect the Cherenkov photons produced in air by charged, locally superluminal particles in atmospheric showers. For reasons explained below, they have a low duty cycle (about 1000 to 1500 hours per year) and a small field-of-view (FoV), but they have a high sensitivity and a low energy threshold.

We refer to [1] for the operation principles and for the performance and the main results of these detectors.

A simplified comparison of the characteristics of the Fermi LAT satellite detector, of the IACTs and of the EAS detectors (ground-based), is given in Table 1. The sensitivities of the above described high-energy detectors are shown in Fig. 4.

The construction of the next generation IACT, called the Cherenkov Telescope Array (CTA), has already started.

Table 1. A comparison of the characteristics of Fermi, the IACTs and of the EAS particle detector arrays. Sensitivity computed over one year for Fermi and the EAS, and over 50 hours for the IACTs.

Quantity	**Fermi**	**IACTs**	**EAS**
Energy range	20 MeV–200 GeV	100 GeV–50 TeV	400 GeV–100 TeV
Energy res.	5–10%	15–20%	$\sim$ 50%
Duty Cycle	80%	15%	> 90%
FoV	$4\pi/5$	$5° \times 5°$	$4\pi/6$
PSF	0.1°	0.07°	0.5°
Sensitivity	1% Crab (1 GeV)	1% Crab (0.5 TeV)	0.5 Crab (5 TeV)

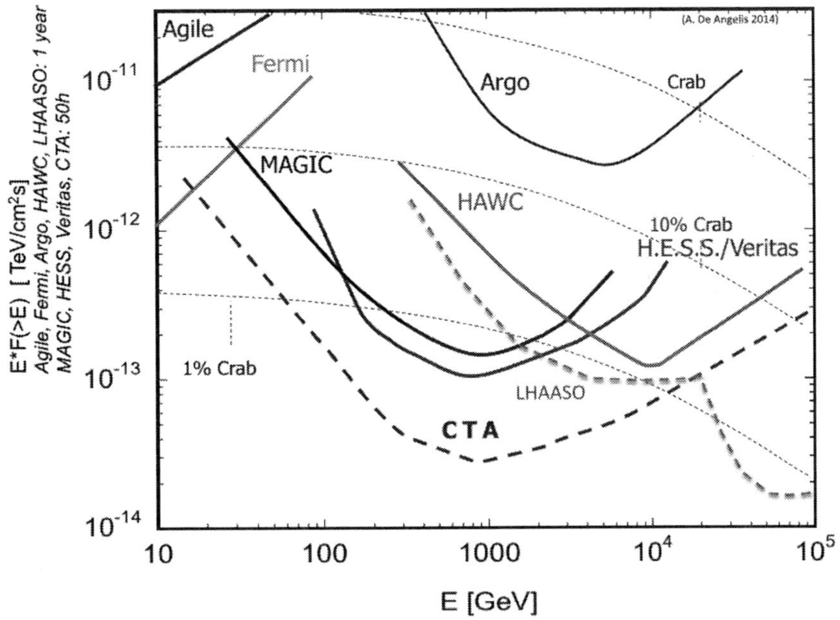

Figure 4. Sensitivities of some present and future HE gamma detectors, measured as the minimum intensity source detectable at 5 σ. The performance for EAS and satellite detector is based on one year of data taking; for Cherenkov telescopes it is based on 50 hours of data. From [1].

2.3 The near future of VHE astrophysics with Cherenkov telescopes: CTA

The CTA [2] is a future instrument for VHE gamma astrophysics that is expected to provide an order of magnitude improvement in sensitivity over existing instruments (Fig. 4).

CTA is a robust collaboration, including more than 1000 scientists from all around the world, and large part of its funding (above 200 MEUR) is granted. CTA is included in the 2008 roadmap of the European Strategy Forum on Research Infrastructures (ESFRI). It is one of the "Magnificent Seven" experiments of the European strategy for astroparticle physics published by ASPERA, and highly ranked in the strategic plan for European astronomy of ASTRONET. In addition CTA is a recommended project for the next decade in the US National Academies of Sciences Decadal Review.

An array of tens of telescopes will detect gamma-ray induced showers over a large area on the ground, increasing the efficiency and the sensitivity, while providing a much larger number of views of each cascade. This will result in both improved angular resolution and better suppression of charged cosmic-ray background events.

In the present design scenario, CTA will be deployed in two sites. The southern hemisphere array, possibly in Paranal, will consist of three types of telescopes with different mirror sizes, in order to cover the full energy range. In the northern hemisphere array, possibly in the Canary island of La Palma, the

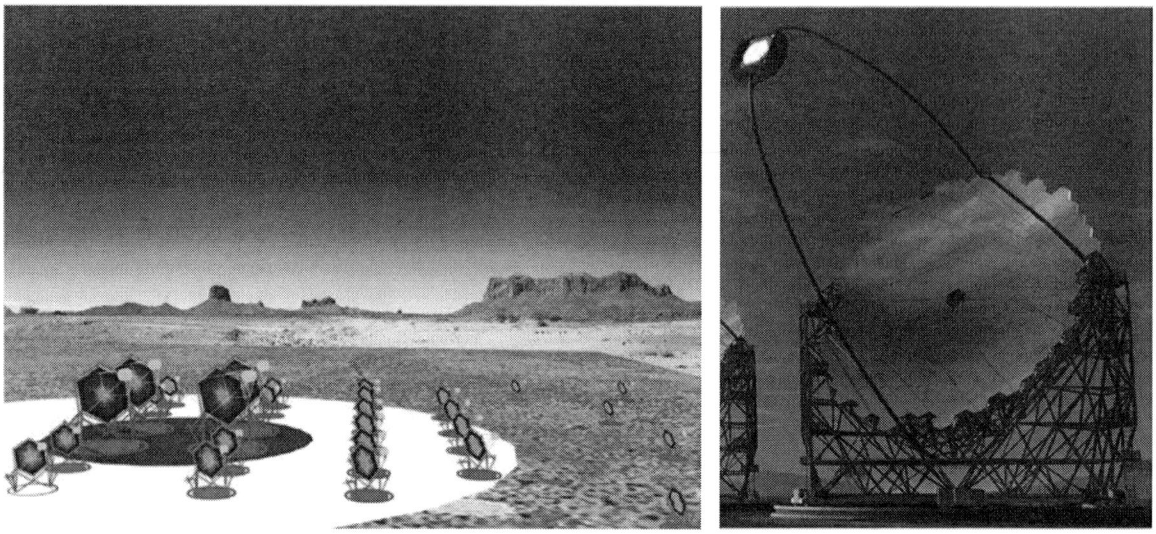

Figure 5. Left: possible layout of the CTA. Right: project of the large-size telescope (LST). Credit: CTA Collaboration.

two larger telescope types would be deployed.

- The low energy (the goal is to detect showers starting from an energy of 20 GeV) instrumentation will consist of four 23 meter telescopes with a FoV of about 4–5 degrees.
- The medium energy range, from around 100 GeV to 1 TeV, will be covered by some 15–25 telescopes of the 12 meter class with a FoV of 6–8 degrees. An alternative design uses smaller reflectors (abot 9 m) with a secondary optics (Schwarzschild-Couder design, [3]).
- The high energy instruments, operating above 10 TeV, will consist of a large number (of the order of 50) of small (4–6 meter diameter) telescopes with a FoV of around 10 degrees. Also in this case single-reflector designs are in competition with Schwarzschild-Couder design.

Prototyping is advanced for all these telescopes, but there is still room for technical improvement (SiPM in particular).

The southern CTA will cover about three square kilometers of land with around 60 telescopes that will monitor all the energy ranges in the center of the Milky Way's galactic plane. The northern CTA will cover one square kilometer and be composed of some 20 telescopes. These telescopes will be mostly targeted at extragalactic astronomy.

The telescopes of different sizes will be disposed in concentrical circles, the largest in the center (Fig. 5). Different modes of operation will be possible: deep field observation; pointing mode; scanning mode.

CTA is expected to surpass in performance the present IACTs around 2020.

2.4 The near future of EAS detectors

There is a lot of activity on the development of EAS detector, since they are relatively cheap (of the order of 30 MEUR) and we know now what is the correct size for the detection of a relevant number of sources.

2.4.1 *HAWC+*

An upgrade of the HAWC high-energy gamma-ray observatory with a sparse array of small outrigger tanks is being investigated [4]. For showers where the core falls outside the array there are ambiguities in the reconstruction between the core position, the shower angle and the shower size or energy. An outrigger array can determine the core position for showers falling outside the main array elevating the ambiguities and making these showers well reconstructable. A gain of 3–4 in sensitivity for gammas above 10 TeV can be obtained over what is presently achieved, and such a detector can be built in 2–3 years. Funds are already available.

2.4.2 *LHAASO*

LHAASO (Large High Altitude Air Shower Observatory) is planned to be located at about 4400 m asl and latitude 30° N in the Daochen site, China. It will consist of the following major components [5, 6]:

- 1 km^2 array (LHAASO-KM2A), including 5600 scintillator detectors, with 15 m spacing, for electromagnetic particle detection.
- An overlapping 1 km^2 array of 1200, 36 m^2 underground water Cherenkov tanks, with 30 m spacing, for muon detection (total sensitive area 40,000 m^2).
- A close-packed, surface water Cherenkov detector facility with a total area of 90,000 m^2 (LHAASO-WCDA), four times that of HAWC.
- 24 wide field-of-view air Cherenkov (and fluorescence) telescopes (LHAASO-WFCTA).
- 452 close-packed burst detectors, located near the center of the array, for detection of high energy secondary particles in the shower core region (LHAASO-SCDA).

A configuration corresponding to 25% of the final detector might be ready by 2019–2020. It is crucial that the completion of the first 25% of the detector (nominally foreseen for 2018) is not too much delayed with respect to the extension of HAWC, otherwise most of its scientific impact might be scooped.

2.4.3 *HiSCORE*

HiSCORE (Hundred*i Square-km Cosmic ORigin Explorer), located in the Tunka Valley near Lake Baikal at about 3200 m asl and latitude 51° N, will consist of an array of wide-angle ($\Omega \sim 0.6$–0.85 sr) light-sensitive detector stations, distributed over an area of the order of 100 km^2. A HiSCORE detector station consists of four photomultiplier tubes, each equipped with a light-collecting Winston cone of 30° half-opening angle [7].

The primary goal of this non-imaging Cherenkov detector is gamma-ray astronomy in the 10 TeV to several PeV range. A prototype array of 9 wide-angle optical stations, spread on a $300 \times 300\,m^2$ area, has been deployed since October 2013, and technical tests are underway. A 1 km^2 engineering array has been deployed and is in commissioning.

The sensitivity of the HiSCORE standard configuration to point-sources is shown in Fig. 6. The curve has been calculated for 1000 h of exposure time, corresponding to roughly 1.4 years of serendipitous mode operation (see [7] for details).

HiSCORE will be possibly incorporated in TAIGA (Tunka Advanced Instrument for cosmic ray physics and Gamma Astronomy), a hybrid detector system including in particular IACTs for sensitivity to photons of lower energies.

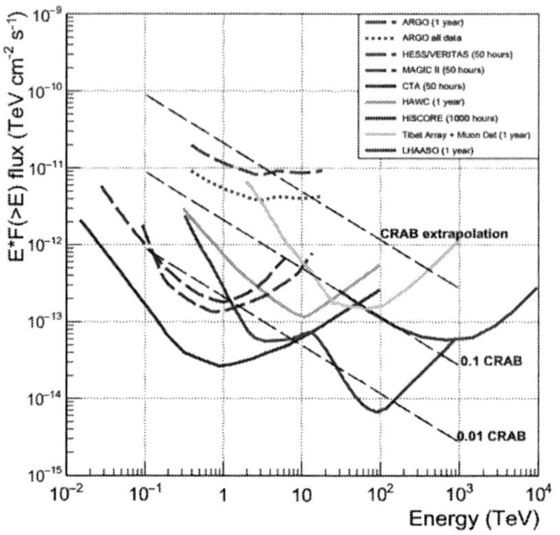

Figure 6. Sensitivity of ground-based γ-ray detectors to a Crab-like point source, not considering the attenuation of the source at large energies (from [6]).

2.4.4 Possible EAS experiments in the Southern hemisphere

Several proposals are being formulated now for a EAS detector at South, which would be the key to study the VHE emission from the Galaxy. HAWC has demonstrated that present EAS technology can be competitive with the technology of large-area coverage by small Cherenkov telescopes.

In particular among the proposals are:

- a Southern site for HAWC, HAWC-South, to be located in the Atacama desert [8];
- LATTES, a hybrid water Cherenkov-RPC detector to be located in Southern America [9].

3. A view in a longer term

Let us analyze the possible future of gamma-ray astrophysics in, say, 15 years from now.

Gamma rays can be divided in 5 different energy regions, subjectively chosen from the requirements of detection techniques and of main scientific issues. These energy regions are: (a) Till a few (~30) MeV; (b) GeV: from 30 MeV to 30 GeV: (c) Sub-TeV: from 30 GeV to 300 GeV; (d) TeV: from 300 GeV to 30 TeV; (e) PeV: from 30 TeV to a few PeV.

3.1 The region till a few MeV

This region has important implications on the science at the TeV and above, since a good knowledge of the spectra in the MeV region can constrain the fit to the emitted spectra at high energies, thus allowing:

- to evidence additional contributions from new physics (dark matter in particular);
- to estimate cosmological absorption (due to EBL or to possible interactions with axion-like fields [10]).

On top of this, The 0.3-30 MeV energy range is important per se, but experimentally difficult to study. It requires an efficient instrument working in the Compton regime with an excellent background subtraction, and possibly with sensitivity to the measurement of polarization. Since COMPTEL, which operated two decades ago, no space instrument obtained extra-solar gamma-ray data in the few MeV range; now we are able to build an instrument one-two orders of magnitudes more sensitive than COMTEL based on Silicon technology, state-of-the-art analog readout, and efficient data acquisition.

Several proposals (ASTROGAM [11], COMPAIR [12], ...) have been made, and convergence is likely for an experiment to be launched around 2025.

3.2 The GeV region

It is difficult to think for this century of an instrument for GeV photons improving substantially the performance of the Fermi LAT: the cost of space missions is such that the size of Fermi cannot be reasonably overcome with present technologies. New satellites in construction (like the Russian-Italian mission GAMMA400 [13] and the Chinese mission HERD [14]) improve some of the aspects of Fermi, e.g., calorimetry. For sure a satellite in the GeV region with sensitivity comparable with Fermi will be needed.

3.3 The sub-TeV and TeV regions

CTA appears to have no rivals for the gamma astrophysics in the Sub-TeV and TeV (from a few GeV to a few TeV) energy regions. These are crucial regions for fundamental physics, and for astronomy; they include also the region where sensitivity is maximum.

CTA will be probably upgraded including state-of-the art photon detection devices of higher efficiencies with respect to the present ones.

There is some room for new concepts of Cherenkov telescopes, in particular:

- a possible array of 4 or 5 very large detectors at 5000 m in order to boost the sensitivity around in the sub-TeV region (in particular from 20 GeV to 100 GeV) [15]. This would boost the science potential in particular in sectors related to fundamental physics (dark matter, photon propagation from far sources);
- a large-FoV detector to guarantee good coverage of the sky in the TeV region [16].

3.4 The PeV region

Due to the opacity of the Universe to gamma rays, less than a handful of sources could be visible in the Northern sky, and less than a dozen in the Southern sky, all galactic. The experiments in the Northern hemisphere (HAWC with its likely extension to HAWC+, LHAASO, TAIGA/HiSCORE if the extension up to 10 km^2 will be funded) provide an appropriate coverage of the Northern sky.

The situation in the Southern hemisphere has room for improvement. An EAS detector in the South might give substantial input with respect to the knowledge of the gamma sky, and of possible Pevatrons in the Galaxy, and outperform in this sense the SSTs of CTA. Several proposals are being formulated now, and they should join. A large detector in Southern America could compete in sensitivity with the small-size telescopes of CTA-South already at 100 TeV, offering in addition a serendipitous approach.

References

[1] A. De Angelis and M. Pimenta, *Introduction to particle and astroparticle physics* (Springer, Heidelberg, 2015)

[2] `https://www.cta-observatory.org/`

[3] V. Vassiliev et al., Astrop. Phys. **28**, 10 (2007)

[4] A. Sandoval for the HAWC Collaboration, arXiv:1509.04269, Presented ICRC2015, The Hague, The Netherlands
[5] M. Zha et al., Int. J. Mod. Phys. **10**, 147 (2012)
[6] G. Di Sciascio, arXiv:1503.05512, presented at RICAP 2014, Noto, Italy
[7] M. Tluczykont et al., Astrop. Phys. **56**, 42 (2014)
[8] Michael DuVernois for the HAWC Collaboration, PoS(ICRC2015)967, Presented at ICRC2015, The Hague, The Netherlands
[9] G. Matthiae, AIP Conf. Proc. 1693 (2015) 060007 and Proc. MG14, http://w2srvg9.icra.it/upload/archivio/HE1-938MA857IO.pdf
[10] A. De Angelis, M. Roncadelli and O. Mansutti, Phys. Rev. D **76**, 121301 (2007)
[11] http://astrogam.iaps.inaf.it/
[12] A. Moiseev et al., arXiv:1508.07349
[13] http://gamma400.lebedev.ru/indexeng.html
[14] http://herd.ihep.ac.cn/
[15] F. Aharonian et al., arXiv:astro-ph/0006163
[16] J. Cortina et al., http://arxiv.org/abs/1507.02532

EPJ Web of Conferences **116**, 11013 (2016)
DOI: 10.1051/epjconf/201611611013

Opportunities within ASTERICS

Rob van der Meer[a] and Giuseppe Cimò for the ASTERICS Consortium

ASTRON, Dwingeloo, The Netherlands

Abstract. ASTERICS[1], The Astronomy ESFRI and Research Infrastructure Cluster project, brings together astronomers and astroparticle physicists of 22 institutes in Europe to help Europe's world-leading observatories work together to find common solutions to their Big Data challenges, their interoperability and scheduling, and their data access, searching for cross-cutting solutions with mutual and wide-ranging benefit to all concerned. ASTERICS is a four year project, funded through the European Union's Horizon 2020 Framework Programme. The facilities supported by ASTERICS include SKA, CTA, KM3NeT, E-ELT. ASTERICS aims to open up multi messenger astronomy to all scientists and the public through the Virtual Observatory and the citizen science work. I will draw a picture of the landscape in which ASTERICS operates and the possible interaction with the Very Large Volume Neutrino Telescope community. Attention will be given to emerging opportunities for the Neutrino community and how these can be recognised or created.

1. Introduction

The astronomy community and the astroparticle community are building for the future. The European Strategy Forum for Research Infrastructures (ESFRI) Roadmap identifies new Research Infrastructures (RI) of pan-European interest corresponding to the long term needs of the European research communities, covering all scientific areas. At the moment, four large ground based Astronomy and astroparticle physics facilities are on the ESFRI Roadmap 2010 [1] and are being prepared or under construction. Each of these four facilities will be capable of delivering transformational science, moving boundaries in resolution, accumulating surface or volume, or completely new science. During the development and exploration of these facilities, the developers and users will encounter major challenges on data handling, and with the large science and engineering groups they will find solutions to all of these. It should be possible to increase the scientific outcome and decrease the development time by joining forces.

ASTERICS (Astronomy ESFRI and Research Infrastructure Cluster) is a project in the Horizon 2020 Work Programme INFRADEV-4-2014/2015 Call – "Implementation and operation of cross-cutting services and solutions for clusters of ESFRI and other relevant research infrastructure initiatives". ASTERICS brings together for the first time astronomers and astroparticle physicists of 22 institutes from six European countries to help Europe's world-leading observatories work together to find common

[a] e-mail: meer@astron.nl

[1] ASTERICS project website: www.asterics2020.eu

solutions. The project, coordinated by ASTRON, The Netherlands, is funded 15 million Euros over the course of four years, beginning May 1st, 2015.

2. The ASTERICS project

2.1 Objectives

The focus of ASTERICS is on four ESFRI facilities, each with its own window on the universe:

- SKA, Square Kilometre Array, *radio waves*
- CTA, Cherenkov Telescope Array, *gamma rays*
- KM3NeT, Kilometre cube Neutrino Telescope, *neutrinos*
- E-ELT, European Extremely Large Telescope, *visible light.*

The main objective of ASTERICS is to join forces. This will result in finding common solutions to data challenges, interoperability and scheduling, and data access, searching for cross-cutting solutions with mutual and wide-ranging benefit to many facilities and user groups, gathering critical mass and creating economy of scale and exploiting complementarities. Besides the above four, ASTERICS also has close links to facilities that want to be on the ESFRI roadmap, like the Einstein Telescope (ET). Since no data is coming out of the focus facilities yet, the data from complementary, pathfinder and precursor facilities, like LOFAR, Euclid, LSST, VIRGO, LIGO, eVLBI, HESS, MAGIC, ANTARES, IceCube will be used for comparing data formats, design new data structures, and analysis software, benchmark hardware, and test prototypes. ASTERICS aims to open up multi messenger astronomy to all scientists and the public through the Virtual Observatory (VO), as described in Sect. 2.3.4, and citizen science work, as described in Sect. 2.3.2, where members of the general public are encouraged to undertaken scientific work in collaboration with or under the direction of professional scientists and scientific institutions. Following from these objectives, there will be plenty of opportunities for project partners and linked facilities to contribute to the search for common solutions and to profit from solutions from other fields. The project results will be widely distributed. It is in the interest of all scientific communities to actively participate and recognise, or create, emerging opportunities.

2.2 Added value

The ESFRI facilities are very well capable of producing interesting science results. However, the added value of ASTERICS will be to create the capability to move the science frontiers beyond the expectations within the science cases of the facilities, and lift the outcome far above the results of the single facilities. To raise the single ESFRI facilities to the level of intended added value, ASTERICS will start with coordinating the individual requirements, to disseminate results, educate and train the developers, manager and users of facilities. The close and direct involvement of ESFRI project and pathfinder staff is essential in realising the major ASTERICS ambitions. There are several opportunities for the VLVnT community to work together with the other facilities to exchange information and to showcase their knowledge and experience with the precursor experiments. The intended added value of ASTERICS compared to the individual ESFRI facilities is summarised in paragraphs below.

2.2.1 Design requirements

The ESFRI facilities will have their own design requirements aimed for optimal science return of the own facility.

ASTERICS aims at:

- enabling interoperability between facilities
- minimising fragmentation of research developments, stimulating collaboration between or mergers of several small groups working on parts of the research of a much larger challenge, into larger (virtual) teams
- encouraging cross-fertilisation, where ideas of one research groups inspires another research group to find solutions for unrelated issues
- developing joint multi-wavelength/multi-messenger capabilities
- opening-up the next generation of observatories to discovery and usage (and re-usage) of data and data reduction and analysis software
- searching for generic approach to the data challenge.

2.2.2 Technology choices

The ESFRI facilities will (have to) go for mature technology choices at some decision points.

ASTERICS aims at:

- identifying areas of rapid technology development
- creating complementary and synergic approaches across the project
- finding significant added value.

2.2.3 Observations

The ESFRI facilities will optimise their own observations. They will have exceptional science cases per facility.

ASTERICS aims at:

- coordinating and harmonising joint and efficient scheduling, operation and interoperability
- multi-wavelength / multi-messenger observations
- engaging with society
- enlarging the Citizen science capacity and usage
- achieving the Science 2.0 goals (open access to all data, analysis tools, publication platforms).

2.2.4 Data analysis

The ESFRI facilities and their users will focus on the analysis of their own data.

ASTERICS aims at:

- data integration across ESFRI facilities
- open standards
- adapting Virtual Observatory framework and tools
- e-science infrastructures
- data mining
- co-development for software and hardware
- software libraries.

2.3 Organisation

The activities within ASTERICS are organised in five work packages. The starting points are the data challenges. There are activities around data collection at the facilities, data access for users, standardisation for data and meta data, interoperability, usability of analysis tools, re-use of data and software, awareness increase of users, general public, and policy makers. Besides reaching out to research communities like the VLVnT community at their conferences and meetings, also interest groups are targeted, e.g. during the Astronomical Data Analysis Software & Systems (ADASS) conference by F. Pasian [5]. The specific tasks and activities of the five ASTERICS work packages are described in the following paragraphs.

2.3.1 WP1: ASTERICS Management Support Team (AMST)

The ASTERICS Management Support Team (AMST) facilitates the project organisation and governance. It coordinates the communication between internal and external stakeholders and coordinates the work to be done in the project. The project is built on four ESFRI facilities originating from four completely different science disciplines that have developed their fields separately since their origin (optical astronomy, gamma ray physics, neutrino physics, and radio astronomy). One major task of the AMST is to morph these four pillars of the project into one big family, while conserving the individual characteristics of the science disciplines and to accelerate the progress by enthusiastic collaboration.

2.3.2 WP2: Dissemination, Engagement and Citizen Science (DECS)

The first task of WP2 is to disseminate the project results to the stakeholders and to promote ASTERICS and the facilities it serves to as large an audience as possible, by creating high quality branding and promotional outreach material. Possible audiences are: scientific and technical communities, academia, private industry, other public research centres, SME's, policy makers, general public. The second task is to engage the public in mass participation events through the Citizen Science platform. This platform serves two purposes: entertaining and educating the general public in an (inter)active way, and giving the scientist a flexible tool by using a much larger number of eyes and brains. To make optimal use of the Citizen Science platform, scientist should ask themselves: "What could I do with a million eyes, to improve my scientific return on the data I have collected?" To develop relevant questions for the VLVnT community, researchers are invited to participate in the citizen science brainstorming workshops.

2.3.3 WP3: Observatory E-environments Linked by common ChallengeS (OBELICS)

The work package OBELICS (Observatory E-environments Linked by common ChallengeS) aims at co-development of the robust, scalable, flexible handling and exploitation of huge data streams and distributed petascale database systems. The work package aims to enable interoperability and software re-use in the three fields of data generation, integration, and analysis. The work package is divided into three tasks, each focussing on one of these fields.

2.3.4 WP4: Data Access, Discovery and Interoperability (DADI)

DADI (Data Access, Discovery and Interoperability) is a prolongation of the Virtual Observatory (VO) projects that started fifteen years ago, and in particular of the series of Euro-VO projects supported by the European Commission (Genova et al., 2015, [2]). The goal is to make the ESFRI and pathfinder projects data available for discovery and usage by the whole astronomical community, interoperable in a homogeneous international framework, and accessible with a set of common tools. In order to optimise evolving the VO, the work is organised around a sequence of actions: train and support ESFRI and pathfinder staff, collect requirements and feedback from the users, adapt the VO framework to the collected requirements and train again.

The DADI work to optimize usage of the data from large projects through the VO is in line with the priority of the IVOA[2] (International Virtual Observatory Alliance). The VO is a framework

[2] `http://www.ivoa.net`

that allows data services following common rules, using common tools. ASTERICS wants to support implementation of the IVOA framework in the ESFRI projects and their pathfinders and to contribute to the development of the VO standards and tools as well. The VO is the interface between the domain specific and generic infrastructure and it works on new and archive data. This was demonstrated in a visualisation of IceCube-40 string data with VO tools by Abbasi et al. [3].

During the ASTERICS project there are recurring events where the partner projects and interested scientists can interact to improve the VO and their own data analysis. These events are: the DADI technology Forums, to inform about technological developments and build collaborations, the annual ASTERICS European Schools (with open call to European early career scientists, Post-doc, 50 participants), the two ESFRI Forum and training events (network, share lessons learnt, discuss requirements, training), and the two Data Provider Forums (open to all European Data providers). These meetings are useful also for the VLVnT community by providing input for optimising the VO and for learning how to use the VO tools.

2.3.5 WP5: CLEOPATRA (Connecting Locations of ESFRI Observatories and Partners in Astronomy for Timing and Real-time Alerts)

CLEOPATRA (Connecting Locations of ESFRI Observatories and Partners in Astronomy for Timing and Real-time Alerts) aims specifically at synergetic observing modes, and fast and reliable access to large data streams. Tasks in the work package are: technology development for fibre connectors, relaying alerts, data streaming software, data dissemination, and advanced scheduling algorithms. This builds on WRE (White Rabbit Ethernet) and the EC EXPReS/NEXPRES projects. The scheduling, triggering and timing can be of interest to the VLVnT community for the Astrophysical Multimessenger Observatory Network (AMON), presented at this VLVnT workshop by Douglas Cowen [4].

3. Conclusion

The close and direct involvement of ESFRI project and pathfinder staff is essential in realising the major ASTERICS ambitions. There are several opportunities for the VLVnT community to work together with the other facilities to exchange information and to showcase their knowledge and experience with the precursor experiments. All work packages have meetings to contact users and projects for input/output. Everyone is encouraged to use these meetings to engage with the project and to maximise the output to the community. Keep an eye on `www.asterics2020.eu` for the schedule of meetings.

STERICS is a project funded by the European Commission under the Horizon2020 programme (id 653477). All the partner institutions and individual participants are gratefully acknowledged for their work in the project.

References

[1] ESFRI Roadmap 2010, `http://www.esfri.eu/roadmap-archive` (2010)
[2] F. Genova et al., `arXiv:1506.06567`, DOI: 10.1016/j.ascom.2015.03.001, Astronomy and Computing **11**, 181 (2015)
[3] R. Abbasi et al., `arXiv:1104.5187`, Phys. Rev. D **84**, 082001 (2011)
[4] D. Cowen et al., this proceedings (EDP Sciences, to be published)
[5] F. Pasian, M.A. Garrett, F. Genova, G. Lamanna, S. Serjeant, A. Szomoru, R. van der Meer, Astronomical Data Analysis Software and Systems XXV, *ASP Conference Series*, in press, 2016

AUTHOR INDEX

AUTHOR INDEX

AUTHOR INDEX

EDP Sciences – Web of Conferences
17, Avenue du Hoggar
Parc d'Activité de Courtabouf
BP 112
F-91944 Les Ulis Cedex A
France

ISBN 978-1-5108-2268-9

CURRAN ASSOCIATES INC.
proceedings.com